Signal Processing in Magnetic Resonance Spectroscopy with Biomedical Applications

Dževad Belkić
Karolinska Institute
Stockholm, Sweden

Karen Belkić
Karolinska Institute
Stockholm, Sweden

CRC Press is an imprint of the
Taylor & Francis Group, an **informa** business
A TAYLOR & FRANCIS BOOK

CRC Press
Taylor & Francis Group
6000 Broken Sound Parkway NW, Suite 300
Boca Raton, FL 33487-2742

Printed in the United States of America on acid-free paper
10 9 8 7 6 5 4 3 2 1

International Standard Book Number: 978-1-4398-0644-9 (Hardback)

Library of Congress Cataloging-in-Publication Data

Belkic, Dž (Dževad)
Signal processing in magnetic resonance spectroscopy with biomedical applications / authors, Dževad Belkic, Karen Belkic.
p. cm.
Includes bibliographical references and index.
ISBN 978-1-4398-0644-9 (hardcover : alk. paper)
1. Nuclear magnetic resonance spectroscopy. 2. Signal processing. I. Belkic, Karen, 1952- II. Title.

QP519.9.N83B45 2010
616.07'548--dc22 2009039096

Visit the Taylor & Francis Web site at
http://www.taylorandfrancis.com

and the CRC Press Web site at
http://www.crcpress.com

Contents

About the Authors

Dževad Belkić is a theoretical physicist. He is professor of Mathematical Radiation Physics at the Karolinska Institute in Stockholm, Sweden. His current research activities are in atomic collision physics, radiation physics, radiobiology, magnetic resonance physics and mathematical physics.

In atomic heavy-particle collision physics, his past and current work encompasses many problems reflecting major challenges in this research field, such as theory of charge exchange and ionization at high non-relativistic energies. He is one of the world's leading experts on the Coulomb asymptotic convergence problem, distorted wave representations and perturbation expansions methods. He is known for furthering the powerful and versatile continuum distorted wave method and its derivatives that were advantageously exported to ion-atom and photon-atom collisions and also found their useful applications in medical physics.

In radiation physics, Dževad Belkić works on the passage of fast electrons and multiply charged ions through tissue, as needed in radiation therapy in medicine. Here he furthers both deterministic methods through the Boltzmann equation and stochastic simulations via Monte Carlo computations. In radiobiology, his research entails mathematical modeling for cell survival, with the main emphasis on mechanistic approaches by including the chief pathways for survival of cells under irradiation during radiotherapy.

In magnetic resonance physics, Dževad Belkić works on magnetic resonance spectroscopy with the main applications to medical diagnostics, aided critically by high-resolution parametric signal processors that go beyond the conventional shape estimations of spectra and fitting approaches to quantification. The leading processor here is his fast Padé transform for exact spectral analysis of generic time signals and unequivocal signal-noise separation via Froissart doublets or pole-zero cancellations in response functions.

In mathematical physics, he works on many problems including the derivation of analytical expressions for scattering integrals or bound-free form factors, for rational response functions in signal processing, and for coupling parameters in the nearest neighbor approximation, which is one of the most frequently used methods in physics and chemistry.

Dževad Belkić has published more than 180 scientific works, which have received over 2500 citations. His three books "Principles of Quantum Scattering Theory", "Quantum Mechanical Signal Processing and Spectral Analysis" and "Quantum Theory of High-Energy Ion-Atom Collisions" (the latter being one of the top selling physics books) were published by the Institute of Physics Publishing and Taylor & Francis in 2003, 2004 and 2008, respectively.

He has received numerous international awards for his scientific research, including the triple guest professorship in Atomic Physics from the Nobel Foundation and the Royal Swedish Academy of Sciences.

Karen (maiden name Edinger) Belkić is a clinical scientist with a PhD in neuroscience and physician specialist in internal medicine. She is adjunct professor of preventive medicine at the University of Southern California School of Medicine, Institute for Health Promotion and Disease Prevention Research. She is also affiliated to the Oncology and Pathology Department of the Karolinska Institute, where she holds a scientific tenure position. Her research activities are broad within several areas including preventive medicine, diagnostics and rehabilitation. In stress research and in molecular imaging, she has two published books as single author and over 75 full length papers in peer-reviewed journals with over 1500 external citations. She has twice been co-editor of special topical issues in peer-reviewed international journals.

Her work has strived to bridge clinical and basic scientific domains, seeking to answer to both these callings, by addressing difficult questions raised by multi-disciplinary, translational research. Her major goal has been to find non-invasive, sensitive and specific tools that can identify initial and often reversible changes, at a stage when timely intervention could be most effective. Her interest and expertize is in early detection, risk assessment and prevention with a focus on cancer and heart disease, with attention to psychosocial factors and the potential mediating mechanisms. Her current scientific activity is focused upon improvement of early cancer detection through *in vivo* magnetic resonance by enhancing the diagnostic information obtained by applying modern advances in signal and imaging processing to signals encoded from patients with cancers, and comparing these to findings from non-malignant tissue. Younger women at high risk for breast and ovarian cancer are a particular target group.

Karen Belkić has taken a broad view, looking not just at the immediate (i.e., proximal) markers of risk, but taking into account the more distal, and potentially key, determinants of disease. Thus, she has been very interested in how the environment (especially the work environment) impacts upon target organs, often mediated by the central nervous system. Within this framework, Karen Belkić has developed multi-level models. These incorporate, *inter alia*, non-linear, parametric methods in signal processing in relation to multiple physiological time signals for functional diagnostic testing. Karen Belkić is the originator of the widely implemented "Occupational Stress Index", a practical diagnostic tool for assessment and subsequent modification of the work environment.

She is involved in risk assessment and determinants of adherence to cancer screening guidelines among vulnerable groups as well as design, implementation and testing of interventions for patients with cancer to return to health-promoting work conditions. Karen Belkić has a special interest in pedagogy, in particular to help medical students and physicians at various levels of training acquire an appreciation of the importance of signal processing for medical diagnostics and to be able to identify situations in which this could be of critical significance in the clinical context, especially for cancer diagnostics.

Preface

In this book, we strive to bridge the gap between the optimal approaches to mathematical signal processing for magnetic resonance spectroscopy and their clinical relevance. This is especially vital for clinical oncology diagnostics, for which accurate quantification is urgently needed in order to achieve timely cancer detection. Clearly, we are aiming at a broad multi-disciplinary audience. Thus, we provide a mathematically rigorous presentation of the principles of signal processing, emphasizing cutting edge theory-based innovations that can optimize magnetic resonance spectroscopy. This was made possible by widening the horizons of signal processing through finding its natural framework in a larger and well-established theory – quantum physics. We also provide an in-depth examination of the application of optimized signal processing through the fast Padé transform to several specific problem areas in oncologic diagnostics such as ovarian cancer, breast cancer and prostate cancer.

The book is organized along two main parts. The first of these is theoretical and analytical, together with detailed proof-of-principle results. We begin by pointing out that from the mathematical vantage point, the quantification problem entails using the encoded time signals, or equivalently, free induction decay curves to reconstruct the true number of harmonic transients together with the pairs of spectral parameters that are complex-valued physical quantities. These are the frequencies and the corresponding amplitudes. Moreover, every time signal has a structure which is quantifiable by a relatively limited number of parameters. In practice, neither this structure nor the related parameters of the studied time signals are known prior to signal processing. The task of spectral synthesis is to reveal or reconstruct these unknowns from the given time signal, and this is achieved by performing spectral decomposition. The motivation for this entire book is the inadequacy of conventional Fourier-based analytical methods to solve this quantification problem. A key question is: *Why is this so*? The fundamental reason is that Fourier analysis ignores the structure of the investigated time signal. We go on to explain that there are two kinds of spectral shapes in signal processing. These are the component shape spectra for each separate resonance and the total shape spectrum as the sum of all the separate component shape spectra, resulting from interference patterns of all the individual resonances. The fast Fourier transform as a non-parametric processor can give only the envelope or total shape spectra. In contrast, parametric estimators can generate the component as well as the total shape spectra. This is because they can obtain the peak positions, widths, heights and phases of individual physical resonances. The main distinction between these two kinds of processors is in the type of information extracted from the input time signal. Non-parametric processors can generate only qualitative information, which is used to create graphs of spectral shapes. However, the quantitative data of vital importance are provided by parametric estimators through unfolding the hidden spectral structure of the apparent envelope. Parametric methods can reconstruct the quantitative features of resonances (peak position, width, height and phase) that are needed to estimate metabolite concentrations. With such processors, these key resonance parameters are retrieved first. Thereafter, the spectra can be constructed in any mode, if desired. This is a completely different methodology from fitting estimators that require the envelope spectrum before attempts to quantify the encoded data. These two distinct approaches – ambiguous fittings versus unambiguous quantum-mechanical estimation are anticipated to give substantially different results, especially with respect to closely overlapping resonances. Through a review of the quantum-mechanical concept of resonances in scattering and spectroscopy, the Padé approximant is seen to be an integral part of the description of the physics of resonance phenomena. The Padé approximant is ideally suited to yield the proper theoretical predictions and physical interpretation of resonance data that can be measured in experiments.

We present an in-depth analysis of the role of quantum mechanics in signal processing. By identifying the quantification problem in signal processing as quantum-mechanical spectral analysis, the key door was opened for using a highly-developed mathematical apparatus to overcome the otherwise insurmountable difficulties of the fast Fourier transform as well as fitting recipes and the like. It is through this direct connection of signal processing with quantum physics that a veritable *paradigm shift* has been established, and the stage set for the emergence of the most powerful and versatile spectral analyzer – the fast Padé transform. From the standpoint of mathematical modeling, the fast Padé transform is capable of extracting the missing information from the analyzed time signals, because it has more degrees of freedom via the use of two polynomials in the form of their ratio rather than only one such polynomial encountered in the fast Fourier transform. Among all the existing signal processors, the fast Padé transform is particularly advantageous because of its self-contained cross-validation. This is because the fast Padé transform, as a system function which is formulated to model an optimal response to generic external perturbations, possesses two complementary versions. These equivalent variants are initially defined in two entirely different convergence regions located inside and outside the unit circle. They correspond to the causal and anti-causal filters. Such an intrinsic cross-validation within the fast Padé transform avoids the need to compare different processors that might be in agreement, but nevertheless might not accurately reconstruct the spectral parameters. The causal and anti-causal versions of the fast Padé transform have a joint conceptual design within the same general Padé mathematical methodology as a system function represented by the corresponding Green function.

Under most general circumstances, including magnetic resonance spectroscopy, the physical considerations invariably lead to a polynomial quotient, as in the fast Padé transform, for the response function, which represents a frequency spectrum in signal processing. This is expected, since the formalism of the Green function is completely equivalent to the Schrödinger equation by which quantum physics becomes applicable to any system, including biological systems. Green functions can be computed for any physical system, irrespective of its structural complexity insofar as the auto-correlation functions, or equivalently, time signal points are available. Precisely such time signals are found in magnetic resonance spectroscopy. The fast Padé transform converges when the reconstructed frequencies and amplitudes become stable. This point of stabilization is a *veritable signature* of the exact number of resonances. With further increase of the partial signal towards the full signal length, i.e., beyond the point at which full convergence was reached, all the fundamental frequencies and amplitudes remain constant.

Protection against contamination by spurious resonances is provided within the fast Padé transform itself, since each pole from spurious resonances stemming from the denominator polynomial coincides with the corresponding zero of the numerator polynomial. This leads to pole-zero cancellation in the polynomial quotient of the fast Padé transform. Such a feature can be used to differentiate between spurious and genuine content of the signal. Since the unphysical poles and zeros always appear as pairs in the fast Padé transform, they are viewed as doublets. These are termed *Froissart doublets* after Froissart who discovered this extremely useful phenomenon, unique to the Padé methodology.

Chapters 3, 6 and 7 are a rigorous presentation of the proof of principle results applying the fast Padé transform to magnetic resonance time signals. In chapter 7 the performances of the fast Fourier transform and the said two variants of the fast Padé transform are compared using time signals encoded *in vivo* via magnetic resonance spectroscopy at 4T and 7T from the occipital grey brain region of a healthy volunteer. These time signals were long with excellent signal-to-noise ratio. The fast Padé transform yields a remarkably steady, as well as rapid convergence of the total shape spectrum, much faster than the fast Fourier transform. This is sharply contrasted to most other non-linear parametric estimators that wildly oscillate before they eventually stabilize and only then perhaps converge.

Detailed error analysis is performed for these time signals. It is thereby shown that the mentioned variants of the fast Padé transform are sufficient to establish consistency, without the need for external confirmation against the fast Fourier transform or any other estimator. The causal and anti-causal versions of the fast Padé transform with the initial convergence regions inside and outside the unit circle yield, respectively, the upper and lower bounds to envelopes of the shape spectra separated from each other by residual values of the order of the background noise.

In chapter 3 the benchmark results are presented showing that the fast Padé transform provides exact quantification of time signals from magnetic resonance spectroscopy and thereby metabolite concentrations are reliably and unequivocally obtained. Validation is given for the causal and anti-causal computational algorithms by which the fast Padé transform yields quantitative spectral parameters. This is done without fitting and the solution is *unique*. Further, the fast Padé transform outperforms all other parametric estimators. Confidence in the fast Padé transform is built systematically by considering theoretically generated as well as experimentally encoded time signals. We present the computations using the fast Padé transform to reconstruct spectral parameters for time signals that closely match encoded free induction decay curves on clinical scanners via magnetic resonance spectroscopy from the human brain. Included in these successful reconstructions are not only isolated and closely overlapping resonances, but also those which are nearly degenerate. These latter resonances cannot possibly be detected via the total shape spectrum or error analysis through residual spectra or difference spectra – the difference between the input and the modeled spectra. Only the parametric analysis provided by, e.g., the fast Padé transform can detect and also exactly quantify such resonances, which are often of major clinical importance.

In chapter 6, we perform further computations using the fast Padé transform to reconstruct these spectral parameters. Therein it is shown that full convergence is reached with a fraction of the full signal length by reconstructing the exact numerical values of all the fundamental frequencies and amplitudes within 12-given decimal places. This confirms the achievement of machine accuracy (i.e., numerically exact) in the reconstruction of the spectral parameters, from which metabolite concentrations are then readily deduced. Moreover, the fast Padé transform is shown to unambiguously distinguish genuine from spurious peaks in spectra via pole-zero cancellations or Froissart doublets. We demonstrate the exact reconstruction of the true number of genuine harmonics for noise-free as well as noisy time signals with the pole-zero canonical forms of the fast Padé transform. The clinical significance of this capability of the fast Padé transform is that, in practice, the physical time signals are always corrupted with noise and the major problem is to identify the genuine resonances with fidelity.

The second part of the book begins with chapter 8, first giving a state-of-the-art review of the achievements to date applying magnetic resonance spectroscopy and spectroscopic imaging in the area of neuro-oncology. This is followed by an analysis of the major limitations and dilemmas with magnetic resonance spectroscopy and chemical shift imaging in neuro-oncology with respect to reliance upon conventional Fourier-based data analysis. Then we explore how the accurate extraction of clinically-relevant metabolite concentrations for neuro-diagnostics via magnetic resonance spectroscopy could be achieved via Padé-processing, and including an example of such in a clinician-friendly format. This is both a graphic and a numerical overview which facilitates repeated cross-checking, to help the clinician acquire a deeper grasp of the method, which could build acumen in interpretation of patterns typical of malignancy versus benign pathologies.

Chapters 9, 10 and 11 are direct applications of the fast Padé transform to three problem areas in cancer diagnostics through magnetic resonance spectroscopy: ovarian, breast and prostate cancer, respectively. These problem areas were chosen because of their urgent clinical and public health importance. Each of these chapters begins with a state-of-the-art review of diagnosis of these three types of malignancies, with a focus upon modalities from magnetic resonance.

Compared to the fast Fourier transform, the fast Padé transform applied to magnetic resonance time signals as encoded *in vitro* from benign and malignant ovarian fluid yields dramatically improved signal-to-noise ratio and highly accurate determination of key metabolite concentrations for identifying ovarian cancer. In applications to magnetic resonance signals as encoded *in vitro* from breast cancer, fibroadenoma and normal breast tissue, the fast Padé transform demonstrates the ability to unequivocally resolve and precisely quantify extremely closely lying resonances, including phosphocholine, a marker of malignant transformation of the breast. The fast Padé transform is also shown to yield unequivocal resolution and exact quantification of numerous overlapping resonances including multiplets of metabolites that distinguish cancerous prostate, normal glandular and stromal prostate. We show that the fast Padé transform is optimally suited to resolve and quantify the abundant overlapping resonances, encompassing multiplets in this very difficult area of signal processing in magnetic resonance spectroscopy within prostate cancer diagnostics. Only relatively short time signals were needed for this achievement, and this is a major advantage because free induction decay data become heavily corrupted with noise at the long total acquisition times required by the fast Fourier transform, which lacks interpolation and extrapolation features. Herein, we have once again shown that the fast Padé transform can unequivocally disentangle physical from spurious content of the studied time signals. The many multiplets characteristic of the spectra of healthy and cancerous prostate represent a major challenge for signal processing. Padé optimization has been shown here to meet with success in these challenges.

Readers with an expertize is signal processing, mathematical physics and/or quantum mechanics will likely want to read this book in the order in which it appears, beginning with the theory and analysis, together with the proof-of-principle results for magnetic resonance studies applying the fast Padé transform, and finally its application and potential in urgently needed clinical areas. Perhaps for some such readers this trajectory will be a spur to involvement in further research in this area, as well as being a realization of the remarkable practicality of these theoretical advances. Medical physicists may choose to begin with the proof-of-principle results. They could then progressively branch out in both directions: towards the theory as well as the clinical applications.

Those whose primary expertize is quite distant from the realm of mathematics such as biomedical researchers and clinical practitioners will most likely wish to start with the clinical chapters as motivation. Then, they may proceed to the proof-of-principle results, and finally, the theoretical portions which hopefully can challenge some preconceived assumptions, and thereby offer intellectual gratification. We recommend that this latter readership pay particular attention to section 1.4 in which we address the question: *Why is this topic relevant for biomedical researchers and clinical practitioners*? In striving to answer this question, it is our hope that this book will be part of the overall efforts via "translational" research to help biomedical researchers and clinicians more actively and confidently engage in more fundamental areas such as optimization of molecular imaging through magnetic resonance, as a vital part of the battle against the scourge of cancer.

Dr. Dževad Belkić
Professor of Mathematical Radiation Physics
Karolinska Institute
Stockholm, Sweden
June 2009

Dr. Karen Belkić, MD
Adjunct Professor of Preventive Medicine
University of California School of Medicine
Institute for Health Promotion and
Disease Prevention Research
Los Angeles, USA

Also affiliated to:
Department of Oncology and Pathology
Karolinska Institute
Stockholm, Sweden
June 2009

Acknowledgments

The authors thank the King Gustav the 5th Jubilee Foundation, the Swedish Cancer Society Research Fund, the Karolinska Institute Research Fund as well as the Signe and Olof Wallenius Research Foundation for support. They also appreciate a fruitful cooperation with Mr. John Navas, Ms Amber Donley and Ms Suzanne Lassandro from Taylor & Francis.

1

Basic tasks of signal processing in spectroscopy

Magnetic Resonance Spectroscopy (MRS) and Magnetic Resonance Spectroscopic Imaging (MRSI) are among the leading and most rapidly developing non-invasive diagnostic modalities in medicine [1]–[7]. While MRS focuses upon a single voxel, the combination of MRS and Magnetic Resonance Imaging (MRI) yields MRSI, which provides information from multiple voxels. Thus, the hybrid nature of MRSI lies in its merging of the spectroscopic and the morphological information of MRS and MRI, respectively. An auxiliary terminology for MRSI is occasionally used via Chemical Shift Imaging (CSI). However, this is not an appropriate alternative nomenclature, since it makes merely an implicit reference to the most important, *spectroscopic* aspect of this diagnostic modality.

Irrespective of whether using one or the other competing acronyms, MRSI or CSI, the current practice with Magnetic Resonance (MR) in medicine reduces the rich quantitative interpretation of the spectroscopic information from MRSI/CSI mainly to the qualitative aspect of imaging. This is done by overlaying a few metabolite concentration maps in different colors upon anatomical images of the scanned tissue. Such presentations unnecessarily simplify MRSI by conveying a misleading message that this spectroscopic imaging can sufficiently be understood and used merely in terms of a color-coded MRI scan. Even when the conventional attempts are made to show the spectroscopic information from MRSI, this is done by displaying exclusively Fourier envelope spectra throughout a scanned tissue. Again this is only a qualitative description. The color-coded concentration maps superimposed on MRI scans stem from fitting Fourier envelopes to determine the peak areas. Since any such fitting is ill-conceived, via least-square adjustments of free parameters to the hidden local components of the Fourier global shapelines, it necessarily leads to biased and non-unique estimates of the sought information. This could hardly constitute a clinically appealing avenue, since instead of taking advantage of spectroscopic complementarity to imaging, the currently implemented MRSI with its fitted Fourier envelopes offers to physicians new dilemmas, ambiguities and uncertainties. Such severe deficiencies are inherited and imported directly from MRS. Therefore, if these hampering drawbacks could be overcome first in MRS, this would dramatically improve the prospect of MRSI for becoming a reliable and invaluable clinical modality in modern diagnostics.

One of the main goals of the present book is to show how MRS can potentially establish its theoretically anticipated, but thus far unrealized high-rank status. This can be accomplished through performing data processing by an exclusive reliance upon *ab initio* spectral analyses of proven validity from the realm of quantum mechanics as the most successful physics theory. The proposed strategy is expected to shed new light upon MRSI by providing the most adequate, unequivocal quantitative tabular and graphical data of concentrations of many MR-detectable and clinically informative metabolites. Such analytical chemistry type spectroscopic data contain not only concentrations, but also chemical shifts and relaxation times of all the reconstructed physical metabolites. These detailed data would be impoverished if they were reduced only to chemical shift color-coding of MRI scans. Rather, they require a novel look at MRSI whose proper interpretation necessitates spectroscopic skills from mainstream mathematical physics and chemistry. Indeed, such a comprehensive strategy is dictated by the inter-disciplinarity of MR phenomena. It is through this fresh avenue that imaging and spectroscopy could be synergistically integrated within magnetic resonance into MRSI and, as such, would have a greater chance of attaining its full potential of becoming an unprecedented diagnostic modality.

The experimentally measured or encoded data from MRS and MRSI are time signals, or alternatively, free induction decay (FID) curves. These data then undergo spectral analysis. This is achieved by solving an inverse problem called quantification, entailing the reconstruction of quantitative physical and biochemical information from the examined tissue. The sought data include the number of metabolites, their chemical shifts, relaxation times as well as their abundance, i.e., concentrations. Within the realm of medical diagnostics, the obtained results are typically compared to normative data.

From the mathematical standpoint, the quantification problem involves the use of encoded time signals to reconstruct the true number of harmonic transients and the pairs of spectral parameters that are complex-valued physical quantities – the frequencies and the corresponding amplitudes. These quantities are the sole elements of the fundamental harmonics that comprise the FID under study. Each harmonic transient possesses a resonant frequency, relaxation time, as well as intensity and phase. These four parameters completely characterize the underlying normal-mode damped oscillations in the FID. From the reconstructed complex frequencies and amplitudes these four real-valued parameters can be identified directly, and thereby the peak areas of the associated resonance profiles in the corresponding spectrum can be obtained. The relative concentrations of various metabolites can then also be obtained, since these are proportional to the computed peak areas. Here, constants of proportionality are the concentrations of certain chosen reference metabolites, e.g., water or some other molecules.

The total number of harmonic transients is also a quantifiable parameter, and can be reconstructed during spectral analysis of the FID. Quantification is recognized as one of the main problems from quantum theory of resonances and spectroscopy [5]. Thus, the general quantum-mechanical relaxation formalism can be profitably utilized to solve the quantification problem from MRS and MRSI.

To this aim, we employ quantum-mechanical signal processing and spectral analysis via the Green function. This is implemented algorithmically through the fast Padé transform (FPT) [5]–[37]. The frequency spectrum is thereby obtained as the unique quotient of two polynomials. This form of a rational quantum response function to the external perturbation (static and gradient magnetic fields, radio-frequency pulses) is dictated both by the quantum origin of MRS and MRSI, as well as by the resolvent form, i.e., the operator Padé approximant of the Green operator/matrix of the system under study.

The FPT includes two different versions of the Green function, with the outgoing and incoming boundary conditions inside and outside the unit circle of the complex harmonic variable z. The former and the latter variants of the FPT correspond, respectively, to the causal and anti-causal Padé-z transform from the signal processing literature, as well as from mathematical statistics [5]. In this book, we show *inter alia* how the FPT solves the quantification problem exactly by extracting the sought spectral parameters. This is done using synthesized noise-free and noise-contaminated FIDs generated theoretically, but based upon entirely realistic data as typically encoded via MRS. Experimentally measured time signals from MRS are studied here, as well.

In both MRS as well as MRSI, it is very important to avoid encoding long FIDs, particularly when employing clinical scanners. This is because the exponentially damped envelopes of these FIDs become completely embedded in noise at long total acquisition times T. We emphasize that the fast Fourier transform (FFT) requires long T to augment its frequency resolution which is equal to $2\pi/T$. Therefore, reliance upon the FFT leads to a conundrum, with two diametrically opposing requisites. Namely, attempts at resolution enhancement entail long T, whereas this leads to increased noise which hampers the sought improvement in spectral quality.

With the FFT, it is typically seen that attempts are made to solve this problem via a "patching" procedure by using shorter bandwidths or window sizes. These attempts engender further problems, such as spectral deformations via Gibbs ringing as well as diminished resolution.

The more fundamental reason why these attempted strategies within the FFT have been unsatisfactory lies in the fact that the Fourier analysis completely ignores the actual structure of the encoded time signal. Namely, with the FFT all time signals encoded with the same T have the same resolution $2\pi/T$, regardless of their internal structure. As such, the FFT is limited solely to shape estimations [5]. Yet, each FID has its structure and this is quantifiable by specific parameters. Typically, neither this structure nor the related parameters of the encoded time signal are known before the signal is processed. Spectral synthesis must reveal or reconstruct these unknowns from the encoded time signal. This is accomplished by performing spectral decomposition which, in turn, avoids the need for long T. In parametric estimation, in contrast to Fourier non-parametric processing, resolution is not determined solely by T.

As noted, the structural parameters are the fundamental frequencies and the associated amplitudes, as the elements from which the attenuated harmonics from the investigated FID are built. These nodal frequencies are complex-valued because external perturbations set the investigated system into motion via exponentially damped harmonic oscillations. With the passage of time, these oscillations must be attenuated thus exhibiting decreased intensity of the signal from the system, since the total time duration of the signal is always finite ($T < \infty$) in any realistic experiment. In other words, no process which generates time signals as the system's response to external excitations could possibly last forever, and this is the origin of the said attenuation causing the system to be dissipative. Dissipative systems are associated with non-Hermitean "Hamiltonians" which have complex fundamental frequencies $\{\omega_k\}$ whose real and imaginary parts describe the cosinusoidal oscillations and exponentially damped amplitudes, respectively. Gamow's [38]–[43] complex energies E_k or frequencies ω_k are common in scattering theory physics for transient (metastable, decaying) states such as resonances. Their real parts have the customary meaning of energy or frequency observables (experimentally measurable quantities) as in Hermitean quantum mechanics, whereas the imaginary parts are proportional to the inverses of the individual k th state's half-lifetime, as the measure of the Heisenberg uncertainty stemming from the decaying nature of the considered state.

An important group of algorithms for spectral decomposition are those in the category known as convergence accelerators and analytical continuators of Fourier series and sequences. It should be pointed out that for a given bandwidth, the FFT converges in a linear fashion, at a slow rate $1/N$ with increasing signal length N. This convergence can be accelerated significantly. The FPT is a convergence accelerator which has at least a quadratic convergence $1/N^2$ with a systematically augmented N. Furthermore, near full convergence as a function of the signal length, the FPT displays "spectral resolving power", which denotes a remarkable exponential convergence rate to the exact values of the sought genuine parameters. This feature translates directly into two simultaneous advantages: resolution enhancement and shortening the FID. With this, the so-called Fourier dichotomy that "a resolution enhancement cannot be obtained without prolonging T", no longer is the case. Operationally, this advantage of the FPT is related to its modeling of the spectrum by a quotient of two unique polynomials, rather than the single polynomial from the FFT.

The same result can alternatively be achieved via analytical continuation rather than convergence acceleration. Cauchy introduced this powerful concept of analytical continuation, by which a divergent sequence is made to converge. If, e.g., a series in powers of the harmonic variable z (as the Green function representing the exact spectrum) is convergent outside the unit circle ($|z| > 1$), it will automatically diverge inside the unit circle ($|z| < 1$). An analytical continuator renders the Green function well-defined inside the unit circle and, therefore, throughout the complex frequency plane with the exception of singular points (poles). In addition to being a convergence accelerator, the FPT is also an analytical continuator. In this book, we present several illustrations with FIDs from MRS as cross-validation, to demonstrate that both convergence acceleration and analytical continuation by the FPT yield exactly the same true spectral parameters.

Prior to quantification of FIDs from MRS and MRSI, clearly, the total number of resonances is unknown. This parameter should actually be added to the usual list of unknown quantities that are the complex frequencies and amplitudes [5]. Similarly to the sought fundamental frequencies and amplitudes, the total number of resonances needs to be reliably determined from the spectrally analyzed time signal. Thus far, in signal processing within MRS and MRSI, the usual practice has been to surmise *ad hoc* about the number of resonances. This is done by fitting the Fourier-reconstructed spectral line-shapes via pre-assigned expansion functions or some model *in vitro* spectra. These fittings are neither unique nor objective. The consequence is either under- or over-fitting by under- or over-estimating the true number of physical resonances. In other words, either some genuine resonances could easily be missed, or some spurious ones might be registered. Neither of these contingencies is acceptable, particularly for medical diagnostics.

The total number of resonances is handled by the FPT on the same footing as the fundamental complex frequencies and amplitudes. Here, all three quantities are considered to be the unknowns of the quantification problem. The total number of resonances will not be subject to guessing any longer nor need this parameter be defined in advance. Thus, with Padé-based quantification in MRS and MRSI, the exact number of genuine resonances is determined in a completely reliable manner, as in the case of the reconstructed frequencies and amplitudes. This remarkable feature of the FPT in providing exact retrieval of the true number of physical resonances is due to the fact that this method is the only exact filter for FIDs built from transient harmonics in the form of damped complex exponentials with stationary and/or non-stationary polynomial-type amplitudes. These FIDs are precisely predicted by the quantum-mechanical description of magnetic resonance phenomena. Moreover, such data are encoded via MRS/MRSI when stringent experimental conditions are fulfilled. These conditions are, for instance, proper suppression of the giant water resonance in, e.g., neuro-diagnostics, adequate shimming of the static magnetic field to reduce inhomogeneities, etc. To unequivocally prove these statements, it is essential to apply the FPT to exactly solvable model problems with all controllable, i.e., precisely known input data. Therefore, using synthesized FIDs, we provide illustrations on precisely how the FPT unambiguously determines the exact total number of resonances as well as all the fundamental frequencies and amplitudes.

More specifically, the exact number of resonances is reconstructed employing the powerful concept of Froissart doublets [44], a phenomenon unique to the Padé polynomial quotient, which is the form of the spectrum in the FPT. To make a link to the field of signal processing, it should be noted that the equivalent name "pole-zero cancellation" for Froissart doublets was independently used in studies on discrete time systems described by the PzT [45]. Froissart reported his finding in 1969 in a Symposium Proceeding in Mathematics [44]. Unluckily, his results have not appeared subsequently as an *in extenso* paper in a regular journal to be accessible to a wider audience. Fortunately, already in 1970 – 1972 his method was cited and further analyzed by Basdevant [46, 47], as well as in 1973/1974 by Gammel and Nuttall [48]–[50]. These initial studies dealt with Froissart doublets within the Padé approximant (PA) [51] based upon some Taylor series having the expansion coefficients perturbed by random noise. Subsequently, from 1978 to the current times, the most thorough investigations on Froissart doublets, or equivalently, Froissart "noise functions" as well as on the so-called Froissart polynomials, were accomplished by Gilewicz [52]–[57] who developed a more formal mathematical basis for the whole subject. Froissart doublets are within the realm of non-analytic or quasi-analytic "noise functions" [58]–[61] because they closely mimic random noise, which is encountered in all computations with finite precision numerical arithmetic (round-off-errors), as well as in experimental measurements. This aspect was examined by Barone [62] and illustrated with the help of Monte Carlo simulations. Froissart doublets within the decimated Padé approximant [5, 14] were used by O'Sullivan *et al.* [63, 64] in spectral analysis of experimentally measured noisy time signals from reverberant acoustic environments, including recorded music (the so-named music transposition) where the measured FIDs are reliably modeled by sums of complex damped harmonics with stationary amplitudes. A multi-variate version of Froissart phenomena was formulated by Becuwe and Cuyt [65] and numerically illustrated in the special case of two dimensions.

Within MRS, we have systematized the concept of Froissart doublets in two variants of the Padé approximant, the FPT$^{(+)}$ and FPT$^{(-)}$, by establishing a novel and long-awaited strategy of exact signal-noise separation (SNS), as an unequivocal twofold signature through the extreme closeness of spurious poles to zeros and, simultaneously, via the smallness of the corresponding amplitudes [11, 24, 30, 31, 34]. All these mentioned studies on Froissart doublets represent only a selected small part of the otherwise large bibliography which can be found in Ref. [6]. Froissart doublets will be revisited in depth with detailed illustrations in chapters 3, 5 and 6.

Noise is ubiquitous and, to re-emphasize, it appears in time signals that are measured experimentally or generated theoretically. As stated, the origin of noise in computations is in finite arithmetics, and it manifests itself through random round-off-errors. Noise in measured FIDs is usually due to many factors that can be systematic errors, statistical uncertainties, random perturbations, etc. Irrespective of its multifaceted origin, for reliable interpretation of measured time signals and their spectra, it is of prime importance to devise an adequate method for distinguishing, with certainty, the true from false information which are mixed together in all experiments. Failure to do precisely this is one of the major causes for potentially incorrect conclusions, even when meticulous attention has been paid to perform the given measurement. Froissart doublets within the Padé approximant come to rescue the situation by providing the most robust method to date for unequivocally distinguishing genuine from spurious information, as has systematically been demonstrated within MRS [11, 24, 30, 31, 34]. Given that the Padé methodology is unprecedentedly spread over many areas, as reviewed recently in Refs. [5, 34], wider applications of the powerful and versatile Froissart concept of signal-noise separation are anticipated across interdisciplinary research fields of basic and applied sciences as well as technologies.

Fourier analysis, being linear, treats noise and physical signal on the same footing and, therefore, cannot distinguish one from the other. Fourier grid frequencies, being real, are all located precisely on the circumference $|z| = 1$ of the unit circle in the complex plane of the harmonic variable z. The same locations along $|z| = 1$ represent the natural boundary of noise to which noise frequencies tend to distribute themselves with probability 1, except at most for a nearly zero-area set (i.e., an open set of points of measure smaller than δ, where δ is an infinitesimally small positive number) [58, 59, 61]. Thus, maximal mixing of noise and physical signal occurs when using the FFT. This explains why the Fourier analysis fails for time signals with poor signal-to-noise ratio (SNR). By implication, such a feature automatically advocates against using maximally noise-polluted total shape spectra from the FFT for post-processing via fitting as an attempt to solve the quantification problem in MRS and elsewhere.

On the other hand, the non-linearity of the Padé analysis helps distinguish the genuine from noisy part of the signal, such that the latter can be suppressed. The mechanism for this is simple and direct for, e.g., the $\mathrm{FPT}^{(+)}$ which pushes all the noise poles to the unit disc boundary $|z| = 1$ (which thus acts as a noise attractor), and simultaneously keeps all the physical poles strictly in the interior of the unit circle $|z| < 1$. Hence the cleanest signal-noise separation. Specifically, using the concept of Froissart doublets in the FPT, non-physical poles are identified by their very tight pairing with the corresponding zeros (strong pole-zero couplings), and the ensuing negligibly small amplitudes. Discarding such spurious Froissart doublets as defective resonances *de facto* purifies the genuine information, and this is how noise suppression is achieved in the FPT. In general, should some "defect poles" ("stray poles") be found by the FPT, they will automatically be accompanied by the corresponding "defect zeros" ("stray zeros") which would annul the ensuing spurious Froissart resonances through the vanishingly small amplitudes/residues. Thus, in the Padé spectrum, any defective part stemming from over-estimating the exact number K of resonances disappears altogether in the end through either implicit or explicit cancellations of the found defect poles and defect zeros.

Defect poles and zeros are also encountered in non-spectral problems, e.g., in approximations of a given function by the Padé approximant. This can best be seen when using a finite-order (say K') diagonal Padé approximant $[K'/K']_f(z)$ to represent a function $f(z)$ which has an infinite number of known true poles and zeros. Invariably, a number $L' < K'$ of Padé-detected poles and zeros will be close to their original counterparts from $f(z)$, whereas the remaining $M' \equiv K' - L' > 0$ ones will be fake. Good agreement will be found between the PA and the original function in the domain containing L' true poles and zeros. Near the fake pairs, $[K'/K']_f(z)$ will considerably deviate from $f(z)$. Such deviations will widely and unpredictably change when the order is increased from K' to $K'' > K'$. The number of true poles and zeros will increase from L' to $L'' > L'$, and so will the number M'' of the fake ones ($M'' \equiv K'' - L' > 0$). In general, the numerical values of true poles and zeros will improve for K'' relative to K'. Simultaneously, the distances between new fake poles and zeros will diminish, as this is precisely how the concept of "convergence in measure" manifests itself [61]. Eventually, with the Padé order becoming infinitely large, the PA will converge almost everywhere to $f(z)$ except at a countable number of tightly packed fake poles and zeros. However, even in this simple application of the PA aimed at obtaining the approximation $[K/K]_f(z) \approx f(z)$, the found fake poles and zeros are not a totally useless defect. Quite the contrary, we may say that their existence is even necessary for any finite order $K < \infty$. It is precisely through these fake poles and zeros, appearing as Froissart pairs/doublets, that the Padé approximant $[K/K]_f(z)\,(K < \infty)$, despite its finite number of poles and zeros, is nevertheless able to mimic quite well the function $f(z)$ which has an infinite number of poles and zeros. Such fake poles and zeros, i.e., Froissart doublets, being artefacts, cannot be found in $f(z)$ and, moreover, they induce randomness which is the noisy part of the approximation $[K/K]_f(z)$. Yet, we see that such random artefacts can be useful. In some other examples, it is precisely through lining up of Froissart poles and zeros that the PA is able to closely mimic functions with cuts and branch points. Thus, in general, it is because of Froissart doublets that the PA, as a meromorphic function, having poles and zeros as its only singularities, is capable of providing good approximations to non-meromorphic functions with cuts and/or branch points.

In signal processing, by increasing the running order K' of the diagonal PA, one always finds a number of genuine and spurious (Froissart) resonances for a time signal which has exactly K true harmonics. Further, one can illustrate, as will be done in the present book, that within finite arithmetics, the optimal Padé spectral analysis is achieved in reconstructing a large number of input parameters for all the genuine resonances if K' is considerably larger than K. The resulting $K' - K > 0$ Froissart doublets help achieve machine accuracy (e.g., 12-digit input, 12-digit output for the retrieved true resonances), while simultaneously annihilating themselves in the final absorption spectra through the pole-zero cancellations. By contrast, if one could be able to perform spectral analysis within exact arithmetics, the Kth order Padé approximant $[K/K]_G(\omega)$ to the input K th rank Green function $G(\omega)$, built from a time signal with K exponentially damped complex harmonics, would suffice for a complete reconstruction with infinitely many decimal places. This would occur due to the lack of round-off errors in the exact arithmetics in which all the computations are hypothetically done with infinite precision. However, in reality, one uses finite, i.e., inexact arithmetics with approximate computations leading to inevitable round-off errors, which appear as random noise with the ensuing Froissart doublets, even for idealized noise-free input data. Crucially, noise in the input data is manifested in the Padé-created output results also through Froissart doublets with a pattern of forming easily identifiable defect pairs similarly to those due to round-off computational noise.

The most important issue to study in system theory is robustness of systems. A system is said to be robust if it is stable. Stability of a system is judged by its sensitivity to external perturbations. One of the major destabilizing factors of any system's performance is noise. Random noise is very sensitive to even the slightest changes of its environment. The reason for this is the lack of coherence of sets of random elements that do not have anything in common to hold them together because they are uncorrelated. Thus, it is imperative to eliminate or at least significantly reduce the amount of random noise in any investigated system. To achieve this goal of paramount fundamental and practical importance, one has to answer the key question: how to tell the difference between what is false and true or spurious and genuine? Padé tells us in which mathematical form the response function should be for the system to act robustly to external perturbations. Froissart furnishes a way to get rid of random noise. As such, the fast Padé transform and the denoising Froissart filter (DFF) act synergistically to stabilize the system's performance. Hence the sought robustness.

Hereafter, the terms genuine and spurious are used to mean physical and unphysical, respectively, in the precise sense of non-zero and zero norms (or equivalently, finite and infinite normalization constants) of the associated eigen-functions. Spurious poles and zeros in Froissart doublets literally coalesce onto each other and, consequently, this leads to the corresponding zero amplitudes. Amplitudes, being residues, are proportional to the pole-zero distances. Norms are proportional to amplitudes. Zero-valued spurious amplitudes, like those in Froissart doublets, correspond to eigen-functions with zero norms or infinite normalization constants and, therefore, are associated with improper, unphysical states. Conversely, all genuine resonances are described by eigen-functions with finite normalization constants or non-zero norms and, as such, can describe physical states of the investigated system.

For an adequate answer to the posed question, one must first acquire the necessary knowledge about the ways systems respond to external perturbations. Here, e.g., system theory teaches us from practical considerations that the most robust response function of a given system to generic external perturbations is provided by rational functions in the form of polynomial quotients, i.e., the Padé approximant. Precisely the same response function, which is equivalently called the Green function, is one of the major methods in quantum mechanics, as well. It is remarkable that these two seemingly distant disciplines share the common Padé approximant as one of their main, multifaceted work strategies.

In fact, many among the existing major algorithms are equivalent or can be reduced to the Padé approximant, e.g., Frobenieus' normal forms, Krylov-Danielevskii expansion method, tridiagonal Jacobi $J-$matrix, Shanks' transform, Wynn's $\varepsilon-$algorithm, auto-regressive moving average, decimated signal diagonalization, continued fractions, Lanczos nearest neighbor approximation (or minimal iterations), Stiltjes' power moments, Löwdin's partitioning and inner projections, Willer's modified moments, Rutishauser's quotient-difference, Gordon's product-difference, Haydock-Heine-Cullum-Willoughby's recursive residue generation method, variation-iteration method, Schwinger's variational principle, the so-called N/D (numerator/denominator) method, Fredholm's determinants, method of finite-rank separable potentials, etc. [5]. Moreover, the Padé approximant P/Q has one unparalleled advantage over all these or other unmentioned methods in signal processing. This is the availability of the closed, analytical expressions derived by Belkić [5, 34] for the general expansion coefficients of the numerator P and denominator Q polynomials of any degree, continued fractions, as well as for the parameters of the quotient-difference, product-difference and Lanczos' algorithms (coupling constants, eigen-functions, including their norms, etc.). This is of paramount importance because Padé-based signal processing from explicit formulae can be used to benchmark the corresponding numerical algorithms for otherwise mathematically ill-conditioned inverse problems, such as the quantification problem in MRS and elsewhere.

On a more general note, in addition to unifying so many apparently different computational algorithms into the general Padé formalism of rational functions, the Padé approximant has an unprecedented status in theoretical physics for certain special reasons. This is the case because the Padé approximant is known to have rescued the working status of several entire theories. The most illustrative examples that lend support to this assertion are the Brillouin-Wigner perturbation theory and the theory of strong interactions in elementary particle physics. The Rayleigh-Ritz perturbation theory has good convergence properties. It is a well-established framework, which for non-degenerate states predicts accurate discrete energies with estimates for the related bounds. To degenerate states, however, one needs to apply the Brillouin-Wigner perturbation theory, which is plagued by divergences that cast doubt upon the usefulness and the prospect of the whole theory. Here, the Padé approximant played a direct role to resum the divergent perturbation expansions and, moreover, to find bounds to the exact energies of degenerate systems. In the theory of strong interactions, the only working method is the perturbation expansion, for which much effort has to be invested for computations of larger-order Feynman diagrams. But when such hard-earned perturbative corrections are finally summed up, divergent results are invariably obtained. This occurs because the coupling constant for strong interaction is large and, as such, does not represent a good perturbation parameter for series expansions. As a consequence, the perturbative solutions of renormalizable field theories in strong interactions were severely limited to mainly understanding certain analytical properties of the scattering $S-$matrix, without yielding any quantitative predictions of observables [66]–[68]. Further, from experimental measurements, a great many resonances were available awaiting proper interpretations of mass spectra via quantitative predictions of positions and widths of these unstable particle states. As such, the said perturbation divergences made the entire theory of strong interactions unphysical but, here again, the Padé approximant salvaged this situation by resumming divergent series to the correct results. Moreover, experimental data on resonant scattering phenomena were and continue to be properly interpreted by the Padé approximant which gave accurate predictions for positions and widths of peaks in spectra for masses of metastable particles [69]–[72].

Overall, we emphasize that although the present book will primarily be focused on spectral analysis of biomedical time signals from MRS in all the concrete illustrations, the expounded general methodology remains applicable as well to MRSI, and all other fields where FIDs can be described by linear combinations of complex attenuated exponentials.

1.1 Challenges with quantification of time signals

At present, versatile MR modalities in medical diagnostics are widely considered as one of the fastest expanding fields of the cross-disciplinary research in medical physics. Nevertheless, the most recent vivid progress particularly in MRS and MRSI, also uncovered the fact that the indispensable signal processing meets with huge challenges in reliably extracting information about biochemical and physical functionality of the most clinically relevant metabolites of the investigated tissue. This type of information could be obtained by MRS and MRSI under the assumption of achieving high-quality, both in encoding and signal processing. As to signal processing, it is mandatory to be able to carry out unequivocal, accurate and robust quantifications of concentration of metabolites of the examined tissue. As mentioned, quantification consists of spectral decomposition of the encoded FID into its harmonic constituents. Such a synthesis is achieved mathematically once the encoding has been performed, because quantitative information cannot be extracted directly from the encoded raw FID without resorting to signal processing.

In the framework of MRS, the signal processor which has been most frequently used is the FFT. The FFT is employed to efficiently convert the encoded FID to the frequency domain with a spectrum of resonances. However, such a transformation is merely a qualitative description of metabolites of the studied tissue. Each metabolite has a molecular structure which might contain more than one resonance. This often leads to a number of congested peaks in a typical spectrum of metabolites. Among such spectral structures, unambiguous quantification of overlapping resonances presents a major challenge for adequate interpretation of encoded data. Attractiveness for processing FIDs by the FFT within MRS is largely due to the stability, efficiency and robustness of automatic computations via the fast $N\ln_2 N$ algorithm for any fixed signal length N whenever $N = 2^s$ $(s = 0, 1, 2, ...)$. For a given FID, the FFT spectrum has a simple stick structure and, as such, it exists only at the Fourier grid frequencies $2\pi k/T \, (0 \leq k \leq N-1)$. Here, T as the total acquisition time is given by $T = N\tau$, whereas $\tau > 0$ is the sampling rate, or equivalently, dwell time, which is equal to the inverse of the selected bandwidth. Clearly, for a theoretically synthesized FID, T has the same significance as the total acquisition time in measurement, but actually it is called the total duration of the FID. Naturally, in a larger interval, no encoded FID is zero for any realistically selected T. Nevertheless, the common practice with the FFT is to artificially double the original FID length by adding zeros, and this is called zero filling or zero padding. As a result, the length of the ensuing FFT spectrum is also doubled. The supplementary N frequencies in the FFT spectrum give a simple sinc-interpolation characterized by wiggling side lobes around each genuine resonance. A drawback of such an outcome, especially for any two closely spaced resonances, is that the sinc side lobes coupled with truncation artefacts might interfere constructively or destructively to yield extraneous peaks or dips. Any extraneous or spurious spectral peak is a false and unphysical structure which is not a part of the true information contained in the investigated FID. This sinc-interpolation cannot be systematically improved, as empirical practice shows that no enhanced spectral quality is achieved by introducing additional zeros into the FID beyond the first doubling of N.

The mentioned truncation artefacts in the FFT are spectral deformations known as Gibbs oscillations that stem from the experimental impossibility to encode infinitely long FIDs. On the other hand, only FIDs of infinite length can give the exact Fourier coefficients. Therefore, the FFT itself can give merely some approximate Fourier coefficients for any encoded FID. In encoding an FID, the first selected quantity is the bandwidth. The second selected quantity is the signal length N, so that T is automatically fixed by $T = N\tau$. However, the unavoidable truncation in the time domain is critically determined by τ itself. This occurs because τ is the only quantity determining the Nyquist frequency $1/(2\tau)$ as the largest frequency which can be sampled for a fixed bandwidth. The Nyquist frequency sets the upper limit to the frequency content of the FID prior to encoding. Hence, it is such a limit which leads to unavoidable truncation errors in the time domain, since no FID can be limited simultaneously in the two conjugate domains – time and frequency. The role of a concrete value of N appears on the level of indicating the actual extent of the truncation error. A sufficiently large N means that the total truncation error will be dominated by noise. Gibbs oscillations are reduced for a long FID, but the resulting signal-to-noise ratio is diminished. A better SNR is obtained for a shorter FID, but this enhances Gibbs oscillations. Thus, in the FFT, a longer FID reduces spectral deformations, but also worsens SNR. This trouble is unsolvable using only the FFT because of its fundamental drawback – the lack of extrapolation features.

As noted, the resolution in the FFT is defined by the minimal separation $\omega_{\rm min}$ between any two adjacent frequencies, $\omega_{\rm min} = 2\pi/T$. Thus, all the FIDs with the same T, irrespective of their internal structure, will have the same resolution which is fixed prior to processing by the FFT. In other words, the resolving power in the FFT is pre-determined by the separation between any two adjacent Fourier grid points $2\pi k/N \, (0 \leq k \leq N-1)$. The minimal frequency separation $\omega_{\rm min}$ is also called the Rayleigh bound or the Fourier uncertainty principle.

In signal processing, two kinds of spectral shapes are encountered. These are the component shape spectra and the corresponding total shape spectrum. The component shape spectra are the spectra for every separate resonance, and they can be generated only via parametric estimators. The total shape spectrum represents the sum of all the separate component shape spectra. The total shape spectrum is also the envelope spectrum. This stems from the fact that a constructive and destructive interference of all the individual resonances yields the total spectral shape.

As stated, the FFT as a non-parametric processor can give only envelope spectra. For this reason, the FFT and other non-parametric processors are called envelope estimators. By contrast, parametric estimators are able to yield both component and total shape spectra, since they can obtain the peak positions, widths, heights and phases of individual physical resonances. The main difference between these two categories of processors is in the kind of information extracted from the input FID. Non-parametric processors can give merely qualitative information, which is an apparent information seen on graphs of spectral shapes. However, more important quantitative information can be obtained by parametric estimators through unfolding the hidden spectral structure of the envelope. Crucially, parametric methods can reconstruct the quantitative features of resonances (peak position, width, height, phase) as the essential ingredient for reliable estimates of concentrations and relaxation times of the clinically most relevant metabolites of the examined tissue.

The spectral parameters of resonances that determine chemical shifts, relaxation times and concentrations of metabolites are the complex frequencies and the corresponding complex residues (amplitudes) as the main constituents of the damped harmonics from the associated FID. Surprisingly, with all its drawbacks, fitting is still most frequently employed in MRS for estimation of these critically important spectral parameters. Here, envelope spectra from the FFT are fitted by least square (LS) adjustments of some assumed resonance line-shapes such as Lorentzians, Gaussians or their sum. Similar LS techniques are also employed in the time domain by fitting a sum of damped complex exponentials or Gaussians or their product to a given FID. Examples of such fittings are the "variable projection" (VARPRO), the so-called "advanced method for accurate, robust and efficient spectral fitting" (AMARES), etc. [73]–[75]. As another option to these fittings of every individual peak, similar LS adjustments of the given *in vivo* FFT spectrum as a whole are also frequently practiced with MRS data. This can be done via, e.g., the "linear combination of model" (LCModel) *in vitro* spectra [76]. The LCModel uses a separately encoded FID on a phantom to set up a basis of "model *in vitro* spectra" that are subsequently fitted to the studied *in vivo* spectrum which is subjected to quantification. Such model metabolites from *in vitro* spectra can alternatively be pre-selected from the corresponding data banks, if available, for every concrete case. In practice, no sufficiently matching data bases exist for patients, and pre-encoding with phantoms is invariably not close enough to *in vivo* FIDs. These fittings can use some prior information to constrain the variations of the adjusted parameters during non-linear minimization of the LS residuals. Such residuals are defined as the difference between the observed envelopes of spectra or FIDs and the corresponding modeled data. In practice, these fittings are set up by means of iterations. In particular, the LCModel uses the Levenberg-Marquardt non-linear fitting algorithm [77] and gives some error assessments of the Kramer-Rao type or the like.

The most serious fault of all the existing fitting recipes is non-uniqueness, as the same envelope spectrum can be freely adjusted to have any subjectively chosen number of resonances. In practice, all curve fitting algorithms are sensitive to the choice of the number of components and extremely sensitive to even small errors in the input data. Moreover, in every LS minimization based upon non-linear fitting techniques, such as VARPRO, AMARES, LCModel, etc., the found minimum is, in fact, local, rather than global. This means that there is more than one minimum having statistically the same χ^2 and LS residuals, within the prescribed threshold for the sought accuracy. In other words, different minimae can give nearly the same residual or error spectra (defined as the model function minus the input data), but the ensuing collections for predictions of spectral parameters could be very different and there is no criterion to state which of the sets of the estimates for the sought parameters is correct (if any). In order to patch such fundamental inconsistencies, some prior information is customarily used in LS fittings in MRS, but this could easily lead to biased estimates in quantification even when the assessed Kramer-Rao bounds decrease, as pointed out in Refs. [8]–[11]. This is partially due to the usage of non-orthogonal expansion sets. With such sets, any change of the weighted sum of the squares of the LS residuals, caused by alterations of one or more adjustable parameters, could largely be compensated by independent variations of all the remaining free parameters [78]. To recapitulate, all fitting algorithms are sensitive to the choice of the number of components to be fitted and are also extremely sensitive to even small errors in the input data[1]. Various fitting procedures used in the time and frequency domains within MRS were reviewed in Refs. [79]–[83]. However, all such fittings constitute a naive approach to spectral analysis due the usage of some *ad hoc* mathematical formulae that do not stem from any adequate physical description of the underlying dynamics. This basic inadequacy is usually patched by employing some prior information in order to artificially constrain the otherwise completely arbitrary variations of adjustable parameters.

Conceptually different from fitting, there exists another strategy for signal processing in MRS. This is quantum-mechanical spectral analysis which converts the quantification problem from MRS into the Schrödinger eigenvalue problem of the evolution matrix or the Hankel (data) matrix. Here, diagonalization of such matrices yields the spectral parameters. Rigorously the same results can alternatively be obtained without diagonalization by solving the corresponding secular equation through rooting the characteristic polynomial. Being parametric and non-iterative, this kind of estimator can unequivocally reconstruct the unknown complex frequencies and amplitudes of each physical resonance. With such processors, these main resonance parameters are retrieved first, and it is only afterwards that the corresponding component and total shape spectra can be constructed, if desired. In other words, this represents a completely different methodology from fitting estimators that need the pre-computed envelope spectrum prior to their attempts to quantify the encoded data. These two distinctly different approaches, i.e., ambiguous fittings and unambiguous quantum estimation are anticipated to give appreciably different results, particularly regarding closely overlapping resonances. Experience shows that spectra in MRS are abundant with tightly overlapped resonances. These latter spectral structures are often of primary clinical importance.

An example of quantum-mechanical parametric estimators is the Hankel-Lanczos Singular Value Decomposition (HLSVD) which is in frequent use within MRS [74]. The HLSVD arranges the encoded raw FID points $\{c_n\}$ as a data matrix of the Hankel form $\{c_{i+j+1}\}$, which is subsequently diagonalized to yield the fundamental complex frequencies for the reconstructed harmonics from the time signal. The associated complex amplitudes are extracted in the HLSVD subsequently through a description of the studied FID by a linear combination of complex damped exponentials with time-independent amplitudes. Such a description yields a system of linear equations for the sought amplitudes. This latter system uses all the found frequencies (genuine and spurious). Such a mix of frequencies is disadvantageous, since the admixture of even a slight amount of extraneous frequencies could severely undermine the reliability of the estimates for the amplitudes in the HLVSD.

[1]Here, in the context of fitting by, e.g., VARPRO, AMARES, etc., within MRS, "components" are the constituent resonances in a total shape spectrum. More generally, "components" could be, e.g., the individual species in a chemical or any other compound/sample or the individual compartments in a multi-compartment tissue/system.

Another drawback of the HLSVD is the fact that from the onset it is designed to work only with Lorentzian line-shapes, despite the abundant appearance of non-Lorentzians in all spectra encountered in MRS with encoded FIDs. In practice, the HLSVD is implemented for an over-determined system of linear equations. In such a system, the number of equations is larger than the number of the unknowns. Consequently, the redundant input information from the data matrix leads to the so-called singular eigen-values. These are the nearly zero eigen-values that cause the inverse of the diagonalized Hankel matrix to become almost singular. Removing such singular values reduces the rank of the investigated matrix, and this occurrence is an integral part of the procedure called the singular value decomposition (SVD).

The fast Padé transform is a powerful and versatile quantum-mechanical processor for parametric estimations which simultaneously lifts both of the mentioned restrictions of the HLSVD. The FPT can be set up in both the time and frequency domain of estimations. This is achievable through several numerical algorithms and algebraic methods that invariably give the spectrum in the form of the quotient of two uniquely determined frequency-dependent polynomials. Such rational polynomials can be in several different forms, e.g., the most general non-diagonal $P_L/Q_K (K \geq L)$, diagonal P_K/Q_K, para-diagonal P_{K-1}/Q_K, etc. For example, in the case of the diagonal quotient where both polynomials have the same degree K, the algebraic condition for a strictly determined system is given by $2K = N$. In other words, here, at least $2K$ FID points are needed to obtain K frequencies and K amplitudes, since in a determined system, the number of equations is equal to the number of unknown quantities.

While staying in the same computational framework, the FPT comprises both the usual Padé approximant and the causal Padé z−transform (PzT). The PA is a well-known method from numerical analysis and the theory of approximations whose most important branch is the class of rational functions. Similarly, the PzT is also well-known from signal processing and statistical mathematics [84, 85]. The PA and the PzT variants of the FPT are defined with their initial convergence regions lying outside and inside the unit circle in the complex harmonic variable plane, or the z−plane. The acronyms for these two versions of the FPT are $\mathrm{FPT}^{(-)}$ and $\mathrm{FPT}^{(+)}$, respectively.

It is important to emphasize that the $\mathrm{FPT}^{(+)}$ does not diverge outside the unit circle and neither does the $\mathrm{FPT}^{(-)}$ inside the unit circle. This crucial feature is secured by the universal Cauchy principle of analytical continuation. Such a principle applied to the $\mathrm{FPT}^{(+)}$ prolongs its initial convergence region, which is inside the unit circle, to the complementary domain outside the unit circle. Likewise, the same Cauchy principle when used in connection with the $\mathrm{FPT}^{(-)}$, extends its initial convergence region, which is outside the unit circle, to the complementary domain inside the unit circle. Thus, in the case of both the $\mathrm{FPT}^{(+)}$ and $\mathrm{FPT}^{(-)}$, their respective initial and extended convergence regions cover the entire complex frequency plane, with the exception of poles as points of singularity.

Unification of these two computationally different algorithms into a single concept and methodology yields the clause 'Padé transform' in the acronym FPT. Further, the adjective 'fast' in the FPT stands to indicate that the envelope spectra from the FPT can optionally be computed by the fast $N(\ln_2 N)^2$ Euclid algorithm [86, 87]. In its role of a transform, the FPT does not need to reconstruct the spectral parameters first. Such an application of the FPT can be implemented through, e.g., acceleration of Fourier sequences by the well-known Wynn $\varepsilon-$non-linear recursion. The Wynn algorithm has been invented to alleviate a cumbersome, direct evaluation of quotients of Hankel determinants that enter the definition of the Shanks transform (ST) [5]. The Shanks transform is an extension of the Aitken Δ^2-extrapolation from one to an arbitrary number of harmonics (further details are given in chapter 2). When a sequence or a series to be non-linearly transformed is comprised of partial sums, as is the case in the acceleration of Fourier sequences by the Wynn algorithm in the frequency domain, the Shanks transform is identical to the Padé approximant. In such a case, the Wynn $\varepsilon-$algorithm is simply a recursive generation of the Padé approximant in a frequency-by-frequency sweeping throughout a chosen window in the spectrum or in the whole Nyquist range. In the present context of MRS, the Wynn recursion is extremely efficient if applied to many frequencies because the Fourier sequences to be accelerated are very short, containing barely a dozen pre-computed FFT spectra of increasing length, $N = 2^s$ $(s = 0, 1, ...)$ [12]. These sequences are short, since the full length of a typical FID encoded at clinical scanners by means of MRS for a chosen bandwidth does not ordinarily go beyond $N = 2048$, and this corresponds to 2^s for $s = 11$.

By construction, the FPT is set up to work for both Lorentzian (non-degenerate) and non-Lorentzian (degenerate) spectra. The latter degenerate spectra include peaks due to multiple roots of the denominator polynomial Q_K in the defining quotient from the FPT. Crucially, being a rational response function, the FPT is simultaneously an interpolator and extrapolator. A particularly important feature of the FPT is extrapolation, since it leads directly to resolution enhancement. Moreover, unlike the resolution $\omega_{\rm min} = 2\pi/T$ in the FFT, resolution $\omega_{\rm ave}$ in the FPT is not restricted critically by T. Instead $\omega_{\rm ave}$ depends upon the average density of resonances in the given frequency window. In most situations, we have $\omega_{\rm ave} < \omega_{\rm min}$. Such a circumstance permits resolution enhancement below the Rayleigh-Fourier limit $2\pi/T$ [5, 13].

Among users of fitting prescriptions for *in vivo* MRS, there is a perception that, in general, successful quantification should include some kind of prior information. This is certainly untrue for non-fitting algorithms such as the FPT or HLSVD. The real reason for which some prior information is used at all in fitting recipes should be clarified. Ample evidence proves that fittings in MRS customarily employ certain chosen prior information (e.g., a fixed ratio among the heights or widths of some resonant peaks or the like), exclusively to constrain the possibly wide variations of the otherwise free adjustable parameters. Without such constraints, fittings usually yield inadequate reconstructions since, as mentioned, changing some free parameters could easily lead to large and unphysical variations in the other adjustable parameters. Such a basic defect of fitting is then generalized by the users of fittings to state that all the adequate algorithms for quantification should introduce prior information. This generalization is misleading. In principle, certain physically motivated prior information could be imposed upon the Padé estimation, if needed. In fact, the adequate prior information to the Schrödinger eigen-problem is already inherently introduced by the proper initial or boundary condition. Moreover, according to the main working postulate of quantum mechanics, the reconstructed total Schrödinger eigen-state contains the whole information about the studied system. Hence, there is no need to impose any other prior information, since the correct solution of the Schrödinger eigen-problem leads to the true eigen-spectrum with the sought complex-valued fundamental frequencies and amplitudes. Should some of these spectral parameters obey certain relationships that are known to exist prior to quantification, such relationships would also be inherently ingrained in the investigated FID and, hence, extractable from the retrieved Schrödinger eigen-spectrum with no imposed constraints. In parametric methods, prior information is used through a structure of the given time signal. The FPT is a parametric estimator which is also a quantum-mechanical eigen-problem solver [5]. This method employs directly or indirectly the Schrödinger equation, which can be solved exactly in a finite-dimensional space through the Krylov or the Lanczos basis set functions [5].

Within the FPT, computation is done by systematically enlarging the dimension of the Schrödinger state space (also carrying the name of Krylov) with the purpose of monitoring and detecting constancy of the retrieved spectral parameters. This represents a procedure of proven validity for demonstrating that the information uncovered from the given FID is complete, without the need to introduce any constraint by some prior information. Therefore, prior information used in fitting techniques becomes superfluous and unsubstantiated by quantum-mechanical spectral analysis, which quantifies MRS data without any fitting, as has previously been demonstrated within the FPT [5, 24] and this will also be illustrated in the present book. Even describing FIDs from MRS by linear combinations of complex damped exponentials with both stationary and non-stationary amplitudes, as done in the FPT, cannot be considered as an imposed, constrained relationship. This comes from the fact that precisely such a relationship follows directly from the two main entities of quantum physics – the exact evolution operator and the related auto-correlation functions.

Recent vigorous advances of the Padé methodology in MRS and MRI have made it evident that the FPT is a very reliable and robust signal processor, which gives the full and intrinsically cross-validated spectral information of great significance for medical diagnostics [16]–[37], as will also be thoroughly reviewed in this book. Such a finding from the MR literature is reminiscent of the well-established status of the general Padé theory over the years across inter-disciplinary scientific and engineering research, including physics, chemistry, biology, signal processing, system theory, circuit theory, speech pattern recognition, etc. A recent review of the versatile use of the PA in such vastly different research branches has been provided in Ref. [5]. Further, relevance of the FPT for medical diagnostics, particularly in clinical oncology has also been been thoroughly analyzed [7]. We re-emphasize that the PA and FPT belong to the same general Padé methodology. It has been shown in quantum chemistry that the FPT can carry out parametric analysis of theoretically generated and experimentally measured FIDs through extremely accurate computation of complex frequencies and amplitudes of the reconstructed physical resonances [5, 28, 34]. Such a conclusion is also relevant to medicine, especially when quantifying FIDs that are encoded from patients via MRS or MRSI. In these reconstructions, the quantification problem is recognized as the harmonic inversion problem and/or the complex-valued power moment problem [5]. The task of this inverse problem is to reconstruct the proper number K of physical resonances, as well as their true fundamental complex frequencies and amplitudes $\{\omega_k, d_k\}$ $(1 \leq k \leq K)$ that build the given FID, which is the only input data available.

Regarding the quantification problem on synthesized FIDs, we will select the input complex frequencies and amplitudes according to other similar data in the MRS literature [88]. In order to secure a robust error analysis and cross-validation of the obtained results, both versions of the FPT via the $\mathrm{FPT}^{(+)}$ and $\mathrm{FPT}^{(-)}$ will be employed throughout. One of the main objectives of the exemplified quantifications in MRS is to assess the overall performance of the FPT for parametric estimation of fully controlled theoretical FID with and without noise. The FPT is scrutinized rigorously against these entry data by testing for the possibility of reconstructing all the spectral parameters exactly. These testings also aim at verifying whether the FPT is stable and reliable while retrieving the total number of resonances and their complex frequencies and amplitudes by exhausting only a portion of the full signal length N.

A successful outcome of such a testing would have far reaching ramifications beyond one-dimensional (1D) MRS and thus also become important for two-dimensional (2D) MRS, 2D MRSI as well as three-dimensional (3D) MRSI and MRI. In particular, for 2D MRS, one of the total acquisition times is a fraction of the corresponding value from 1D MRS. Therefore, whenever the 2D FFT is employed to compute cross-correlation plots using *in vivo* 2D FIDs encoded via 2D MRS, low-resolution results are obtained along the frequency axis associated with the short total acquisition time [89]. In such cases, no appreciable improvement in the results of the FFT is achieved by zero filling of the FID on the short time axis. Hence the need for signal processors that possess extrapolation features. Such features are automatically built into the FPT via modeling the response function by polynomial quotients. These latter rational functions describe adequately and realistically the manner in which the examined physical system responds to external perturbations. As opposed to the FFT, the FPT predicts that the system's answer to these excitations is non-linear. Here, non-linearity of the FPT can turn into an advantage through the possibility of diminishing the influence of noise on the reconstructed data.

Before proceeding to detailed applications of the FPT, some basic definitions with a brief, but thorough discussion of pertinent aspects of the quantum-mechanical origin of resonances will be given next.

1.2 The quantum-mechanical concept of resonances in scattering and spectroscopy

When two nuclei collide with each other, certain particle transformations known as nuclear transmutations can occur. For instance, in the case of scattering of proton (${}^{1}_{1}\mathrm{H}$) with carbon (${}^{12}_{6}\mathrm{C}$) or aluminum (${}^{27}_{13}\mathrm{Al}$), the following nuclear reactions termed radiative capture are possible with emission of photons (γ)

$$ {}^{1}_{1}\mathrm{H} + {}^{12}_{6}\mathrm{C} \longrightarrow {}^{13}_{7}\mathrm{N}^{*} \longrightarrow {}^{13}_{7}\mathrm{N} + \gamma \tag{1.1} $$

$$ {}^{1}_{1}\mathrm{H} + {}^{27}_{13}\mathrm{Al} \longrightarrow {}^{28}_{14}\mathrm{Si}^{*} \longrightarrow {}^{28}_{14}\mathrm{Si} + \gamma. \tag{1.2} $$

In the collisional channel ${}^{1}_{1}\mathrm{H} + {}^{12}_{6}\mathrm{C} \longrightarrow {}^{13}_{7}\mathrm{N}^{*}$ from process (1.1), or in the similar part ${}^{1}_{1}\mathrm{H} + {}^{27}_{13}\mathrm{Al} \longrightarrow {}^{28}_{14}\mathrm{Si}^{*}$ from (1.2), the so-called compound nucleus of nitrogen (${}^{13}_{7}\mathrm{N}^{*}$) or silicon (${}^{28}_{14}\mathrm{Si}^{*}$) can be created in an excited state as formally denoted with the star superscript. Such nuclear excited states are intrinsically unstable and, therefore, they will decay through radiative de-excitations like ${}^{13}_{7}\mathrm{N}^{*} \longrightarrow {}^{13}_{7}\mathrm{N} + \gamma$ or ${}^{28}_{14}\mathrm{Si}^{*} \longrightarrow {}^{28}_{14}\mathrm{Si} + \gamma$, as per (1.1) or (1.2), respectively.

It is a well-established fact that at a fixed incident energy E, cross sections $Q(E)$ for resonant scattering of particles are adequately described by the one- and many-level Breit-Wigner (BW) formulae [90]. This latter expression appears as either a single Lorentzian

$$ Q(E) \approx Q_1 \propto \frac{\Gamma_1^2/4}{(E - E_1)^2 + \Gamma_1^2/4} \tag{1.3} $$

or a linear combination of any number of such Lorentzians

$$ Q(E) \approx Q_K \propto \sum_{k=1}^{K} \frac{\Gamma_k^2/4}{(E - E_k)^2 + \Gamma_k^2/4} \tag{1.4} $$

where E_k is the resonance energy and Γ_k is the resonance width[2]. In its

[2]Throughout this book, we will use the units in which the reduced Planck constant is equal to unity ($\hbar = 1$) such that the energy and frequency are permitted to be interchangeably employed as synonyms, $E = \hbar\omega = \omega$.

original appearance, Breit and Wigner [90] derived their dispersion formula or resonance formula (1.3) with the help of a perturbation method of the type of the Weisskopf-Wigner theory of dispersion of light by atoms.

A more direct derivation of the BW formula from the viewpoint of a quantum mechanical scattering wave function for particle scatterings was made by Siegert [91]. He also employed the perturbation formalism of quantum scattering theory. Crucially, his presentation of decaying states of a transient compound system consisting of a projectile and a target is very plausible. This is dictated by the nature of the initial problem in which one seeks to describe an experimental fact that at certain total energies of the compound system, the cross sections for a collision between a projectile and a target acquire unusually large values. Such a situation is precisely within the category of the universal phenomenon called the resonance effect, with similar appearances irrespective of its origin (mechanical, quantum-mechanical, acoustic and the like).

Hence, it is natural to expect that the observed enhancement could be inferred directly from the notion of the cross section $Q(E)$. By definition, cross section $Q(E)$, in the units of area for a given collisional event, is given by a ratio between the outgoing and incoming particle flux $\Phi^{(+)}$ and $\Phi^{(-)}$

$$Q(E) \propto \frac{\Phi^{(+)}}{\Phi^{(-)}}. \tag{1.5}$$

This implies that $Q(E) \longrightarrow \infty$ if $\Phi^{(-)} \longrightarrow 0$. As such, cross section $Q(E)$ becomes singular with the disappearance of the incoming wave $\Phi^{(-)}$. Hence, a compound state of the projectile-target system can be perceived as a state describing annihilation of the incoming wave. Stated differently, a compound state is a state whose total energy makes the pertinent cross section singular, i.e., $Q(E)$ becomes enormously large. Hence the resonance effect. Within this framework, the target captures (absorbs) the projectile to form a temporarily existing compound system. These compound states are unstable or metastable and, thus, prone to decay once the decay time $1/\Gamma_k$ has elapsed. The inverse resonance width $1/\Gamma_k$ is the half lifetime of the kth metastable state. An alternative way of conceiving this picture could be to say that the compound system relaxes with the relaxation time $1/\Gamma_k$.

This key identification in the definition of $Q(E)$ yields the signature of the resonance effect as the singularities of the cross sections at certain values of complex energies $E'_k (k = 1, 2, 3, ...)$ of the compound system. Siegert's hypothesis of a complex energy $E'_k = E_k - i\Gamma_k/2$ of the compound system at the resonance stems from the occurrence that these states possess a definite width $\Gamma_k \neq 0$, since they are decaying (radiative) states.

In contrast to genuine or true discrete states of negative energies with zero widths, resonance states have positive energies. Therefore, a resonance state is described by a wave packet localized on the positive part of the real energy axis. The real and the imaginary parts of the complex resonance energy represent, respectively, the position and the width of the corresponding peak in the cross section $Q(E)$ as a function of the incident energy E. This resonant maximum in $Q(E)$ at $E \approx E_k$ is manifested through a peak which is superimposed on top of a smoothly varying background described by a regular function $f(E)$. Since they are located in the complex energy plane, the resonance energies could be generally classified as sharp and broad. The mentioned singularities in $Q(E)$ associated with E_k near the real energy axis (small Γ_k) generate the well-delineated (isolated) resonances. In the energy spectrum, they are seen as sharp peaks. The area underneath of such peaks represents the resonant cross sections

$$Q_k(E) \equiv Q(E_k). \tag{1.6}$$

In such a case, the total cross section $Q(E)$ near the resonant energy in the single- and multi-level BW formulae can be approximated by

$$\begin{aligned} Q(E) &\approx f(E) + Q_1(E) \\ Q(E) &\approx f(E) + \sum_{k=1}^{K} Q_k(E) \end{aligned} \tag{1.7}$$

respectively. The non-singular function $f(E)$ is associated with a background which could contain the contributions from broad resonances with large widths Γ_k. As stated, the physical interpretation of the BW formula is that it describes absorption of an incident particle (or a wave) by a target. Due to these circumstances, $Q_k(E)$ is proportional to the absorption spectrum $A(E)$, which represents the real part of the complex-valued spectrum $F(E)$

$$F(E) \equiv i\frac{C_1}{(E - E_1) + i\Gamma_1/2} \equiv A_1(E) + iD_1(E) \tag{1.8}$$

where C_1 is a constant. The imaginary part of $F(E)$, denoted by $D_1(E)$, is the dispersion spectrum

$$\begin{aligned} A_1(E) &= C_1\frac{\Gamma_1/2}{(E - E_1)^2 + \Gamma_1^2/4} \\ D_1(E) &= C_1\frac{E - E_1}{(E - E_1)^2 + \Gamma_1^2/4}. \end{aligned} \tag{1.9}$$

Hence, for a single-level BW formula, the function $F(E)$ is recognized as the simplest PA to $Q(E)$ with a numerator and denominator polynomial of the degree 0 and 1, respectively, as symbolized by $[0/1]_Q(E)$

$$F(E) = [0/1]_Q(E). \tag{1.10}$$

Similarly, the many-level BW formula for $F(E)$ is obviously the PA of the order $[(K-1)/K]$ to $Q(E)$

$$F(E) = i\sum_{k=1}^{K} \frac{C_k}{(E - E_k) + i\Gamma_k/2} \tag{1.11}$$

$$F(E) = \frac{P_{K-1}(E)}{Q_K(E)} = [(K-1)/K]_Q(E) \tag{1.12}$$

where $P_{K-1}(E)$ and $Q_K(E)$ are the numerator and denominator polynomial of degree $K-1$ and K, respectively. This complex-valued spectrum can also be written as a sum of its multi-level BW absorption and dispersion spectra $A(E)$ and $D(E)$

$$F(E) = i\sum_{k=1}^{K} \frac{C_k}{(E-E_k)+i\Gamma_k/2} \equiv A(E) + iD(E) \tag{1.13}$$

$$A(E) = \sum_{k=1}^{K} A_k(E) \equiv \sum_{k=1}^{K} C_k \frac{\Gamma_k/2}{(E-E_k)^2+\Gamma_k^2/4} \tag{1.14}$$

$$D(E) = \sum_{k=1}^{K} D_k(E) \equiv \sum_{k=1}^{K} C_k \frac{E-E_k}{(E-E_k)^2+\Gamma_k^2/4}. \tag{1.15}$$

Naturally, the same BW formulae for one or many levels are also directly applicable to spectroscopy.

The above outlines clearly show that the Padé approximant is an integral part of the adequate description of the physics of resonance phenomena. Thus, from the outset, the PA is optimally suited to yield the proper theoretical predictions and physical interpretation of resonance data that are measured in experiments. This is indeed the case not only in physics, but also across interdisciplinary research fields [5]. This briefly expounded quantum-mechanical concept of resonances is deeply rooted in one of the most fundamental quantities in physics, termed the scattering or $S-$matrix [92, 93]. The $S-$matrix is employed to map the incoming to the outgoing total scattering wave function $\Psi^{(-)}$ and $\Psi^{(+)}$, respectively, for the initial and final state of the whole system under consideration [93]

$$\Psi^{(+)}(E) = S(E)\Psi^{(-)}(E). \tag{1.16}$$

The incoming and outgoing particle fluxes $\Phi^{(-)}$ and $\Phi^{(+)}$ can be computed from $\Psi^{(+)}$ and $\Psi^{(-)}$, respectively. This implies that the cross section $Q(E)$ from (1.5) can be extracted from $S(E)$. Hence, the elements of the $S-$matrix contain all the physical properties of cross sections, including resonances. Importantly, $S(E)$ is an analytic function in the complex energy plane with singularities represented by poles and branch cuts[3]. In particular, the poles of $S(E)$ lead to the Lorentzian-shaped resonances in $Q(E)$. Clearly, a function which possesses poles ought to be a rational function, and such is $S(E)$. Hence, it is not surprising that the matrix elements of the $S-$matrix are the Padé

[3]It is well-known from the Cauchy analysis that any analytical function can be represented by a sum of its poles [44, 47].

approximants, as the simplest rational functions in the form of the unique quotient of two polynomials from (1.11). The overall goal of this succinct reminder is to illuminate the key feature of the Padé approximant, i.e., the fact that this method forms a natural integral part of the leading mathematical formulations and implementations of quantum mechanics.

1.3 Resonance profiles

As stated, collision theory predicts that on physical grounds the most natural spectral line-shapes are given by Lorentzian profiles. The term Lorentzian for a bell-shaped profile of a spectral line stems from optics where Lorentz [94] used a dispersion-type function of the form (1.3). Alternatively, in mathematics, a Lorentzian functional form is called the Cauchy distribution.

There exist other line-shapes that are frequently used, e.g., the Gauss and Voigt profiles [95]. In particular, the Voigt profile f_V is a convolution integral of a Lorentzian $(\gamma/\pi)/[\gamma^2 + (y-\omega)^2]$ by a Gaussian $[1/(\sigma\sqrt{\pi})]\exp(-y^2/\sigma^2)$

$$f_V(\omega) = \frac{\gamma}{\pi}\frac{1}{\sigma\sqrt{\pi}}\int_{-\infty}^{\infty}\frac{\mathrm{e}^{-y^2/\sigma^2}}{\gamma^2+(y-\omega)^2}\mathrm{d}y. \tag{1.17}$$

As such, the Voigt function can be perceived as a broadened Lorentzian, or equivalently, a narrowed Gaussian depending upon which of the two constituents (Gaussian or Lorentzian) is taken to be the primary profile. Originally, the Voigt profile was introduced in optics with the purpose of accounting for the Doppler broadening (via Gaussians) of the primary line-shapes given by dispersion functions (Lorentzians) to describe absorption spectra [95]–[97]. After this initial appearance, many applications of the Voigt profile emerged in diverse fields, including MRS [98, 99]. Being a symmetric function, the Voigt frequency profile is intermediate to a Gaussian and a Lorentzian. More specifically, the Voigt profile is reduced to a Gaussian or a Lorentzian function in the pertinent limiting cases of the two widths. For instance, not too far from the center $\omega \approx \omega_1$, the Lorentz and Gauss distributions look very much alike. Nevertheless, the discrepancy between these two distributions becomes appreciable away from the center where a Lorentzian decreases more slowly than a Gaussian when the value of the frequency variable ω is augmented.

From the mathematical viewpoint, a Lorentz distribution represents the one-sided Fourier integral of an exponentially damped oscillatory wave function. Likewise, from the physics standpoint, exposing an oscillator (e.g., a pendulum, an atom, a molecule, etc.) to a given external influence will produce damped harmonic oscillations of a type of forced excitations as a response to the applied perturbation. This response function due to exponentially attenuated harmonic oscillations represents a pure Lorentzian in the absence of any other external disturbance. Importantly, a convolution (folding) of two or more Voigtians gives a single Voigt frequency profile. This is similar to the well-known fact that a convolution of two or more Gaussians (Lorentzians) gives a single Gaussian (Lorentzian). In plasma physics, the Doppler broadening of spectral lines originates from statistical velocity distributions of emitting atoms. By reference to the Boltzmann thermodynamical theory of gases, it follows that the average velocity and plasma temperature are directly proportional to each other. When gas temperatures are far from the Bose-Einstein condensation, the velocity distribution of atoms is Maxwellian and, therefore, the said Doppler broadening will appear as a Gaussian profile. Alternatively, in some other applications, the Voigt function can be viewed as a primary Gaussian profile, which is narrowed through a convolution with a Lorentzian and the basis for this is the occurrence of collisions among atoms.

In MRS, the basic forms of line-shapes are also rooted in pure Lorentzians that stem from absorption of radio-frequency photons. However, whenever there are inhomogeneities of static magnetic fields, the original Lorentzian profiles could appropriately be modified by Gaussian broadening of the basic Lorentzian profiles. Such a physically motivated convolution of the Lorentzian leads to the Voigt profile. In practical applications, the Voigt folding employed in MRS has often been simplified by a distribution in the form of a linear combination of Lorentzians and Gaussians with some constant fitting coefficients [98, 99]. Alternatively, the Voigt convolution integral can be computed directly. This might be especially necessary if this convolution is used many times in the course of an iterative analysis, e.g., in various least square estimations of spectra. In particular, it is well-known that the Voigt convolution integral represents the real part of the complex-valued error-function or the probability function. The error-function of complex variables can be generated efficiently to within machine accuracy by the continued fraction (CF) algorithm proposed by Gautschi [100]. This latter algorithm enables the Voigt frequency profile to be used extensively without any simplification as a sum of a Gaussian and a Lorentzian. Continued fractions are the special cases (diagonal and para-diagonal) of the general Padé approximant.

It should be noted that there exists also the Voigt time profile. This function is defined by the inverse Fourier integral of the Voigt frequency profile. Explicitly, the result of such an inversion is represented by the twofold exponential function $\exp(-\gamma t - \sigma^2 t^2/4)$ where t is time ($t \geq 0$). This stems from the convolution theorem according to which the Fourier integral of two functions convolved in the frequency domain is effectively reduced (in the time domain) to the ordinary product of the two inverse Fourier transforms. In particular, the one-sided inverse Fourier integral over frequencies (from zero to infinity) of a Lorentzian $\propto 1/(\gamma^2 + \omega^2)$ is given by $\propto \exp(-\gamma t)$, whereas that of a Gaussian $\propto \exp(-\omega^2/\sigma^2)$ is $\propto \exp(-\sigma^2 t^2/4)$. This yields the time-domain Voigt time profile in the above-quoted form, $\propto \exp(-\gamma t - \sigma^2 t^2/4)$. Hence, in the case of static field inhomogeneities, for an MR spectrum with more than one peak, the corresponding time signal $c(t)$ could be described by a linear combination of the fundamental Voigt time profiles $\exp\left(-\gamma_k t - \sigma_k^2 t^2/4\right)$

$$c(t) = \sum_{k=1}^{K} d_k e^{-\gamma_k t - \sigma_k^2 t^2/4}. \tag{1.18}$$

Here, $\gamma_k \equiv \omega_k$ and d_k are complex fundamental frequencies and amplitudes, whereas the damping factors $\sigma_k^2/4$ mimic the magnetic field inhomogeneities.

We will now proceed to address the relevance of signal processing through MRS as it applies to medical diagnostics.

1.4 Why is this topic relevant for biomedical researchers and clinical practitioners?

At first glance, this book might seem far afield from the concerns and purview of biomedical researchers and clinical practitioners, including especially oncologists. However, with a second look coupled with reflection about the *actual* barriers to reliable applications of MRS and MRSI for cancer diagnostics, a different perspective can emerge.

Nowadays, many clinical MR scanners offer the possibility of automatically generating MR spectra. This opportunity is naturally welcomed by diagnosticians who increasingly recognize that the anatomic information provided by MRI, while invaluable, is still insufficient for unequivocal identification and characterization of malignancy and its clear distinction from various benign pathologies. Indeed, in the recent period, the role of *in vivo* MRS for cancer diagnostics has been highlighted in e.g. Refs. [7], [101]–[113], especially for neuro-oncology [101, 108], [113]–[115] and prostate cancer [101]–[104]. Breast cancer detection and prediction of response to chemotherapy have also been improved by MRS [101, 106, 107, 111], [116]–[118]. Unfortunately, however, when one examines these automatically-generated spectra, innumerable dilemmas arise. The interpretation of these spectra is frequently shrouded by confusion and ambiguity. In actual clinical practice, the outcome of this endeavor is all too often frustration, and retreat to the familiar realm of customary anatomical diagnostic modalities. As a consequence, among the potentially most valuable diagnostic information is left by the wayside. In fact, MRS has yet to become a standard diagnostic tool in the area where it is needed the most, for clinical oncology, including especially cancer screening and surveillance.

Technological improvements have been sought in attempts to solve these dilemmas. Needless to say, meticulous attention to coil design, shimming, water suppression, as well as the use of appropriate encoding sequences are a prerequisite for proper acquisition of MR time signals. Albeit at higher cost, higher field MR scanners with stronger static magnetic fields obviously provide better SNR, thereby yielding improved spectral quality.

Notwithstanding the importance of these technical, i.e., hardware considerations, and the need for further advances in this area, the critical limitation of current applications of MRS and MRSI is *directly* related to reliance upon the conventional signal processing method, the fast Fourier transform and the accompanying post processing via fitting and/or peak integrations. The more advanced mathematical methods that are the focus of this book are the vital remedy. The strategic importance of robust and uniform data processing of MRS signals has been strongly emphasized [107, 119] at, e.g., the expert meeting on MRS for oncology, held recently by the U.S. National Cancer Institute [107], as well as at a special conference in November 2006 on Data Processing in MR Spectroscopy and Imaging by the International Society for Magnetic Resonance in Medicine.

One may wonder: how could mathematics play such a critical role in medical diagnostics? This is because data encoded directly from patients by means of existing imaging techniques, e.g., computerized tomography (CT), positron emission tomography (PET), single photon emission tomography (SPECT), ultrasound (US), as well as MRI [12, 120] and MRS [5, 8] are not at all amenable to direct interpretation, which therefore need mathematics via signal processing.

The starting point for grasping the basics of signal processing in medical diagnostics is the concept of "conjugate variables". Unfortunately, this concept is rarely explained in an intellectually satisfying manner for those whose primary expertize is distant from the realm of mathematics and physics. Rather, far too prematurely, technical terminology is usually introduced, and this is done with inadequate definition. The result is often hazy thinking about such important relationships as that between k-space (momentum space) and the image obtained from the MR scanner. We would like to emphasize the tremendous intellectual gratification when we grasp that this relationship is closely analogous to that between time signals and their spectral representation.

Biomedical researchers and clinicians should readily appreciate that diagnostically important information can sometimes be difficult to quantify, and may not even be apparent in the domain in which it is recorded. Familiar illustrations include slow activity on the electroencephalogram, 60-120 Hz late potentials and heart rate variability in the 0.15 to 0.4 Hz range on the electrocardiogram, to name a few. The reason for which spectral analysis of these physiological signals is justified is that time and frequency are complementary representations or conjugate variables. We also proceed to another set of conjugate variables, momentum and position, from whence the k space representation is transformed into an MR image in the more familiar coordinate spatial representation. The next logical step is to examine the mathematical procedures needed to achieve this transformation.

In MRS, the encoded data are heavily packed time signals that decay exponentially in an oscillatory manner. These time domain data are not directly interpretable. The corresponding total shape spectrum is obtained by mathematical transformation of the FID into its complementary representation in the frequency domain. This total shape spectrum provides qualitative information, but *not* the quantitative one about the actual number of metabolites that underlie each peak or the relative strength of individual components, their abundance, etc. At best, the FFT takes us only to this second step. More information is needed before the metabolites can be identified and their concentrations reliably determined, and from the total shape spectrum alone this can only be guessed (see section 3.1.4).

This undetected information can be extracted by novel and self-contained data analysis, namely the FPT, which we have introduced into MRS [5, 16] with detailed implementations reported in Refs. [8]–[11] and [17]–[34]. In this book, we review the "proof of principle" evidence establishing that the FPT meets the most stringent criteria imposed by clinical disciplines such as oncology for MRS, as outlined in Refs. [10, 11] and [18]–[20]. The high resolution and stability of the FPT have been clearly confirmed in our studies of MR total shape spectra [8, 9], thereby overcoming one of the major hindrances to wider application of MRS in clinical oncology. However, as stated, total shape spectra do not provide the information needed to determine how many metabolite resonances are actually present in the tissue and in which concentrations. It is this information which is essential for improving the diagnostic yield and accuracy of MRS in oncology. We demonstrate that the FPT provides *exact quantification* of MR signals and thereby metabolite concentrations are reliably and unequivocally obtained with an intrinsic and robust error analysis [10, 11, 20, 34].

We have emphasized that there is an urgent need for accurate quantification to determine metabolite concentrations, so that MRS can be better used to detect and characterize cancers, with clear distinction from non-malignant processes. This is clearly illustrated by applying the FPT to time signals that were generated according to *in vitro* MRS data as encoded from (a) malignant and benign ovarian lesions [27, 29] as reviewed in chapter 9, (b) breast cancer, fibroadenoma and normal breast tissue [32], as analyzed in chapter 10 and (c) for cancerous prostate, normal stromal and glandular prostate [33], as presented in chapter 11. We chose these problem areas because of their urgent clinical importance.

This approach was made possible by widening the horizons of signal processing through finding its natural framework in a larger and well-established theory – quantum physics [5]. By identifying the quantification problem in signal processing as quantum-mechanical spectral analysis, the key door was opened for using a highly-developed mathematical apparatus to overcome the otherwise insurmountable difficulties of the FFT, fittings and other similar techniques [10, 11, 34]. It is through this direct connection of signal processing with quantum physics that a veritable *paradigm shift* has been established, and the stage set for the emergence of the most powerful and versatile spectral analyzer – the FPT.

2

The role of quantum mechanics in signal processing

In quantum mechanics, the dynamics of physical systems that evolve in time are described by the Schrödinger equation $\hat{\mathrm{U}}(\tau)|\Upsilon_k) = u_k|\Upsilon_k)$ where $u_k = \exp(-i\omega_k\tau)$. This represents the eigen-value problem of the evolution operator $\hat{\mathrm{U}} = \exp(-i\hat{\Omega}\tau)$ where $\hat{\Omega}$ is the system operator which generates the dynamics. To secure decay to zero of u_k at the infinite time, we must have $\mathrm{Im}(\omega_k) < 0$. Physically, $\hat{\Omega}$ is the energy operator associated with the Hamiltonian which is the total energy (kinetic + potential) of the whole system. For resonances with complex energies, the operator $\hat{\Omega}$ is non-Hermitean, $\hat{\Omega}^\dagger \neq \hat{\Omega}$. Given a 'Hamiltonian' $\hat{\Omega}$ of the studied system, direct spectral analysis deals with extraction of the spectral set with all the eigen-frequencies and eigen-functions $\{\omega_k, \Upsilon_k\}$. Such a task is usually accomplished by solving the Schrödinger eigen-value problem via, e.g., diagonalization of $\hat{\Omega}$ by employing certain basis functions from a set which is complete, or locally complete. For practical purposes, all one needs are the complex frequencies $\{\omega_k\}$ and the corresponding complex amplitudes $\{d_k\}$. By definition, the amplitudes follow from the squared projection of the full state vector Υ_k onto the initial state Φ_0 of the system, $d_k = (\Phi_0|\Upsilon_k)^2$. In particular, the absolute values $\{|d_k|\}$ are the intensity of spectral line-shapes, whereas $\{\mathrm{Re}(\omega_k)\}$ and $\{\mathrm{Im}(\omega_k)\}$ are, respectively, equal or proportional to the positions and widths of the resonances in the spectrum of the investigated system. Hereafter, the notation $\mathrm{Re}(z)$ and $\mathrm{Im}(z)$ will be used to label the real and imaginary parts of the complex quantity z, respectively. Thus, by extracting all the spectral parameters $\{\omega_k, d_k\}$ for the given system operator $\hat{\Omega}$, quantum mechanics can examine the structure of matter on any concrete level (nuclear, atomic, molecular, etc.). This is known as the direct quantification problem.

The rationale for calling the quantity $\hat{\mathrm{U}}$ the evolution operator can at once be appreciated by inspecting the time-dependent Schrödinger equation $(i\partial/\partial t)\Phi(t) = \hat{\Omega}\Phi(t)$. 'Hamiltonian' operator $\hat{\Omega}$ is stationary for conservative systems, in which case the solution $\Phi(t)$ of the Schrödinger equation can be generated simply through the relation $\Phi(t) = \hat{\mathrm{U}}(t)\Phi(0)$. In other words, $\Phi(t)$ at an arbitrary time $t > 0$ is obtained by subjecting $\Phi(0)$ to the action of the operator $\hat{\mathrm{U}}(t)$. Therefore, given the operator $\hat{\Omega}$, the state of the system $\Phi(t)$ at any instant t will be known if the initial state $\Phi(0) = \Phi(t=0) \equiv \Phi_0$ is specified at $t = 0$. This is the origin of determinism of quantum mechanics. This mapping justifies the name of the time evolution operator for $\hat{\mathrm{U}}$. Spectral problems can alternatively be solved by reliance upon another powerful quantum-mechanical formalism called the Green operator $\hat{\mathrm{G}}(u)$ which is defined as the resolvent operator, $\hat{\mathrm{G}}(u) = (u\hat{1} - \hat{\mathrm{U}})^{-1}$.

All physical systems develop in time via their correlated dynamics. Correlations between any two states of the system are contained in the evolution operator $\hat{\mathrm{U}}(t)$ which connects $\Phi(0)$ with $\Phi(t)$. Since operators are not observables, i.e., they cannot be measured directly in experiments, certain related scalar quantities are needed. For instance, the projection technique can be used to project one state onto the other via the scalar or inner product. Thus, to correlate $\Phi(t)$ and $\Phi(0)$ with the ensuing scalar result, one can make the inner product of these two states, $C(t) \equiv (\Phi(0)|\Phi(t))$. The obtained quantity $C(t)$ is known as the auto-correlation function, because it correlates the same system to itself via two different states $\Phi(t)$ and $\Phi(0)$. Furthermore, using $\Phi(t) = \hat{\mathrm{U}}\Phi(0)$, it follows that $C(t)$ basically correlates the state $\Phi(0)$ to the same vector $\Phi(0)$ 'weighted' with the evolution operator $\hat{\mathrm{U}}$, such that $C(t) = (\Phi_0|\hat{\mathrm{U}}(t)|\Phi_0)$.

The chief working hypothesis of quantum mechanics asserts that the entire information about any given system is ingrained in the wave function of the system. These stationary and non-stationary wave functions are the state vectors Υ and $\Phi(t)$, respectively. The mathematical formula which encapsulates this postulate is given by the requests of the existence of the global (or at least the local) completeness relation $\sum_{k=1}^{K} \hat{\pi}_k = \hat{1}$, where $\hat{\pi}_k$ is the projection operator $\hat{\pi}_k = |\Upsilon_k)(\Upsilon_k|$. The term 'completeness' refers explicitly to the complete information. To interpret the completeness relation, one resorts to the probabilistic meaning of quantum mechanics and all its entities. Thus, the state vector $|\Upsilon_k)$ and its dual counterpart $(\Upsilon_k|$ are the probability densities to find the system in the state described by these wave functions. Plausibly, knowing everything about the system under study is equivalent to achieving the maximum probability, which is equal to 1. This is precisely what is stated by the completeness relation in its operator form. The same statement can also be expressed in the scalar form by putting the operator completeness relation in a matrix element between the initial states $|\Phi_0)$ and its dual counterpart $(\Phi_0|$. The result reads as $\sum_{k=1}^{K} d_k = 1$ where d_k is the scalar amplitude d_k given by $d_k = (\Phi_0|\hat{\pi}_k|\Phi_0) = (\Phi_0|\Upsilon_k)^2$ which agrees with the above-quoted formula for d_k. Therefore, one of the ways of checking the completeness relation is to verify whether the sum of all the amplitudes $\{d_k\}$ $(1 \le k \le K)$ is equal to unity. It should also be noted that every form (operational or scalar) of the completeness relation implicitly contains the whole information about the fundamental frequencies $\{\omega_k\}$, since Υ_k and d_k correspond to ω_k.

The quantities that hold the complete information, the states vectors Υ and $\Phi(t)$, can be derived from the system operator $\hat{\Omega}$. This, in turn, means that the entire sought information is also present in $\hat{\Omega}$, or equivalently, in $\hat{\mathrm{U}}(t)$. The total evolution operator $\hat{\mathrm{U}}(t)$ itself is the major physical content of $C(t)$.

Hence, the stated quantum-mechanical postulate on completeness implies that everything one could possibly learn about any considered system is also contained in the auto-correlation function $C(t)$. Despite the fact that the same full information is available from $\Upsilon, \Phi(t), \hat{\Omega}, \hat{\mathrm{U}}(t)$ and $C(t)$, the auto-correlation functions are more manageable in practice, since they are observables. As a scalar, the quantity $C(t)$ has a functional form which is defined by its dependence upon the independent variable t. Hence, since $C(t)$ ingrains the whole information about the studied generic system, it is critically important to have the quantum-mechanical prediction for the shape of the auto-correlation function. To obtain the explicit form of the auto-correlation function $C(t)$ for varying time t, one can use the spectral representation of $\hat{\mathrm{U}}(t)$. This representation is easily obtained via multiplication of the defining exponential $\exp{(-i\hat{\Omega}t)}$ for $\hat{\mathrm{U}}(t)$ by the unity operator $\hat{1}$, which is taken from the completeness relation. The end result is $\hat{\mathrm{U}}(t) = \sum_{k=1}^{K} \exp{(-i\omega_k t)}\hat{\pi}_k$. Therefore, preservation of the entire information from the system *imposes* the form of the spectral representation of the evolution operator given by the sum of K damped complex exponentials with the operator amplitudes $\hat{\pi}_k$. Substituting such a representation for $\hat{\mathrm{U}}(t)$ into $C(t)$ yields $C(t) = \sum_{k=1}^{K} d_k \exp{(-i\omega_k t)}$. Thus, the obtained shape of the quantum-mechanical auto-correlation functions represents a linear combination of K complex damped exponentials with the scalar amplitudes d_k. Crucially, this is not a fitting model for $C(t)$ introduced by hand. Rather, it is the shape of the auto-correlation function *demanded* by the form $\exp{(-i\hat{\Omega}t)}$ of the quantum-mechanical evolution operator $\hat{\mathrm{U}}(t)$. The derived form of $C(t)$ corresponds precisely to time signals $c(t)$ in many research branches. Crucially, such time signals are measured in experiments across interdisciplinary fields.

The outlined procedure establishes the equivalence between time signals and auto-correlation functions from quantum mechanics. This leads to a direct link between quantum mechanics and signal processing as two otherwise separated branches. Such a connection is of fundamental importance, because a correct mathematical modeling in signal processing can only be done with a proper physical description of the investigated phenomenon. As seen, quantum mechanics amply fulfills this strict requirement.

2.1 Direct link of quantum-mechanical spectral analysis with rational response functions

In quantum physics, spectral analysis in the time domain employs the non-stationary Schrödinger equation [5]

$$i\frac{\partial}{\partial t}|\Phi(t)) = \hat{\Omega}|\Phi(t)). \tag{2.1}$$

Supposing that the beginning of counting the time from an instant t_0 is taken to be zero ($t_0 = 0$), the boundary condition to (2.1) reads as $|\Phi(0)) = |\Phi_0)$. As before, $|\Phi_0)$ is the known initial state of the whole system with the dynamics generated by the total 'Hamiltonian' $\hat{\Omega}$. Although it is not necessary to normalize the state vector $|\Phi_0)$, the associated norm ought to be finite and non-zero

$$||\Phi_0||^2 \equiv (\Phi_0|\Phi_0) \equiv C_0 \neq 0 \qquad |C_0| < \infty. \tag{2.2}$$

Thus, generally, the constant C_0 is not unity and, furthermore, the norm (2.2) could be a complex number. The norm $(\Phi_0|\Phi_0)$ must be non-zero for both mathematical and physical reasons. Mathematical, because this norm is a part of the normalized state $|\Phi_0)$ as a denominator of the quotient $|\Phi_0)/||\Phi_0||$, which prohibits division by zero. Physical, because the norm $||\Phi_0||$ could be zero provided that, e.g., $|\Phi_0)$ is the zero state vector $|0)$. Time propagation of the state $|0)$ will give again the same zero state which is, as such, of no interest due to the lack of any information about the system. As mentioned earlier, to encompass resonances that correspond to complex-valued frequencies, the 'Hamiltonian' $\hat{\Omega}$ should be a non-Hermitean linear operator. To make it transparent, non-Hermiticity of the operator $\hat{\Omega}$ will be indicated in the definition of the scalar product as the non-Hermitean symmetric inner product via

$$(\chi|\psi) = (\psi|\chi). \tag{2.3}$$

As seen, no conjugation through the usual star superscript is put on either of the two state vectors $(\chi|$ or $|\psi)$. Such a symmetry property of the inner product is symbolically highlighted by employing small soft round brackets $|\psi)$ and $(\chi|$ in lieu of the conventional Dirac notation $\langle\chi|$ and $|\psi\rangle$ encountered in Hermitean Hamiltonians. The customary Dirac bra-ket symbolism and the soft brackets are inter-related by $\langle\chi|\psi\rangle = \langle\psi^*|\chi^*\rangle$ and $\langle\chi^*|\psi\rangle = (\chi|\psi)$.

As emphasized, one of the major working hypotheses of quantum mechanics is determinism, which prescribes the time evolution of the state of the system as follows. Whenever the wave packet $|\Phi(0))$ of the studied arbitrary system is well-prepared, i.e., well-controlled, at the initial time $t = 0$ and, moreover, if a subsequent time development is dictated exclusively by the given dynamics (with interactions being the chief constituent), then at any later instant t, the state $|\Phi(t))$ shall be fully determined, i.e., completely known [5]. In the Schrödinger picture of quantum mechanics, wave functions are non-stationary, whereas operators are stationary. Thus, for a given stationary dynamical operator $\hat{\Omega}$, which is exactly what is needed for the definition of a conservative system, the solution of (2.1) becomes

$$|\Phi(t)) = e^{-i\hat{\Omega}t}|\Phi_0) \qquad |\Phi_0) \equiv |\Phi(0)). \tag{2.4}$$

The application of an exponential operator in (2.4) can be conceived by means of the corresponding Maclaurin series expansion [5], or alternatively, by its cited spectral representation which is also called the spectral decomposition.

The mechanism of creating and keeping the whole information about the system through the state vector $|\Phi(t)$ from (2.4) is rooted in the dynamics generator $\hat{\mathrm{U}}(t)$ or the time evolution operator. This is also called the time relaxation operator which is given by

$$\hat{\mathrm{U}}(t) = \mathrm{e}^{-i\hat{\Omega}t} \qquad \therefore \qquad |\Phi(t)) = \hat{\mathrm{U}}(t)|\Phi_0). \tag{2.5}$$

According to (2.4) and (2.5), if $\hat{\Omega}$ and $|\Phi_0)$ are given, the determinism of quantum mechanics predicts that the state $|\Phi(t))$ of the entire system will be completely known. Stated equivalently, the full information which could conceivably be obtained about any given system under investigation will be available, provided that one knows how to extract such information from the time evolution operator (2.5).

As mentioned, the stationary state $|\Upsilon_k)$ of any studied system satisfies the time independent Schrödinger equation

$$\hat{\Omega}|\Upsilon_k) = \omega_k|\Upsilon_k) \qquad 1 \leq k \leq K. \tag{2.6}$$

The solution of this equation can be found with the help of the expansion method by which $|\Upsilon_k)$ is developed in terms of a selected basis set functions. This is followed by computation of the matrix elements of $\hat{\Omega}$ with these basis functions. Such a procedure converts (2.6) into a system of linear equations that are solved to obtain the K eigen-values $\{\omega_k\}$. Substitution of the obtained eigen-values into the same system of linear equations yields the K eigen-states $\{|\Upsilon_k)\}$. Thus, even for a direct problem (2.6) these brief outlines indicate that the explicit knowledge of $\hat{\Omega}$ is not necessary. As seen, all one needs are the matrix elements of $\hat{\Omega}$, and not this operator itself. Note that hereafter K is a non-negative integer which is the total number of the genuine physical states of the considered system. Each given $\hat{\Omega}$ possesses a well-defined and fixed number K which is usually unknown prior to analysis. However, an adequately designed analysis and its computer implementation ought to be capable of finding the unique value of K, since surmising this number would be unacceptable for any valuable practical purpose. Any guessing of K inevitably leads to either over-estimation or under-estimation of the true number of physical states. Thus, in over-estimation, spurious unphysical states that are forbidden to the investigated system would be detected. In under-estimation, some of the genuine states would be missed. As such, both over-estimation and under-estimation are anathema to a proper theory as well as to practice. Of course, the same criticism against guessing K applies also to the corresponding inverse problem, such as harmonic inversion for quantification in MRS. Specifically, in the quantification problem, $\hat{\Omega}$ is completely unknown. Nevertheless, this is not an obstacle, since only the matrix elements of $\hat{\Omega}$ are required to solve (2.6). These matrix elements, $C(t) = (\Phi_0|\exp(-i\hat{\Omega}t)|\Phi_0)$, are recognized as quantum-mechanical auto-correlation functions $C(t)$. It is remarkable that precisely these auto-correlation functions arc measured directly in experiments as free induction decay curves or time signals $c(t)$, as encountered throughout Nuclear Magnetic Resonance (NMR), Ion Cyclotron Resonance Mass Spectroscopy (ICRMS), MRS, etc. This is the basis of the earlier mentioned equivalence of fundamental importance between the auto-correlation functions and time signals, $C(t) = c(t)$. Moreover, it is this equivalence which renders the inverse quantification problem exactly solvable by numerical means. Such an accomplishment is made by reduction of the originally non-linear inverse problem of quantification to the same above-outlined linear algebra of the direct eigen-value problem (2.6).

The eigen-functions $\{|\Upsilon_k)\}$ form a basis set of complete orthonormalized state vectors so that, up to a normalization constant, we can write

$$(\Upsilon_{k'}|\Upsilon_k) \propto \delta_{k,k'} \tag{2.7}$$

where $\delta_{k,k'}$ is the Kronecker $\delta-$symbol, $\delta_{k,k} = 1$ and $\delta_{k,k'} = 0\,(k' \neq k)$. Here, $||\Upsilon_k||$ represents the norm of the eigen-state $|\Upsilon_k)$

$$||\Upsilon_k||^2 = (\Upsilon_k|\Upsilon_k). \tag{2.8}$$

Given an un-normalized total wave function $|\widetilde{\Upsilon}_k)$ of the system, its normalized counterpart $|\Upsilon_k)$ can be cast into the usual form

$$|\Upsilon_k) = N_k|\widetilde{\Upsilon}_k) \tag{2.9}$$

where N_k is a normalization constant. This latter constant can be determined from the requirement that the flux of the particles described in terms of the wave function Υ_k is fixed to be, e.g., a unit flux

$$(\Upsilon_k|\Upsilon_k) = ||\Upsilon_k||^2 = 1 \qquad \therefore \qquad (\Upsilon_{k'}|\Upsilon_k) = \delta_{k,k'}. \tag{2.10}$$

Inserting (2.9) into (2.10) gives the standard expression for the normalization constant

$$N_k = \frac{1}{||\widetilde{\Upsilon}_k||} \tag{2.11}$$

$$\therefore \quad |\Upsilon_k) = \frac{1}{||\widetilde{\Upsilon}_k||}|\widetilde{\Upsilon}_k). \tag{2.12}$$

Employing (2.6), the Cayley-Hamilton theorem leads to the following important equation for any operator analytic function $f(\hat{\Omega})$

$$f(\hat{\Omega})|\Upsilon_k) = f(\omega_k)|\Upsilon_k). \tag{2.13}$$

Here, the eigen-functions $|\Upsilon_k)$ are the same in (2.6) and (2.13), whereas the eigen-values ω_k and $f(\omega_k)$ are associated with the former and the latter equation, respectively.

As discussed, the main working postulate of quantum mechanics is that the full information about any studied system is contained in the set of the total wave functions $\{|\Upsilon_k)\}$. This hypothesis coheres with (2.6), since $\hat{\Omega}$ stores the whole information about the considered system. As stated, this is known as completeness of the quantum-mechanical description of general phenomena in nature, and transcribed mathematically by the closure relation

$$\sum_{k=1}^{K} \hat{\pi}_k = \hat{1} \qquad \hat{\pi}_k = |\Upsilon_k)(\Upsilon_k| \qquad 1 \leq k \leq K \tag{2.14}$$

where $\hat{\pi}_k$ is the conventional projection operator.

Clearly, equation (2.14) is also the spectral decomposition of the unity operator, $\hat{1}$. This allows the derivation of the corresponding spectral representation of $f(\hat{\Omega})$ via multiplication of both sides of equation (2.13) by $(\Upsilon_k|$. The subsequent usage of the closure relation (2.14) gives

$$f(\hat{\Omega}) = \sum_{k=1}^{K} f(\omega_k)\hat{\pi}_k. \tag{2.15}$$

For $f(\hat{\Omega}) = \mathrm{e}^{-i\hat{\Omega}t}$, the general formula (2.15) leads to the spectral decomposition of the evolution operator $\hat{\mathrm{U}}(t)$ via

$$\hat{\mathrm{U}}(t) = \sum_{k=1}^{K} \mathrm{e}^{-i\omega_k t}\hat{\pi}_k \qquad \mathrm{Im}(\omega_k) < 0. \tag{2.16}$$

Hence, the *exact* spectral representation of the time evolution operator is given precisely by the sum of K complex damped exponentials with the amplitudes in the form of the projectors $\hat{\pi}_k$. Since the operators $\hat{\Omega}$ and $\hat{\mathrm{U}}(t)$ are linear, the same conclusion also extends to a scalar variant of (2.16). This follows by taking the matrix element of $\hat{\mathrm{U}}(t)$ between the states $(\Phi_0|$ and $|\Phi_0)$

$$C(t) \equiv (\Phi_0|\hat{\mathrm{U}}(t)|\Phi_0). \tag{2.17}$$

The obtained quantum-mechanical quantity $C(t)$ represents the continuous auto-correlation function. Its physical meaning is understood by the argument which runs as follows. As seen in (2.5), the state $|\Phi(t))$ at time t is generated by propagation of the initial wave packet $|\Phi_0)$ from $t = 0$ to t through $\hat{\mathrm{U}}(t)$. Given a fixed $|\Phi_0)$ at $t = 0$, there will be a non-zero chance of detecting the system in the state $|\Phi(t))$ at the subsequent time $t > 0$, if the overlap between these two wave packets is not zero. In quantum mechanics, this type of overlap is defined via the scalar product of $|\Phi(t))$ and $(\Phi_0|$. The result is the matrix element $(\Phi_0|\Phi(t))$ which is $C(t)$ from (2.17) by reference to (2.5). The auto-correlation function $C(t)$ indicates the degree of correlations between the states $|\Phi(t = 0))$ and $|\Phi(t \neq 0))$ in the course of the exposure of the system to the action of the dynamical operator $\hat{\Omega}$. The operator $\hat{\Omega}$ itself is responsible for the difference between $|\Phi(t))$ and $|\Phi_0)$, as seen in (2.4). When the dynamics are turned off by setting $\hat{\Omega} = \hat{0}$, the system would indefinitely remain in the initial state $|\Phi_0)$. This yields $C(t) = C_0 = (\Phi_0|\Phi_0) \neq 0$ at any time t, as per (2.2). At any two times t' and t with $t < t'$, the state $|\Phi(t'))$ can be viewed as a delayed copy of $|\Phi(t))$. Hence, the overlap $(\Phi(t')|\Phi(t))$ taken at $t < t'$ indeed gives a measure of correlation between the state $|\Phi(t))$ and its delayed copy $|\Phi(t'))$. The special case $t' = 0$ gives the auto-correlation function, $C(t) = (\Phi(0)|\Phi(t))$.

Once (2.16) is inserted into (2.17), the following expression is obtained

$$C(t) = \sum_{k=1}^{K} \mathrm{e}^{-i\omega_k t} d_k \tag{2.18}$$

where $\mathrm{Im}(\omega_k) < 0$, as in (2.16). During the derivation of (2.18), the quantity d_k is identified as the matrix element of the projection operator $\hat{\pi}_k$ taken over the initial state

$$d_k \equiv (\Phi_0|\hat{\pi}_k|\Phi_0) = (\Phi_0|\Upsilon_k)^2 \tag{2.19}$$

where symmetry (2.3) of the inner product is used. The result $d_k = (\Phi_0|\Upsilon_k)^2$ is the residue linked to the eigen-frequency ω_k.

The usefulness of the spectral representation (2.16) can be seen by introducing the Green operator as the operator-valued Fourier integral of the evolution operator

$$\hat{\mathrm{G}}(\omega) \equiv \int_0^\infty \mathrm{d}t\, \mathrm{e}^{i\omega t} \hat{\mathrm{U}}(t). \tag{2.20}$$

The operator integral (2.20) can be reduced to the ordinary integration by substituting the spectral representation of $\hat{\mathrm{U}}(t)$ into (2.20). The resulting scalar integral is carried out via

$$\int_0^\infty \mathrm{d}t\, \mathrm{e}^{i\omega t} \hat{\mathrm{U}}(t) = \sum_{k=1}^{K} \hat{\pi}_k \int_0^\infty \mathrm{d}t\, \mathrm{e}^{i(\omega-\omega_k+i\epsilon)t} = \sum_{k=1}^{K} \hat{\pi}_k (\omega - \omega_k + i\epsilon)^{-1}.$$

Here, $\epsilon \longrightarrow 0^+$ is an infinitesimally small positive number known as the Dyson exponential damping factor, which is introduced to secure convergence at the upper limit $t = \infty$. As such, the final result is understood with a limit being taken for ϵ to approach zero through positive numbers, as indicated by the plus superscript on zero, 0^+. Thus, the Green operator (2.20) becomes

$$\hat{\mathrm{G}}(\omega) = -i \sum_{k=1}^{K} \frac{\hat{\pi}_k}{\omega - \omega_k + i\epsilon} \qquad \mathrm{Im}(\omega_k) < 0 \qquad \epsilon \longrightarrow 0^+. \tag{2.21}$$

Now, in principle, the $\epsilon-$damping might safely be set to zero (as will be done in the sequel), since ω_k is already a complex number with $\mathrm{Im}(\omega_k) < 0$, as per (2.16). The sum over k in (2.21) can explicitly be performed to yield the equivalent exact expression

$$\hat{\mathrm{G}}(\omega) = -i \frac{\hat{\mathrm{P}}_{K-1}(\omega)}{Q_K(\omega)}. \tag{2.22}$$

The numerator $\hat{\mathrm{P}}_{K-1}(\omega)$ and denominator $Q_K(\omega)$ are an operator and a scalar polynomial of degree $K-1$ and K, respectively. Hence, expression (2.22) is a mixed operator-scalar Padé approximant for the Green operator $\hat{\mathrm{G}}(\omega)$.

Operators do not possess their direct physical meanings in quantum mechanics. However, their scalar products taken over physical states could have physical meaning. Hence, it is important especially for comparison with experimental data, to see the physical meaning of the overlap integrals involving certain physical states, such as Φ_0, weighted with operators $\hat{\mathrm{U}}(t)$ or $\hat{\mathrm{G}}(\omega)$. As to $\hat{\mathrm{U}}(t)$, the answer is already known from (2.17), where the observable $C(t)$ is the auto-correlation function given by the overlap $(\Phi_0|\hat{\mathrm{U}}(t)|\Phi_0)$. Likewise, the significance of the Green operator $\hat{\mathrm{G}}(\omega)$ is in the fact that its matrix element taken over, e.g., the initial state Φ_0 is the Green function $\mathrm{G}(\omega)$. This Green function is the exact frequency spectrum of the system

$$\mathrm{G}(\omega) \equiv (\Phi_0|\hat{\mathrm{G}}(\omega)|\Phi_0). \tag{2.23}$$

Substituting (2.21) into the rhs of (2.23), employing (2.19) and ignoring the ϵ−factor, we have

$$\mathrm{G}(\omega) = -i\sum_{k=1}^{K}\frac{d_k}{\omega - \omega_k}. \tag{2.24}$$

Either by carrying out the sum over k on the rhs of (2.24), or by putting (2.22) into (2.23) invariably yields

$$\mathrm{G}(\omega) = -i\frac{P_{K-1}(\omega)}{Q_K(\omega)}. \tag{2.25}$$

This quotient is explicitly the scalar Padé approximant for the Green function $\mathrm{G}(\omega)$ and the same is implicitly true for the Heaviside representation (2.24).

In an alternative derivation, inserting (2.21) into the rhs of (2.23) and identifying the matrix element $(\Phi_0|\hat{\mathrm{U}}(t)|\Phi_0)$ as $C(t)$ from (2.18) gives

$$\mathrm{G}(\omega) = \int_0^\infty \mathrm{d}t\, \mathrm{e}^{i\omega t} C(t). \tag{2.26}$$

This demonstrates that the Green function $\mathrm{G}(\omega)$ is the Fourier integral of the auto-correlation function $C(t)$ similarly to their operator-valued counterparts $\hat{\mathrm{G}}(\omega)$ and $\hat{\mathrm{U}}(t)$ from (2.20). Due to the invertability of the Fourier integrals, $\hat{\mathrm{U}}(t)$ and $C(t)$ can be retrieved exactly from $\hat{\mathrm{G}}(\omega)$ and $\mathrm{G}(\omega)$, respectively

$$\hat{\mathrm{U}}(t) = \frac{1}{2\pi}\int_0^\infty \mathrm{d}\omega\, \mathrm{e}^{-i\omega t}\hat{\mathrm{G}}(\omega) \tag{2.27}$$

$$C(t) = \frac{1}{2\pi}\int_0^\infty \mathrm{d}\omega\, \mathrm{e}^{-i\omega t}\mathrm{G}(\omega). \tag{2.28}$$

When (2.21) is inserted into (2.28), the result (2.18) for the auto-correlation function $C(t)$ is obtained as the sum of K damped complex exponentials. This is enabled by passing to the complex frequency plane and employing the Cauchy residue theorem to perform the integral exactly through the sum of poles at $\omega = \omega_k$. Functions $\mathrm{G}(\omega)$ and $C(t)$ are associated with the same operator $\hat{\Omega}$ as the only source of the whole information about the system. Such a property and the possibility of switching from $C(t)$ to $\mathrm{G}(\omega)$ or the other way around is a guarantee that the information is fully preserved when passing from the time to the frequency domain or *vice versa*. The spectral resonance parameters from (2.24) are the position, width, height and phase of the kth peak that are equal or proportional to $\mathrm{Re}(\omega_k), \mathrm{Im}(\omega_k), |d_k|, \mathrm{Arg}(\omega_k)$, respectively. By definition (2.19), the residue d_k is a measure of the extent of the squared projection of the state $|\Upsilon_k)$ onto $(\Phi_0|$. In other words, the amplitudes $\{d_k\}$ are the weights that contain information about the extent of the contributions from individual fundamental frequencies $\{\omega_k\}$ to $C(t)$ in (2.18). Furthermore, the magnitudes $\{|d_k|\}$ are the intensities of the harmonics $\{\exp(-i\omega_k t)\}$ that are the principal components in the auto-correlation function $C(t)$. Moreover, $\phi_k = \mathrm{Arg}(d_k)$ is the kth phase of complex-valued $C(t)$. The real-valued quantities of a particular interest, such as the magnitude $|\mathrm{G}(\omega)|$, power $|\mathrm{G}(\omega)|^2$, absorption $\mathrm{Re}(\mathrm{G}(\omega))$ and dispersion $\mathrm{Im}(\mathrm{G}(\omega))$ spectra are all available as soon as the Green function $\mathrm{G}(\omega)$ is constructed. This illuminates the key role of the Green function.

The summation from (2.21) represents the Heaviside partial fraction representation of the Green operator $\hat{\mathrm{G}}(\omega)$. This sum is also the spectrum in terms of a mixed operator-scalar valued function of frequency ω. The outlined derivation makes no approximation and, therefore, the obtained Heaviside representation is the exact spectrum. Furthermore, the sum over k in (2.21) can be performed exactly via

$$\sum_{k=1}^{K} \frac{\hat{\pi}_k}{\omega - \omega_k} = \sum_{k=1}^{K} \frac{|\Upsilon_k)(\Upsilon_k|}{\omega - \omega_k} = \sum_{k=1}^{K} |\Upsilon_k)(\Upsilon_k|(\omega\hat{1} - \hat{\Omega})^{-1} = \hat{1}(\omega\hat{1} - \hat{\Omega})^{-1}$$

where the ϵ−factor is omitted. Here, we employed the following Green eigenvalue problem where ω appears as a parameter

$$\hat{\mathrm{G}}(\omega)|\Upsilon_k) = \lambda_k(\omega)|\Upsilon_k) \qquad \lambda_k(\omega) = \frac{1}{\omega - \omega_k} \tag{2.29}$$

in accordance with the relationship (2.13) for the particular choice $f(\hat{\Omega}) = -i(\omega\hat{1} - \hat{\Omega})^{-1}$. Therefore, we finally arrive at the result

$$\hat{\mathrm{G}}(\omega) = -i\hat{1}(\omega\hat{1} - \hat{\Omega})^{-1} = -i(\omega\hat{1} - \hat{\Omega})^{-1}\hat{1} \equiv -i(\omega\hat{1} - \hat{\Omega})^{-1} \tag{2.30}$$

where the unity operator $\hat{1}$ can be placed either before or after the inverse of the Schrödinger operator $\omega\hat{1} - \hat{\Omega}$ depending upon whether the Green eigenvalue problem is used for $|\Upsilon_k)$ or $(\Upsilon_k|$. Of course, both states $|\Upsilon_k)$ and $(\Upsilon_k|$ are simultaneously present in the projector $\hat{\pi}_k$. Here, we use the name Schrödinger operator for $\omega\hat{1} - \hat{\Omega}$ because this operator originates from the Schrödinger equation $\hat{\Omega}|\Upsilon) = \omega|\Upsilon)$ which can alternatively be written as the secular or characteristic equation $(\omega\hat{1} - \hat{\Omega})|\Upsilon) = \hat{0}$. The inverse operator $(\omega\hat{1} - \hat{\Omega})^{-1}$ is in a special category called resolvent operators. Naturally, (2.30) can exist on its own as the definition of the Green operator from the outset without any derivation. Therefore, if (2.30) is considered as the definition of $\hat{G}(\omega)$, it is useful to replace the unity operator by the closure (2.14). The subsequent use of the Green eigen-value problem (2.29) yields the spectral decomposition (2.21) without recourse to the Fourier integral (2.26).

Importantly, when using the Green function from (2.30), which is obtained by the above derivation, then $\hat{G}(\omega)$ is recognized as the operator Padé approximant (OPA) [121] (there is also the matrix Padé approximant (MPA) [122]). Generally, the left $\hat{A}_L\hat{B}_K^{-1}$ and the right $\hat{B}_K^{-1}\hat{A}_L$ variants of the OPA yield the same OPA through $\hat{A}_L\hat{B}_K^{-1} = \hat{B}_K^{-1}\hat{A}_L$. This allows one to drop the adjective left or right associated with OPA. In fact, the equality between the left and the right OPA is easily established from their definitions that are of the same type as the scalar PA, except for the operator-valued polynomial coefficients in $\hat{A}_L$ and $\hat{B}_K$. These notions also apply to (2.30) because of the presence of the unity operator for $\hat{A}$ (the other operator $\hat{B}$ is $\omega\hat{1} - \hat{\Omega}$). The OPA structure of the exact Green operator, which is a more fundamental quantity than the associated scalar, implies that in every realization, the corresponding exact Green function will necessarily be the scalar PA. Hence, it is not surprising that the main building block of, e.g., the Heisenberg picture of quantum mechanics, the scattering operator or the $S-$operator must be in the mathematical form of a rational function, as shown first by Hu [123].

The above analysis demonstrates that the Padé approximant is actually elevated to the status of being an inherent, fundamental formalism of quantum mechanics *per se* from the outset. As such, the PA represents an inseparable part of the conceptual foundation of quantum mechanics, and not only an exceptionally successful computational method. This remark is made in the abstract theoretical framework with an underlying total generality and with no reference to any particular algorithmic methodology. This status of the Padé approximant justifies the previous concrete derivations using a number of the leading quantum-mechanical methods, that the Padé approximant (operator-valued and/or scalar) is equivalent to the Schwinger variational principle, variational-iteration methods and the Fredholm determinant, to name only a few of the leading theories [5]. In particular, the establishment of the variational nature of the Padé approximant is the critical feature of this method especially related to the stability of spectral estimation as well as to the feasibility of obtaining the upper and lower bounds to the computed spectral parameters of generic systems.

Regarding the computational performance, judicious implementation of the theoretical formalism of the Green function is significant in itself, on top of being a veritable alternative to the Schrödinger equation, especially for spectral analysis and signal processing. Furthermore, it is advantageous to solve the Green eigen-value problem (2.29) instead of the original Schrödinger equation (2.1) because $\hat{\mathrm{G}}(\omega)$ from the former equation is a bound operator, whereas $\hat{\Omega}$ from the latter one is an unbound operator[1]. This means that the eigen-values $\{\omega_k\}$ obtained from the Green eigen-problem can simultaneously possess the lower and upper bounds or limits, whereas only the lower bounds could be provided by the Schrödinger equation. In the Green eigen-problem (2.29), frequency ω is considered as a scaling parameter, which can be taken to be $\omega = \omega_0$, where ω_0 is any selected fixed number. Upon finding the solution of (2.29) for $\omega = \omega_0$, the sought eigen-frequencies $\{\omega_k\}$ can be deduced from the eigen-values $\lambda_k(\omega_0)$ through the following relationship

$$\omega_k = \omega_0 - \frac{1}{\lambda_k(\omega_0)}. \tag{2.31}$$

Advantageously, the corresponding eigen-functions $\{|\Upsilon_k)\}$ are the same in the Green and the Schrödinger eigen-value problem. This feature has already been encountered previously in the case of the general equation (2.13).

2.2 Expansion methods for signal processing

2.2.1 Non-classical polynomials

Herein, we will employ real orthogonal polynomials $\{Q_r(x)\}$ as basis functions to establish an expansion method for computation of the most important physical quantities, such as Green functions, density of states, integrated density of states, etc. Such polynomials are introduced on the real axis in the interval $[a, b]$ which can be finite or infinite. The expansion coefficients of these polynomials are also real-valued. However, since $Q_r(x)$ is an analytic function, the quantity $Q_r(z)$ exists for a complex variable z in the limit $\mathrm{Im}(z) \longrightarrow 0$. For a real x and a complex z we have

$$Q_n^*(x) = Q_n(x) \qquad Q_n^*(z) = Q_n(z^*) \tag{2.32}$$

where the star superscript denotes the usual complex conjugation. By $\mathrm{d}\sigma(x)$ we denote a non-negative Riemann measure on $[a, b]$

$$\mathrm{d}\sigma(x) = \mathrm{W}(x)\mathrm{d}x \tag{2.33}$$

[1] An operator is said to be bound if and only if its norm is limited both from below and above. However, an operator bound only from below (as is actually the case with most Hamiltonian operators) is an unbound operator.

where the weight function $\mathrm{W}(x) > 0$ has at least $n+1$ points of increase. Next, in the space of polynomials $\{Q_r(x)\}$, one can define the asymmetric scalar/inner product with respect to the measure $\mathrm{d}\sigma(x)$ via

$$\langle Q_m(x)|Q_n(x)\rangle \equiv \int_a^b Q_m^*(x)Q_n(x)\mathrm{d}\sigma(x). \tag{2.34}$$

Of course, asymmetric scalar products are customary in the conventional formulation of quantum mechanics defined on the field of Hermitean operators. Clearly, the star in $Q_m^*(x)$ is superfluous in (2.34) because $Q_m^*(x) = Q_m(x)$ according to (2.32). We shall keep the star superscript on all the 'bra' vectors $(f(x)|$ even for a real function $f(x)$ in order to indicate the use of the asymmetric inner product. For the value $Q_n(z)$ with a complex z, it is obvious that the star superscript on such a polynomial cannot be ignored. Such circumstances can be relevant for resolvents of even Hermitean operators (Hamiltonians) encountered in quantum mechanics. Further, we assume that polynomials $\{Q_n(x)\}$ are orthonormalized with respect to the measure $\mathrm{d}\sigma(x)$

$$\int_a^b Q_n^*(x)Q_m(x)\mathrm{d}\sigma(x) = \mathrm{w}_n\delta_{nm} \tag{2.35}$$

where the coefficients $\{\mathrm{w}_n\}$ are the weights ($\mathrm{w}_n > 0$). The latter weights should not be confused with the standard Christoffel numbers $\{w_n\}$ from numerical integrations [5]. We shall also use the completeness relation

$$\delta(x-x_0) = \sum_{n=0}^{\infty} \mathrm{W}_n(x)Q_n^*(x)Q_m(x_0) \qquad \mathrm{W}_n(x) = \frac{\mathrm{W}(x)}{\mathrm{w}_n}. \tag{2.36}$$

This relationship formally retains its form for operator-valued functions

$$\delta(E\hat{1} - \hat{\mathrm{H}}) = \sum_{n=0}^{\infty} \mathrm{W}_n(E)Q_n^*(E)Q_m(\hat{\mathrm{H}}) \tag{2.37}$$

where E is the real energy and $\hat{\mathrm{H}}$ is the Hermitean Hamiltonian operator ($\hat{\mathrm{H}}^\dagger = \hat{\mathrm{H}}$) of the investigated physical system. In (2.37), the Dirac δ-operator represents the so-called density operator for the total energy E and Hamiltonian $\hat{\mathrm{H}}$. There exists the following operator relationship, in the limit $\epsilon \longrightarrow 0^+$

$$\frac{1}{(E+i\epsilon)\hat{1} - \hat{\mathrm{H}}} = \mathcal{P}\left(\frac{1}{E\hat{1} - \hat{\mathrm{H}}}\right) - i\pi\delta(E\hat{1} - \hat{\mathrm{H}}) \tag{2.38}$$

where the symbol $\mathcal{P}$ denotes the usual principle value. Using this expression, the density operator becomes

$$\delta(E\hat{1} - \hat{\mathrm{H}}) = -\frac{1}{\pi}\mathrm{Im}\hat{\mathrm{G}}(z) \tag{2.39}$$

where $\hat{\mathrm{G}}(z)$ is the Green operator

$$\hat{\mathrm{G}}(z) = \left(z\hat{1} - \hat{\mathrm{H}}\right)^{-1}. \tag{2.40}$$

For a Hermitean Hamiltonian $\hat{\mathrm{H}}$, it follows

$$\hat{\mathrm{G}}^*(z) = \hat{\mathrm{G}}(z^*). \tag{2.41}$$

The existence of the inverse operator $\hat{\mathrm{G}}(z)$ is guaranteed if z is not an eigen-energy E_k from the spectrum of $\hat{\mathrm{H}}$. This will be the case for a Hermitean Hamiltonian $\hat{\mathrm{H}}$ provided that

$$z = E + i\epsilon \qquad z^+ = E + i0^+ \tag{2.42}$$

where ϵ is an infinitesimally small positive number. On the other hand, the resolvent $\hat{\mathrm{G}}^+(z)$ can be viewed as the standard Fourier integral of the time evolution operator

$$\hat{\mathrm{G}}(E) = -i\int_{-\infty}^{\infty} \mathrm{d}t\theta(t)\hat{\mathrm{U}}(t)\mathrm{e}^{iEt} \tag{2.43}$$

$$\hat{\mathrm{U}}(t) = \mathrm{e}^{-i\hat{\mathrm{H}}t} \tag{2.44}$$

where $\theta(t)$ is the usual Heaviside step function and t is the real time variable.

As it stands, $\hat{\mathrm{G}}(z)$ is the operator Padé approximant, i.e., the OPA, because this resolvent is defined by the product of the unity operator and $(z\hat{1} - \hat{\mathrm{H}})^{-1}$, and formally, $\hat{\mathrm{G}}(z) = \hat{1}(z\hat{1} - \hat{\mathrm{H}})^{-1} = (z\hat{1} - \hat{\mathrm{H}})^{-1}\hat{1}$. Therefore, diagonal or off-diagonal matrix elements of the Green function $\hat{\mathrm{G}}$ are the customary scalar Padé approximant. This follows from the corresponding Schrödinger eigenvalue problem

$$\hat{\mathrm{H}}|\Upsilon_k\rangle = E_k|\Upsilon_k\rangle \tag{2.45}$$

where $|\Upsilon_k\rangle$ is the eigen-state corresponding to the eigen-energy E_k. Employing completeness of the set $\{|\Upsilon_k\rangle\}$, in the form of the spectral decomposition of the unity operator, i.e., $\hat{1} = \sum_{k=1}^{K} |\Upsilon_k\rangle\langle\Upsilon_k|$, we have from (2.40)

$$\hat{\mathrm{G}}(z) = \sum_{k=1}^{K} \frac{|\Upsilon_k\rangle\langle\Upsilon_k|}{z - E_k}. \tag{2.46}$$

Non-negative integer K is finite or infinite, depending on the number of eigenstates E_k in the spectrum of $\hat{\mathrm{H}}$. The associated diagonal ($r = s$) and off-diagonal ($r \neq s$) matrix elements of $\hat{\mathrm{G}}(z)$ taken over the states $|\Phi_s\rangle$ and $\langle\Phi_r|$ can be extracted from (2.46) as

$$\mathrm{G}_{r,s}(z) \equiv \langle\Phi_r|\hat{\mathrm{G}}|\Phi_s\rangle = \sum_{k=1}^{K} \frac{d_k^{(r,s)}}{z - E_k} \equiv \frac{A_{K-1}(z)}{B_K(z)} \tag{2.47}$$

$$d_k^{(r,s)} = \langle \Phi_r | \Upsilon_k \rangle \langle \Upsilon_k | \Phi_s \rangle. \tag{2.48}$$

A simplification occurs in the diagonal case when the residue (2.48) is reduced to $d_k^{(s)} = |\langle \Phi_s | \Upsilon_k \rangle|^2$ where $d_k^{(s,s)} \equiv d_k^{(s)}$. We see from (2.47) that $\mathrm{G}_{r,s}(z)$ is the ratio of two polynomials $A_{K-1}(z)/B_K(z)$, which is the Padé approximant, as in (2.25), and that was set to prove (QED). This indicates that the Padé approximant is the most natural direct method for obtaining the Green function. Namely, by construction, the Padé approximant is exact whenever the function to be modeled is itself a quotient of two polynomials. This is the case with the Green function (2.47), which is a polynomial quotient, since it originates from the matrix element of the operator Padé approximant represented by the resolvent operator (2.40).

Due to their orthonormality, the polynomials $\{Q_r(x)\}$ fulfill the standard three-term recursion relation

$$\left.\begin{array}{c} \beta_{n+1} Q_{n+1}(x) = (x - \alpha_n) Q_n(x) - \beta_n Q_{n-1}(x) \\ Q_{-1}(x) = 0 \qquad Q_0(x) = 1 \end{array}\right\}. \tag{2.49}$$

When (2.49) is multiplied by $Q_m^*(x)$ and the product integrated over x from a to b with the help of (2.33) and (2.35), the following result is obtained

$$\alpha_n = \frac{1}{\mathrm{w}_n} \int_a^b Q_n^*(x) x Q_n(x) \mathrm{d}\sigma(x). \tag{2.50}$$

Similarly, multiplication of (2.49) by $Q_{n-1}(x)$ yields the expression

$$\beta_n = \frac{1}{\mathrm{w}_{n-1}} \int_a^b Q_{n-1}^*(x) x Q_n(x) \mathrm{d}\sigma(x). \tag{2.51}$$

The formulae (2.50) and (2.51) can be taken as the definitions of the parameters α_n and β_n. To check these relations, we first eliminate the term $xQ_n(x)$ from (2.50) and (2.51). Afterwards, using (2.35), (2.50) and (2.51) the identities $\alpha_n \equiv \alpha_n$ and $\beta_n \equiv \beta_n$ are obtained (QED). Hence, α_n from (2.50) represents the diagonal element of the position operator $\hat{x}$ in the coordinate representation. Similarly, in this latter representation, the same operator $\hat{x}$ contains the parameter β_n from (2.51) as the coupling constant between the nth and $(n-1)$st polynomials. We can arrive at an alternative interpretation of β_n. To this end, we initially eliminate $xQ_{n-1}^*(x)$ from (2.51) employing (2.49) so that $xQ_{n-1}^*(x) = \beta_n Q_n^*(x) + \alpha_{n-1} Q_{n-1}^*(x) + \beta_{n-1} Q_{n-2}^*(x)$. This is followed by the application of the orthonormality condition (2.35) with the final result

$$\begin{aligned} \beta_n &= \frac{1}{\mathrm{w}_{n-1}} \int_a^b |Q_n(x)|^2 \mathrm{d}\sigma(x) = \frac{1}{\mathrm{w}_{n-1}} \int_a^b Q_n^2(x) \mathrm{d}\sigma(x) \\ &= \frac{1}{\mathrm{w}_{n-1}} ||Q_n||^2 \Longrightarrow \beta_n > 0. \end{aligned} \tag{2.52}$$

Here, β_n becomes proportional to the squared norm of $Q_n(x)$ and, hence, all the parameters $\{\beta_n\}$ are strictly positive. Furthermore, since the diagonal case of the matrix element (2.35) is the norm $\|Q_n\|^2 = \int_a^b |Q_n(x)|^2 \mathrm{d}\sigma(x) = \mathrm{w}_n$, we can derive a very simple formula for β_n from (2.52) via

$$\beta_n = \frac{\mathrm{w}_n}{\mathrm{w}_{n-1}}. \tag{2.53}$$

Thus, the entire procedure of a numerical construction of the orthonormalized polynomials $\{Q_r(x)\}$ with the self-generating coefficients $\{\alpha_r, \beta_r\}$ is reminiscent of the well-known Lanczos algorithm when applied to the quantum-mechanical position operator $\hat{x}$ [5]. Moreover, the Lanczos algorithm for physical state vectors can be devised by replacing operator $\hat{x}$ by the Hamiltonian $\hat{\mathrm{H}}$ in the scalar recursion (2.49). The polynomials $\{Q_n(x)\}$ are called the polynomials of the first kind. Every such polynomial $Q_n(x)$ of the degree n has n zeros $\{x_k\}$ $(1 \leq k \leq n)$. Due to the mentioned assumptions on $\mathrm{W}(x)$, all the quantities $\{x_k\}$ are real-valued simple zeros ($x_k \neq x_{k'}$ for $k' \neq k$) and belong to the interval (a, b).

We also need the polynomials of the second kind denoted by $\{P_r(x)\}$ that are tightly linked to the basis functions $\{Q_r(x)\}$. Formally, the polynomials $P_r(x)$ satisfy the same recursive relation from (2.49), but the initial conditions are different

$$\left.\begin{array}{c} \beta_{n+1}P_{n+1}(x) = (x - \alpha_n)P_n(x) - \beta_n P_{n-1}(x) \\ P_0(x) = 0 \qquad P_1(x) = 1 \end{array}\right\}. \tag{2.54}$$

Otherwise, both recursions (2.49) and (2.54) are numerically stable. Alternatively, the polynomials $Q_n(x)$ and $P_n(x)$ can be introduced by their power series representations

$$Q_n(x) = \sum_{r=0}^{n} q_{r,n-r}x^r \qquad P_n(x) = \sum_{r=0}^{n-1} p_{r,n-r}x^r. \tag{2.55}$$

Here, the expansion coefficients $p_{r,n-r}$ and $q_{r,n-r}$ can also be computed recursively by the same kind of relations from (2.49) and (2.54)

$$\left.\begin{array}{c} \beta_{n+1}p_{n+1,n+1-r} = p_{n,n+1-r} - \alpha_n p_{n,n-r} - \beta_n p_{n-1,n-1-r} \\ p_{n,-1} = 0 \qquad p_{n,m} = 0 \qquad (m > n) \qquad p_{0,0} = 1 \qquad p_{1,1} = 1 \end{array}\right\} \tag{2.56}$$

$$\left.\begin{array}{c} \beta_{n+1}q_{n+1,n+1-r} = q_{n,n+1-r} - \alpha_n q_{n,n-r} - \beta_n q_{n-1,n-1-r} \\ q_{n,-1} = 0 \qquad q_{n,m} = 0 \qquad (m > n) \qquad q_{0,0} = 1. \end{array}\right\}. \tag{2.57}$$

It can be shown that the zeros $\{\tilde{x}_k\}$ of $P_n(x)$ interlace with the zeros x_k of $Q_n(x)$

$$x_1 < \tilde{x}_1 < x_2 < \tilde{x}_2 < \cdots < x_n < \tilde{x}_n. \tag{2.58}$$

In the mathematical literature, this is known as the Cauchy–Poincaré interlacing theorem, which was rediscovered in quantum chemistry as the Hylleraas–Undheim theorem [124].

Because of the scaled initial conditions relative to (2.49), the degree of the polynomial $P_n(x)$ is by 1 smaller than that of $Q_n(x)$. Thus, the degrees of the polynomials $Q_n(x)$ and $P_n(x)$ are n and $n-1$, respectively, as per (2.55). There is a relationship between the polynomials $Q_n(x)$ and $P_n(x)$ by means of the following integral

$$P_n(z) = \frac{\beta_1}{\mu_0}\int_a^b \frac{Q_n(z)-Q_n(x)}{z-x}\mathrm{d}\sigma(x) \tag{2.59}$$

$$P_n(z^*) = P_n^*(z) \tag{2.60}$$

where μ_0 is the simplest case ($n=0$) of the power moment

$$\mu_n = \int_a^b x^n \mathrm{d}\sigma(x). \tag{2.61}$$

These polynomials can be linked to the standard Gaussian quadratures. To this end, we introduce the Stieltjes integral $S(z)$ via

$$S(z) = \int_a^b \frac{\mathrm{d}\sigma(x)}{z-x} \qquad S(z^*) = S^*(z). \tag{2.62}$$

As usual, the Stieltjes integral can be represented by the Padé approximant $\mathcal{R}_n(u)$ via

$$S(z) \approx \mathcal{R}_n(z) \tag{2.63}$$

$$\mathcal{R}_n(x) \equiv \frac{\mu_0}{\beta_1}\frac{P_n(x)}{Q_n(x)}. \tag{2.64}$$

To estimate the remainder $S(z)-\mathcal{R}_n(z)$ in (2.63), we define the error term $\mathcal{E}_n(z)$, which is a measure of the approximation $S(z)\approx\mathcal{R}_n(z)$

$$S(z) = \mathcal{R}_n(z) + \mathcal{E}_n(z). \tag{2.65}$$

It is possible to derive an expression for the error or remainder $\mathcal{E}_n(z)$ by substituting (2.65) into (2.59) and employing (2.64)

$$\begin{aligned}
P_n(z) &= \frac{\beta_1}{\mu_0}\left\{Q_n(z)\int_a^b \frac{\mathrm{d}\sigma(x)}{z-x} - \int_a^b \frac{\mathrm{d}\sigma(x)}{z-x}Q_n(x)\right\}\\
&= \frac{\beta_1}{\mu_0}\left[Q_n(z)[\mathcal{R}_n(z)+\mathcal{E}_n(z)] - \int_a^b \frac{\mathrm{d}\sigma(x)}{z-x}Q_n(x)\right]\\
&= \frac{\beta_1}{\mu_0}\left\{Q_n(z)\left[\frac{\mu_0}{\beta_1}\frac{P_n(z)}{Q_n(z)} + \mathcal{E}_n(z)\right] - \int_a^b \frac{\mathrm{d}\sigma(x)}{z-x}Q_n(x)\right\}\\
P_n(z) &= P_n(z) + \frac{\beta_1}{\mu_0}\left[Q_n(z)\mathcal{E}_n(z) - \int_a^b \frac{\mathrm{d}\sigma(x)}{z-x}Q_n(x)\right]
\end{aligned}$$

$$\therefore \qquad \mathcal{E}_n(z) = \frac{\pi_n(z)}{Q_n(z)} \tag{2.66}$$

$$\pi_n(z) \equiv \int_a^b \frac{\mathrm{d}\sigma(x)}{z-x}Q_n(x). \tag{2.67}$$

While arriving to the expression (2.66), a division by the constant quotient β_1/μ_0 was made. This is allowed, since $\beta_1 \neq 0$ and $\mu_0 \neq 0$. Employing (2.32) and (2.67), we have

$$\pi_n^*(z) = \pi_n(z^*). \tag{2.68}$$

This is reminiscent of the relation (2.41). Moreover, (2.68) can also be obtained from the complex conjugate counterpart of (2.67) by using (2.32) and (2.41). The derived expression (2.66) for the error $\mathcal{E}_n(z)$ is remarkable because the function $Q_n(z)\mathcal{E}_n(z)$ coincides precisely with the Hilbert transform of the polynomial $Q_n(x)$ in the vector space defined by the measure $\mathrm{d}\sigma(x)$ and the asymmetric scalar product for $x \in [a,b]$.

The polynomial quotient $\mathcal{R}_n(x)$ from (2.64) is a meromorphic function. Therefore, $\mathcal{R}_n(x)$ can be given by its spectral representation, which is a linear combination of the Heaviside partial fractions

$$\mathcal{R}_n(x) = \sum_{k=1}^{n} \frac{w_k}{x - x_k}. \tag{2.69}$$

Here, $\{x_k\}$ are the zeros of the polynomials $Q_n(x)$, whereas $\{w_k\}$ are the corresponding Christoffel numbers. Explicitly, $\{w_k\}$ can be obtained by means of the Cauchy residue theorem applied to $\mathcal{R}_n(x)$

$$w_k = \lim_{x \to x_k} (x - x_k)\mathcal{R}_n(x) \tag{2.70}$$

$$w_k = \frac{\mu_0}{\beta_1}\frac{P_n(x_k)}{Q_n'(x_k)} = \frac{\mu_0}{\beta_n}\frac{1}{Q_{n-1}(x_k)Q_n'(x_k)} \tag{2.71}$$

$$P_n(x_k) = \frac{\beta_1}{\beta_n Q_{n-1}(x_k)} \qquad Q_n(x_k) = 0 \tag{2.72}$$

where $Q_n'(x) = (\mathrm{d}/\mathrm{d}x)Q_n(x)$. The formula (2.71) can also be derived through the Christoffel–Darboux formula [5]

$$\sum_{m=0}^{n-1} Q_m(x)Q_m(y) = \beta_n \frac{Q_n(x)Q_{n-1}(y) - Q_{n-1}(x)Q_n(y)}{x - y}. \tag{2.73}$$

Summation of all the terms $\{w_k/(x - x_k)\}$ on the rhs of (2.69) leads to the quotient of the two polynomials $(\mu_0/\beta_1)P_n(x)/Q_n(x)$. This result represents the Padé approximant (2.64) from which we started. Hence consistency.

The formulae (2.65) and (2.66) show that it is possible to compute the error $\mathcal{E}_n(z)$ with an absolute certainty whenever the Stieltjes integral (2.62) is represented by the rational polynomial $\mathcal{R}_n(z)$ from (2.64). This can be achieved by the exact Gauss-type numerical quadrature rule

$$\int_a^b \frac{\mathrm{d}\sigma(x)}{z - x} = \lim_{n \to \infty} \sum_{k=1}^{n} \frac{w_k}{z - x_k}. \tag{2.74}$$

This expression is of the same kind as the customary formula encountered in the well-known Gauss numerical integrations with the help of the classical polynomials. The only difference is that we employ the non-classical Lanczos polynomials in (2.74). Naturally, in all realistic computations, the infinite upper limit in the sum over n from (2.74) is replaced by a finite cut-off number K which is the order or rank of the quadrature rule. Thus, when the development in (2.74) is truncated to the first K terms, the following approximation is obtained

$$\int_a^b \frac{\mathrm{d}\sigma(x)}{z-x} \approx \sum_{k=1}^{K} \frac{w_k}{z-x_k}. \tag{2.75}$$

Importantly, even with only K terms retained in the Heaviside partial fractions, the final result for $S(z)$ can still be exact if the error $\mathcal{E}_K(z)$ is taken into account via

$$\begin{aligned} S(z) \equiv \int_a^b \frac{\mathrm{d}\sigma(x)}{z-x} &= \sum_{k=1}^{K} \frac{w_k}{z-x_k} + \mathcal{E}_K(z) \\ &= \frac{\mu_0}{\beta_1} \frac{P_K(z)}{Q_K(z)} + \mathcal{E}_K(z). \end{aligned} \tag{2.76}$$

This analysis can be extended to the corresponding exact Gauss-type quadrature for the case of a general function $f(x)$ in place of $(z-x)^{-1}$

$$\int_a^b f(x)\mathrm{d}\sigma(x) = \lim_{n\to\infty} \sum_{k=1}^{n} w_k f(x_k) + \mathcal{E}_n \tag{2.77}$$

$$\mathcal{E}_n = \frac{1}{Q_n(z)} \int_a^b \mathrm{d}\sigma(x) f(x) Q_n(x). \tag{2.78}$$

Employing the binomial expansion for $(z-x)^{-1} = \sum_{n=0}^{\infty} x^n z^{-n-1}$ in (2.67), we have

$$\pi_n(z) = \sum_{m=0}^{\infty} \mu_{n,m} z^{-m-1} \tag{2.79}$$

where $\mu_{n,m}$ is the modified moment [5]

$$\mu_{n,m} = \langle Q_n(x)|x^m\rangle = \int_a^b Q_n^*(x) x^m \mathrm{d}\sigma(x) \qquad \mu_{n,0} = \mu_n. \tag{2.80}$$

By means of the orthogonality (2.34), the following important feature of $\mu_{n,m}$ is deduced

$$\begin{aligned} \langle Q_n(x)|x^m\rangle &= 0 \qquad m = 0,1,\ldots,n-1 \\ \therefore \qquad \mu_{n,m} &= 0 \qquad m \le n-1. \end{aligned} \tag{2.81}$$

This property makes the first n terms of series in (2.79) equal to zero and this reduces $\pi_n(z)$ to

$$\pi_n(z) = \sum_{m=n}^{\infty} \mu_{n,m} z^{-m-1} = \sum_{m=0}^{\infty} \mu_{n,n+m} z^{-n-m-1}. \tag{2.82}$$

The expansion coefficients $\mu_{n,m}$ from (2.82) can be computed recursively via

$$\beta_{n+1}\mu_{n+1,m} = \mu_{n,m+1} - \alpha_n\mu_{n,m} - \beta_n\mu_{n-1,m} \qquad \mu_{0,0} = 1 \tag{2.83}$$

where the parameters $\{\alpha_n\}$ and $\{\beta_n\}$ are the same Lanczos coupling constants from (2.49). Employing (2.55) and (2.61), we can rewrite (2.81) as

$$\sum_{r=0}^{n} q_{r,n-r}\mu_{r+m} = 0 \qquad (0 \le m < n) \tag{2.84}$$

or in the alternative matrix counterpart

$$\begin{pmatrix} \mu_0 & \mu_1 & \mu_2 & \cdots & \mu_{n-1} \\ \mu_1 & \mu_2 & \mu_3 & \cdots & \mu_n \\ \mu_2 & \mu_3 & \mu_4 & \cdots & \mu_{n+1} \\ \vdots & \vdots & \vdots & \ddots & \vdots \\ \mu_{n-1} & \mu_n & \mu_{n+1} & \cdots & \mu_{2n-2} \end{pmatrix} \begin{pmatrix} q_{0,n} \\ q_{1,n-1} \\ q_{2,n-2} \\ \vdots \\ q_{n-1,1} \end{pmatrix} = - \begin{pmatrix} \mu_n \\ \mu_{n+1} \\ \mu_{n+2} \\ \vdots \\ \mu_{2n-1} \end{pmatrix}. \tag{2.85}$$

It is seen that the $n \times n$ matrix on the lhs of (2.85) represents the usual Hankel matrix $\mathbf{H}_n(\mu_0) \equiv \{\mu_{i+j}\}\ (0 \le i \le n-1\,,\, 0 \le j \le n-1)$. Given the first $2n$ moments $\{\mu_r\}$, the system (2.85) of n linear inhomogeneous equations can be solved by taking $q_{n,n-r}$ as the unknown. In particular, the solution of the system (2.85) exists and it is unique for the non-zero value of the Hankel determinant

$$\det \mathbf{H}_n(\mu_0) \neq 0. \tag{2.86}$$

The relation (2.86) can be proven by supposing that, e.g., the opposite is valid, $\det \mathbf{H}_n(\mu_0) = 0$, and establishing the contradiction afterwards. The relation $\det \mathbf{H}_n(\mu_0) = 0$ means that there exists a linear dependence among the rows of $\det \mathbf{H}_n(\mu_0)$. Linear dependence signifies that there is a non-zero column vector $\mathbf{y} = \{y_r\}\ (0 \le r \le n-1)$ such that $\mathbf{H}_n(\mu_0)\mathbf{y} = \mathbf{0}$

$$\begin{pmatrix} \mu_0 & \mu_1 & \mu_2 & \cdots & \mu_{n-1} \\ \mu_1 & \mu_2 & \mu_3 & \cdots & \mu_n \\ \mu_2 & \mu_3 & \mu_4 & \cdots & \mu_{n+1} \\ \vdots & \vdots & \vdots & \ddots & \vdots \\ \mu_{n-1} & \mu_n & \mu_{n+1} & \cdots & \mu_{2n-2} \end{pmatrix} \begin{pmatrix} y_0 \\ y_1 \\ y_2 \\ \vdots \\ y_{n-1} \end{pmatrix} = \begin{pmatrix} 0 \\ 0 \\ 0 \\ \vdots \\ 0 \end{pmatrix} \tag{2.87}$$

or alternatively

$$\mathbf{y}^T \mathbf{H}_n(\mu_0) \qquad \mathbf{y} \neq \mathbf{0} \tag{2.88}$$

where $\mathbf{y}^T$ is the transpose matrix of $\mathbf{y}$. Next, a straightforward calculation with the help of (2.88) yields

$$
\begin{aligned}
0 = \mathbf{y}^T \mathbf{H}_n(\mu_0)\mathbf{y} &= \sum_{r=0}^{n-1} y_r \sum_{s=0}^{n-1} y_s \mu_{r+s} \\
&= \sum_{r=0}^{n-1} y_r \left\langle \sum_{s=0}^{n-1} y_s x^s \middle| x^r \right\rangle = \left\langle \sum_{s=0}^{n-1} y_s x^s \middle| \sum_{r=0}^{n-1} y_r x^r \right\rangle = \left\| \sum_{s=0}^{n-1} y_s x^s \right\|^2 \\
\therefore \qquad & \left\| \sum_{s=0}^{n-1} y_s x^s \right\|^2 = 0. \qquad (2.89)
\end{aligned}
$$

Since this zero norm of the polynomial $\sum_{s=0}^{n-1} y_s x^s$ is not permitted, the obtained result (2.89) contradicts the assumption $\det \mathbf{H}_n(\mu_0) = 0$. Hence, the opposite is true, which proves (2.86) (QED).

When we first replace n by $n+1$ in (2.67) and subsequently multiply the ensuing formula by β_{n+1}, the following expression is obtained with the use of the recursion (2.49)

$$\beta_{n+1}\pi_{n+1}(z) = (z - \alpha_n)\pi_n(z) - \beta_n \pi_{n-1}(z). \qquad (2.90)$$

This indicates that the denominator $\{\pi_n(z)\}$ of the error $\mathcal{E}_n(z)$ from (2.66) can be constructed using the recursion (2.49). Therefore, as the Lanczos algorithm proceeds, the error at each step can be computed in concert with the same kind of recursion, as is clear from (2.49) and (2.90). Such an efficient error analysis is of great practical value in versatile applications of the Lanczos algorithm.

Because the set $\{Q_r(x)\}$ is a basis, it is possible to expand, e.g., the Green operator $\hat{\mathrm{G}}(z)$ from (2.40) in terms of $Q_n(x)$ as

$$\hat{\mathrm{G}}(z) = \sum_{n=0}^{\infty} \gamma_n(z) Q_n(\hat{\mathrm{H}}) \qquad \gamma_n(z) \equiv \frac{\eta_n(z)}{\mathrm{w}_n} \qquad (2.91)$$

where $\{\eta_n(z)\}$ are the expansion coefficients. The explicit expression for the coefficients $\eta_n(z)$ can be derived by the standard projection technique. This procedure shows that the general coefficient $\eta_n(z)$ is given by the Hilbert transform of $Q_n^*(x)$

$$\eta_n(z) = \int_a^b \frac{\mathrm{d}\sigma(x)}{z - x} Q_n^*(x). \qquad (2.92)$$

Moreover, by comparing (2.67) with (2.92) and using (2.32), we have

$$\eta_n(z) = \pi_n(z). \qquad (2.93)$$

It should be noted that the coefficients $\{\eta_n(z)\}$ in (2.91) are independent of the Hamiltonian $\hat{\mathrm{H}}$, which is here present merely via $Q_n(\hat{\mathrm{H}})$. Thus, as soon as the basis $\{Q_r\}$ is numerically generated (or chosen from the family of classical polynomials), the assembly of the coefficients $\{\eta_n(E)\}$ will remain the same for different systems if they are taken at the same energy E [125]–[131].

Alternatively, the integral representation (2.59) of the polynomial $P_n(z)$ can also be employed to obtain the expansion coefficient $\eta_n(z)$. This is achieved by replacing z by z^* in (2.59) yielding $P_n(z^*)$ which is, by reference to(2.60), equal to $P_n^*(z)$

$$\begin{aligned}
P_n(z^*) = P_n^*(z) &= \frac{\beta_1}{\mu_0}\left\{Q_n(z^*)\int_a^b \frac{\mathrm{d}\sigma(x)}{z^*-x} - \int_a^b \frac{\mathrm{d}\sigma(x)}{z^*-x}Q_n(x)\right\} \\
&= \frac{\beta_1}{\mu_0}\left\{Q_n(z)\int_a^b \frac{\mathrm{d}\sigma(x)}{z-x} - \int_a^b \frac{\mathrm{d}\sigma(x)}{z-x}Q_n^*(x)\right\}^* \\
&= \frac{\beta_1}{\mu_0}\left\{Q_n(z)\int_a^b \frac{\mathrm{d}\sigma(x)}{z-x} - \eta_n(z)\right\}^* \\
\therefore \qquad \eta_n(z) &= S(z)Q_n(z) - \frac{\mu_0}{\beta_1}P_n(z) \qquad (2.94)
\end{aligned}$$

where $S(z)$ is the Stieltjes integral (2.62). Employing (2.32) in a direct comparison of (2.66) with (2.92) or inserting (2.92) into (2.94), we obtain

$$\eta_n(z) = Q_n(z)\mathcal{E}_n(z). \qquad (2.95)$$

Thus, the expansion coefficient $\eta_n(z)$ in the series for the Green operator (2.92) is given by the product of the polynomial $Q_n(z)$ and the error term $\mathcal{E}_n(z)$ from the representation of the Stieltjes integral (2.62) by the Padé approximant $\mathcal{R}_n(z)$ or from its equivalent Heaviside partial fractions (2.76).

In the method of propagation of wave packets in the Schrödinger picture of quantum mechanics, the evolution operator $\hat{\mathrm{U}}(t)$ is employed. Following the outlined procedure for $\hat{\mathrm{H}}$, the evolution operator $\hat{\mathrm{U}}(t)$ can also be expanded in the polynomial basis $\{Q_n(x)\}$ via

$$\hat{\mathrm{U}}(t) = \sum_{n=0}^{\infty} \tilde{\eta}_n(t)Q_n(\hat{\mathrm{H}}). \qquad (2.96)$$

Here, the expansion coefficients $\tilde{\eta}_n(t)$ are related to $\eta_n(E)$ from (2.91) by the Fourier integral

$$\eta_n(E) = -i\int_{-\infty}^{\infty} \mathrm{d}t\,\theta(t)\tilde{\eta}_n(t)\mathrm{e}^{iEt}. \qquad (2.97)$$

When the set $\{Q_n(x)\}$ is taken to be within the classical polynomials [77, 132], the Stieltjes integral (2.62) with a known weight function $\mathrm{W}(x)$ can be calculated explicitly in the analytical form.

The expressions (2.49) and (2.54) can be used in the corresponding operator forms

$$\left.\begin{array}{c}\beta_{n+1}Q_{n+1}(\hat{\mathrm{H}}) = (\hat{\mathrm{H}} - \alpha_n\hat{1})Q_n(\hat{\mathrm{H}}) - \beta_nQ_{n-1}(\hat{\mathrm{H}}) \\ Q_{-1}(\hat{\mathrm{H}}) = \hat{0} \qquad Q_0(\hat{\mathrm{H}}) = \hat{1}\end{array}\right\}. \qquad (2.98)$$

For a given physical state $|\Phi_n\rangle$, which could be an eigen-vector of a part of $\hat{\rm H}$, or an orbital from a basis set, two state vectors $|\Psi_n\rangle$ and $|\psi_n^{(s)}\rangle$ can be introduced as

$$|\Psi_n\rangle = \hat{\rm G}(z)|\Phi_n\rangle \qquad |\psi_n^{(s)}\rangle = Q_n(\hat{\rm H})|\Phi_s\rangle. \tag{2.99}$$

Given a fixed s, the vector $|\psi_n^{(s)}\rangle$ can be obtained by a repeated application of (2.98). This procedure yields the Lanczos algorithm for physical states

$$\beta_{n+1}|\psi_{n+1}^{(s)}\rangle = (\hat{\rm H} - \alpha_n\hat{1})|\psi_n^{(s)}\rangle - \beta_n|\psi_{n-1}^{(s)}\rangle. \tag{2.100}$$

Both diagonal and off-diagonal matrix elements of the Green operator can be computed in this manner

$$\langle\Phi_r|\hat{\rm G}(z)|\Phi_s\rangle = \sum_{n=0}^{\infty}\gamma_n(z)I_{r,n,s} \qquad I_{r,n,s} = \langle\Phi_r|\psi_n^{(s)}\rangle. \tag{2.101}$$

We could also calculate the matrix elements of any number of Green operators that might be multiplied by some other operators. This gives the expression

$$|\Psi_s\rangle = \hat{\rm G}(z)|\Phi_s\rangle = \sum_{n=0}^{\infty}\gamma_n(z)|\psi_n^{(s)}\rangle. \tag{2.102}$$

In particular, it can be deduced from here that

$$\langle\Phi_r|\hat{\rm G}(z)\hat{\rm V}\hat{\rm G}(z)|\Phi_s\rangle = \langle\Psi_r|\hat{\rm V}|\Psi_s\rangle. \tag{2.103}$$

By reliance upon the quantum-mechanical trace, Tr, of the Green function, the density of states (DOS), as denoted by $\rho(E)$, can be introduced via

$$\rho(E) = -\frac{1}{\pi}{\rm Im}({\rm Tr})\{\hat{\rm G}(z^+)\}. \tag{2.104}$$

On the other hand, by definition, the trace represents the sum of the diagonal elements of $\delta(E\hat{1} - \hat{H})$

$$\rho(E) = \sum_{s=0}^{L}\langle\Phi_s|\delta(E\hat{1} - \hat{H})|\Phi_s\rangle. \tag{2.105}$$

Substituting (2.37) into (2.104) gives the following expression for $\rho(E)$

$$\rho(E) = \sum_{n=0}^{\infty}W_n(E)Q_n(E)I_n \qquad I_n = \sum_{s=0}^{L}I_{s,n,s} \tag{2.106}$$

where L is the number of the retained vectors from the set $\{|\Phi_s\rangle\}$. Another important quantity is the integrated density of state (IDOS), as denoted by $N(E)$, which can be obtained through integration of $\rho(x)$

$$N(E) = \int_{-\infty}^{E}{\rm d}x\rho(x) = \sum_{n=0}^{\infty}I_n\int_{-\infty}^{E}W_n(x)Q_n(x){\rm d}\sigma(x). \tag{2.107}$$

Since the set $\{|\psi_r)\}$ represents a basis, it can be employed to expand the complete Schrödinger state vector $|\Upsilon_k)$ of $\hat{\mathrm{H}}$ from (2.45). Furthermore, the total state vector can be written at any energy E, and not only E_k, through the following development

$$|\Upsilon(E)) = \sum_{n=0}^{\infty} B_n(E)|\psi_n^{(s)}). \tag{2.108}$$

This is valid for any vector $|\Phi_s)$, which is present in $|\psi_n^{(s)})$ through the state $|\psi_n^{(s)}) = Q_n(\hat{\mathrm{H}})|\Phi_s)$. Therefore, $|\Upsilon(E))$ is independent of $|\Phi_s)$ at any energy E. Moreover, the following expression for the expansion coefficients $B_n(E)$ can be obtained

$$B_n(E) = \mathrm{W}_n(E)Q_n(E). \tag{2.109}$$

As such, the exact Schrödinger eigen-vector $|\Upsilon(E))$ of $\hat{\mathrm{H}}$ can be generated by propagation of the state $|\Phi_s)$

$$|\Upsilon(E)) = \mathcal{P}(E, \hat{\mathrm{H}})|\Phi_s) \tag{2.110}$$

$$\mathcal{P}(E, \hat{\mathrm{H}}) = \sum_{n=0}^{\infty} \mathrm{W}_n(E)Q_n(E)Q_n(\hat{\mathrm{H}}). \tag{2.111}$$

For example, the vector $|\Phi_s)$ can be viewed as the initial state of the studied system. More generally, an arbitrary state can be used for $|\Phi_s)$ including a random state. Comparison of the completeness relation (2.37) with (2.111) gives the relation

$$|\Upsilon(E)) = \delta(E\hat{1} - \hat{\mathrm{H}})|\Phi_s). \tag{2.112}$$

The eigen-states $|\Upsilon_k) \equiv |\Upsilon(E_k))$, associated with the eigen-value E_k, can be extracted from the expansion (2.108) taken at $E = E_k$

$$|\Upsilon_k) = \sum_{n=0}^{\infty} B_n(E_k)|\psi_n^{(s)}) = \sum_{n=0}^{\infty} \mathrm{W}_n(E_k)Q_n(E_k)Q_n(\hat{\mathrm{H}})|\Phi_s). \tag{2.113}$$

By way of a summary, let us enumerate the key properties of the expansion method devised using non-classical polynomials that are intertwined with rational polynomials as analyzed in this chapter (see also [127, 128]):

- Diagonal and off-diagonal elements can be computed, not only for the Green function itself, but also for the product of any number of Green functions with other quantum-mechanical operators,

- The energy resolution in the sought spectrum is controlled by the expansion order, which is equal to the number of retained polynomials,

- It can be used for systems with discrete energies as well as for resonances and scattering,
- It provides the integrated density of states, i.e., the IDOS, eigen-values E_k and the associated eigen-vectors $|\Upsilon_k\rangle$ of the Schrödinger eigen-value problem without ever solving this problem explicitly,
- The computational and storage costs are attractive, since they scale linearly with the size of the considered system,
- It is suitable for parallel processing.

2.2.2 Classical polynomials

For the reason of having a more general presentation, the expansion method from sub-section 2.2.1 is concerned with the space of non-classical, Lanczos polynomials $\{P_n(x), Q_n(x)\}$. Nevertheless, the whole analysis and the corresponding conclusions also hold true for classical polynomials. This is the case because all the classical polynomials are orthogonal polynomials and, therefore, they satisfy the same three-term recursion (2.49) with the unchanged definition of the coupling constants α_n and β_n from (2.50) and (2.51). The sole difference is that α_n and β_n can be obtained in their analytical forms for all the classical polynomials. Moreover, the Christoffel numbers $\{w_n\}$ for classical polynomials are computed from the same formula (2.72). Furthermore, the expression (2.53) can be checked analytically to be valid.

Remarkably, for classical polynomials, the error or remainder $\mathcal{E}_n(z)$ from (2.65) can also be calculated analytically. For instance, if $Q_n(x)$ is the Chebyshev polynomial $T_n(x)$ of the first kind, then the polynomial $P_n(x)$ is the Chebyshev polynomial $U_n(x)$ of the second kind

$$T_n(z) = \cos(n\varphi) \qquad U_n(z) = \frac{\sin([n+1]\varphi)}{\sin\varphi} \qquad \varphi = \cos^{-1}(z). \tag{2.114}$$

In such a case, we have from (2.66) that

$$\mathcal{E}_n^{(\mathrm{C})}(z) = \frac{\pi_n^{(\mathrm{C})}(z)}{T_n(z)} \tag{2.115}$$

where $\pi_n^{(\mathrm{C})}(z)$ is given by

$$\pi_n^{(\mathrm{C})}(z) = \frac{1+\delta_{n,0}}{2}\pi\left[U_{n-1}(z) - \frac{i}{\sqrt{1-z^2}}T_n(z)\right] \tag{2.116}$$

with $\delta_{n,0}$ being the Kronecker δ-symbol. It is also possible to derive a simple analytical expression for $\mathcal{E}_n(z)$ if $Q_n(x)$ is the Hermite $H_n(x)$, Jacobi $P_n^{(\alpha,\beta)}(x)$, Legendre and Laguerre polynomial. For instance, we can write

$$\mathcal{E}_n^{(\mathrm{H})}(E) = \frac{\pi_n^{(\mathrm{H})}(E)}{H_n(E)} \qquad \mathcal{E}_n^{(\mathrm{J})}(E) = \frac{\pi_n^{(\mathrm{J})}(E)}{P_n^{(\alpha,\beta)}(E)} \tag{2.117}$$

where the superscripts H and J stand for the Hermite and Jacobi polynomials. In (2.117), the following expressions for $\pi_n^{(\mathrm{H})}(E)$ and $\pi_n^{(\mathrm{J})}(E)$ exist [128]

$$\pi_n^{(\mathrm{H})}(E) = -\frac{1}{n!\sqrt{2\pi}}[i\pi w(E/\sqrt{2})H_n(E) - \mathcal{H}_n(E)] \tag{2.118}$$

$$\pi_n^{(\mathrm{J})}(E) = -\frac{2^{\alpha+\beta+1}B(\alpha+1,\beta+1)}{1-E}$$
$$\times\, {}_2F_1\left(1,\beta+1;\alpha+\beta+2;\frac{2}{1-E}\right)P_n^{(\alpha,\beta)}(E) - Q_n^{(\alpha,\beta)}(E) \tag{2.119}$$

$$w(z^+) = \frac{i}{\pi}\int_0^\infty \mathrm{d}x\frac{\mathrm{e}^{-x^2}}{z^+ - x} = \mathrm{e}^{-x^2/2}[\mathrm{erf}(iE/\sqrt{2}) - 1] \tag{2.120}$$

where z^+ is defined by (2.42) and (α,β) are the fixed parameters from the Jacobi polynomial, $w(z)$ is the w-function which is related to the error-function (erf) or the probability function, $B(\alpha,\beta)$ is the beta-function and ${}_2F_1(a,b;c;z)$ is the Gauss hypergeometric function [132, 133]. For machine accurate and fast computations of the w-function of a complex variable by means of continued fractions, the algorithm of Gautschi [100] is optimal. Further, in (2.118) and (2.119), $\mathcal{H}_n(E)$ and $Q_n^{(\alpha,\beta)}(E)$ are the Hermite and Jacobi polynomials of the second kind. They stem from (2.54) where the constants α_n and β_n are the same as in the conventional recursions for $H_n(E)$ and $P_n^{(\alpha,\beta)}(E)$ [133], but with a set of different initial conditions given by $\mathcal{H}_0(E)=0$, $\mathcal{H}_1(E)=1$ and $Q_0^{(\alpha,\beta)}(E)=0$, $Q_1^{(\alpha,\beta)}(E)=1$. The Legendre and Chebyshev polynomials are obtained from $P_n^{(\alpha,\beta)}(x)$ as two special cases with $\alpha=0=\beta$ and $\alpha=-1/2=\beta$, respectively. Therefore, we can write, e.g., $\mathcal{E}_n^{(\mathrm{L})}(E) = \pi_n^{(\mathrm{L})}(E)/P_n^{(0,0)}(E)$ where $\pi_n^{(\mathrm{L})}(E)$ for the Legendre polynomials is available from (2.119) for $\alpha=0=\beta$.

Advantageously, the integral in (2.107) can be carried out analytically for classical polynomials. Thus, for the Hermite $H_n(x)$ and Jacobi $P_n^{(\alpha,\beta)}(x)$ polynomials the specification $N^{(\mathrm{H})}(E)$ and $N^{(\mathrm{J})}(E)$ can, respectively, be deduced in the forms [128]

$$N^{(\mathrm{H})}(E) = M\left[1+\frac{1}{\sqrt{\pi}}\mathrm{erf}(E/\sqrt{2\pi})\right] - \frac{1}{\sqrt{2\pi}}\sum_{n=1}^{\infty}\frac{I_n}{n!}\mathrm{e}^{-E^2/2}H_{n-1}(E) \tag{2.121}$$

$$N^{(\mathrm{J})}(E) = M\frac{B_{E'}(\alpha+1,\beta+1)}{B(\alpha+1,\beta+1)} - \frac{1}{2}\sum_{n=1}^{\infty}\frac{I_n}{n}\mathrm{W}_n(E)P_{n-1}^{(\alpha+1,\beta+1)}(E) \tag{2.122}$$

$$\mathrm{W}_n(x) = \frac{\mathrm{W}(x)}{\mathrm{w}_n} \qquad \mathrm{W}(x) = (1-x)^\alpha(1+x)^\beta \tag{2.123}$$

$$\mathrm{w}_n = \frac{2^{\alpha+\beta+1}}{2n+1}\frac{\Gamma(n+\alpha+1)\Gamma(n+\beta+1)}{n!\Gamma(n+\alpha+\beta+1)}. \tag{2.124}$$

Here, $E'=(1+E)/2$, $\Gamma(a)$ is the gamma function and $B_\gamma(\alpha,\beta)$ is the incomplete beta function [133].

In practice, the expansion methods with the classical orthogonal polynomials, including the Chebyshev polynomials as the simplest, were employed frequently in solving quantum-mechanical problems that necessitate extensive computations of Green functions and other related physical quantities [134, 135]. Such a method, which is also known as 'the recursive orthogonal polynomial expansion method' (ROPEM) [128], was used with the Chebyshev, Jacobi and Legendre polynomials. Naturally, all the chief characteristics listed at the end of sub-section 2.2.1 for non-classical polynomials are also applicable to classical polynomials in the setting of the ROPEM [128]. Specifically in the case of the ROPEM with the Legendre polynomials (and possibly with the other classical polynomials), the truncation error, caused by using a finite number M of terms in the series development, was shown to be proportional to M^{-1} for IDOS and to $M^{-3/2}$ for the tight-binding chain models with continuous spectra [128]. In a number of recent applications of the expansion methods for large physical systems, the ROPEM was found to be accurate, stable and efficient. This should be a good motivation for further applications of the ROPEM.

Nevertheless, compared with the ROPEM, there is a very important advantage of a more general expansion method constructed from non-classical Lanczos polynomials elaborated in sub-section 2.2.1. In many applications from physics and chemistry, the studied functions to be expanded on a basis must often obey the given prescribed initial or boundary conditions. For instance, continuum wave functions are required to behave at large distances as a sinusoidal function which contains a phase shift for a short and/or a long-range Coulombic potential. Likewise, discrete wave functions must have exponentially decaying behavior at the asymptotically large distances ($x \longrightarrow \infty$). Moreover, the boundary conditions at the origin are usually imposed on those functions that describe physical states by requiring $f(x) \longrightarrow 0$ as $x \longrightarrow 0$. Further, many functions employed in physics frequently possess essential singularities within the interval of interest $[a, b]$. For these and other reasons, it is advantageous that the needed asymptotic behaviors of the investigated function are included in the employed basis set from the outset. Such prior knowledge can readily be implemented for non-classical polynomials, but *not* with the classical ones. For example, if a solution of second-order differential equations should have a singularity at $x \approx a$, all classical polynomials would inevitably undergo large cancellations (with the undesirable consequence of considerable round-off errors) to satisfy the required boundary condition. Namely, for the reason of completeness of the employed basis set, any classical polynomial of degree n must contain a free term, which is a constant, i.e., a factor independent of x. However, even a small admixture of an irregular solution of the second-order differential equation from this example could yield a large loss of accuracy when classical polynomials are used to expand the regular solution. By contrast, this kind of obstacle is easily alleviated with the non-classical Lanczos polynomials $\{Q_n(x)\}$ by simply modifying their recursion relation so as to start with x instead of a constant [136, 137].

2.3 Recurrent time signals and their generating fractions as spectra with no recourse to Fourier integrals

We recall that one of the equivalent ways to introduce classical polynomials is to use their so-called generating functions. For example, the Chebyshev polynomials of the second kind $U_n(x)$ can be introduced as the general expansion coefficient in the following development

$$f(x,y) \equiv \frac{1}{1-2xy+y^2} = \sum_{n=0}^{\infty} U_n(x)x^n \tag{2.125}$$

$$U_n = \frac{\sin(n\cos^{-1}x)}{\sin(\cos^{-1}x)}. \tag{2.126}$$

Here, $1/(1-2xy+y^2)$ is the generating function of $U_n(x)$. The generating function $f(x,y)$ is a rational function, which is a polynomial quotient and, as such, it represents the PA of the order $[0/1]$ in the variable x for a fixed y or of order $[0/2]$ in y for a fixed x. The simplest polynomial in x is obtained when all its expansion coefficients are set to zero and simultaneously the coefficient of the highest power is taken as one. This is the monomial or power function, $M_n(x) = x^n$, and its generating function is $1/(x-y)$ according to the binomial expansion

$$g(x,y) \equiv \frac{1}{x-y} = \sum_{n=0}^{\infty} M_n(x)y^{-n-1} \qquad M_n(x) = x^n. \tag{2.127}$$

This generating function $g(x,y)$ is also a rational function. Moreover, $1/(x-y)$ is the PA of order $[0/1]$ in the variable x for a fixed y or likewise in y for a fixed x. Such reasoning can be extended to the operator-valued functions. For example, the Green operator $\hat{\mathrm{G}}(u,\hat{\mathrm{U}}) \equiv \hat{\mathrm{G}}(u) = (u\hat{1}-\hat{\mathrm{U}})^{-1}$ is the generating function of the operator monomial $M_n(\hat{\mathrm{U}}) = \hat{\mathrm{U}}^n$ via

$$\hat{\mathrm{G}}(u) = \frac{1}{u\hat{1}-\hat{\mathrm{U}}} = \sum_{n=0}^{\infty} M_n(\hat{\mathrm{U}})u^{-n-1} \qquad M_n(\hat{\mathrm{U}}) = \hat{\mathrm{U}}^n. \tag{2.128}$$

Likewise, the Green function $\mathcal{R} \equiv (\Phi_0|\hat{\mathrm{G}}(u)|\Phi_0)$ is a generating function for the auto-correlation functions or signal points $(\Phi_0|\{u\hat{1}-\hat{\mathrm{U}}\}^{-1}|\Phi_0) = \sum_{n=0}^{\infty} c_n u^{-n-1}$ where $c_n = (\Phi_0|\hat{\mathrm{U}}^n|\Phi_0)$. To get an insight into the functional dependence of the generating function $\mathcal{R}(u)$ for the c_n's, one can use the Schrödinger eigen-value problem for the evolution operator $\hat{\mathrm{U}}$ viz $\hat{\mathrm{U}}|\Upsilon_k) = u_k|\Upsilon_k)$. The eigen-vectors $\{|\Upsilon_k)\}$ form a basis and the completeness can be expressed through the spectral decomposition of the unity operator $\hat{1} = \sum_{k=0}^{K} \hat{\pi}_k$

where $\hat{\pi}$ is the projection operator $\hat{\pi} = |\Upsilon_k)(\Upsilon_k|$. Here, positive integer K can be finite or infinite depending on the number of the eigen-values u_k of $\hat{\mathrm{U}}$. Thus, it follows $\mathcal{R}(u) = (\Phi_0|\{u\hat{1} - \hat{\mathrm{U}}\}^{-1}|\Phi_0) = \sum_{k=1}^{K}(\Phi_0|\{u\hat{1} - \hat{\mathrm{U}}\}^{-1}|\Upsilon_k)(\Upsilon_k|\Phi_0) = \sum_{k=1}^{K}(\Phi_0|\{u - u_k\}^{-1}|\Upsilon_k)(\Upsilon_k|\Phi_0) = \sum_{k=1}^{K}(\Phi_0|\Upsilon_k)(\Upsilon_k|\Phi_0)/(u - u_k)$ so that

$$\mathcal{R}(u) = (\Phi_0|\{u\hat{1} - \hat{\mathrm{U}}\}^{-1}|\Phi_0) = \sum_{k=1}^{K} \frac{d_k}{u - u_k} \tag{2.129}$$

with $d_k = (\Phi_0|\Upsilon_k)^2$ where the symmetry $(\Phi_0|\Upsilon_k) = (\Upsilon_k|\Phi_0)$ is employed. It follows from (2.129) that $\mathcal{R}(u)$ is the scalar PA of the order $[(K-1)/K]$. This is expected, since the corresponding resolvent (2.128) is the operator PA, i.e., $\hat{1}(u\hat{1} - \hat{\mathrm{U}})^{-1} = (u\hat{1} - \hat{\mathrm{U}})^{-1}\hat{1}$. Thus, the generating fraction[2] of the auto-correlation functions or signal points c_n is the PA of the order $[(K-1)/K]$

$$\frac{A_{K-1}(z)}{B_K(z)} = \sum_{n=0}^{\infty} c_n z^n \qquad z = \frac{1}{u} \tag{2.130}$$

$$\frac{a_0 + a_1 z + a_2 z^2 + \cdots + a_{K-1} z^{K-1}}{b_K + b_{K-1} z + b_{K-2} z^2 + \cdots + b_0 z^K} = \sum_{n=0}^{\infty} c_n z^n \tag{2.131}$$

$$A_{K-1}(z) = \sum_{r=0}^{K-1} a_r z^r \qquad B_K(z) = \sum_{r=0}^{K} b_{K-r} z^r. \tag{2.132}$$

In (2.125), the given expansion coefficients $U_n(x)$ of the Taylor series in y retrieves exactly the generating function $f(x,y)$, where all the coefficients of the denominator polynomial are known. However, in (2.131) when the c_n's are given, the form of the generating fraction is known to be the PA $A_{K-1}(z)/B_K(z)$, but the coefficients $\{a_r\}$ and $\{b_r\}$ of the numerator and denominator polynomials are unknown. Of course, they can be uniquely determined through multiplication of (2.131) by $B_K(z)$ and via the subsequent equating the coefficients of the like powers. We re-emphasize that the generating fraction of the signal points always has the degree of the denominator polynomial by one unit higher than that of the numerator, as per (2.131).

We shall now consider the inverse problem of reconstructing the auto-correlation functions $\{c_n\}$ when their generating fraction $A_{K-1}(z)/B_K(z)$ from the lhs of (2.129) is given. No Fourier integral will be used. Instead, only the purely algebraic method originated by Prony [138] will be employed. With

[2]The term 'generating function' is customarily used for ordinary polynomials, but for rational polynomials, the corresponding nomenclature 'generating fraction' of Prony [138] seems to be more transparent.

this goal, we shall develop the generating fraction in a power series expansion. Then, by using the method of undetermined coefficients, we can write

$$\frac{a_0 + a_1 z + \cdots + a_{K-1} z^{K-1}}{b_K + b_{K-1} z + \cdots + b_0 z^K} = c_0 + c_1 z + \cdots + c_K z^K + \cdots . \tag{2.133}$$

The constants $c_0, c_1, \ldots, c_K, \ldots$, and so on from the rhs of (2.133) are the undetermined coefficients. In order to find their values when the a_k's and b_k's are known, we multiply both sides of (2.133) by $b_K + b_{K-1}u + \cdots + b_0 u^K$. Then passing all the ensuing terms to the same side and ordering them regarding the power of z will yield the result

$$\left.\begin{array}{llll} b_K c_0 + & b_K c_1 z + & b_K c_2 z^2 + & b_K c_3 z^3 \; + \cdots \\ -a_0 & + b_{K-1} c_0 z & + b_{K-1} c_1 z^2 & + b_{K-1} c_2 z^3 + \cdots \\ & - \; a_1 z & + b_{K-2} c_0 z^2 & + b_{K-2} c_1 z^3 + \cdots \\ & & - \; a_2 z^2 & + b_{K-3} c_0 z^3 + \cdots \\ & & & - \; a_3 z^3 \quad + \cdots \\ & & & \qquad - \cdots \end{array}\right\} = 0. \tag{2.134}$$

In (2.134), every vertical band of coefficients of the same power of z should be zero and this yields the following equations [138]

$$\begin{aligned} c_0 &= \frac{a_0}{b_K} \\ c_1 &= -\frac{1}{b_K}[b_{K-1} c_0] + \frac{a_1}{b_K} \\ c_2 &= -\frac{1}{b_K}[b_{K-1} c_1 + b_{K-2} c_0] + \frac{a_2}{b_K} \\ c_3 &= -\frac{1}{b_K}[b_{K-1} c_2 + b_{K-2} c_1 + b_{K-3} c_0] + \frac{a_3}{b_K} \\ &\;\;\vdots \\ c_{K-1} &= -\frac{1}{b_K}[b_{K-1} c_{K-2} + b_{K-2} c_{K-3} + b_{K-3} c_{K-4} + \cdots + b_1 c_0] + \frac{a_{K-1}}{b_K} \end{aligned} \tag{2.135}$$

$$\begin{aligned} c_K &= -\frac{1}{b_K}[b_{K-1} c_{K-1} + b_{K-2} c_{K-2} + b_{K-3} c_{K-3} + \cdots + b_0 c_0] \\ &\;\;\vdots \\ c_m &= -\frac{1}{b_K}[b_{K-1} c_{m-1} + b_{K-2} c_{m-2} + b_{K-3} c_{m-3} + \cdots + b_0 c_{m-K}] \\ &\;\;\vdots \\ c_{m+K} &= -\frac{1}{b_K}[b_{K-1} c_{m+K-1} + b_{K-2} c_{m+K-2} + b_{K-3} c_{m+K-3} + \cdots + b_0 c_m]. \end{aligned} \tag{2.136}$$

It is seen that the isolated terms $a_0/b_K, a_1/b_K, a_2/b_K, \ldots$, given by all the scaled coefficients $\{a_k/b_K\}$ $(k = 0, 1, 2, ...)$ of the numerator of the generating fraction are present only in (2.135) for determination of the c_n's from c_0 to c_{K-1}. Starting from c_K, any coefficient c_m is always expressed by means of the K preceding coefficients $c_{m-1}, c_{m-2}, c_{m-3}, \ldots, c_{m-K}$, each of which is individually multiplied by the respective coefficients $b_{K-1}, b_{K-2}, b_{K-3}, \ldots, b_0$ of the denominator polynomial of the generating fraction. Finally, the sum of the latter products is divided by b_K. In particular, we shall single out the last equation from the string (2.136) as

$$b_0 c_n + b_1 c_{n+1} + b_2 c_{n+2} + \cdots + b_K c_{n+K} = 0. \tag{2.137}$$

Since here the constant coefficients $\{b_r\}$ $(0 \le r \le K)$ are known, we can recognize (2.137) as a difference equation whose solution is given by the complete integral as the geometric sequence

$$c_n = \sum_{k=1}^{K} d_k u_k^n. \tag{2.138}$$

The u_k's are the roots of the characteristic polynomial $L_K(u)$ due to (2.138)

$$L_K(u) = b_0 + b_1 u + b_2 u^2 + \cdots + b_K u^K. \tag{2.139}$$

Here, the d_k's are some arbitrary integration constants which can be determined by imposing the K boundary conditions to (2.137). These K initial conditions can always be reduced to the request of passing the continuous curve $c(t)$ through a fixed number K of given points $t_n = n\Delta t \equiv n\tau$. In this way, the values of $t_0, t_1, t_2, \ldots, t_K$ will be associated with $c_0, c_1, c_2, \ldots, c_K$, respectively. This procedure gives K linear equations

$$\left.\begin{aligned} c_0 &= d_1 + d_2 + d_3 + \cdots + d_K \\ c_1 &= d_1 u_1 + d_2 u_2 + d_3 u_3 + \cdots + d_K u_K \\ c_3 &= d_1 u_1^2 + d_2 u_2^2 + d_3 u_3^2 + \cdots + d_K u_K^2 \\ &\ \vdots \\ c_K &= d_1 u_1^K + d_2 u_2^K + d_3 u_3^K + \cdots + d_K u_K^K \end{aligned}\right\}. \tag{2.140}$$

From here, we can extract the values of the d_k's that will acquire their forms

$$d_1 = c_0 \tag{2.141}$$

$$\left.\begin{aligned} d_1 &= \frac{-u_2 c_0 + c_1}{u_1 - u_2} \\ d_2 &= \frac{-u_2 c_0 + c_1}{u_2 - u_1} \end{aligned}\right\} \quad K = 2 \tag{2.142}$$

$$\left.\begin{aligned} d_1 &= \frac{u_2 u_3 c_0 - (u_2 + u_3) c_1 + c_2}{(u_1 - u_2)(u_1 - u_3)} \\ d_2 &= \frac{u_1 u_3 c_0 - (u_1 + u_3) c_1 + c_2}{(u_2 - u_1)(u_2 - u_3)} \\ d_3 &= \frac{u_1 u_2 c_0 - (u_1 + u_2) c_1 + c_2}{(u_3 - u_1)(u_3 - u_2)} \end{aligned}\right\} \quad K = 3 \tag{2.143}$$

$$\left.\begin{aligned}
d_1 &= \frac{-u_2u_3u_4c_0 + (u_2u_3 + u_2u_4 + u_3u_4)c_1 - (u_2 + u_3 + u_4)c_2 + c_3}{(u_1 - u_2)(u_1 - u_3)(u_1 - u_4)}\\
d_2 &= \frac{-u_1u_3u_4c_0 + (u_1u_3 + u_1u_4 + u_3u_4)c_1 - (u_1 + u_3 + u_4)c_2 + c_3}{(u_2 - u_1)(u_2 - u_3)(u_2 - u_4)}\\
d_3 &= \frac{-u_1u_2u_4c_0 + (u_1u_2 + u_1u_4 + u_2u_4)c_1 - (u_1 + u_2 + u_4)c_2 + c_3}{(u_3 - u_1)(u_3 - u_2)(u_3 - u_4)}\\
d_4 &= \frac{-u_1u_2u_3c_0 + (u_1u_2 + u_1u_3 + u_2u_3)c_1 + (u_1 + u_2 + u_3)c_2 + c_3}{(u_4 - u_1)(u_4 - u_2)(u_4 - u_3)}
\end{aligned}\right\} K = 4 \tag{2.144}$$

etc. We see that, in general, irrespective of the actual value of the number K, the numerator of d_k is obtained by first taking all the roots of the characteristic equation $L_K(u_k) = 0$ except u_k and forming all possible products of the $K-1$ roots $\{u_r\}$ $(r \neq k)$. Afterwards, the sums of these products are formed and multiplied by $c_0, c_1, c_2, \ldots, c_{K-2}$, respectively. Subsequently, each of the resulting terms is multiplied by -1 or $+1$ dependent on whether K is even or odd. Finally, all these terms are summed up and the constant c_{K-1} is added at the end of the procedure to give the numerator of d_k, as per (2.141)–(2.144). Likewise, to obtain the denominator of d_k, we first subtract successively from u_k all the other roots $\{u_r\}$ $(r \neq k)$. Then, the product of all these differences of the roots $u_r - u_s$ $(r \neq s)$ yields the denominator of d_k, as is clear from (2.141)–(2.144).

Returning now to the K equations of the string (2.135), we see that they represent the terms that are of the general form

$$c_r = \frac{a_{r-1}}{b_r} - \frac{1}{b_r}\sum_{k=1}^{K-1} b_k c_{k-1}. \tag{2.145}$$

To gain an insight into this relationship, we introduce the following characteristic polynomial

$$J_K(u) = b_0u^K + b_1u^{K-1} + b_2u^{K-2} + b_3u^{K-3} + \cdots + b_K. \tag{2.146}$$

Comparing the polynomials $L_K(u)$ and $J_K(u)$, it follows

$$L_K(u) = J_K(u^{-1}). \tag{2.147}$$

This means that if u_k is the kth root of $L_K(u)$, then the kth root of $J_K(u^{-1}) = 0$ is u_k^{-1}

$$L_K(u) = b_K(u-u_1)(u-u_2)\cdots(u-u_K) = \frac{b_0}{b_K} + \frac{b_1}{b_K}u + \frac{b_2}{b_K}u^2 + \cdots + u^K \tag{2.148}$$

$$\begin{aligned}
J_K(u) &= b_0\left(\frac{1}{u_1} - u\right)\left(\frac{1}{u_2} - u\right)\cdots\left(\frac{1}{u_K} - u\right)\\
&= u^K + \frac{b_1}{b_0}u^{K-1} + \frac{b_2}{b_0}u^{K-2} + \cdots + \frac{b_K}{b_0}.
\end{aligned} \tag{2.149}$$

The rhs of (2.149) is the denominator polynomial of the generating fraction from (2.133). Then substituting the lhs of (2.149) into (2.133) and writing the resulting polynomial quotient via its partial fractions, we have

$$\frac{a_0 + a_1 u + a_2 u^2 + \cdots + a_{K-1} u^{K-1}}{b_0 (z_1 - u)(z_2 - u) \cdots (z_K - u)} \equiv \frac{\tilde{d}_1}{z_1 - u} + \frac{\tilde{d}_2}{z_2 - u} + \cdots + \frac{\tilde{d}_K}{z_K - u} \tag{2.150}$$

where $\{\tilde{d}_k\}$ are some constants to be determined, and

$$z_k = \frac{1}{u_k}. \tag{2.151}$$

If each partial fraction on the rhs of (2.150) is expanded into its own binomial series, we will have

$$\left.\begin{aligned} \frac{\tilde{d}_1}{z_1 - u} &= \tilde{d}_1 u_1 (1 + u_1 u + u_1^2 u^2 + u_1^3 u^3 + \cdots + u_1^m u^m + \cdots) \\ \frac{\tilde{d}_2}{z_2 - u} &= \tilde{d}_2 u_2 (1 + u_2 u + u_2^2 u^2 + u_2^3 u^3 + \cdots + u_2^m u^m + \cdots) \\ \frac{\tilde{d}_3}{z_3 - u} &= \tilde{d}_3 u_3 (1 + u_3 u + u_3^2 u^2 + u_3^3 u^3 + \cdots + u_3^m u^m + \cdots) \\ &\vdots \\ \frac{\tilde{d}_K}{z_K - u} &= \tilde{d}_K u_K (1 + u_K u + u_K^2 u^2 + u_K^3 u^3 + \cdots + u_K^m u^m + \cdots) \end{aligned}\right\}^{+} = \sum_{n=0}^{\infty} c_n u^n. \tag{2.152}$$

Here, when we sum up these K equations, we shall obtain the generating fraction on the lhs of (2.152), as per (2.149) and (2.150). Consequently, the sum of K series from the rhs (2.152) ought to be equal to the Taylor expansion $c_0 + c_1 u + c_2 u^2 + \cdots + c_m u^m + \cdots$, as per (2.133) and this is symbolically indicated via $\sum_{n=0}^{\infty} c_n u^n$ after the curly bracket. The superscript + associated with the curly bracket in (2.152) is written to point at the said summation of the K equations. This procedure leads to

$$\sum_{k=1}^{K} \frac{\tilde{d}_k}{z_k - u} = \frac{A_{K-1}(u)}{B_K(u)} \tag{2.153}$$

$$\sum_{k=1}^{K} \tilde{d}_k u_k \sum_{r=0}^{\infty} (u_k u)^r = \sum_{n=0}^{\infty} c_n u^n. \tag{2.154}$$

After equating the coefficients of the like power of u in (2.154) and setting

$$u_k \tilde{d}_k = d_k \tag{2.155}$$

we arrive at

$$\left.\begin{aligned} c_0 &= d_1 + d_2 + d_3 + \cdots + d_K \\ c_1 &= d_1 u_1 + d_2 u_2 + d_3 u_3 + \cdots + d_K u_K \\ c_2 &= d_1 u_1^2 + d_2 u_2^2 + d_3 u_3^2 + \cdots + d_K u_K^2 \\ &\ \vdots \\ c_n &= d_1 u_1^n + d_2 u_2^n + d_3 u_3^n + \cdots + d_K u_K^n \\ &\ \vdots \end{aligned}\right\}. \tag{2.156}$$

Hence, the obtained general term of the form

$$c_n = d_1 u_1^n + d_2 u_2^n + d_3 u_3^n + \cdots + d_K u_K^n = \sum_{k=1}^{K} d_k u_k^n \tag{2.157}$$

coincides with the complete integral (2.138) of the Kth order difference equation (2.137). This sets the prescription for the usage of the first K results $c_0, c_1, c_2, \ldots, c_{K-1}$ deduced from the generating fraction. As mentioned, these results $c_0, c_1, c_2, \ldots, c_{K-1}$ contain the coefficients $a_0, a_1, a_2, \ldots, a_{K-1}$ of the numerator polynomial of the generating fraction. In order to find all the terms $c_0, c_1, c_2, \ldots,$ of the rhs of (2.154) to the recurring order K, where c_n is the general member of the Taylor series $\sum_{n=0}^{\infty} c_n u^n$, it is sufficient to have only the K starting values $\{c_r\}\,(0 \le r \le K-1)$ that should be given in advance. By means of these K initial c_n's, one could continue the sequence forward or backward using the difference equation (2.137). In this way, one can either predict the new c_n's beyond the originally given $\{c_r\}\,(0 \le r \le K-1)$ or retrieve all the preceding c_n's by recurring with the descending value of the suffices. In other words, if the K terms are given $c_{s+1}, c_{s+2}, \ldots, c_{s+K}$, one could prolong/extrapolate this latter sequence by means of the relationship

$$\left.\begin{aligned} c_{s+K+1} &= -\frac{1}{b_K}(b_{K-1}c_{s+K} + b_{K-2}c_{s+K-1} + \cdots + b_0 c_{s+1} \\ c_{s+K+2} &= -\frac{1}{b_K}(b_{K-1}c_{s+K+1} + b_{K-2}c_{s+K} + \cdots + b_0 c_{s+2} \\ &\ \vdots \end{aligned}\right\}. \tag{2.158}$$

Thus the new c_n's are predicted using the previous ones, so in general

$$c_{s+K+m} = -\frac{1}{b_K}\sum_{r=0}^{K-1} b_r c_{r+s+m} \qquad \text{(predicting the new } c_n\text{'s)}. \tag{2.159}$$

Alternatively, the calculation could also be carried out by recurring towards the origin in a descending order with the aid of the expressions

$$\left.\begin{aligned} c_s &= -\frac{1}{b_0}(b_1 c_{s+1} + b_2 c_{s+2} + \cdots + b_K c_{s+K} \\ c_{s-1} &= -\frac{1}{b_0}(b_1 c_s + b_2 c_{s+1} + \cdots + b_K c_{s+K-1} \\ &\vdots \end{aligned}\right\}. \tag{2.160}$$

This recursion recovers the preceding c_n's and the general formula is

$$c_{s-m} = -\frac{1}{b_0}\sum_{r=1}^{K} b_r c_{r+s-m} \qquad \text{(recovering the old } c_n\text{'s)}. \tag{2.161}$$

With such an established pattern, the set $\{c_n\}$ is said to form a forward or backward self-recurring sequence whose general term c_n is given by the geometric progression (2.138). The sum and the product of the two recurrent sequences of the order K and K' also represent a recurrent sequence of the order $K + K'$ and KK', respectively. Importantly, all the recurrent sequences are summable, i.e., they are convergent to the correct limit.

The three main expressions (2.133), (2.159), (2.161) from this exposition were derived by Prony [138] in 1790 and they are identical to those defining the Padé approximant [51] from 1892. Based upon his course of Mathematical Analysis via "Suite de Leçons d'Analyse de Prony" [138], Prony developed in 1795 [139] a versatile method called later after his name for approximating functions by linear combinations of exponentials. This is the well-known Prony method [139], which marks the birth of what later became the whole field of signal processing. Rational approximations in the forms of polynomial quotients, known as the Padé approximants, were used by a number of mathematical giants before Padé [51] dating back to the period 1730 – 1870 (Bernoulli, Stirling, Euler, Lambert, Lagrange, Jacobi, Hankel, Frobenius, Darboux, Laguerre and Chebyshev). Padé [51], through his doctoral thesis, was widely credited to be the first to carry out the most systematic study of these rational polynomials. Unfortunately, it does not seem to be widely known that the published work of Prony [138, 139], who, due to his great achievements became Baron de Prony, was also extremely systematic in investigations of what later was termed the Padé approximant.

2.4 The fast Padé transform for quantum-mechanical spectral analysis and signal processing

Irrespective of whether the analysis is carried out in the time or frequency domain, the FPT appears as an optimal estimator especially in MRS. This is evident from the following two reasons. For any given time signal c_n, represented by a sum of the K damped complex exponentials $c_n = \sum_{k=1}^{K} d_k u_k^n$ with, e.g., stationary amplitudes $\{d_k\}$, the exact spectrum reads as $\sum_{k=1}^{K} d_k/(u - u_k)$ at any $u = \exp(-i\omega\tau)$. Here, the sum can be carried out analytically to yield the polynomial quotient $P_{K-1}(u)/Q_K(u)$, which is recognized as the para-diagonal FPT. A like rationale also holds true for a more general time signal in the form of a linear combination of attenuated complex exponentials with non-stationary polynomial-type amplitudes. Hence, the FPT is optimal for processing time signals modeled as described. This is expected, since the FPT is the exact theory for a function which is itself a rational polynomial.

From an algorithmic viewpoint, the FPT gives the unique quantification in MRS by solving the characteristic equation $Q_K(u) = 0$. The results of such rooting are the retrieved K values of the complex harmonics $\{u_k\}$ from the input time signal. Furthermore, the corresponding amplitudes $\{d_k\}$ are reconstructed with no extra labor in computations by using the Cauchy residue formula $d_k \propto P_{K-1}(u_k)[(\mathrm{d}/\mathrm{d}u)Q_K(u)]_{u=u_k}$ for all the distinct roots ($u_p \neq u_q\,, p \neq q$). Once the spectral parameters $\{u_k, d_k\}$ are obtained in this way, the associated non-degenerate Lorentzian spectrum in the complex mode is constructed by means of the Heaviside partial fractions $\sum_{k=1}^{K} d_k/(u - u_k)$. Such an analysis extends easily to the case of multiple roots of $Q_K(u)$ to describe fully overlapping (degenerate) resonances.

On the other hand, if no model is pre-supposed for c_n, the resulting exact spectrum is the Green function given by the Maclaurin expansion $G(u) = \sum_{n=0}^{\infty} c_n u^{-n-1}$. Here again the FPT is optimal, since its non-diagonal polynomial quotient $P_L(u)/Q_K(u)$ has the best contact with $G(u)$ through the first $L+K$ expansion terms of the Maclaurin series. This follows from the exact accord between the partial sum $\sum_{n=0}^{L+K} c_n u^{-n-1}$ in $G(u) = \sum_{n=0}^{\infty} c_n u^{-n-1}$ which is truncated at $n = L + K$ and the original Maclaurin series of $P_L(u)/Q_K(u)$ via the first $L+K$ terms, as per the definition of the FPT, i.e., $P_L(u)/Q_K(u) - \mathrm{G}(u) = \mathcal{O}(u^{-L-K-1})$. As usual, the symbol $\mathcal{O}(u^{-L-K-1})$ represents the remainder which is itself a series in terms of $u^{-L-K-2-n} (n \geq 0)$ with calculable expansion coefficients. For consistency, the outlined two circumstances (with and without an explicit modeling of the examined FID) need to be interrelated.

With this goal, we substitute the modeled time signal $c_n = \sum_{k=1}^{K} d_k u_k^n$ into the exact model-free spectrum $\mathrm{G}(u) = \sum_{n=0}^{\infty} c_n u^{-n-1}$. The obtained infinite Maclaurin sum from the Green function can be performed exactly by means of the geometric series with the result $\mathrm{G}(u) = \sum_{k=1}^{K} d_k/(u - u_k)$, which represents the para-diagonal FPT, i.e., $P_{K-1}(u)/Q_K(u)$. Thus, the general spectrum $\mathrm{G}(u)$ without an explicit modeling of the time signal is reduced, when the signal is afterwards assumed to be model-dependent, to the result which coincides with the corresponding finding from a different and independent analysis. Hence consistency of the two representations.

Of course, according to section 2.3, the Padé frequency spectrum in the form

$P_{K-1}(u)/Q_K(u)$ from the stated representation $G(u) = \sum_{n=0}^{\infty} c_n u^{-n-1} \approx P_{K-1}(u)/Q_K(u)$ can be considered as a generating fraction, which upon inversion yields the geometric progression model $c_n = \sum_{k=1}^{K} d_k u_k^n$ for the time signal.

Moreover, if spectral analysis is performed directly in the time domain via, e.g., the Shanks algorithm, the FPT is found to be the *exact* filter for a general time signal represented by the most conventional form of a sum of complex damped exponentials with stationary and/or time-dependent amplitudes. Overall, the discussed arguments for modeling in the time and frequency domain highlight the optimal suitability of the FPT for MRS.

2.5 Padé acceleration and analytical continuation of time series

Considering all the currently available processors, the FPT is especially important because of its self-contained cross-validation. Specifically, the FPT, as a system function for modeling an optimal response to general external perturbations, has two complementary variants. As noted earlier, these represent the equivalent versions $\text{FPT}^{(+)}$ and $\text{FPT}^{(-)}$ that are initially defined in two totally different convergence regions located inside ($|z| < 1$) and outside ($|z| > 1$) the unit circle, respectively [18, 19].

By design, the fast Padé transform performs equally well in the time and frequency domains. For instance, in the frequency domain, the $\text{FPT}^{(+)}$ and $\text{FPT}^{(-)}$ possess a common starting platform given by the same input spectrum in the form of the following truncated power series representation of the finite-rank Green function

$$G_N(z^{-1}) = \frac{1}{N} \sum_{n=0}^{N-1} c_n z^{-n} \qquad z = \mathrm{e}^{i\omega\tau}. \tag{2.162}$$

Here, $\{c_n\}(0 \le n \le N-1)$, are the expansion coefficients that represent the signal points, where N is the total length. As discussed, the time signal from the Green function in (2.162) is equivalent to the quantum-mechanical auto-correlation function. By reference to the theory of the $z-$transform, the Green function (2.162) is introduced by the expansion in powers of the inverse z^{-1} of the harmonic variable z.

The superscripts $\pm$ in $\text{FPT}^{(\pm)}$ indicate the power ± 1 of the expansion harmonic variable $z^{\pm 1}$. Also in use are the usual conventions $z^{+1} \equiv z^{+} = z$ and $z^{-1} \equiv z^{-} = 1/z$. In the limit $N \longrightarrow \infty$, the series (2.162) converges outside and diverges inside the unit circle, $|z| > 1$ and $|z| < 1$, respectively. The initial convergence regions of the $\text{FPT}^{(+)}$ and the $\text{FPT}^{(-)}$ are $|z| < 1$ and

$|z| > 1$. According to the main feature of the Padé approximant for the input power expansions of the form (2.162), both response functions or spectra in the $\mathrm{FPT}^{(\pm)}$ represent the unique rational polynomials $P_L^{\pm}(z^{\pm 1})/Q_K^{\pm}(z^{\pm 1})$ in their respective variables $z^{\pm 1}$

$$G_N(z^{-1}) - G_{L,K}^{\pm}(z^{\pm 1}) = \mathcal{O}(z^{\pm L+K\pm 1}) \tag{2.163}$$

where

$$G_{L,K}^{\pm}(z^{\pm 1}) = \frac{P_L^{\pm}(z^{\pm 1})}{Q_K^{\pm}(z^{\pm 1})} \qquad\qquad L+K \leq N. \tag{2.164}$$

As before, the symbols $\mathcal{O}(z^{\pm L+K\pm 1})$ are remainders that are themselves the Maclaurin series starting with the expansion terms $z^{\pm L+K\pm 1}$. The conventional meaning of the $\mathcal{O}(z^{\pm L+K\pm 1})$ symbols from (2.163) becomes clear when the Green-Padé functions $G_{L,K}^{\pm}(z^{\pm 1})$ are expanded in their own Maclaurin series. Then the ensuing series agree exactly with the partial sum G_{L+K} containing the first $L+K \leq N$ terms of the original Green function G_N. However, the remaining $N-L-K$ terms from G_N and $G_{L,K}^{\pm}$ are different and they are all contained in the symbols $\mathcal{O}(z^{\pm L+K\pm 1})$ from (2.163). Therefore, for brevity, (2.163) can take the form

$$G_N(z^{-1}) = G_{L,K}^{\pm}(z^{\pm 1}) \tag{2.165}$$

with the understanding that the left and the right sides are equal only up to the neglected terms of the orders $\mathcal{O}(z^{\pm L+K\pm 1})$.

By construction, the rational polynomials $G_{L,K}^{+}(z)$ and $G_{L,K}^{-}(z^{-1})$ are the Padé approximants to the finite-rank Green function G_N and, as such, they are also called the Green-Padé functions [5]. In such functions, $P_L^{\pm}(z^{\pm 1})$ and $Q_K^{\pm}(z^{\pm 1})$ represent the numerator and denominator polynomial of the degree L and K

$$P_L^{\pm}(z^{\pm 1}) = \sum_{r=r^{\pm}}^{L} p_r^{\pm} z^{\pm r} \tag{2.166}$$

$$Q_K^{\pm}(z^{\pm 1}) = \sum_{s=0}^{K} q_s^{\pm} z^{\pm s} \tag{2.167}$$

where $r^{+} = 1$ and $r^{-} = 0$. Because the Green functions $G_N(z^{-1})$ and $G_{L,K}^{-}(z^{-1})$ possess the same independent variables z^{-1}, it is evident that the $\mathrm{FPT}^{(-)}$ is the standard PA, which is denoted by the conventional symbol $[L/K]_{G_N}(z^{-1})$. On the other hand, in contrast to the input $z-$transform $G_N(z^{-1})$, the complementary Green-Padé function $G_{L,K}^{+}(z)$ is introduced in terms of the variable z and, as such, the $\mathrm{FPT}^{(+)}$ is recognized as the causal Padé transform, PzT.

2.6 Description of the background contribution by the off-diagonal fast Padé transform

Mathematically, the values of the degrees L and K of the polynomials $P_L^{\pm}(z^{\pm 1})$ and $Q_K^{\pm}(z^{\pm 1})$ in the general Green-Padé functions $G_{L,K}(z^{\pm 1})$ from (2.164) can be any positive integer. Of course, the sum $L+K$ cannot exceed the total number of available signal points, as indicated in (2.164). Physically, the input FID has a definite number of harmonics, and this number is equal precisely to K which is not known prior to spectral analysis. However, this does not mean at all that K should be guessed as regularly done in all fitting recipes throughout MRS, and beyond, as well as in the HLSVD. Quite the opposite practice is recommended, by which the unknown true value K is unambiguously reconstructed from the input FID together with retrieval of the unknown fundamental frequencies and amplitudes, as accomplished exactly in the FPT.

Two cases are particularly important in versatile applications of the FPT. These are the diagonal ($L = K$) and para-diagonal ($L = K - 1$) variants of the fast Padé transform as denoted by

$$G_K^{\pm}(z^{\pm 1}) \equiv G_{K,K}^{\pm}(z^{\pm 1}) \qquad \bar{G}_K^{\pm}(z^{\pm 1}) \equiv G_{K-1,K}^{\pm}(z^{\pm 1}) \tag{2.168}$$

$$G_K^{\pm}(z^{\pm 1}) = \frac{P_K^{\pm}(z^{\pm 1})}{Q_K^{\pm}(z^{\pm 1})} \qquad \bar{G}_K^{\pm}(z^{\pm 1}) = \frac{\bar{P}_{K-1}^{\pm}(z^{\pm 1})}{Q_K^{\pm}(z^{\pm 1})}. \tag{2.169}$$

Here, the diagonal Green-Padé functions $G_K^{\pm}(z^{\pm 1})$ can be given by a sum of a constant term and the corresponding para-diagonal form, labeled as $\tilde{G}_K^{\pm}(z^{\pm 1})$

$$G_K^{\pm}(z^{\pm 1}) = b_K^{\pm} + \tilde{G}_K^{\pm}(z^{\pm 1}) \qquad \tilde{G}_K^{\pm}(z^{\pm 1}) = \frac{\tilde{P}_{K-1}^{\pm}(z^{\pm 1})}{Q_K^{\pm}(z^{\pm 1})} \tag{2.170}$$

$$\tilde{P}_{K-1}^{\pm}(z^{\pm 1}) = \sum_{r=r^{\pm}}^{K-1} \tilde{p}_r^{\pm} z^{\pm r} \qquad \tilde{p}_r^{\pm} = p_r^{\pm} - b_K^{\pm} q_r^{\pm} \qquad b_K^{\pm} = \frac{p_K^{\pm}}{q_K^{\pm}}. \tag{2.171}$$

All absorption spectra constructed from FIDs encoded via MRS are invariably seen to exhibit quite pronounced backgrounds beneath the main physical resonances. These backgrounds are usually interpreted as stemming from large macromolecules that all the fitting techniques from MRS treat roughly by adjusting some 3–4 expansion coefficients of *ad hoc* 3rd or 4th degree least-square polynomials. Such background polynomials are subsequently patched in an artificial way to the employed basis set for the remaining resonances in the fitted spectra. As opposed to this empirical procedure, the FPT is capable of describing these spectral backgrounds from MRS with full accuracy and does so quite naturally by reliance upon the off-diagonal Green-Padé functions $G_{L,K}(z^{\pm 1})$ for $L > K$. This is achieved by using the relationships

$$\frac{P_L^{\pm}(z^{\pm 1})}{Q_K^{\pm}(z^{\pm 1})} = B_{L-K}^{\pm}(z^{\pm 1}) + \frac{A_K^{\pm}(z^{\pm 1})}{Q_K^{\pm}(z^{\pm 1})}. \tag{2.172}$$

Functions $B_{L-K}^{\pm}(z^{\pm 1})$ are the polynomials of degree $L - K > 0$ and they can describe the background contributions to typical spectra from MRS. The remaining genuine resonances are described by the diagonal Green-Padé functions $A_K^{\pm}(z^{\pm 1})/Q_K^{\pm}(z^{\pm 1})$ where $A_K^{\pm}(z^{\pm 1})$ are the numerator polynomials of the same degree K as the denominators $Q_K^{\pm}(z^{\pm 1})$. Of course, this latter ratio of polynomials could also be replaced by the para-diagonal Green-Padé functions $\tilde{A}_{K-1}^{\pm}(z^{\pm 1})/Q_K^{\pm}(z^{\pm 1})$. In such a case, the degree of the new polynomial for the background should be increased accordingly by 1

$$\frac{P_L^{\pm}(z^{\pm 1})}{Q_K^{\pm}(z^{\pm 1})} = \tilde{B}_{L-K+1}^{\pm}(z^{\pm 1}) + \frac{\tilde{A}_{K-1}^{\pm}(z^{\pm 1})}{Q_K^{\pm}(z^{\pm 1})}. \tag{2.173}$$

The sole purpose of writing the expressions (2.172) and (2.173) is to indicate that a modeling of the background due to macromolecules is achieved naturally by a simple division in the Padé quotients. This division can factor out a background polynomial and the remaining quotient is responsible for the description of the physical resonances. When a given spectrum to be modeled possesses a rolling and smoothly varying background, as encountered in MRS, one would not first extract $P_L^{\pm}(z^{\pm 1})/Q_K^{\pm}(z^{\pm 1})$ for $L > K$ and then perform divisions as done in, e.g., (2.172) to single out macromolecules from the contributions due to other resonances. Rather, from the outset, one should set up the Padé approximants with the distinct contributions from the background and the remaining spectrum via the model

$$G_N(z^{-1}) \approx B_{L-K}^{\pm}(z^{\pm 1}) + \frac{A_K^{\pm}(z^{\pm 1})}{Q_K^{\pm}(z^{\pm 1})} \tag{2.174}$$

with the diagonal polynomial ratio, or with the corresponding para-diagonal quotient

$$G_N(z^{-1}) \approx \tilde{B}_{L-K+1}^{\pm}(z^{\pm 1}) + \frac{\tilde{A}_{K-1}^{\pm}(z^{\pm 1})}{Q_K^{\pm}(z^{\pm 1})}. \tag{2.175}$$

In such a procedure, the expansion coefficients of the background polynomials $B_{L-K}^{\pm}(z^{\pm 1})$ and those of the remaining two polynomials in the quotients $A_K^{\pm}(z^{\pm 1})/Q_K^{\pm}(z^{\pm 1})$ can be extracted simultaneously from the system of linear equations that are implicitly present, e.g., in (2.174). By this design of the FPT, the background for large macromolecules and the main physical resonances are adequately treated on the same footing without any artificial patching of one contribution to the other.

2.7 Diagonal and para-diagonal fast Padé transform

In the rest of this book, we will illustrate only the diagonal and para-diagonal forms of the $\mathrm{FPT}^{(\pm)}$. The computational algorithm of the $\mathrm{FPT}^{(\pm)}$ begins by extracting the Padé polynomials $P_K^{\pm}(z^{\pm 1})$ and $Q_K^{\pm}(z^{\pm 1})$ from the given set of the input time signal points $\{c_n\}$. This is prescribed already in the definition of the $\mathrm{FPT}^{(\pm)}$ from (2.169) by which the expansion coefficients $\{p_r^{\pm}\}$ and $\{q_s^{\pm}\}$ of $P_K^{\pm}(z^{\pm 1})$ and $Q_K^{\pm}(z^{\pm 1})$ can be extracted. To this end within the $\mathrm{FPT}^{(+)}$ or $\mathrm{FPT}^{(-)}$, we first multiply (2.165) for $L = K$ by $Q_K^+(z)$ or $Q_K^-(z^{-1})$, and then compare the coefficients of the same powers of the expansion variables. For the $\mathrm{FPT}^{(+)}$, the results of this procedure are the two systems of linear equations for the polynomial coefficients $\{q_s^+\}$ and $\{p_r^+\}$

$$\sum_{s=1}^{K} q_s^+ c_{m+s} = -c_m \qquad (0 \le m \le I) \tag{2.176}$$

$$p_k^+ = \sum_{r=0}^{K-k} c_r q_{k+r}^+ \qquad (1 \le k \le K) \tag{2.177}$$

where $I = N - K - 1$. Similarly, in the case of the $\mathrm{FPT}^{(-)}$, we obtain the two systems of linear equations for the expansion coefficients $\{q_s^-\}$ and $\{p_r^-\}$

$$\sum_{s=1}^{K} q_s^- c_{K+m-s} = -c_{K+m} \qquad (1 \le m \le I) \tag{2.178}$$

$$p_k^- = \sum_{r=0}^{k} c_r q_{k-r}^- \qquad (0 \le k \le K). \tag{2.179}$$

In (2.176) and (2.178), the expansion coefficients $q_0^{\pm}$ can be set to unity. The equivalent matrix forms of the systems of equations (2.176)–(2.179) will be given in chapter 4. Notice that the expansion coefficients $\{q_s^-\}$ and $\{q_s^+\}$ of the polynomial $Q_K^-(z^{-1})$ and $Q_K^+(z)$ have the meaning of the forward and backward prediction coefficients in the $\mathrm{FPT}^{(-)}$ and $\mathrm{FPT}^{(+)}$, respectively. Thus, when all the K coefficients $\{q_s^-\}$ are determined, (2.178) can be employed to predict the new, unavailable time signal points c_n for $n > N$ from a linear combination of the known input data $\{c_n\}\,(0 \le n \le N-1)$. This is the essence of the powerful extrapolation feature of the $\mathrm{FPT}^{(-)}$ directly in the time domain. Likewise, if all the K coefficients $\{q_s^+\}$ are computed, (2.176) can be used to recur backwards and retrieve the old time signals points, as in (2.159) and (2.161). Such retrieved signal points fully agree with the given input data $\{c_n\}\,(0 \le n \le N-1)$ and this is the proof that the computed coefficients $\{q_s^+\}$ are accurate. Explicit computations of the polynomial coefficients $\{p_r^+\}$ and $\{q_s^+\}$ are simplified by the occurrence that the two systems of equations (2.176) and (2.177) are decoupled. The same remark holds true for $\{p_r^-\}$ and $\{q_s^-\}$ from the two systems of equations (2.178) and (2.179). Hence, in the $\mathrm{FPT}^{(+)}$ or $\mathrm{FPT}^{(-)}$, only the system (2.176) or (2.178) for the coefficients $\{q_s^+\}$ or $\{q_s^-\}$ needs to be solved, respectively. In other words, if the sets $\{q_s^\pm\}$ are obtained, it is not necessary to solve the linear equations from (2.177) and (2.179) for the remaining coefficients $\{p_r^+\}$ or $\{p_r^-\}$. In fact, for the known sets $\{q_s^\pm\}$, the equations (2.177) and (2.179) themselves *are* the analytical results for $\{p_r^+\}$ and $\{p_r^-\}$, respectively. In practice, regarding the systems of linear equations (2.176) and (2.178) for the coefficients $\{q_s^+\}$ and $\{q_s^-\}$, the standard MATLAB algorithms can be used with the optional implementation of the conventional SVD for a refinement of the obtained solutions.

By construction, the Green-Padé spectra $G_K^\pm(z^{\pm1})$ are meromorphic functions and, as such, the only singularities these functions could possibly have are their poles. For this reason, the poles of these functions are the same as the corresponding zeros of the denominator polynomials $Q_K^\pm(z^{\pm1})$. This enables the $\mathrm{FPT}^{(\pm)}$ to determine all the fundamental complex frequencies $\{\omega_k^\pm\}(1 \le k \le K)$, counted with their possible multiplicities, by solving the pertinent characteristic equations

$$Q_K^\pm(z^{\pm1}) = 0. \tag{2.180}$$

These equations represent the well-known secular equations for the roots in terms of the harmonic variables $z_k^\pm$

$$z_k^\pm \equiv \mathrm{e}^{\pm i\omega_k^\pm \tau} \qquad\qquad \mathrm{Im}(\omega_k^\pm) > 0. \tag{2.181}$$

Physical, genuine roots have $\mathrm{Im}(\omega_k^\pm) > 0$, meaning that all the K poles $\{z_k^+\}$ and $\{z_k^-\}$ are located inside and outside the unit circle, respectively. The fundamental frequencies $\omega_k^\pm$ are deduced from $z_k^\pm$ via

$$\omega_k^\pm = \mp(i/\tau)\ln(z_k^\pm) \qquad\qquad \ln(z_k^\pm) = \ln(|z_k^\pm|) + \mathrm{Arg}(z_k^\pm). \tag{2.182}$$

In the case of non-degenerate resonances, a fixed fundamental frequency is associated with only one amplitude. As discussed, these resonances yield pure Lorentzian line-shapes in the corresponding spectrum. Within the Padé spectral analysis, such amplitudes are the Cauchy residues of the associated Green-Padé functions. For instance

$$d_k^\pm = \lim_{z^{\pm1}\longrightarrow z_k^{\pm1}} \{(z^{\pm1} - z_k^{\pm1})\tilde{G}_K^\pm(z^{\pm1})\} = \frac{\tilde{P}_{K-1}^\pm(z_k^\pm)}{\left[(\mathrm{d}/\mathrm{d}z^{\pm1})Q_K^\pm(z^{\pm1})\right]_{z^{\pm1}=z_k^\pm}}. \tag{2.183}$$

Here, the kth amplitude corresponds only to the kth frequency. Hence non-degeneracy. The Padé reconstructions for the true phases ϕ_k of the kth resonance are given by $\phi_k^{\pm} \equiv \text{Arg}(d_k^{\pm})$. The absolute values of the amplitudes, $|d_k^{\pm}|$, are not the only spectral parameters that determine the height of the corresponding resonance peak. This becomes clear from the ersatz spectrum

$$\mathcal{E}_K(\omega) = -i \sum_{k=1}^{K} \frac{|d_k|}{\omega - \omega_k}. \tag{2.184}$$

As before, this Heaviside partial fraction representation can explicitly be summed up to produce a quotient of a numerator and denominator polynomial of degree $K-1$ and K, respectively. As such, $\mathcal{E}_K(\omega)$ is seen to be a para-diagonal FPT in the variable ω [5]. For the sake of brevity, the superscripts $\pm$ in (2.184) are dropped. The absorption ersatz spectrum corresponding to (2.184) is a sum of K pure Lorentzians

$$\text{Re}\left(-i \sum_{k=1}^{K} \frac{|d_k|}{\omega - \omega_k}\right) = \sum_{k=1}^{K} \frac{|d_k|\text{Im}(\omega_k)}{[(\omega - \text{Re}(\omega_k)]^2 + [\text{Im}(\omega_k)]^2}. \tag{2.185}$$

From here, the height h_k of the k th Lorentzian peak is given by the expression $|d_k|\text{Im}(\omega_k)/\{[(\omega - \text{Re}(\omega_k)]^2 + [\text{Im}(\omega_k)]^2\}$ taken at $\omega = \text{Re}(\omega_k)$ via

$$h_k \equiv \lim_{\omega \longrightarrow \text{Re}(\omega_k)} \frac{|d_k|\text{Im}(\omega_k)}{[(\omega - \text{Re}(\omega_k)]^2 + [\text{Im}(\omega_k)]^2} = \frac{|d_k|}{\text{Im}(\omega_k)}. \tag{2.186}$$

Thus, the height h_k of the k th Lorentzian peak is determined by the two spectral parameters, $|d_k|$ and $\text{Im}(\omega_k)$, and not by $|d_k|$ alone, as further illustrated through the figures in chapter 3. The briefly outlined procedure completes the determination of all the fundamental complex frequencies and the associated complex amplitudes. Such reconstructed spectral data yield the sought four peak parameters (position, width, height, phase) for each of the K Lorentzian resonances in the spectra computed by both variants of the FPT.

Given the input Maclaurin sum (2.162), the uniqueness of the FPT guarantees that the $\text{FPT}^{(+)}$ and $\text{FPT}^{(-)}$ must reconstruct the identical frequencies and amplitudes $\{\omega_k^+, d_k^+\}$ and $\{\omega_k^-, d_k^-\}$. Furthermore, these retrieved spectral parameters ought to be equal to the true complex frequencies and amplitudes $\{\omega_k, d_k\}$ from the input FID

$$\omega_k^+ = \omega_k^- = \omega_k \qquad\qquad d_k^+ = d_k^- = d_k. \tag{2.187}$$

Such equalities will be fulfilled if the values for all the reconstructed spectral parameters $\{\omega_k^{\pm}, d_k^{\pm}\}$ have converged as a function of the increased number of signal points, as shown in chapter 3. Moreover, during this self-contained cross-checking, the FPT can also discriminate with certainty between the genuine and spurious resonances by relying upon the concept of Froissart doublets [44] or pole-zero cancellations [45].

The overall meaning of the explained intrinsic cross-validation in the $\text{FPT}^{(\pm)}$ is in avoiding altogether the customary need to compare different processors.

As discussed, even if different signal processors mutually agree, this still does not necessarily imply that they retrieved the true spectral parameters. The $\mathrm{FPT}^{(+)}$ and $\mathrm{FPT}^{(-)}$ possess a joint conceptual design for a system response function in the Green-Padé functions $G_K^+(z)$ and $G_K^-(z^{-1})$ within the same general mathematical Padé methodology. These versions from the FPT are computationally complementary and, therefore, they could rightly be regarded as two distinct strategies for the same problem. As stated, the main difference between these two variants is in that the $\mathrm{FPT}^{(-)}$ accelerates slowly convergent series [18], whereas the $\mathrm{FPT}^{(+)}$ converts divergent series into convergent ones by the principle of the Cauchy analytical continuation [19]. Specifically, for the diagonal case $[K/K]_{G_N}(z^{-1})$, the optimal performances of the $\mathrm{FPT}^{(-)}$ and $\mathrm{FPT}^{(+)}$ are expected for $N = 2K$ and $N > 2K$, respectively. The two requirements, $N = 2K$ and $N > 2K$ yield, respectively, the algebraically determined and over-determined system of linear equations for the expansion coefficients $p_r^{\pm}$ and $q_s^{\pm}$ of the polynomials $P_K^{\pm}(z^{\pm 1})$ and $Q_K^{\pm}(z^{\pm 1})$. Determination of these expansion coefficients, as the critical part of computations, must be performed with high accuracy to provide maximal precision for the reconstructed fundamental frequencies and amplitudes. Machine accuracy can be achieved by the outlined algorithm of the FPT, as done in our computations with the results that are illustrated in chapter 6.

After highly accurate extraction of all the fundamental frequencies and amplitudes $\{\omega_k^{\pm}, d_k^{\pm}\}$ from the input FID, the Green-Padé spectra can be generated in their complex modes defined by the Heaviside partial fractions

$$\tilde{G}_K^{\pm}(z^{\pm 1}) = \frac{\tilde{P}_{K-1}^{\pm}(z^{\pm 1})}{Q_K^{\pm}(z^{\pm 1})} = \sum_{k=1}^{K} \frac{d_k^{\pm}}{z^{\pm 1} - z_k^{\pm}}. \tag{2.188}$$

One can also establish an alternative representation of the same spectrum by using only the roots of the numerator and denominator polynomials. To this end, we additionally need to solve the equation

$$\tilde{P}_{K-1}^{\pm}(z^{\pm 1}) = 0 \tag{2.189}$$

which yields the roots denoted by $\{\tilde{z}_k^{\pm}\}$ $(1 \le k \le K-1)$

$$\tilde{z}_k^{\pm} \equiv \mathrm{e}^{\pm i\tilde{\omega}_k^{\pm}\tau}. \tag{2.190}$$

Thus, employing the sets of the roots $\{z_k^{\pm}\}$ and $\{\tilde{z}_k^{\pm}\}$ it becomes feasible to express $\tilde{G}_K^{\pm}(z^{\pm 1})$ in the canonical forms that are fully equivalent to (2.188)

$$\frac{\tilde{P}_{K-1}^{\pm}(z^{\pm 1})}{Q_K^{\pm}(z^{\pm 1})} = \frac{p_{K-1}^{\pm}}{q_K^{\pm}} \frac{\prod_{k=1}^{K-1}(z^{\pm 1} - \tilde{z}_k^{\pm})}{\prod_{k=1}^{K}(z^{\pm 1} - z_k^{\pm})}. \tag{2.191}$$

By computing the Cauchy residues of (2.191), we are led to the following expressions for the Padé amplitudes

$$d_k^{\pm} = \frac{p_{K-1}^{\pm}}{q_K^{\pm}} \frac{\prod_{k'=1}^{K-1}(z_k^{\pm} - \tilde{z}_{k'}^{\pm})}{\prod_{k'=1,k'\neq k}^{K}(z_k^{\pm} - z_{k'}^{\pm})} \tag{2.192}$$

that are equivalent to (2.183). Such a procedure is valid only for the purely Lorentzian line-shapes encountered in a non-degenerate spectrum with non-coincident fundamental frequencies. By this algorithm we can also reconstruct the corresponding FID as a sum of K attenuated complex harmonics with time-independent amplitudes $d_k^{\pm}$

$$c_n^{\pm} = \sum_{k=1}^{K} d_k^{\pm}\, \mathrm{e}^{in\omega_k^{\pm}\tau} \qquad c_n^{\pm} \approx c_n \qquad \mathrm{Im}(\omega_k^{\pm}) > 0. \tag{2.193}$$

The expounded concept ought to be modified if one or more confluent (coincident) fundamental frequencies are present in the input FID. In such a case, more than one amplitude corresponds to the same given frequency. This is properly described by inclusion of multiple roots of the characteristic polynomials $Q_K^{\pm}(z^{\pm 1})$. The corresponding line-shapes are non-Lorentzians and they are associated with the degenerate part of the entire spectrum. Let us assume that there are $J \leq K$ degenerate fundamental frequencies in the input FID. Further, let the kth degenerate frequency have the multiplicity m_k whose maximal value is denoted by M_k

$$M_k = \max\{m_k\} \qquad\qquad M_1 + M_2 + \cdots + M_J = K. \tag{2.194}$$

The inequality $1 \leq m_k \leq M_k$ with $m_k > 1$ implies that the kth root of $Q_K^{\pm}$ is repeated m_k times. The special case $m_k = 1$ is associated with non-degeneracy. When all the multiplicities m_k of the kth root of $Q_K^{\pm}(z^{\pm 1})$ are included, the Cauchy residues of the quotient $\tilde{P}_K^{\pm}(z^{\pm 1})/Q_K^{\pm}(z^{\pm 1})$ become

$$D_{k,m_k}^{\pm} = \frac{\tilde{P}_{K-1}^{\pm}(z_k^{\pm})}{[(\mathrm{d}/\mathrm{d}z^{\pm 1})^{m_k} Q_K^{\pm}(z^{\pm 1})]_{z^{\pm 1} = z_k^{\pm}}}. \tag{2.195}$$

These expressions extend (2.183) to the case when there are fundamental frequencies that coincide exactly with each other. The associated expressions that generalize (2.188) to the case of simultaneous presence of simple and multiple roots of $Q_K^{\pm}(z^{\pm 1})$ in the $\mathrm{FPT}^{(\pm)}$ are represented by the mixed complex spectra

$$\tilde{G}_K^{\pm}(z^{\pm 1}) = \frac{\tilde{P}_{K-1}^{\pm}(z^{\pm 1})}{Q_K^{\pm}(z^{\pm 1})} = \sum_{k=1}^{J} \sum_{m_k=1}^{M_k} \frac{D_{k,m_k}^{\pm}}{(z^{\pm 1} - z_k^{\pm})^{m_k}}. \tag{2.196}$$

These results permit the exact reconstruction of the general form of the FID for degenerate and non-degenerate fundamental frequencies. The degenerate part of such an FID has the confluent harmonics given by a linear combination of attenuated complex exponentials with coefficients that are non-stationary (time-dependent) amplitudes $d^{\pm}_{k,n}$. Such a time dependence of the amplitudes is given by a polynomial of degree M_k so that

$$c^{\pm}_{n} = \sum_{k=1}^{J} d^{\pm}_{k,n}\, \mathrm{e}^{in\omega^{\pm}_{k}\tau} \qquad d^{\pm}_{k,n} = \sum_{m_k=1}^{M_k} D^{\pm}_{k,m_k}\, (n\tau)^{m_k-1} \tag{2.197}$$

where $\mathrm{Im}(\omega^{\pm}_{k}) > 0$. Hence, the simple analytical formulae also exist in the $\mathrm{FPT}^{(\pm)}$ for degenerate amplitudes in the case of non-Lorentzian spectral line-shapes with coincident resonances having exactly the same real parts of the corresponding complex fundamental frequencies. This outlined analysis demonstrates that the two Green-Padé versions, the $\mathrm{FPT}^{(+)}$ and $\mathrm{FPT}^{(-)}$, are able to treat both Lorentzian and non-Lorentzian spectra on the same footing. Non-degenerate and degenerate spectra correspond to FIDs represented by linear combinations of K damped complex exponentials with stationary and non-stationary amplitudes, respectively.

2.8 Determination of the exact number K of resonances

Maximally accurate retrieval of the unknown true number K of physical resonances in a given FID is of primary importance. In the FPT, this number K is determined exactly. Moreover, K can be reconstructed in either the time or frequency domain analysis, by employing the Shanks transform [5] and Froissart doublets [44], respectively. The main outlines of the mathematical basis upon which this is achieved will be presented in this section (for a number of complementary aspects, see chapter 5).

2.8.1 Exact Shank's filter for finding K, including the fundamental frequencies and amplitudes: the use of Wynn's recursion

In the case of (2.193) or (2.197), the Shanks transform, which is denoted by $e_K(c_n)$, is defined as [5]

$$e_k(c_n) \equiv \frac{\mathrm{H}_{k+1}(c_n)}{\mathrm{H}_k(\Delta^2 c_n)} = \frac{1}{\mathrm{W}_k^2}\frac{\mathrm{H}_{k+1}(c_n)}{\mathrm{H}_k(c_n)} \tag{2.198}$$

$$\mathrm{W}_k = \prod_{k'=1}^{k} (z_{k'} - 1) \neq 0 \tag{2.199}$$

where $\Delta^2 c_n = \Delta c_{n+1} - \Delta c_n$, $\Delta c_n = c_{n+1} - c_n$ and $\mathrm{H}_n(c_s)$ is the Hankel determinant

$$\mathrm{H}_n(c_s) = \begin{vmatrix} c_s & c_{s+1} & c_{s+2} & \cdots & c_{s+n-1} \\ c_{s+1} & c_{s+2} & c_{s+3} & \cdots & c_{s+n} \\ c_{s+2} & c_{s+3} & c_{s+4} & \cdots & c_{s+n+1} \\ \vdots & \vdots & \vdots & \ddots & \vdots \\ c_{s+n-1} & c_{s+n} & c_{s+n+1} & \cdots & c_{s+2n-2} \end{vmatrix}. \tag{2.200}$$

The Hankel determinants are not computed directly from their definition (2.200) for large dimensions. This would be numerically impractical for higher ranks due to the factorially growing number of multiplications. Rather, the $e_k(c_n)$'s are computed recursively by the Wynn $\varepsilon-$algorithm [5]

$$\varepsilon_n^{(m+1)} = \varepsilon_{n+1}^{(m-1)} + \frac{1}{\varepsilon_{n+1}^{(m)} - \varepsilon_n^{(m)}}$$
$$\varepsilon_n^{(-1)} = 0 \qquad \varepsilon_n^{(0)} = c_n \tag{2.201}$$

where c_n is present in $\varepsilon_n^{(m)}$ via the second initial condition, $\varepsilon_n^{(0)} = c_n$. Upon reaching convergence in (2.201), the ST can simply be extracted from the relationship

$$e_m(c_n) = \varepsilon_n^{(2m)}. \tag{2.202}$$

The Wynn recursion (2.202) itself was established from a recursion which links four neighboring elements in the table of Padé approximants of varying order. As such, the recursive algorithm for the Shanks transform is an efficient algorithm of the general Padé methodology [5].

If a given FID is comprised of precisely K non-degenerate or degenerate attenuated complex harmonics, (2.198) becomes

$$e_K(c_0) = 0. \tag{2.203}$$

This finding is the signature for detection of the exact K in the time domain spectral analysis. Regarding the ST, the quotient form (2.198) itself serves as the definition of the necessary and sufficient condition for the time signal c_n to possess exactly K fundamental transients $\{z_k\}$ $(1 \le k \le K)$. Such a condition is the simultaneous fulfillment of the following two relations

$$\mathrm{H}_{K+1}(c_n) = 0 \qquad \mathrm{H}_K(c_n) \neq 0. \tag{2.204}$$

The remaining two sets of the system's parameters, that are the fundamental harmonics and the corresponding amplitudes $\{z_k, d_k\}$, can also be found by continuing the work on spectral analysis directly in the time domain. The explicit formulae for this purpose are [5]

$$z_k = \frac{e_{k-1}(c_{n+1})}{e_{k-1}(c_n)} \qquad n \longrightarrow \infty \tag{2.205}$$

$$d_{k+1} = \frac{1}{\mathrm{R}_k^2 z_{k+1}^n} \frac{\mathrm{H}_{k+1}(c_n)}{\mathrm{H}_k(c_n)} \qquad n \longrightarrow \infty \tag{2.206}$$

$$\mathrm{R}_k = \prod_{k'=1}^{k} (z_{k+1} - z_{k'}). \tag{2.207}$$

This expression for R_k is deduced by simplifying the original formula $\mathrm{R}_k = \mathrm{V}_{k+1}/\mathrm{V}_k$ derived in Ref. [5] where V_k is the Vandermond determinant

$$\mathrm{V}_k = \begin{vmatrix} 1 & 1 & 1 & \cdots & 1 \\ z_1 & z_2 & z_3 & \cdots & c_k \\ z_1^2 & z_2^2 & z_3^2 & \cdots & z_k^2 \\ \vdots & \vdots & \vdots & \ddots & \vdots \\ z_1^{k-1} & z_2^{k-1} & z_3^{k-1} & \cdots & z_k^{k-1} \end{vmatrix} = \prod_{s=1}^{k} \prod_{s'=1, s'\neq s}^{k} (z_s - z_{s'}) \neq 0\,. \tag{2.208}$$

We see that the same quotient of the Hankel determinants $\mathrm{H}_{k+1}(c_n)/\mathrm{H}_k(c_n)$ appears in the Shanks transform $e_k(c_n)$ and in the amplitude d_{k+1}. Therefore, the necessary and sufficient condition (2.193) that c_n has precisely K components can equivalently be written as $d_{K+1} = 0$, or more generally

$$d_{K+m} = 0 \qquad (m = 1, 2, 3, ...). \tag{2.209}$$

An important implication of this result is that, whenever the input time signal c_n has K harmonics $\{z_k\}\,(1 \leq k \leq K)$, the exact computation which eventually finds some higher-order transients $\{z_{K+m}, d_{K+m}\}\,(m = 1, 2, 3, ...)$ must produce zero-valued amplitude $d_{K'} = 0\,(K' > K)$, as per (2.209). Let a computation find an estimate $\tilde{c}_n$ with $K + m\,(m > 0)$ harmonics for the input time signal c_n which, however, possesses only K harmonics

$$\tilde{c}_n = \sum_{k=1}^{K+m} d_k\, z_k^n \qquad (m = 1, 2, 3, ...). \tag{2.210}$$

Then, the result (2.209) will apply to yield the reduction formula which, in turn, reconstructs the exact c_n

$$\begin{aligned} \tilde{c}_n &= \sum_{k=1}^{K+m} d_k\, z_k^n \\ &= \sum_{k=1}^{K} d_k\, z_k^n + \left\{ d_{K+1} z_{K+1}^n + d_{K+2} z_{K+2}^n + \cdots + d_{K+m} z_{K+m}^n \right\} \\ &= \sum_{k=1}^{K} d_k\, z^n = c_n \qquad \text{(QED)}. \end{aligned} \tag{2.211}$$

In other words, the higher-order harmonics $\{z_{K+m}, d_{K+m}\}\,(m = 1, 2, 3, ...)$ in the estimate $\tilde{c}_n$ ought to be interpreted as being spurious in the sense that the original time signal c_n simply does not possess such components, $\{z_{K'}, d_{K'}\}\,(K' > K)$. We see that, in the end, all such spurious components $\{z_{K+1}, d_{K+1}; z_{K+2}, d_{K+2}; z_{K+3}, d_{K+3}; ...\}$ of the estimate $\tilde{c}_n$ for the input time signal c_n with K harmonics $\{z_1, d_1; z_2, d_2; z_3, d_3; ...; z_K, d_K\}$ are washed out from the output data by their zero-valued intensities. This leads straight to the exact result $\tilde{c}_n = c_n$, as per (2.211).

2.8.2 Exact number K and the existence of the solution of ordinary difference equations

At this point, we make an apparent digression, and write the Kth order ordinary difference equation (OΔE) similarly to (2.137) and (2.138), except for the $z-$variable, $z = \mathrm{e}^{i\omega\tau}$

$$\sum_{k=1}^{K} q_k c_{n+k} = 0 \qquad \therefore \qquad c(t) = \sum_{k=1}^{K} d_k \mathrm{e}^{in\omega_k\tau} \tag{2.212}$$

where $\{q_k\}_{k=1}^{K}$ are the given constant coefficients (not all of which are zero). This is the discretized version of the associated ordinary differential equation (ODE) defined by

$$\sum_{k=1}^{K} q_k \left(-i\frac{\mathrm{d}}{\mathrm{d}t}\right)^k c(t) = 0 \qquad \therefore \qquad c(t) = \sum_{k=1}^{K} d_k \mathrm{e}^{i\omega_k t}. \tag{2.213}$$

All the constants $\{d_k\}\,(1 \le k \le K)$ are obtained from the K boundary conditions to (2.212) or (2.213) [5]. The ω_k's are deduced from the roots $z_k = \mathrm{e}^{i\omega_k\tau}$ of the characteristic or secular equation

$$\sum_{k=1}^{K} q_k z^k = 0 \qquad \therefore \qquad z_k = \mathrm{e}^{i\omega_k\tau} \tag{2.214}$$

which is automatically generated by (2.212). Crucially, the necessary and sufficient condition for the existence of the solution c_n of (2.212) in the unique form $c_n = \sum_{k=1}^{K} d_k \mathrm{e}^{in\omega_k\tau}$ is given precisely by (2.204). Hence, our temporary excursion to the OΔE was not a digression after all. Nowhere have we mentioned the Padé approximant in this passage to the differential calculus, and yet the nature of the invoked dynamics has naturally driven us to the PA through the explicit appearance of the Padé denominator polynomial $Q_K(z)$ disguised in the secular equation (2.214), which can equivalently be written as $Q_K(z) \equiv \sum_{k=1}^{K} q_k z^k = 0$, as in (2.180). It is remarkable that different mathematical strategies in the time domain cohere while jointly determining the exact number K of harmonics by the same condition (2.204).

2.8.3 The role of linear dependence as spuriousness in determining K within the state space-based perspective of signal processing

The investigated FID from its geometric progression $c_n = \sum_{k=1}^{K} d_k\, z_k^n$ can also be viewed from a state space-based perspective in signal processing. This can be done by introducing a $K-$dimensional linear vector space $\mathcal{K}$ with the symmetric inner product (2.3) and a basis $\{\Upsilon_k\}$ $(1 \le k \le K)$ which is a complete set of the solutions of the stationary Schrödinger equation (2.6). The corresponding non-stationary solution $|\Phi(t))$ of the time-dependent Schrödinger equation (2.1) is given by (2.4). Its discretized version is $|\Phi_n) = \hat{\mathrm{U}}^n|\Phi_0)$ where $\hat{\mathrm{U}}$ is the time evolution operator $\hat{\mathrm{U}} \equiv \hat{\mathrm{U}}(\tau) = \exp{(-i\hat{\Omega}\tau)}$. Here, the continuous time variable t is discretized via $t = n\tau\, (0 \le n \le N-1)$, where τ is the sampling rate. The two equivalent representations Φ and Υ can be related to each other. This is done by using the spectral decomposition (2.16) of the time evolution operator in Φ_n to arrive at the expansion $\Phi_n = \sum_{k=1}^{K} d_k^{1/2} u_k^n |\Upsilon_k)$, where $u_k = \exp{(-i\omega_k\tau)}$ and d_k is the amplitude/residue from (2.19), i.e., $d_k = (\Phi_0|\Upsilon_k)^2$. To cohere with the notation from the other sub-sections in this section, we shall switch from the $u-$ to the $z-$variable, in which case Φ_n will be replaced by Φ'_n where $|\Phi'_n) = \exp{(in\hat{\Omega}\tau)}|\Phi_0)$ and

$$\Phi'_n = \sum_{k=1}^{K} d_k^{1/2} z_k^n |\Upsilon_k) \qquad d_k = (\Phi'_0|\Upsilon_k)^2. \tag{2.215}$$

This $K-$dimensional state vector Φ'_n is defined through its K components or "coordinates"

$$\begin{aligned}
\Phi'_0 &= \{d_1^{1/2}, d_2^{1/2}, d_3^{1/2}, \cdots, d_K^{1/2}\} \\
\Phi'_1 &= \{d_1^{1/2} z_1, d_2^{1/2} z_2, d_3^{1/2} z_3, \cdots, d_K^{1/2} z_K\} \\
\Phi'_2 &= \{d_1^{1/2} z_1^2, d_2^{1/2} z_2^2, d_3^{1/2} z_3^2 \cdots, d_K^{1/2} z_K^2\} \\
&\;\;\vdots \\
\Phi'_n &= \{d_1^{1/2} z_1^n, d_2^{1/2} z_2^n, d_3^{1/2} z_3^n, \cdots, d_K^{1/2} z_K^n\}.
\end{aligned} \tag{2.216}$$

In other words, since $\{\Upsilon_k\}_{k=1}^{K}$ is a basis in the space $\mathcal{K}$, any function from this space including $\Phi'_n \in \mathcal{K}$ can be expanded in terms of the elements Υ_k such that the kth component/coordinate of Φ'_n is $d_k^{1/2} u_k^n$ as in (2.215), or equivalently

$$\Phi'_n = d_1^{1/2} z_1^n \Upsilon_1 + d_2^{1/2} z_2^n \Upsilon_2 + d_3^{1/2} z_3^n \Upsilon_3 + \cdots + d_K^{1/2} z_K^n \Upsilon_K. \tag{2.217}$$

Taking the scalar product between $|\Phi'_n)$ and $(\Phi'_m|$ via $(\Phi'_m|\Phi'_n)$, setting the normalization of Υ_k to unity through $||\Upsilon_k|| \equiv 1$ and using the orthogonalization $(\Upsilon_{k'}|\Upsilon_k) = \delta_{k,k'}$, it follows from (2.216) and (2.217)

$$S_{n,m} \equiv (\Phi'_m|\Phi'_n) = \sum_{k=1}^{K} d_k z_k^{n+m}. \tag{2.218}$$

The rhs of (2.218) is recognized as the geometric series representation of the time signal point c_{n+m} so that

$$c_{n+m} = (\Phi'_m|\Phi'_n) = S_{n,m}. \tag{2.219}$$

Thus, in the basis $\{\Upsilon_k\}\,(1 \le k \le K)$, the time signal c_{n+m} appears as the matrix element of vectors $|\Phi'_n)$ and $(\Phi'_m|$. Since, in general, $c_{n+m} \neq 0$ for arbitrary non-negative integers n and m, functions $|\Phi'_n)$ and $(\Phi'_m|$ are not orthogonal to each other and, as they stand, they do not form a basis set. The matrix element between $|\Phi'_n)$ and $(\Phi'_m|$ from (2.219) represents, in fact, the so-called overlap $S_{n,m} \equiv (\Phi'_m|\Phi'_n)$. This overlap $S_{n,m}$ is nothing but the signal point c_{n+m} by virtue of (2.219). Note that the set of K vectors from (2.216) could be orthogonalized by using, e.g., the Gram-Schmidt or Lanczos orthogonalization. However, orthogonalization is a convenience, but not a necessity in the expansion methods.

Although they do not represent a basis, the $K-$dimensional vectors $\{\Phi'_1, \Phi'_2, \Phi'_3, ..., \Phi'_K\}$ from (2.216) can be easily shown to be linearly independent[3]. By contrast, the augmented set of vectors $\{\Phi'_1, \Phi'_2, \Phi'_3, ..., \Phi'_K, \Phi'_{K+1}\}$ is linearly dependent and so is any other sequence with more added $\Phi'-$terms, $\{\Phi'_1, \Phi'_2, \Phi'_3, ..., \Phi'_{K+m}\}\,(m = 1, 2, ...)$. This follows from the well-known fact that the necessary and sufficient condition for a set of vectors to be linearly dependent is that the Gram determinant with the elements $S_{n,m} = (\Phi'_m|\Phi'_n)$ is equal to zero

$$\det\{S_{n,m}\} = 0 \qquad (n, m = 1, 2, ..., K + 1). \tag{2.220}$$

Due to (2.218), the lhs of (2.220) is equal to the Hankel determinant $\mathrm{H}_{K+1}(c_0)$. Therefore (2.220) can be rewritten as $\mathrm{H}_{K+1}(c_0) = 0$

$$\det\{S_{n,m}\} = \mathrm{H}_{K+1}(c_0) = 0 \qquad (n, m = 1, 2, ..., K + 1). \tag{2.221}$$

This coincides with the condition (2.204). Therefore, we see that the previous checkings for the fulfillment of the condition (2.204), by which the exact number K of the harmonic components in the time signal c_n can be determined, is entirely equivalent to verifying whether the vectors $\{\Phi'_1, \Phi'_2, \Phi'_3, ..., \Phi'_K, \Phi'_{K+1}\}$ are linearly dependent. Hence, this is yet another way of arriving to the same criterion (2.204) for determining the exact number K of true resonances.

Recall that, by reference to (2.198), zero-valued amplitude d_{K+m} for $m > 1$ represents a signature of spuriousness in the time signal c_n with precisely K

[3]Let $S = \{\xi_1, \xi_2, \xi_3, ..., \xi_M\}$ be a finite set of M distinct elements in a linear space $\mathcal{M}$. The set S is said to be linearly dependent if there exists a final set of scalars $\{a_1, a_2, a_3, ..., a_M\}$ (not all zero) such that $\sum_{m=1}^{M} a_m\xi_m = 0$. Conversely, the set S will be linearly independent, if it is not linearly dependent. In other words, when a set S is linearly independent, then for any choices of distinct elements $\{\xi_1, \xi_2, \xi_3, ..., \xi_M\}$ in S and scalars $\{a_1, a_2, a_3, ..., a_M\}$, the relation $\sum_{m=1}^{M} a_m\xi_m = 0$ could be fulfilled if and only if all the M values of the scalars $\{a_m\}_{m=1}^{M}$ are simultaneously equal to zero, $a_m = 0\,(m = 1, 2, 3, ..., M)$.

components. The mechanism of this signature is in the condition (2.204). The same condition (2.204) determines that the vectors of $K+m$ components $\{\Phi'_1, \Phi'_2, \Phi'_3, ..., \Phi'_{K+m}\}$ for $m \geq 1$ are linearly dependent. On the other hand, the overlap $(\Phi'_m|\Phi'_n)$ of these vectors is equal to the signal point c_{n+m} according to (2.218). Thus, in the representation (2.218), the signal c_{n+m} with K harmonics is viewed as being built from the overlap of the two state vectors $|\Phi'_n)$ and $(\Phi'_m|$, where according to (2.216) each Φ'_r $(r = n, m)$ has K components $\Phi'_r = \{d_1^{1/2} z_1^r, d_2^{1/2} z_2^r, d_3^{1/2} z_3^r, \cdots, d_K^{1/2} z_K^r\}$. Therefore, linear dependence of vectors $\{\Phi'_1, \Phi'_2, \Phi'_3, ..., \Phi'_K, \Phi'_{K+m}\}$ for $m > 1$ represents the same kind of spuriousness as the one due to $d_{K+m} = 0$ for $m > 1$.

In practice, while using the $\text{FPT}^{(-)}$, we search for stabilization of all the genuine spectral parameters and for identification of every spurious resonance. Such a stabilization occurs at those Padé orders K' that are larger than the sought true number K of resonances, $K' \equiv K + m\,(m > 1)$. For $K' > K$, spurious resonances stem from a linear dependence of their state vectors $\{\Phi_s\}$ (spurious states for $K' < K$ will be discussed in chapter 6). Due to zero amplitudes of all the components of a spurious vector $|\Phi'_s) = \sum_{k=1}^{K+m} d_k^{1/2} z_k^s |\Upsilon_k)$ for $m > 1$, its overlap with any genuine (or spurious) state is equal to zero

$$(\Phi'_s|\Phi'_r) = c_{r+s} = 0 \;\; \forall r, s \;\; (\text{for} \;\; \Phi'_r \;\; \text{genuine or spurious} \;\; \& \;\; \Phi'_s \;\; \text{spurious})$$

where $r = s$ is permitted if Φ'_r and Φ'_s are both spurious states. If $\mathcal{A}$ and $\mathcal{B}$ are the two disjoint vector spaces of genuine and spurious states, respectively, then it follows from the preceding expression that

$$\mathcal{A} \perp \mathcal{B} \qquad (\mathcal{A} \text{ genuine } \& \ \mathcal{B} \text{ spurious vector spaces}). \tag{2.222}$$

However, $c_{r+s} = 0$ from $(\Phi'_s|\Phi'_r) = 0$ should not be confused with zero filling from the FFT, in which N zeros $c_n = 0\,(n > N-1)$ are added to the input time signal c_n $(0 \leq n \leq N-1)$ to achieve a sinc-type interpolation. When Φ'_r and Φ'_s coincide, the overlap $(\Phi'_s|\Phi'_s) = c_{2s} = 0$ becomes the squared norm $||\Phi'_s||^2$. Therefore, spurious states have zero norm $||\Phi'_s|| \equiv (\Phi'_s|\Phi'_s)^{1/2} = 0$. A normalized state is given by $N_m \Phi'_m$ where N_m is the normalization constant. An un-normalized physical state Φ'_m is normalized to the unit particle flux via $(N_m\Phi'_m|N_m\Phi'_m) = 1$, which gives $N_m = 1/||\Phi'_m||$. Spurious states $\{\Phi'_s\}$ cannot be normalized, since their normalization constants are equal to infinity

$$\{\Phi'_s\}_{\text{normalized}} = N_s\{\Phi'_s\}_{\text{un-normalized}} \;\; (N_s = \infty \;\; \text{for} \;\; \Phi'_s \;\; \text{spurious}) \tag{2.223}$$

$$N_s \equiv \frac{1}{||\Phi'_s||} \qquad (\,||\Phi'_s|| = 0 \quad \text{for} \;\; \Phi'_s \;\; \text{spurious}). \tag{2.224}$$

Zero norms, or equivalently, infinite normalization constants of spurious states correspond to the zero flux of particles. Thus, spurious states do not describe any physical particle. Overall, the lesson from this sub-section is that linearly dependent vectors or equations bring no new information whatsoever to the analysis. Rather, when they are added to the remaining genuine, linearly independent vectors or equations, spurious Froissart doublets are obtained.

2.8.4 Froissart doublet spuriousness in the frequency domain for finding K

In the frequency domain, the exact Padé-reconstruction of the true number K works with the help of the concept of Froissart doublets

$$z_k^{\pm} = \tilde{z}_k^{\pm} \tag{2.225}$$

where $z_k^{\pm}$ and $\tilde{z}_k^{\pm}$ are the poles and zeros of the Green-Padé transfer function $G_{K'}^{\pm}(z^{\pm 1}) = P_K^{\pm}(z^{\pm 1})/Q_K^{\pm}(z^{\pm 1})$. Because the number K is unknown in advance, the computation in the FPT is carried out by a gradual and systematic augmentation of the order K' in the Green-Padé functions $G_{K'}^{\pm}(z^{\pm 1})$. The changes in K' cause fluctuations in the retrieved functions. However, K' will saturate, i.e., it will reach a value after which such fluctuations in $G_{K'}^{\pm}$ shall disappear within a prescribed level of accuracy. Such a stabilized value of the order K' is the sought K. This phenomenon will be manifested in (2.191) through cancellation of all the terms in the Padé numerator and denominator polynomials if the computation is continued beyond the stabilized value of the order in the Green-Padé functions

$$\frac{P_{K+m}^{\pm}(z^{\pm 1})}{Q_{K+m}^{\pm}(z^{\pm 1})} = \frac{P_K^{\pm}(z^{\pm 1})}{Q_K^{\pm}(z^{\pm 1})} \qquad (m = 1, 2, 3, ...). \tag{2.226}$$

By means of (2.192) and (2.225), Froissart amplitudes are found to be zero

$$d_k^{\pm} = 0 \qquad \text{for} \qquad z_k^{\pm} = \tilde{z}_k^{\pm}. \tag{2.227}$$

Moreover, pole-zero cancellations also take place if the computations discover multiplicities in $\{z_k^{\pm}\}$ even for an FID whose spectrum is known to be non-degenerate. In this case, the same degeneracies also appear in $\{\tilde{z}_k^{\pm}\}$, and this gives the degenerate Froissart doublets whose automatic elimination by means of pole-zero cancellations finally gives a non-degenerate spectrum, as should be from the outset. Pole-zero cancellations can also happen prior to detection of the true K for all the spurious poles that are canceled by the associated spurious zeros. However, what counts is not so much that spurious resonances uncontrollably appear, but rather that in the FPT they can unequivocally be identified through pole-zero pairings as Froissart doublets of zero amplitudes. In sharp contrast, genuine resonances never have zero amplitudes, due to the absence of pole-zero coincidences.

In this way, Froissart doublets emerge as a reliable procedure for deciding whether a reconstructed resonance is genuine or spurious. It should be emphasized that spurious resonances can arise in spectral analysis for both noise-free and noise-corrupted (synthesized or encoded) FIDs. Thus, the concept of Froissart doublets could advantageously be employed as a reliable and robust procedure for separating the physical from non-physical (noisy) information. At the end of the computations, all the Froissart doublets are dropped. Therefore, the list of the reconstructed spectral parameters $\{\omega_k^{\pm}, d_k^{\pm}\}$ will contain only the physical information through the genuine resonances.

2.8.5 Froissart doublets in exact analytical computations

The majority of previous studies consider Froissart doublets exclusively as noise functions. This presumes that Froissart doublets are due only to noise perturbations in the input data. However, spuriousness of Froissart doublets manifested through pole-zero coincidences can also arise in ideal, noiseless data [11, 34]. We already mentioned in chapter 1 that one of the reasons for this is finite arithmetics in numerical computations because of round-off-errors. As discussed, round-off-errors are random and, therefore, they too represent computational noise. Hence, again, noise appears as a cause leading to Froissart doublets even for noiseless input data if finite arithmetics is used on computers.

The key issue to address, however, is whether computational noise is the only cause for Froissart doublets in the case of noiseless input data? In other words, could such doublets occur even when carrying out numerical computations with no round-off-errors using infinite precision arithmetics? If this would be feasible, then computational noise would not be the only reason for Froissart doublets for noise-free data. The answer to this question is in the affirmative, as can be readily shown using symbolic numerical computations with the input data given by integers. However, a cause for spuriousness with such infinite arithmetics might be difficult (if not impossible) to pin-point by numerical means.

If Froissart doublets are bound to happen in both finite and infinite precision numerical computations, could there be any difference, after plotting, at least in their locations relative to the unit circle? There would be a difference which could be made visible after zooming into a minuscule segment of the plot to within machine accuracy (the so-called machine ϵ) of finite arithmetics. Then Froissart doublets from finite arithmetics would be seen as being displaced a bit at the ϵ distance from their counterparts from infinite arithmetics.

Moreover, if Froissart doublets appear despite using infinite precision numerical computations, then they should also appear in purely analytical calculations with no errors whatsoever. This happens to be true for both noise-free and noise-corrupted synthesized time signals, as has recently been demonstrated by Belkić [35] who showed by purely analytical means that in both cases Froissart doublets are due to linear dependence of the system of equations for the Padé denominator polynomial.

As such, Froissart doublets need not necessarily be produced by noise alone in noise-perturbed input data, since these spurious pairs can also appear even when employing the noise-free input data. The only difference one could expect between the results with noiseless and noise-corrupted signals is that, as opposed to the former, the latter Froissart doublets will not be aligned along the unit circle on a regular and smooth arc, but rather they will chaotically bifurcate relative to that noiseless arc, depending on the level of the perturbation. Explicit numerical computations in Refs. [11, 34, 35], as well as those reported in the present book (chapters 3 and 6) have confirmed this description in detail.

In order to inter-connect some of the sub-sections from this section, it is important to ask here whether the phenomenon of Froissart doublets, as manifested through (2.226), is consistent with the concept of spuriousness viewed from the standpoint of linear dependence? This sought consistency can indeed be confirmed by the argument which runs as follows. In general, as mentioned, the discussed linear dependence in mathematical modeling means, vaguely speaking, that something was added which carries no new information. Such an addition is a linear combination of the already included information. Then, naturally, this extra information put into the considered model is redundant and leads to a spurious part in the overall output data of the analysis. This spuriousness need not be due to computational noise nor to noise in the input data, as illuminated by pure analytical means from Ref. [35]. Rather, it is a consequence of the lack of prior knowledge of the exact number K of the constituent components of the investigated time signal. In modeling via the FPT, the fact that we do not know K in advance of the analysis prompts the addition of more superfluous information through increasing the polynomial order. This is what is done in (2.226) by passing from P_K/Q_K to P_{K+m}/Q_{K+m}, which is equivalent to writing $P_{K+m}/Q_{K+m} = \tilde{P}_K A_m/[\tilde{Q}_K B_m]$, where we go beyond K for $m \geq 1$, with A_m and B_m being the two extra polynomials.

Since the unknown exact spectrum is supposed to be P_K/Q_K, by having the surpluses A_m and B_m, we effectively introduce spuriousness. In spite of this latter obstacle, the FPT manages to properly detect the K genuine components by forcing A_m and B_m to coincide with each other, $A_m = B_m \equiv C_m$, so that cancellation takes place in the Padé quotient P_{K+m}/Q_{K+m}. By such a forced confluence of A_m and B_m, the FPT successfully recovers the exact spectrum P_K/Q_K via $P_{K+m}/Q_{K+m} = \tilde{P}_K A_m/[\tilde{Q}_K B_m] = \tilde{P}_K C_m/[\tilde{Q}_K C_m] = \tilde{P}_K/\tilde{Q}_K$, as in (2.226) where, for simplicity, $\tilde{P}_K/\tilde{Q}_K$ is re-labeled as P_K/Q_K.

Moreover, the precise kind of spuriousness brought by A_m and B_m can now be determined by reference to the algorithm which produces these two extra polynomials. To this end, let us take as an example the simplest case $m = 1$, where we enlarge the two systems of equations for the p's and q's (expansion coefficients) of the P's and Q's (numerator and denominator polynomials) from K to $K+1$. By reference to sub-section 2.8.3, this automatically causes the corresponding $(K+1) \times (K+1)$ Gram determinant of the augmented system to be equal to zero, which implies that the system is singular and, hence, linearly dependent.

Therefore, the origin of spuriousness brought into the FPT by the emergence of A_1 and B_1 through augmentation $P_K/Q_K \rightarrow P_{K+1}/Q_{K+1} = \tilde{P}_K A_1/[\tilde{Q}_K B_1]$ is in the linear dependence of the two systems of $K+1$ equations for the polynomial expansion coefficients. A similar argument is extended to the case P_{K+m}/Q_{K+m} with $m > 1$. This is the essence of Froissart doublets for noiseless time signals as mediated by pole-zero cancellation in (2.226). Moreover, it is actually immaterial whether we explicitly computed poles and zeros at all, since the extra polynomials A_m and B_m cancel out any way as a whole from the Padé quotient P_{K+m}/Q_{K+m}.

As discussed, if we limit ourselves to estimation of total shape spectra alone for a given FID with the (unknown) number K, we would use the non-parametric FPT and compute $P_{K'}/Q_{K'}$ at a selected frequency grid by gradually increasing the polynomial degree K' until a saturation occurs at $K' = K + m\,(m = 1, 2, 3, ...)$, as in (2.226). The found constancy in the total shape spectra determines the exact number K of the components in the considered FID. Thus, remarkably, even the non-parametric FPT can determine with certainty at least one parameter: the total number K of the FID's components. Of course, to extract all the components themselves from the FID, the parametric FPT is needed when staying within the Padé methodology.

Given a time signal with K harmonics, we have $2K$ unknowns, K fundamental frequencies $\{\omega_k\}$ and K corresponding amplitudes $d_k\,(k = 1, 2, 3, ..., K)$. If we apply the parametric FPT, we need two systems of linear equations, each yielding K solutions for the expansion coefficients of P_K and Q_K. In this case, the problem of spectral analysis is said to be "determined" by having the $2K$ unknown frequencies and amplitudes to be computed by means of the extracted $2K$ Padé polynomial coefficients that are themselves obtained from their linearly independent equations.

In practice, with the unknown K, we can pick up a running order K' of the FPT. For example, with $K' > K$, we would have an "over-determined" spectral analysis with more equations than unknowns. For, e.g., $K' = K+m\,(m > 1)$, the system of equations for the Padé polynomial coefficients will possess m linearly dependent solutions that would lead to m Froissart doublets. Thus, in the parametric FPT, linear dependence of the system of equations for the Padé polynomial coefficients can be connected with the associated situations where the problem of spectral analysis is over-complete. However, as stated, Froissart doublets regularly also appear in the under-determined case $(K' < K)$ with fewer equations for the polynomial expansion coefficients than the unknown spectral parameters.

Linear dependence of the systems of equations for the expansion coefficients of the Padé polynomials, as a mechanism by which Froissart doublets are produced, also remains in place for noise-corrupted FIDs, as stated. Namely, no estimator is able to model noise in a fully adequate manner, but the FPT attempts to do this by solving a larger system of equations $(K + m', m' > 1)$ than is actually needed for the Padé polynomial coefficients.

Such "larger than needed" systems possess m' linearly dependent equations that lead to Froissart-type spuriousness. In practice, we have $m' > m$ where m is associated with noiseless FIDs. As mentioned, the striking difference between the appearances of Froissart doublets for noise-free and noise-corrupted FIDs is that the former spurious features are distributed in a relatively orderly, regular manner in the complex plane of the harmonic variables (z), whereas the locations of the latter pole-zero pairs are irregular and random, but still near the circumference $|z| = 1$ of the unit circle (for detailed illustrations, see Refs. [11, 34, 35] and the present chapters 3 and 6).

3

Exact quantum-mechanical, Padé-based recovery of spectral parameters

Using the FPT, we shall examine a synthesized noise-free FID in order to reconstruct the input spectral parameters. This FID, reminiscent of time signals typically encountered in MRS [88], is a sum of powers of K damped complex exponentials $\exp(i\omega_k\tau)$ with fundamental frequencies ω_k and stationary amplitudes d_k as in the geometric progression (2.193)

$$c_n = \sum_{k=1}^{K} d_k\, \mathrm{e}^{in\omega_k\tau} \qquad \mathrm{Im}(\omega_k) > 0 \qquad 0 \le n \le N-1. \tag{3.1}$$

For the illustrations in this chapter, we set $K = 25$ and $N = 1024$. Recall that ω is the angular frequency, $\omega = 2\pi\nu$, and ν is the linear frequency. The finite-rank Green function as the system's response function gives the corresponding exact complex-valued spectrum (2.162).

The physics of the representation (3.1) dictates that all the signal points c_n must have finite absolute values, $|c_n| = |\sum_{k=1}^{K} d_k\, z_k^n| < \infty$, where the complex numbers $z_k = \mathrm{e}^{i\omega_k\tau}$ represent the signal poles. This requires that all these latter input fundamental harmonics z_k, that are present in (3.1), must describe the exponentially decaying transients. In other words, each signal pole $z_k = \mathrm{e}^{i\omega_k\tau}$ must be inside the unit circle, $|z_k| < 1$

$$c_n = \sum_{k=1}^{K} d_k\, (z_k)^n \qquad z_k = \mathrm{e}^{i\omega_k\tau} \qquad |z_k| < 1 \tag{3.2}$$

where $|c_n| < \infty$. The condition, $|z_k| = |\mathrm{e}^{i\tau\mathrm{Re}(\omega_k) - \tau\mathrm{Im}(\omega_k)}| < 1$ will be fulfilled by the imposition of the constraint $\mathrm{Im}(\omega_k) > 0\, (1 \le k \le K)$, as in (3.1).

Formally, we can represent the same time signal c_n in terms of the inverses of the signal poles z_k^{-1} rather than the signal poles z_k themselves. However, in this case, as well, we ought to preserve the same sign of the imaginary fundamental frequencies, $\mathrm{Re}(\omega_k) > 0$. This can be done by writing (3.2) as the following identity

$$c_n = \sum_{k=1}^{K} d_k\, \mathrm{e}^{i\omega_k\tau} \equiv \sum_{k=1}^{K} d_k\, (u_k)^{-n} \qquad \mathrm{Re}(\omega_k) > 0 \tag{3.3}$$

where $u_k = \mathrm{e}^{-i\omega_k\tau}$ with $|u_k| > 1$, which is implied by the maintained inequality $\mathrm{Re}(\omega_k) > 0$. In other words, when the harmonic variable z^{-1} is used from the outset, the absolute value $|c_n|$ of every element c_n of the entire set $\{c_n\}$ $(0 \le n \le N-1)$ will still remain finite, $|c_n| < \infty$, but only if we redefine z_k, i.e., the old definition $z_k = \exp{(i\omega_k\tau)}$ from (3.2) cannot remain intact.

The necessary redefinition is $z_k \equiv u_k$, as directly prescribed by (3.3). Simultaneously and advantageously, by this redefinition we return to the old $z-$variable instead of introducing the new $u-$variable. Thus, if we want the fundamental harmonic variables $z_k \equiv \exp{(-i\omega_k\tau)}$ to be the new signal poles, rather than the old ones $z_k = \exp{(i\omega_k\tau)}$, such that in both cases $\mathrm{Re}(\omega_k) > 0$, then the following "new" representation of the FID must consistently be employed

$$c_n = \sum_{k=1}^{K} d_k\,(z_k)^{-n} \qquad z_k \equiv \mathrm{e}^{-i\omega_k\tau} \qquad \mathrm{Re}(\omega_k) > 0 \qquad |z_k| > 1 \tag{3.4}$$

where $|c_n| < \infty$. It is seen here that all the new signal poles $z_k = \mathrm{e}^{-i\omega_k\tau}$ lie outside the unit circle because $|z_k| = |\mathrm{e}^{-i\tau\mathrm{Re}(\omega_k)+\tau\mathrm{Im}(\omega_k)}| > 1$ for $\mathrm{Re}(\omega_k) > 0$.

Working backwards by substituting the redefined signal poles $z_k = \mathrm{e}^{-i\omega_k\tau}$ into the "new" representation $c_n = \sum_{k=1}^{K} d_k\,(z_k)^{-n}$ from (3.4) yields $c_n = \sum_{k=1}^{K} d_k \exp{(in\omega_k\tau)}$, which is the old representation from (3.2), as expected. Likewise, when the old signal poles $z_k = \mathrm{e}^{i\omega_k\tau}$ are inserted into the old representation $c_n = \sum_{k=1}^{K} d_k\,(z_k)^{n}$ from (3.2), the expression (2.193) is obtained. Hence, the two representations (3.2) and (3.4) are, in fact, the same. They are written as two formally different expressions depending whether $z_k = \mathrm{e}^{i\omega_k\tau}$ or $z_k = \mathrm{e}^{-i\omega_k\tau}$ is used for the signal poles.

When displayed as the corresponding Argand plots in the complex $z-$plane, the FID points $\{c_n\}$ in the representation (3.2) are built from the signal poles $z_k = \mathrm{e}^{i\omega_k\tau}$ that all lie inside the unit circle, $|z_k| < 1$. Alternatively, if depicted as the associated Argand plots in the complex $z^{-1}-$plane, the FID points $\{c_n\}$ in the representation (3.4) are generated using the signal poles $z_k = \mathrm{e}^{-i\omega_k\tau}$ that are all located outside the unit circle, $|z_k| > 1$. Suppose that the input signal poles $z_k = \mathrm{e}^{i\omega_k\tau}$ with $|z_k| < 1$ inside the unit circle from the representation (3.2) are all located, e.g., in the first quadrant of the $z-$plane, as in our illustrations. Then, the locations of the alternative input signal poles $z_k = \mathrm{e}^{-i\omega_k\tau}$ from the representation (3.4) will be automatically found in the fourth quadrant of the $z^{-1}-$plane outside the unit circle. This follows from the relationship $z_k^{-1} = 1/z_k$.

When dealing with the input data, the two representations (3.2) and (3.4) are a pure formality, of course. Nevertheless, the above outlines are made in order to clearly trace the origin of the two variants of the fast Padé transform, the $\mathrm{FPT}^{(+)}$ and $\mathrm{FPT}^{(-)}$, that construct the response functions using the harmonic variable z and z^{-1}, respectively. Specifically, using the same FID, the $\mathrm{FPT}^{(+)}$ and $\mathrm{FPT}^{(-)}$ are set to reconstruct the poles z_k^{+} and z_k^{-} as the Padé approximations to the corresponding input data z_k and z_k^{-1} via $z_k^{+} \approx z_k$

and $z_k^- \approx z_k^{-1}$, respectively. Using the Padé-computed signal poles $z_k^{\pm}$ and the corresponding amplitudes $d_k^{\pm}$, one can build the two Padé time signals, one in the FPT$^{(+)}$

$$c_n^+ = \sum_{k=1}^{K} d_k^+ \, (z_k^+)^n \qquad z_k^+ = \mathrm{e}^{i\omega_k^+ \tau} \qquad \mathrm{Re}(\omega_k^+) > 0 \qquad |z_k^+| < 1 \tag{3.5}$$

and the other in the FPT$^{(-)}$

$$c_n^- = \sum_{k=1}^{K} d_k^- \, (z_k^-)^n \qquad z_k^- = \mathrm{e}^{-i\omega_k^- \tau} \qquad \mathrm{Re}(\omega_k^-) > 0 \qquad |z_k^-| > 1 \tag{3.6}$$

where $|c_n^{\pm}| < \infty$. A joint expression for (3.2) and (3.5) was given earlier in (2.193). The representations (3.5) and (3.6) are the Padé approximations to the same input c_n from the two formally different, but otherwise identical expressions (3.2) and (3.4), respectively

$$c_n^+ \approx c_n \qquad c_n^- \approx c_n\,. \tag{3.7}$$

If the input time signal c_n has precisely K harmonics, then for the noiseless case, the two variants of the FPT will exactly reconstruct the input FID via

$$c_n^+ = c_n^- = c_n \qquad \text{(noiseless FID)}. \tag{3.8}$$

On the other hand, when the input signal, containing exactly K resonances, is noise-corrupted, the time-domain residuals $r_n^{\pm}$ obtained by the FPT$^{(+)}$ and FPT$^{(-)}$ via

$$r_n^+ \equiv c_n^+ - c_n \neq 0 \qquad r_n^- \equiv c_n^- - c_n \neq 0 \qquad \text{(noisy FID)} \tag{3.9}$$

will be different ($r_n^+ \neq r_n^-$) and shall represent an approximate reconstruction of the input noise.

In order to formulate the proper theoretical standards, it is vital to reconstruct the spectral parameters from noiseless FIDs, as we shall do in this chapter. These allow validity assessments to be made concerning the designed estimator, such as the FPT in the present case. We re-emphasize that the quantification problem in MRS can be kept under complete control throughout the analysis exclusively via synthesized noise-free FIDs. Thereby, everything that is expected from the theory can be known exactly, and then the remaining part of the investigation can be geared to testing the appropriateness of the employed algorithms. In later steps, the idealized FIDs can be corrupted with random Gaussian noise which can be controlled. Solving the corresponding quantification problem in this latter case can then be used to get a handle on actual data from *in vivo* time signals encoded via MRS.

In noise-free as well as noise-corrupted synthesized FIDs, the spectral parameters are recognized by detecting their stability or constancy with gradually increasing partial signal length $N/M\,(M > 1)$. This expected constancy of the reconstructed spectral parameters is predicted by, e.g., the Hazi-Taylor stabilization method from resonant scattering theory [140]. According to that method, the computed eigen-energies will have constant values when the dimension of the finite-rank Hamiltonian matrix is increased by systematically augmenting the size of the box in which the examined physical system is enclosed. This procedure can be easily implemented in the FPT in order to separate the physical portion of the FID from its noise. As a consequence, the stable transients are distinguished from those transients that are unstable. The former and the latter are identified as true (physical) and spurious (non-physical, noisy or noise-like), respectively. This procedure has been validated using the FPT in ICRMS [13, 14] and NMR [15].

As will be presented in chapter 7, the FPT has been tested using encoded FIDs from MRS. Therein, the superior resolving power and convergence rate of the FPT for total shape spectra are established in relation to the FFT. Comparative analyses within the FPT and FFT are seen to be important when the FFT spectra with the full FID are of high quality so that they could be considered as the gold standard in their own category of envelope spectra. This was the case in the studies from Refs. [8, 9] and [18]-[23] using FIDs of high SNR as encoded via MRS [141] at magnetic field strengths of 4T and 7T from the occipital region of healthy human brain. Among the major conclusions that emerged was that the FPT attains at least two times superior resolution compared to the FFT for the same fraction N/M of the full signal length N. Of particular note, this was true even for $M = 2$ when $N/2$ signal points were used in both the FPT and FFT. These findings essentially mean that the FPT can reach the same resolution as the FFT by using only the first half $N/2$ of the full signal length N. Such a remarkable superiority of the FPT relative to the FFT for total shape spectra computed from time signals encoded with high SNR is expected to be even greater for synthesized FIDs.

Including the envelope spectra in the present chapter helps to link the results reported here with those from chapter 7 as well as from Refs. [8, 9]. This, in turn, would allow another consistency check of our previous conclusions. It should be emphasized, however, that it is necessary to proceed beyond comparisons between the Padé and Fourier total shape spectra. The results of quantification need to be explicitly reported. These results of the FPT for the numerical values of the spectral parameters reconstructed from *in vivo* MRS FIDs were implicitly present in the studies of the total shape spectra. This was the case because the reported total shape spectra [8, 9] were built using the Heaviside partial fraction expansions $\sum_{k=1}^{K} d_k^{\pm} z^{\pm 1}/(z^{\pm 1} - z_k^{\pm 1})$ with $z^{\pm 1} = \exp(\pm i\omega\tau)$ and $z_k^{\pm 1} = \exp(\pm i\omega_k^{\pm}\tau)$. These fractions can only be computed after reconstruction of the complex frequencies $\{\omega_k^{\pm}\}$ and amplitudes $\{d_k^{\pm}\}$ for all the physical resonances. However, the numerical values of the spectral parameters were not explicitly reported in our examination of total shape spectra [8, 9], since the focus was on a comparison of the convergence rates of the FPT and FFT. Such comparisons are only possible using total shape spectra, because these are the only spectra that can be obtained by the FFT. In a total shape spectrum generated through the FPT via the Heaviside representation, the sum is carried out over all the spectral parameters. Thus, the information on the inherently performed quantification is not readily apparent. As a check, the total shape spectrum in the FPT could be computed directly, as well, via a frequency-by-frequency evaluation of the defining polynomial quotient without prior reconstructions of the spectral parameters.

In this chapter, we go beyond presenting the total shape spectra, by giving the explicit numerical values of the reconstructed spectral parameters. In this respect, the most stringent criterion is the exact retrieval of the known complex frequencies and amplitudes. This is possible for a synthesized FID. As mentioned, such time signals are considered in this chapter by choosing their spectral parameters to be consistent with those FIDs that are encoded via MRS. The quantification problems for these synthesized FIDs will presently be solved using the two variants of the FPT, that are initially defined inside and outside the unit circle, namely the $\mathrm{FPT}^{(+)}$ and $\mathrm{FPT}^{(-)}$, respectively.

Strictly speaking, the theory foresees that for synthesized noise-free input FIDs, these two versions of the FPT should generate the same results once convergence has been attained in both variants for all the spectral parameters, as per (2.187). Such a hypothesis is tested herein. This testing is important in light of the fact that with noisy FIDs encoded via MRS, as those studied in Refs. [8, 9], the use of both the $\mathrm{FPT}^{(+)}$ and $\mathrm{FPT}^{(-)}$ is needed as an intrinsic cross-validation of estimations. As per this cross-validation, the only acceptable reconstructions will be those found by both the $\mathrm{FPT}^{(+)}$ and $\mathrm{FPT}^{(-)}$.

It should be pointed out that these two versions of the FPT have different algebraic structure. These are designed to work in the two complementary regions of the complex plane for the harmonic variable z. The $\mathrm{FPT}^{(+)}$ and $\mathrm{FPT}^{(-)}$ operate in the first and the fourth quadrant of the complex $z-$plane, respectively. These two opposite regions are very different with respect to the expected locations of spurious poles. This is the case for those from simulations as well as for encoded FIDs. Thus, for the $\mathrm{FPT}^{(+)}$, we expect that there will be a clear separation of the physical poles (inside the unit circle, $|z| < 1$) from the non-physical ones (outside the unit circle, $|z| > 1$). This separation should facilitate robust and stable estimations by the $\mathrm{FPT}^{(+)}$ whenever there is noise in the examined FIDs. This postulate should be verified, particularly since the $\mathrm{FPT}^{(+)}$ must carry out the numerically difficult and ill-conditioned task of analytical continuation by numerical means by inducing convergence into the input Maclaurin sum (2.162) which diverges inside the unit circle when N is increased.

On the other hand, the $\mathrm{FPT}^{(-)}$ accelerates the already convergent input sum (2.162). This is an easier task from the computational vantage point compared to analytical continuation. Instead, the $\mathrm{FPT}^{(-)}$ has another challenge, namely to separate the genuine from the spurious poles that are intermixed within the same region outside the unit circle.

Viewed together, the major properties of the $\text{FPT}^{(+)}$ and $\text{FPT}^{(-)}$ are complementary. Thus, each of these two versions of the FPT must independently arrive at a trade-off in order to perform optimally. The outlined differences between the $\text{FPT}^{(+)}$ and $\text{FPT}^{(-)}$ justify the use of both versions of the FPT even for noiseless time signals. In such a case, the genuine resonances are known exactly, so that spurious poles can be identifiable with fidelity. A favorable result from the presently-designed testing on a typical synthesized FID will provide an impetus to apply the FPT to distinguish genuine and spurious resonances in encoded time signals from MRS.

Since our aim is quantification of FIDs from MRS, the synthesized time signals are chosen to possess the main physical features of importance to clinical diagnostics. We achieved this similarity by using the numerical values of the complex fundamental frequencies and the amplitudes in the synthesized FID to be reminiscent of the tabulated data as reported in the MRS literature. Furthermore, the FID generated theoretically with these spectral parameters yields absorption total shape spectra similar to a highly resolved spectrum computed from an FID encoded via MRS from a healthy human brain at short echo time (20 ms) at the magnetic field strength B_0 =1.5T [88].

The findings and the detailed interpretation from this chapter are presented in seven sections 3.1–3.7. These results are obtained using a sequence of partial signal lengths N/M ($M = 2 - 32$) as well as the full length N with $M = 1$.

- *The first* section 3.1 presents the tabular input and reconstructed data of all the spectral parameters. It also gives graphic illustrations of the input data and shows the superiority of parametric signal processing in MRS over non-parametric estimations. This includes 4 sub-sections. Three sub-sections are with tabular data: 3.1.1 for the input data (Table 3.1) as well as sub-sections 3.1.2 and 3.1.3 for convergence of the numerical values of all the spectral parameters reconstructed by the $\text{FPT}^{(-)}$ (Tables 3.2 and 3.3). Sub-section 3.1.4 gives the physical interpretations of the graphs of the input data (Fig. 3.1). It also shows the clinical advantages of Padé reconstruction of the component shape spectra and the associated metabolite concentrations that are completely lacking from the corresponding total shape spectra as the only outcome of the Fourier analysis (Fig. 3.2).
- *The second* section 3.2 has sub-section 3.2.1 with the absorption total shape spectra (Fig. 3.3) and sub-section 3.2.2 comparing the convergence rates of these spectra in the $\text{FPT}^{(-)}$ and FFT for varying N/M (Figs. 3.4 and 3.5).
- *The third* section 3.3 continues the analyses in sub-sections 3.3.1 and 3.3.2 for the residual or error spectra (Figs. 3.6 and 3.7) and sub-sections 3.3.3 and 3.3.4 with the consecutive difference spectra (Figs. 3.8 and 3.9). Section 3.3 is a part of the intrinsic error analysis in the $\text{FPT}^{(+)}$ and $\text{FPT}^{(-)}$ without recourse to the FFT or to any other estimator.
- *The fourth* section 3.4 contains the absorption component and total shape spectra (Figs. 3.10 – 3.12). The absorption component shape spectra display each individual resonance. The sum of all these elementary, constituent spec-

tra produces the corresponding absorption total shape spectrum. In Fig. 3.10 the numbers and acronyms are also indicated on a map of MR-detectable metabolites from the human brain. Figures 3.11 and 3.12 address the inadequacy of the usual practice in MRS to rely solely upon the residual spectra in error analysis.

• *The fifth* section 3.5 is presented through 6 sub-sections. These are: subsection 3.5.1 for the distributions of complex frequencies and amplitudes, as well as poles, shown via the harmonic variables for the constituent resonances in the associated complex planes as reconstructed by the $\mathrm{FPT}^{(+)}$ after full convergence (Fig. 3.13), sub-section 3.5.2 which is the same as sub-subsection 3.5.1, except for using the $\mathrm{FPT}^{(-)}$ (Fig. 3.14), sub-sections 3.5.3 and 3.5.4 with the convergence of the reconstructed complex frequencies (Figs. 3.15 and 3.16) and sub-sections 3.5.5 and 3.5.6 for the convergence of the absolute values of amplitudes (Figs. 3.17 and 3.18).

• *The sixth* section 3.6 contains a preview on Froissart doublets (Fig. 3.19) in the $\mathrm{FPT}^{(\pm)}$. Here, the emphasis is in illustrating the main aspect of the exact Froissart-based separation of genuine from spurious resonances in an important part of the full Nyquist range with all the MR-detectable metabolites of the healthy human brain. More details on Froissart doublets are given in chapters 5 and 6.

• *The seventh* section 3.7 provides an in-depth overview of the results from this chapter emphasizing the key importance of exact quantification for magnetic resonance spectroscopy.

3.1 Input data (tabular & graphic) and reconstructed tabular data

3.1.1 Input tabular data for the spectral parameters of 25 resonances

Table 3.1 presents the exact 4-digit input data of all the spectral parameters for 25 complex attenuated exponentials that constitute the presently synthesized FID from (3.1). These parameters are the fundamental frequencies and the associated amplitudes. The real-valued frequencies $\mathrm{Re}(\nu_k)$ are also termed chemical shifts. The quantities $|d_k|$ represent the absolute values or moduli of the corresponding amplitudes d_k. All the individual phases ϕ_k of the amplitudes d_k are chosen to have zero values, so that $d_k = |d_k| \exp{(i\phi_k)} = |d_k|$. These zero values for the phases are customary for spectral analysis of synthesized time signals in MRS [142]. Clearly, however, this is not a restriction for the FPT which can process FIDs with non-zero phases ϕ_k. The decay time constants $\{\lambda_k\}$ that are proportional to the transverse relaxation times T_{2k} are not explicitly given in Table 3.1. These can be deduced from the inverses of the listed imaginary frequency $\{\mathrm{Im}(\nu_k)\}$ via the relationship

$$\lambda_k(\mathrm{s}) = \frac{1}{\mathrm{Im}(\nu_k)(\mathrm{ppm})\nu_L(\mathrm{Hz})} \tag{3.10}$$

where ν_L is the Larmor precession frequency, which is equal to 63.864 MHz at B_0 =1.5T. As per convention in MRS, the frequencies $\mathrm{Re}(\nu_k)$ and $\mathrm{Im}(\nu_k)$ are reported in the dimensionless units of parts per million (ppm) [1], whereas λ_k is expressed in seconds (s) and $|d_k|$ in arbitrary units (au). Furthermore, chemical shifts $\mathrm{Re}(\nu_k)$ given in ppm and hertz (Hz) are related by

$$\mathrm{Re}(\nu_k)(\mathrm{ppm}) = 4.68 - \frac{1}{\nu_L}\mathrm{Re}(\nu_k)(\mathrm{Hz}) \tag{3.11}$$

where 4.68 ppm is the resonance frequency of water. In the case of imaginary frequencies $\mathrm{Im}(\nu_k)$ expressed in ppm and Hz, the following relationship exists

$$\mathrm{Im}(\nu_k)(\mathrm{ppm}) = \frac{1}{\nu_L}\mathrm{Im}(\nu_k)(\mathrm{Hz}). \tag{3.12}$$

These fundamental spectral parameters directly yield the peak area S_k for the kth resonance where $1 \le k \le K$. Furthermore, each peak area S_k is proportional to the dimensionless concentration $\mathrm{C}_k^{\mathrm{met}}$ of the kth metabolite, $\mathrm{C}_k^{\mathrm{met}} \propto S_k$. In this relationship, the proper dimension of the concentration $\mathrm{C}_k^{\mathrm{met}}$ can be introduced by fixing the constant of proportionality as an overall multiplying factor. Such a proportionality constant is chosen to be the concentration of a reference metabolite, $\mathrm{C}_{\mathrm{ref}}$, expressed in units of mM/g per wet weight (ww) of tissue. Various numerical values for $\mathrm{C}_{\mathrm{ref}}$ can be selected for different tissues. For example, for FIDs encoded via MRS in occipital grey matter of the healthy human brain, the reference metabolite is usually NAA (nitrogen acetyl aspartate) or H_2O (water), in which case the usual concentrations are [88, 143]

$$\mathrm{C}_{\mathrm{ref}}(\mathrm{NAA}) = 6\,\mathrm{mM/g\ \ ww} \qquad \mathrm{C}_{\mathrm{ref}}(\mathrm{H_2O}) = 85\,\mathrm{mM/g\ \ ww}\,. \tag{3.13}$$

For the selected $\mathrm{C}_{\mathrm{ref}}$, the concentration $\mathrm{C}_k^{\mathrm{met}}$ of the kth metabolite M_k is deduced from the expression

$$\mathrm{C}_k^{\mathrm{met}} = S_k \mathrm{C}_{\mathrm{ref}}. \tag{3.14}$$

For the Lorentzian line-shape (2.184), the surface S_k is given by the area of a rectangle, $S_k = a_k h_k$, where a_k is proportional to the full-width at the half-maximum (FWHM) of the kth peak, $\mathrm{FWHM}_k = \mathrm{Im}(\omega_k)/2$. Here, h_k is the corresponding height defined as $h_k = |d_k|/[\mathrm{Im}(\omega_k)]$ according to (2.186).

[1]Frequency is expressed in dimensionless units of ppm so that a given resonance always appears at the same position, irrespective of the value of the magnetic field strength B_0.

Therefore, $S_k = |d_k|/2$, which states the well-known fact that the Lorentzian peak area S_k is independent of the resonance width. This reduces (3.14) to the following simple expression for obtaining metabolite concentrations

$$\mathrm{C}_k = \frac{|d_k|}{2}\mathrm{C}_{\mathrm{ref}} \qquad\qquad \mathrm{C}_k \equiv \mathrm{C}_k^{\mathrm{met}}. \tag{3.15}$$

The physical meaning of the concentration $\mathrm{C}_k^{\mathrm{met}}$ is in representing the abundance of the kth metabolite in the scanned tissue. On the other hand, encoding in proton MRS proceeds via detection of time signals emanating from protons that are bound to various chemical compounds/molecules in the examined tissue. Therefore, the concentration $\mathrm{C}_k^{\mathrm{met}}$ is proportional to the number of protons underneath the kth metabolite peak. In various implementations of the existing sequence encodings in clinical scanners for MRS, due to several realistic experimental limitations (echo time delay, TE $\neq 0$, finite repetition times, TR $\neq \infty$, etc.), certain empirical modifications to the metabolite concentrations are usually introduced via division of the peak areas S_k by a product of exponentials of the magnetization relaxation type, $S_k/\{\exp(-\mathrm{TE}/T_{2k})[1-\exp(-\mathrm{TR}/T_{1k})]\}$. Here, T_{1k} and T_{2k} are the kth metabolite relaxation times [143]. In our computations of metabolite concentrations, we shall use (3.15) as it stands, with no recourse to the latter empirical corrections.

The presentation of the results and their analysis is facilitated by referring to resonances according to their numbers k (also denoted by n_k° in the tables) where $1 \leq k \leq K\,(K = 25)$. These numbers are ordered so that the 1st resonance is a lipid and the 25th is water. In proton MRS, the following abbreviations are used for some of the major MR-detectable metabolites, e.g., H_2O, Cho (choline), Cr (creatine), Glu (glutamine), Gln (glutamate), m-Ins (myo-inositol), NAA, Ala (alanine), Lip (lipid), etc., as listed also in Table 3.1. High quality *in vivo* time signals encoded via proton MRS are obtained with water suppression. In such FIDs, the residual water content does not seriously disturb the determination of concentrations of the neighboring metabolites. In order to closely mimic this situation, we set up the presently synthesized FID to contain the residual water metabolite. This resonance will have a small and broad shape determined by the chosen values of the parameters $|d_{25}|$ and $\mathrm{Im}(\nu_{25})$ seen at the chemical shift 4.68 ppm on the 25th line in Table 3.1.

Upon inspection of the values of all the other fundamental parameters for the synthesized time signal from Table 3.1, it can be seen that the corresponding exact absorption spectrum, defined as the real part of the associated complex-valued spectrum, possesses a variety of structures, including isolated, overlapped, tightly overlapped and nearly degenerate resonances. The exact absorption component shape spectra for the clinically most informative frequencies (0 – 5 ppm), will have isolated, but closely-lying resonances, that are associated with the following 10 peaks: $k = 8, 9, 10, 13, 18, 19, 20, 21, 24, 25$ located at chemical shifts 2.261 ppm, 2.411 ppm, 2.519 ppm, 2.855 ppm, 3.481 ppm, 3.584 ppm, 3.694 ppm, 3.803 ppm, 4.271 ppm, 4.680 ppm, respectively.

TABLE 3.1
Four-digit accurate numerical values for all the input spectral parameters: the real $\mathrm{Re}(\nu_k)$ and imaginary $\mathrm{Im}(\nu_k)$ parts of the complex frequencies ν_k, and the absolute values $|d_k|$ of the complex amplitudes d_k of 25 damped complex exponentials from the synthesized time signal (3.1) similar to a short echo time (~20 ms) encoded FID via MRS at the magnetic field strength $B_0 = 1.5\mathrm{T}$ from a healthy human brain [88]. Every phase $\{\phi_k\}$ of the amplitudes is set to zero (0.000), i.e., each d_k is chosen as purely real, $d_k = |d_k| \exp(i\phi_k) = |d_k|$. The letter M_k denotes the kth metabolite.

INPUT DATA for ALL SPECTRAL PARAMETERS of a SYNTHESIZED TIME SIGNAL or FID

n_k^o	$\mathrm{Re}(\nu_k)$ (ppm)	$\mathrm{Im}(\nu_k)$ (ppm)	$\|d_k\|$ (au)	M_k
1	0.985	0.180	0.122	Lip
2	1.112	0.257	0.161	Lip
3	1.548	0.172	0.135	Lip
4	1.689	0.118	0.034	Lip
5	1.959	0.062	0.056	Gaba
6	2.065	0.031	0.171	NAA
7	2.145	0.050	0.116	NAAG
8	2.261	0.062	0.092	Gaba
9	2.411	0.062	0.085	Glu
10	2.519	0.036	0.037	Gln
11	2.675	0.033	0.008	Asp
12	2.676	0.062	0.063	NAA
13	2.855	0.016	0.005	Asp
14	3.009	0.064	0.065	Cr
15	3.067	0.036	0.101	PCr
16	3.239	0.050	0.096	Cho
17	3.301	0.064	0.065	PCho
18	3.481	0.031	0.011	Tau
19	3.584	0.028	0.036	m−Ins
20	3.694	0.036	0.041	Glu
21	3.803	0.024	0.031	Gln
22	3.944	0.042	0.068	Cr
23	3.965	0.062	0.013	PCr
24	4.271	0.055	0.016	PCho
25	4.680	0.108	0.057	Water

In addition, overlapped resonances will appear corresponding to the following 11 peaks: $k = 1, 2$ (0.985 ppm, 1.112 ppm), $k = 3, 4$ (1.548 ppm, 1.689 ppm), $k = 5, 6, 7$ (1.959 ppm, 2.065 ppm, 2.145 ppm), $k = 14, 15$ (3.009 ppm, 3.067 ppm) and $k = 16, 17$ (3.239 ppm, 3.301 ppm). The very closely overlapped resonances will appear as the following 2 peaks: $k = 22, 23$ (3.944 ppm, 3.965 ppm). These latter resonances are separated from each other by 0.021 ppm. Finally, there will be almost degenerate resonances that are comprised of 2 peaks $k = 11, 12$ (2.675 ppm, 2.676 ppm) separated by a mere 0.001 ppm. In the absorption total shape spectrum these two nearly coincident peaks will be completely unresolved and will appear as a single resonance.

It is expected that in the $k = 11 \,\&\, 12$ peaks of near degeneracy all the conventional fitting techniques would fail to detect the smaller peak which is $k = 11$. In fact, there would be no justifiable reason to initialize fitting two peaks for a resonance which appears to be a single structure. Even if a fitting procedure were to begin with the two pre-assigned modeling resonances, disregarding the appearance of a bell-shaped single peak, there would be no justifiable reason for not choosing even a larger number, e.g., 3 or 4 or more small peaks below the dominant resonance $k = 12$ in the $k = 11 \,\&\, 12$ peaks.

This illustrates the fundamental ambiguity of fittings from MRS. This includes such programs as VARPRO, AMARES, LCModel, etc. Specifically, a key weakness of these fittings is their non-uniqueness in handling the inverse quantification problem in MRS. This near-degeneracy $k = 11 \,\&\, 12$ in the presently synthesized FID substantially increases the overall challenge for spectral analysis by any estimator, including the FPT.

3.1.2 Numerical values of the reconstructed spectral parameters at six signal lengths, N/M ($N = 1024, M = 1 - 32$)

Table 3.2 presents the detailed convergence rates of the numerical values of the complex frequencies and amplitudes for the reconstructed resonances by the $\text{FPT}^{(-)}$ using 6 partial signal lengths ($N/32 = 32, N/16 = 64, N/8 = 128, N/4 = 256, N/2 = 512$) as well as the full FID ($N = 1024$). In panels (i), (ii) and (iii) of Table 3.2, the spectral parameters of the detected 10, 14 and 20 resonances are shown at $N/32 = 32, N/16 = 64$ and $N/8 = 128$, respectively. Clearly, these latter findings are approximate values for the corresponding exact input data for the parameters. This occurs because the number of signal points ($N/M \leq 128$) is not sufficient. In contrast, panels (iv), (v) and (vi) of Table 3.2 display the spectral parameters found at $N/4 = 256, N/2 = 512$ and $N = 1024$, respectively. These latter results should be compared with the corresponding input data from Table 3.1. It thereby can be seen that from, e.g., panel (iv) of Table 3.2 for a quarter ($N/4 = 256$) of the full signal length N that the $\text{FPT}^{(-)}$ has retrieved the entire set of 25 resonances with all their exact values for all of the spectral parameters. Furthermore, it is remarkable that these reconstructed spectral parameters remain totally unchanged, even after attaining full convergence, when further signal points

TABLE 3.2
Convergence of the spectral parameters in the $\mathrm{FPT}^{(-)}$ for signal lengths $N/M\,(N = 1024, M = 1 - 32)$.

CONVERGENCE of SPECTRAL PARAMETERS in $\mathrm{FPT}^{(-)}$; FID LENGTH: N/M, N = 1024, M = 1 – 32

(i) N/32 = 32 (15 Missing Peaks)

n_k^o	$\mathrm{Re}(\nu_k^-)$ (ppm)	$\mathrm{Im}(\nu_k^-)$ (ppm)	$\lvert d_k^- \rvert$ (au)
1	1.010	0.206	0.223
–	–	–	–
3	1.517	0.420	0.254
4	1.643	0.097	0.038
–	–	–	–
6	2.064	0.069	0.339
–	–	–	–
–	–	–	–
–	–	–	–
–	–	–	–
–	–	–	–
12	2.637	0.274	0.421
–	–	–	–
–	–	–	–
15	3.055	0.139	0.417
–	–	–	–
17	3.376	0.038	0.018
–	–	–	–
–	–	–	–
–	–	–	–
–	–	–	–
–	–	–	–
23	3.968	0.114	0.222
24	4.093	0.083	0.080
25	4.681	0.106	0.056

(ii) N/16 = 64 (11 Missing Peaks)

n_k^o	$\mathrm{Re}(\nu_k^-)$ (ppm)	$\mathrm{Im}(\nu_k^-)$ (ppm)	$\lvert d_k^- \rvert$ (au)
1	0.989	0.180	0.130
2	1.121	0.241	0.148
3	1.562	0.206	0.195
–	–	–	–
5	2.030	0.012	0.026
6	2.055	0.071	0.376
–	–	–	–
–	–	–	–
9	2.473	0.193	0.313
10	2.590	0.054	0.062
–	–	–	–
–	–	–	–
–	–	–	–
–	–	–	–
15	3.057	0.051	0.160
16	3.237	0.071	0.177
–	–	–	–
–	–	–	–
19	3.565	0.045	0.035
–	–	–	–
21	3.776	0.068	0.075
22	3.941	0.048	0.087
–	–	–	–
24	4.269	0.054	0.016
25	4.680	0.108	0.057

(iii) N/8 = 128 (5 Missing Peaks)

n_k^o	$\mathrm{Re}(\nu_k^-)$ (ppm)	$\mathrm{Im}(\nu_k^-)$ (ppm)	$\lvert d_k^- \rvert$ (au)
1	0.985	0.180	0.122
2	1.112	0.256	0.160
3	1.545	0.169	0.123
4	1.704	0.134	0.051
5	2.012	0.072	0.331
6	2.045	0.042	0.331
7	2.157	0.037	0.041
–	–	–	–
9	2.351	0.015	0.008
10	2.507	0.129	0.196
–	–	–	–
12	2.655	0.048	0.059
13	2.809	0.018	0.001
–	–	–	–
15	3.072	0.053	0.183
16	3.231	0.078	0.207
17	3.367	0.035	0.011
–	–	–	–
19	3.588	0.022	0.025
20	3.699	0.041	0.047
21	3.803	0.027	0.036
22	3.944	0.046	0.084
–	–	–	–
24	4.271	0.055	0.016
25	4.680	0.108	0.057

(iv) N/4 = 256 (Converged)

n_k^o	$\mathrm{Re}(\nu_k^-)$ (ppm)	$\mathrm{Im}(\nu_k^-)$ (ppm)	$\lvert d_k^- \rvert$ (au)
1	0.985	0.180	0.122
2	1.112	0.257	0.161
3	1.548	0.172	0.135
4	1.689	0.118	0.034
5	1.959	0.062	0.056
6	2.065	0.031	0.171
7	2.145	0.050	0.116
8	2.261	0.062	0.092
9	2.411	0.062	0.085
10	2.519	0.036	0.037
11	2.675	0.033	0.008
12	2.676	0.062	0.063
13	2.855	0.016	0.005
14	3.009	0.064	0.065
15	3.067	0.036	0.101
16	3.239	0.050	0.096
17	3.301	0.064	0.065
18	3.481	0.031	0.011
19	3.584	0.028	0.036
20	3.694	0.036	0.041
21	3.803	0.024	0.031
22	3.944	0.042	0.068
23	3.965	0.062	0.013
24	4.271	0.055	0.016
25	4.680	0.108	0.057

(v) N/2 = 512 (Converged)

n_k^o	$\mathrm{Re}(\nu_k^-)$ (ppm)	$\mathrm{Im}(\nu_k^-)$ (ppm)	$\lvert d_k^- \rvert$ (au)
1	0.985	0.180	0.122
2	1.112	0.257	0.161
3	1.548	0.172	0.135
4	1.689	0.118	0.034
5	1.959	0.062	0.056
6	2.065	0.031	0.171
7	2.145	0.050	0.116
8	2.261	0.062	0.092
9	2.411	0.062	0.085
10	2.519	0.036	0.037
11	2.675	0.033	0.008
12	2.676	0.062	0.063
13	2.855	0.016	0.005
14	3.009	0.064	0.065
15	3.067	0.036	0.101
16	3.239	0.050	0.096
17	3.301	0.064	0.065
18	3.481	0.031	0.011
19	3.584	0.028	0.036
20	3.694	0.036	0.041
21	3.803	0.024	0.031
22	3.944	0.042	0.068
23	3.965	0.062	0.013
24	4.271	0.055	0.016
25	4.680	0.108	0.057

(vi) N = 1024 (Converged)

n_k^o	$\mathrm{Re}(\nu_k^-)$ (ppm)	$\mathrm{Im}(\nu_k^-)$ (ppm)	$\lvert d_k^- \rvert$ (au)
1	0.985	0.180	0.122
2	1.112	0.257	0.161
3	1.548	0.172	0.135
4	1.689	0.118	0.034
5	1.959	0.062	0.056
6	2.065	0.031	0.171
7	2.145	0.050	0.116
8	2.261	0.062	0.092
9	2.411	0.062	0.085
10	2.519	0.036	0.037
11	2.675	0.033	0.008
12	2.676	0.062	0.063
13	2.855	0.016	0.005
14	3.009	0.064	0.065
15	3.067	0.036	0.101
16	3.239	0.050	0.096
17	3.301	0.064	0.065
18	3.481	0.031	0.011
19	3.584	0.028	0.036
20	3.694	0.036	0.041
21	3.803	0.024	0.031
22	3.944	0.042	0.068
23	3.965	0.062	0.013
24	4.271	0.055	0.016
25	4.680	0.108	0.057

are included beyond $N/4$. This is shown on panels (v) and (vi) for $N/2 = 512$ and $N = 1024$, respectively. This unique feature of the Padé polynomial quotient, e.g., P_K^-/Q_K^- and the like, is related to pole-zero cancellations or Froissart doublets, as analyzed in section 3.6. A veritable signature of reconstruction of the true number K of resonances is provided by the attained convergence of the spectral parameters. If one continues to increase K in the ratio P_K^-/Q_K^- even after convergence has been achieved, the converged results for the Padé quotient in the $\mathrm{FPT}^{(-)}$ will not change. This is the case because the new poles from Q_{K+m}^- will be exactly the same as the new zeros of P_{K+m}^-. In such a case, the pole-zero cancellation occurs in the Padé quotient yielding $P_{K+m}^-/Q_{K+m}^- = P_K^-/Q_K^-$, as per (2.226), where m is any positive integer. We have verified that this holds true in the present computation, as seen on panels (iv)–(vi) of Table 3.2. Synergistically, the same computation demonstrates that all the amplitudes $\{d_k^-\}$ associated with the poles from the Froissart doublets are identical to zero, as in (2.192). Theoretically, the strict algebraic condition $2K = N$ implies that only 100 FID points should be sufficient for the $\mathrm{FPT}^{(-)}$ to exactly reconstruct all the 25 unknown complex frequencies and 25 complex amplitudes. However, panel (iv) of Table 3.2 reveals that full convergence is attained with the first 256 time signal points. This occurs because the $\mathrm{FPT}^{(-)}$ produces genuine as well as spurious resonances. In other words, in the polynomial quotient P_K^-/Q_K^-, spurious poles and spurious zeros from the denominator and the numerator, respectively, come in pairs as Froissart doublets. Therefore, they cancel each other. Thus, each addition of more time signal points yields new Froissart doublets. However, another process takes place at the same time, namely stabilization of the values of the reconstructed physical spectral parameters. Ultimately, saturation occurs when the total number of genuine resonances stop fluctuating as described in (2.226). At that point, all the spectral parameters become constant for varying partial signal length. This process of stabilization illustrates how the $\mathrm{FPT}^{(-)}$ determines, with certainty, the true total number K of genuine resonances. For the time signal which is currently under study, this stabilization actually takes place by using less than a quarter $N/4 = 256$ of the full FID. In fact, this occurs by exhausting the first 210 signal points (not shown in Table 3.2 where all the signal lengths are of the form 2^s with s being a positive integer). This would give $K = 105$, but the elimination of all the Froissart doublets and other extraneous poles finally yields the exact total number $K = 25$ of the genuine resonances in a Froissart-cleaned spectrum.

3.1.3 Numerical values of the reconstructed spectral parameters near full convergence for 3 partial signal lengths $N_P = 180, 220, 260$

Table 3.3 for the reconstructed spectral parameters from the $\mathrm{FPT}^{(+)}$ (left column) and $\mathrm{FPT}^{(-)}$ (right column) focuses upon a narrow convergence range

TABLE 3.3
Convergence of the spectral parameters in the $\mathrm{FPT}^{(+)}$ (left) and $\mathrm{FPT}^{(-)}$ (right) near full convergence for signal lengths $N_P = 180, 220, 260$.

FAST PADE TRANSFORM: INSIDE and OUTSIDE the UNIT CIRCLE, $\mathrm{FPT}^{(+)}$ (Left) and $\mathrm{FPT}^{(-)}$ (Right) ; N_P = 180, 220, 260

(i) N_P = 180 (1 Missing Peak)

n_k^o	$\mathrm{Re}(\nu_k^+)$ (ppm)	$\mathrm{Im}(\nu_k^+)$ (ppm)	$\lvert d_k^+ \rvert$ (au)
1	0.985	0.180	0.122
2	1.112	0.257	0.161
3	1.548	0.172	0.135
4	1.689	0.118	0.034
5	1.959	0.062	0.056
6	2.065	0.031	0.171
7	2.145	0.050	0.115
8	2.261	0.062	0.094
9	2.411	0.063	0.089
10	2.518	0.036	0.037
–	–	–	–
12	2.675	0.054	0.066
13	2.855	0.014	0.004
14	3.013	0.054	0.058
15	3.065	0.037	0.113
16	3.242	0.053	0.120
17	3.301	0.052	0.045
18	3.482	0.027	0.009
19	3.585	0.028	0.036
20	3.694	0.036	0.042
21	3.803	0.024	0.031
22	3.945	0.042	0.072
23	3.970	0.057	0.010
24	4.271	0.055	0.016
25	4.680	0.108	0.057

(iv) N_P = 180 (1 Missing Peak)

n_k^o	$\mathrm{Re}(\nu_k^-)$ (ppm)	$\mathrm{Im}(\nu_k^-)$ (ppm)	$\lvert d_k^- \rvert$ (au)
1	0.985	0.180	0.122
2	1.112	0.257	0.161
3	1.548	0.172	0.135
4	1.689	0.118	0.034
5	1.959	0.062	0.056
6	2.065	0.031	0.171
7	2.145	0.050	0.115
8	2.261	0.062	0.094
9	2.411	0.063	0.088
10	2.518	0.036	0.037
–	–	–	–
12	2.675	0.054	0.066
13	2.855	0.015	0.004
14	3.011	0.059	0.059
15	3.066	0.037	0.108
16	3.240	0.051	0.103
17	3.299	0.061	0.062
18	3.480	0.030	0.011
19	3.584	0.028	0.035
20	3.694	0.036	0.041
21	3.803	0.024	0.031
22	3.944	0.042	0.067
23	3.964	0.061	0.014
24	4.271	0.055	0.016
25	4.680	0.108	0.057

(ii) N_P = 220 (No Missing Peaks)

n_k^o	$\mathrm{Re}(\nu_k^+)$ (ppm)	$\mathrm{Im}(\nu_k^+)$ (ppm)	$\lvert d_k^+ \rvert$ (au)
1	0.985	0.180	0.122
2	1.112	0.257	0.161
3	1.548	0.172	0.135
4	1.689	0.118	0.034
5	1.959	0.062	0.056
6	2.065	0.031	0.171
7	2.145	0.050	0.116
8	2.261	0.062	0.092
9	2.411	0.062	0.085
10	2.519	0.036	0.037
11	2.675	0.031	0.006
12	2.676	0.061	0.064
13	2.855	0.016	0.005
14	3.009	0.064	0.065
15	3.067	0.036	0.101
16	3.239	0.050	0.096
17	3.301	0.064	0.065
18	3.481	0.031	0.011
19	3.584	0.028	0.036
20	3.694	0.036	0.041
21	3.803	0.024	0.031
22	3.944	0.042	0.068
23	3.965	0.062	0.013
24	4.271	0.055	0.016
25	4.680	0.108	0.057

(v) N_P = 220 (Converged)

n_k^o	$\mathrm{Re}(\nu_k^-)$ (ppm)	$\mathrm{Im}(\nu_k^-)$ (ppm)	$\lvert d_k^- \rvert$ (au)
1	0.985	0.180	0.122
2	1.112	0.257	0.161
3	1.548	0.172	0.135
4	1.689	0.118	0.034
5	1.959	0.062	0.056
6	2.065	0.031	0.171
7	2.145	0.050	0.116
8	2.261	0.062	0.092
9	2.411	0.062	0.085
10	2.519	0.036	0.037
11	2.675	0.033	0.008
12	2.676	0.062	0.063
13	2.855	0.016	0.005
14	3.009	0.064	0.065
15	3.067	0.036	0.101
16	3.239	0.050	0.096
17	3.301	0.064	0.065
18	3.481	0.031	0.011
19	3.584	0.028	0.036
20	3.694	0.036	0.041
21	3.803	0.024	0.031
22	3.944	0.042	0.068
23	3.965	0.062	0.013
24	4.271	0.055	0.016
25	4.680	0.108	0.057

(iii) N_P = 260 (Converged)

n_k^o	$\mathrm{Re}(\nu_k^+)$ (ppm)	$\mathrm{Im}(\nu_k^+)$ (ppm)	$\lvert d_k^+ \rvert$ (au)
1	0.985	0.180	0.122
2	1.112	0.257	0.161
3	1.548	0.172	0.135
4	1.689	0.118	0.034
5	1.959	0.062	0.056
6	2.065	0.031	0.171
7	2.145	0.050	0.116
8	2.261	0.062	0.092
9	2.411	0.062	0.085
10	2.519	0.036	0.037
11	2.675	0.033	0.008
12	2.676	0.062	0.063
13	2.855	0.016	0.005
14	3.009	0.064	0.065
15	3.067	0.036	0.101
16	3.239	0.050	0.096
17	3.301	0.064	0.065
18	3.481	0.031	0.011
19	3.584	0.028	0.036
20	3.694	0.036	0.041
21	3.803	0.024	0.031
22	3.944	0.042	0.068
23	3.965	0.062	0.013
24	4.271	0.055	0.016
25	4.680	0.108	0.057

(vi) N_P = 260 (Converged)

n_k^o	$\mathrm{Re}(\nu_k^-)$ (ppm)	$\mathrm{Im}(\nu_k^-)$ (ppm)	$\lvert d_k^- \rvert$ (au)
1	0.985	0.180	0.122
2	1.112	0.257	0.161
3	1.548	0.172	0.135
4	1.689	0.118	0.034
5	1.959	0.062	0.056
6	2.065	0.031	0.171
7	2.145	0.050	0.116
8	2.261	0.062	0.092
9	2.411	0.062	0.085
10	2.519	0.036	0.037
11	2.675	0.033	0.008
12	2.676	0.062	0.063
13	2.855	0.016	0.005
14	3.009	0.064	0.065
15	3.067	0.036	0.101
16	3.239	0.050	0.096
17	3.301	0.064	0.065
18	3.481	0.031	0.011
19	3.584	0.028	0.036
20	3.694	0.036	0.041
21	3.803	0.024	0.031
22	3.944	0.042	0.068
23	3.965	0.062	0.013
24	4.271	0.055	0.016
25	4.680	0.108	0.057

with 3 partial signal lengths $N_P = 180, 220, 260$. This is suggested by Table 3.2 where all the parameters are found to have converged in the interval $[N/8, N/4] = [128, 256]$. Before full convergence, at the lowest partial signal length considered in Table 3.3 ($N_P = 180$) on panels (i) and (iv), peak $k = 11$ is not detected in the $\text{FPT}^{(\pm)}$. However, full convergence of the entire set of unknowns in both versions of the FPT is reached at $N_P = 260$ on panel (iii) and (vi). Moreover, it is seen on panel (v) that the $\text{FPT}^{(-)}$ converged even at $N_P = 220$. Convergence of the $\text{FPT}^{(\pm)}$ is verified to be maintained at $N_P > 260$ as also implied by Table 3.2.

3.1.4 Graphic presentation of the input data

Figure 3.1 shows the input spectral parameters in the complex planes of frequency and harmonic variables. These are Argand plots $\{\text{Re}(\nu), \text{Im}(\nu)\}$ $\{\text{Re}(z_k), \text{Im}(z_k)\}$ and $\{\text{Re}(z_k^{-1}), \text{Im}(z_k^{-1})\}$ displayed on panels (i), (ii) and (v), respectively. Also shown in panels (iii) and (vi) are the real, $\text{Re}(c_n)$, and imaginary, $\text{Im}(c_n)$, parts of the FID from (3.1) as a function of time. In these latter graphs for $\{\text{Re}(c_n), n\}$ and $\{\text{Im}(c_n), n\}$, integers $n\,(0 \le N-1)$ appear on the abscissa to count time expressed in the units of the sampling rate τ which is itself given in seconds (s). As usual, the continuous time variable t is digitized according to $t \to t_n \equiv \tau n\,(0 \le n \le N-1)$, so that the total duration T of the FID is given by $T = N\tau$. Panel (iv) displays the absolute values $|d_k|$ of the amplitudes d_k as a function of chemical shifts, $\text{Re}(\nu)$ where $|d_k| = d_k$ due to the special choice of the phases $\text{Arg}(d_k) \equiv \phi_k = 0\,(1 \le k \le K)$.

It is seen on panel (i) in Fig. 3.1 that the majority of the input complex frequencies have relatively small imaginary parts $\text{Im}(\nu_k)$, i.e., they are clustered in the proximity of the real axis, $\text{Re}(\nu_k)$. The physical significance of such a distribution is that the resonances associated with these smaller imaginary frequencies $\text{Im}(\nu_k)$ have larger life-times that are characteristic of more stable transients from the inner part [1.959,4.271] ppm of the whole considered spectral band [0.985,4.680] ppm in panel (i). Thus, such frequencies lying close to the real axis in panel (i) will generate narrow spectral line-shapes for the metabolites $n_k^{\circ} = 5 - 24$ (Gaba - PCho) from the middle part of Table 3.1. Further, several frequencies on panel (i) are seen as being quite remote from the real axis. These larger imaginary frequencies $\text{Im}(\nu_k)$ correspond to shorter life-times that are typical of more unstable transients from the outer parts [0.985,1.689] ppm and 4.680 ppm on the right and left wings, respectively, of the entire investigated window [0.985,4.680] ppm in panel (i). Being located deep in the complex plane, these latter frequencies will produce broader spectral line-shapes for the metabolites $n_k^{\circ} = 1-4$ (mobile lipids) and $n_k^{\circ} = 25$ (water) from the top and bottom parts, respectively, of Table 3.1.

The distribution of the input amplitudes for all the metabolites from Table 3.1 is depicted on panel (iv) in Fig. 3.1 for the interval [0.985,4.680] ppm of the chemical shifts, $\text{Re}(\nu)$. In particular, water is seen as having quite a sizable amplitude relative to a number of low-abundant metabolites. This

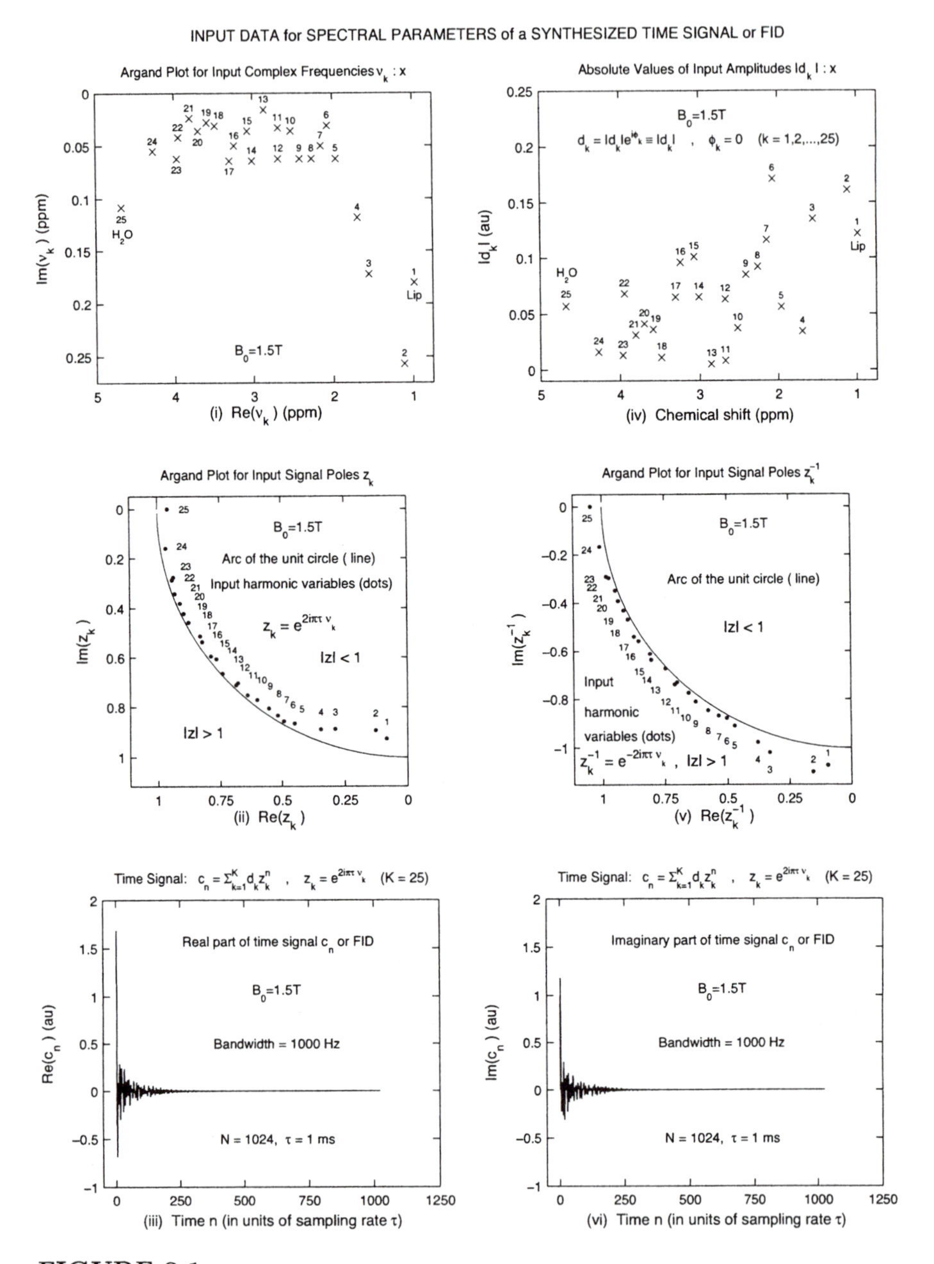

FIGURE 3.1

Argand plot for the input complex frequencies ν_k and a graphical display of the absolute values of amplitudes $d_k = |d_k|$: panels (i) and (iv). Argand plots for the input signal poles or harmonic variables z_k and their inverses z_k^{-1} : panels (ii) and (v). Real $\mathrm{Re}(c_n)$ and imaginary $\mathrm{Im}(c_n)$ part of the input time signal $\{c_n\}$ or FID as a sum of 25 damped complex exponentials with constant amplitudes d_k : panels (iii) and (vi).

mimics a typical, realistic situation seen in the customary presentations of MR-generated spectra of human brain metabolites from experimental encoding after water suppression. The corresponding original FIDs encoded from the brain always generate a single giant resonance due to the fact that brain tissue contains about 75% water. All the other brain metabolites are hardly visible, since they appear as very tiny peaks superimposed on the tail of the exponentially decaying curve of the huge water peak. While water is of primary concern to MRI, which is aimed at detecting the sought proton spin density distributions to depict the anatomy/morphology of the scanned tissue, this molecule is not usually in the focus of MRS. Therefore, regarding the output data, there is every interest to reduce the concentration of the water molecule as much as possible in order to allow the emergence of the other main metabolites that are clinically more informative from the MRS perspective. Customary water suppression can be accomplished in three different ways in clinical scanners [7, 143], and invariably the resulting spectra contain a much smaller, so-called residual water peak, which is simulated by the component $n_k^{\mathrm{o}} = 25$ on panel (iv). As a result of successful water suppression at the end of the experimental encoding, many metabolites pop out from the background, and this is simulated on panel (iv) by a realistic distribution of the input amplitudes $\{|d_k|, \mathrm{Re}(\nu_k)\}$ for metabolites $n_k^{\mathrm{o}} = 1 - 24$ (Lip – PCho) in the interval $\mathrm{Re}(\nu_k) \in [0.985, 4.271]$ ppm. The absolute values $|d_k|$ of the amplitudes d_k are the key spectral parameters for determining the concentrations of metabolites for the given Lorentzian line-shapes, as seen from (3.15).

Argand plots for the input harmonic variables z_k (signal poles) and their inverses z_k^{-1} are shown on panels (ii) and (v) in Fig. 3.1. The former and the latter variables are all lying inside and outside the unit circle, $|z| < 1$ and $|z| > 1$, respectively. As will be displayed later on, the complex quantities z_k and z_k^{-1} are clustered in the 1st and 4th quadrant, respectively. This is not apparent in Fig. 3.1 in which, for convenience, the values on the abscissae and ordinates are used in descending order. The signal poles z_k (or their inverses z_k^{-1}) are tightly packed together and lie close to the inner (or outer) boundary of the unit circle, respectively. They are more or less aligned along the inner or outer arcs. The real and imaginary parts of the FID are seen in panels (iii) and (vi) as heavily oscillating cosinusoids that are exponentially damped with increased time. The earliest sampled signal points exhibit the largest intensities. However, the intensities at larger times quickly become immersed into the zero-valued background for this noiseless FID. Such a situation favors those signal processors, like the FPT, that are capable of reconstructing all the signal components by using the first, relatively short part of the FID with sizable intensities. Of course, this is most relevant for noisy FIDs, since noise dominates the physical signal at larger times.

Further, we present Fig. 3.2 to illustrate the reason for which mathematical methods are absolutely indispensable for MRS and many other fields that rely upon signal processing. The top panel (i) in this figure depicts the time signal from the input data that are encoded via MRS. As discussed, the shown free

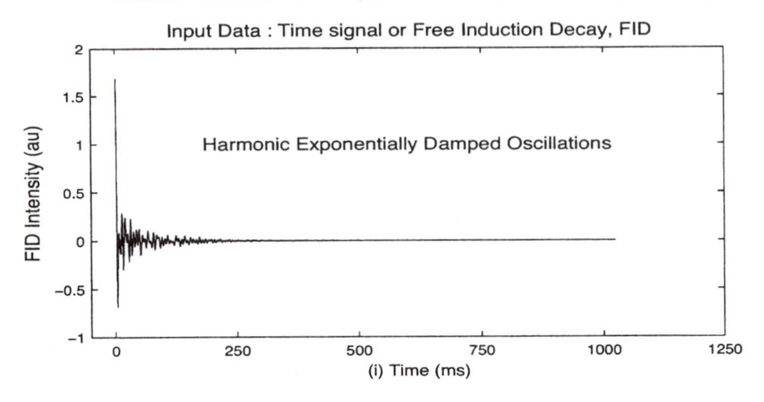

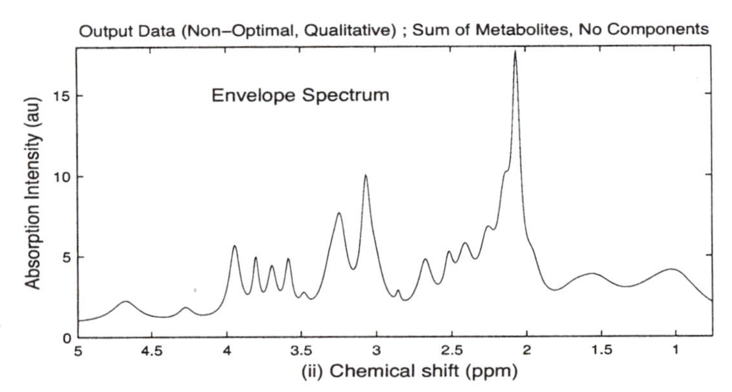

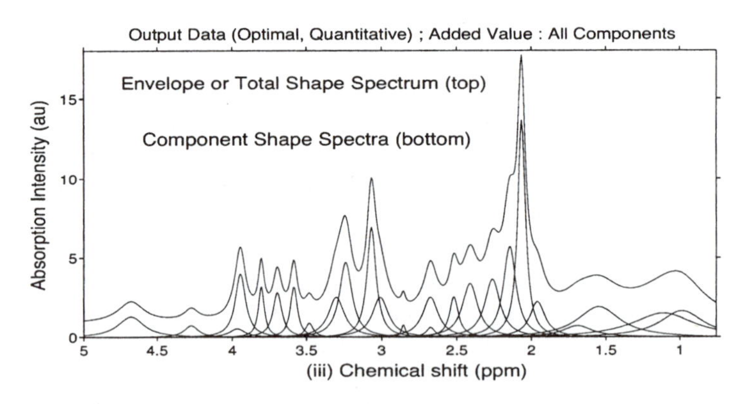

FIGURE 3.2
Time and frequency domain data in signal processing. Top panel (i): the real part of the input FID (the same as in Fig. 3.1 where the imaginary part is also given). Middle panel (ii): absorption total shape spectrum (FFT, FPT). Bottom panel (iii): absorption component (lower curve; FPT) and total (upper curve; FPT, FFT) shape spectra. Panels (ii) and (iii) are generated using both the real and imaginary parts of the FID.

induction decay curve is heavily packed with exponentially decaying oscillations and no other discernable structure appear. Specifically, it is impossible to decipher any clinically meaningful information by inspecting an FID directly in the measured time domain. However, from such a time signal one can compute an MR spectrum which exhibits the definite advantage of displaying a relatively small number of distinct characteristics that are amenable to further analyses and interpretations for clinical purposes. A typical total shape spectrum of this type is shown in the middle panel (ii) in Fig. 3.2 in the absorption mode.

This is obtained by a simple and powerful mathematical transformation of the original time signal into its dual or complementary representation in the frequency domain. The advantage of this passage to the frequency representation is manifested in the emergence of a number of clearly discernable features through the appearance of peaks and valleys. Nevertheless, the total shape spectrum is merely an envelope which, at best, could provide only qualitative information about the overall contribution from the sum of all the constituent resonances, but not the individual components themselves that are seen on panel (iii) in Fig. 3.2, as reconstructed by the $\text{FPT}^{(-)}$. Thus, despite being more revealing than the time signal, the spectral envelope from panel (ii) is still only qualitative as well as inconclusive and, as such, often of limited clinical usefulness. Yet, the FFT, as the most frequently used signal processor in many inter-disciplinary applications, including MRS, is restricted to computations of total shape spectra alone.

Overall, the absorption total shape spectra cannot directly provide the information about any feature of resonances, such as the clinically most important concentrations of the underlying metabolites of the scanned tissue. Indirect information is often guessed from these spectral Fourier-type envelopes by attempting to fit a subjectively preassigned number of resonances hidden beneath each peak structure. These fittings constitute the usual technique encountered in the MRS literature, despite the obvious drawbacks of such a naive approach to spectral analysis. The most serious of these drawbacks is non-uniqueness, which stems from the fact that virtually any chosen number of components could equally well produce an acceptable error in the conventional least-square adjustments to the given spectral envelope. Hence, it would be far more clinically advantageous to have an alternative mathematical transformation, which would use only the original, unedited, raw time signal to first obtain the unique spectral parameters of each peak (position, width, height, phase) and then, if desired, to generate the component as well as total shape spectra in any of the selected modes (absorption, dispersion, magnitude, power). Nevertheless, such spectra with curves, although convenient, are only for visual inspection.

Most important are the numerical values of the retrieved spectral parameters and especially metabolite concentrations. The reason being that, when analyzing clinically encoded FIDs, it is only with these numbers from tables (rather than with envelopes from graphs) that the adequate quantitative assessment could be made as to which metabolites do and which do not have normal concentrations. This together with the presumed availability of metabolite concentrations for the corresponding healthy tissues represent the critical step towards decision making about disease assignment. In aiding these clinical decisions, the mentioned advanced mathematical methods must unambiguously separate the physical from non-physical (noise, noise-like) contents in the input time-domain data, to reconstruct exactly the true number of individual resonances and, finally, to deduce the concentrations of every genuine metabolite.

The signal processor capable of fulfilling all these most stringent physical as well as clinical criteria is the fast Padé transform. This method yields the unique component shape spectra as on panel (iii) in Fig. 3.2 that are generated from the exact envelope. This will be thoroughly documented in the present chapter through maximally accurate reconstructions of all the physical spectral parameters, including their number K. Further, we shall implement exact signal-noise separation by reliance upon the concept of Froissart doublets [44] or pole-zero cancellations [45].

3.2 Absorption total shape spectra

3.2.1 Absorption total shape spectra or envelopes

Figure 3.3 shows, on panels (i) and (ii), the real $\mathrm{Re}(c_n)$ and imaginary $\mathrm{Im}(c_n)$ part, respectively, of the complex-valued synthesized time signal $\{c_n\}(0 \leq n \leq N-1)$ computed from (3.1) by using the spectral parameters given in Table 3.1. The full signal length of this MR time signal is $N = 1024$, the selected bandwidth is 1000 Hz, such that the sampling rate is $\tau = 1$ ms, and the total duration time is $T = N\tau = 1.024$ s. Panel (iii) of Fig. 3.3 displays the initial convergence regions of the $\mathrm{FPT}^{(+)}$ and $\mathrm{FPT}^{(-)}$ located inside and outside the unit circle $|z| < 1$ and $|z| > 1$ in the complex planes of the harmonic variables z and z^{-1}, respectively.

Since the Padé spectra are rational functions given by the quotients of two polynomials, the Cauchy analytical continuation principle lifts the restrictions of the initial convergence regions. Namely, the Cauchy principle extends the initial convergence region from $|z| < 1$ to $|z| > 1$ for the $\mathrm{FPT}^{(+)}$ and similarly from $|z| > 1$ to $|z| < 1$ for the $\mathrm{FPT}^{(-)}$. Thus, both $\mathrm{FPT}^{(+)}$ and $\mathrm{FPT}^{(-)}$ continue to be computable throughout the complex frequency plane without encountering any divergent regions. An exception is the set of the fundamental frequencies of the examined FID that are simultaneously the singular points (poles) of the system's response function.

The small dots seen on panel (iii) depict both the exact input harmonic variables $z_k^{\pm 1} = \exp(\pm i\omega_k\tau)$ and the corresponding Padé counterparts $z_k^{\pm} =$

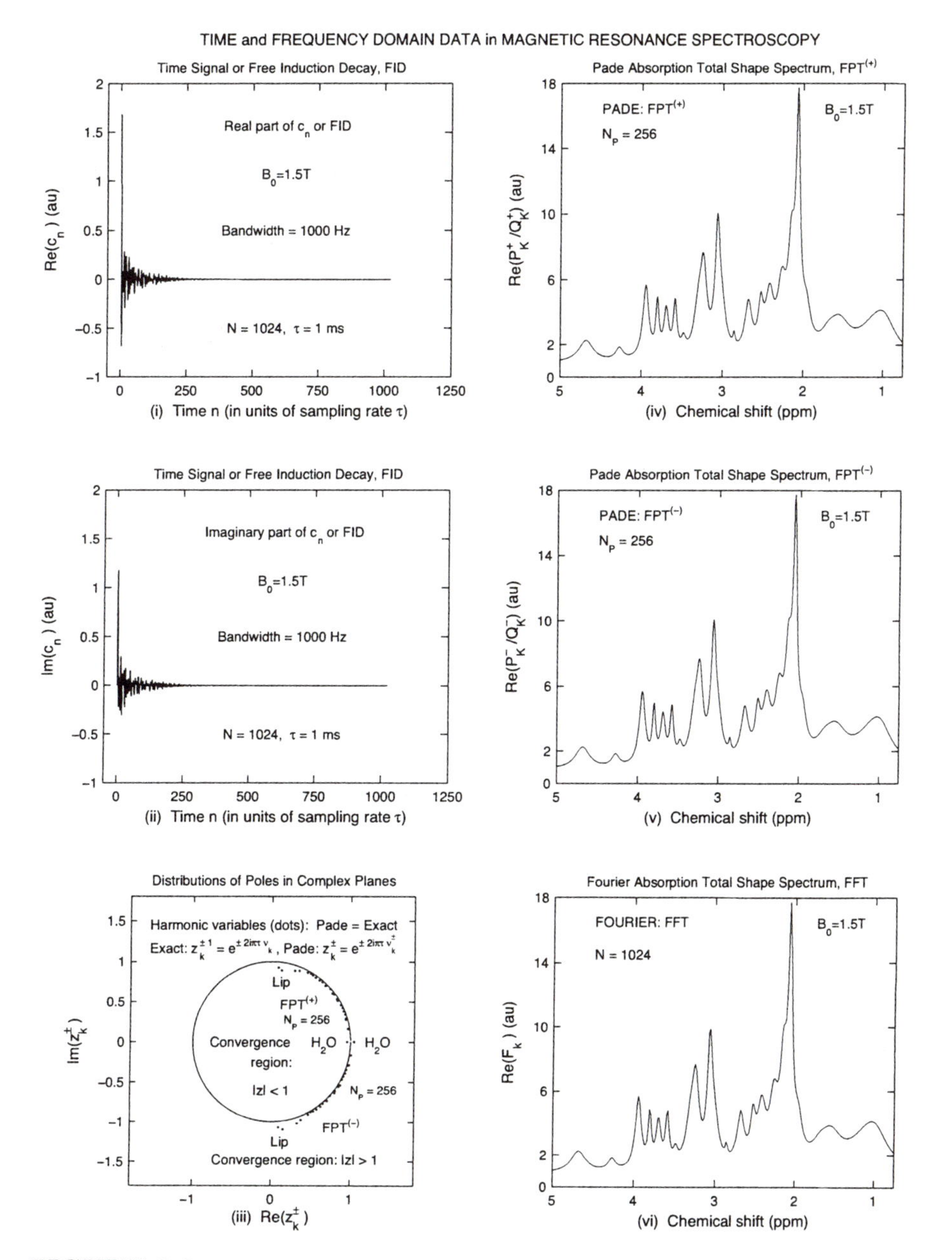

FIGURE 3.3
The synthesized time signal and the corresponding absorption total shape spectra (envelope) in the $\mathrm{FPT}^{(\pm)}$ and FFT. The initial convergence regions in the $\mathrm{FPT}^{(+)}$ and $\mathrm{FPT}^{(-)}$, inside and outside the unit circle, respectively, are shown on panel (iii).

$\exp(\pm i\omega_k^{\pm}\tau)$ reconstructed with $N/4 = 256$ where $\omega_k = 2\pi\nu_k$ and $\omega_k^{\pm} = 2\pi\nu_k^{\pm}$. On panel (iii) the locations of the 1st and the 25th damped harmonics for lipid and water are denoted by Lip and H_2O, respectively. These represent the two endpoints of the complex harmonic variable interval within which all the 25 studied resonances reside. To avoid clutter, numbers for the remaining 23 resonances on both sides of the circumference $|z| = 1$ are not written on panel (iii). These numbers will be shown in Figs. 3.13 and 3.14, whereas the corresponding acronyms for the metabolites shall be depicted in Fig. 3.10. Panels (iv) and (v) in Fig. 3.3 display the two Padé absorption total shape spectra from the Heaviside partial fractions of the FPT$^{(+)}$ and FPT$^{(-)}$, respectively, computed using a quarter signal length ($N/4 = 256$). These results are identical to those obtained with $N/2 = 512$ and $N = 1024$, as is anticipated from the corresponding numerical values of the spectral parameters from panels (iv) – (vi) in Table 3.2.

Panel (vi) in Fig. 3.3 presents the Fourier absorption total shape spectrum evaluated via the FFT using the full FID with $N = 1024$. A comparison of panels (iv)–(vi) in Fig. 3.3 reveals that zero-valued spectra are obtained from the difference between any two selected pairs of these spectra. These differences are termed residual or error spectra and are explicitly shown in Figs. 3.6 and 3.7. Clearly, when the FFT is used in such residuals, subtraction of the Fourier from the Padé spectra is possible only if the FPT is also computed at the Fourier grid points for chemical shifts.

The present finding concerning the error spectra sharpens the corresponding conclusion to be presented in chapter 7 where one half of the entire FID was needed for the FPT to obtain negligible residuals. Herein, the FPT used only the first quarter of the full FID to achieve the resolution of the FFT (which requires the entire time signal) and thereby yields zero-valued residual spectra. It should nevertheless be re-emphasized that in chapter 7, we used encoded FIDs that, despite their excellent SNR, inherently contain noise, whereas the present chapter is dealing with a theoretically constructed noiseless FID.

On all the panels from the right column in Fig. 3.3, the abscissae are the chemical shifts, i.e., $\mathrm{Re}(\nu_k)$ (in ppm), and the ordinates are the intensities (in au) of the structures in the total shape spectra. The Fourier and the Padé absorption total shape spectra are denoted by $\mathrm{Re}(F_k)$ with $0 \le k \le N-1$ and $\mathrm{Re}(P_K^{\pm}/Q_K^{\pm})$, respectively. as the real parts of the corresponding complex-valued spectra, $F \equiv F_k$ and $P_K^{\pm}/Q_K^{\pm} \equiv P_K^{\pm}(z^{\pm 1})/Q_K^{\pm}(z^{\pm 1})$. The Fourier spectrum F_k can be obtained from (2.162), provided that $G_N(z^{-1})$ is evaluated at the Fourier grid points, $F_k = G_N(\exp(-2i\pi k/N))$.

As per the convention in NMR and MRS, chemical shifts in ppm are displayed graphically in descending order when passing from left to right on the abscissa. In the present case, this implies that the largest resonant frequency (water at 4.68 ppm) for resonance $k = 25$ is situated on the extreme left corner of the entire interval for chemical shifts on the abscissa. Such conventions, notations, nomenclature and units will also apply to other figures throughout this book.

3.2.2 Padé and Fourier convergence rates of absorption total shape spectra

Figures 3.4 and 3.5 compare the convergence rates of the absorption total shape spectra from the FFT and $\text{FPT}^{(-)}$ using the FID of the varying length N/M where $M = 1 - 32\,(N = 1024)$. We first analyze Fig. 3.4 which is concerned with 3 partial signal lengths $N/32 = 32, N/16 = 64$ and $N/8 = 128$. The left and the right columns of this figure are for the Fourier and the Padé spectra via panels (i)–(iii) and (iv)–(vi), respectively. In examining the Fourier panels (i) and (ii) together with the Padé panels (iv) and (v) in Fig. 3.4, particularly striking differences are noted between the FFT and $\text{FPT}^{(-)}$ at the two shortest investigated signal lengths $N/32 = 32$ and $N/16 = 64$. The FFT does not provide any spectroscopic information whatsoever on panels (i) and (ii). These two Fourier spectra show merely wide, uninterpretable bumps near 2 ppm and 3 ppm. In sharp contrast, the $\text{FPT}^{(-)}$ on panel (iv) with only $N/32 = 32$ shows several physical resonances, such as the lipid complex between 1 ppm and 2 ppm, NAA near 2 ppm, Cr around 3 ppm as well as close to 4 ppm, PCho in the vicinity of 4.3 ppm and H_2O near 4.7 ppm. Of particular note is that with only 32 signal points out of 1024, these Padé-reconstructed resonance profiles on panel (iv) in Fig. 3.4 are actually quite reasonable. Admittedly, some of these line-shapes in the FPT are rather broad. This is the case because 32 signal points do not provide sufficient accuracy in the estimates for the exact values of the spectral parameters. However, by increasing the input information with $N/16 = 64$ signal points, the performance of the $\text{FPT}^{(-)}$ is improved, as seen on panel (v) in Fig. 3.4. Thus, higher accuracy is achieved in reconstructing the spectral parameters by the FPT at $N/16 = 64$ than at $N/32 = 32$, as is also clear from the associated numerical values listed on panel (ii) in Table 3.2. Furthermore, panel (v) on Fig. 3.4 displays four more resonances that are well-reconstructed by the $\text{FPT}^{(-)}$. These are the peaks corresponding to NAA near 2.7 ppm, Cho close to 3.2 ppm, m-Ins around 3.6 ppm and Glu near 3.8 ppm. Especially in comparison to the spectra from the FFT on panels (i) and (ii) in Fig. 3.4, the results of the $\text{FPT}^{(-)}$ from panels (iv) and (v), albeit not fully complete, display the evidently superior convergence rate in the Padé analysis relative to the FFT. This trend is further displayed on panels (iii) and (vi) at $N/8 = 128$ in Fig. 3.4 for the FFT and $\text{FPT}^{(-)}$, respectively.

Figure 3.5 continues with comparisons of the total shape spectra from Fig. 3.4 using a larger number of signal points: $N/4 = 256, N/2 = 512, N = 1024$. These produce the results presented on panels (i), (ii), (iii) for the FFT and (iv), (v), (vi) for the $\text{FPT}^{(-)}$. Using a quarter ($N/4 = 256$) of the full FID, the FFT on panel (i) on Fig. 3.5 has still not generated adequate estimates for NAA, Cr and Cho that are three diagnostically informative metabolites. This is still the case even with one half ($N/2 = 512$) of the full FID as seen on panel (ii) in Fig. 3.5, since the FFT underestimates the actual height of NAA. It is only at the full signal length ($N = 1024$) that the correct estimates

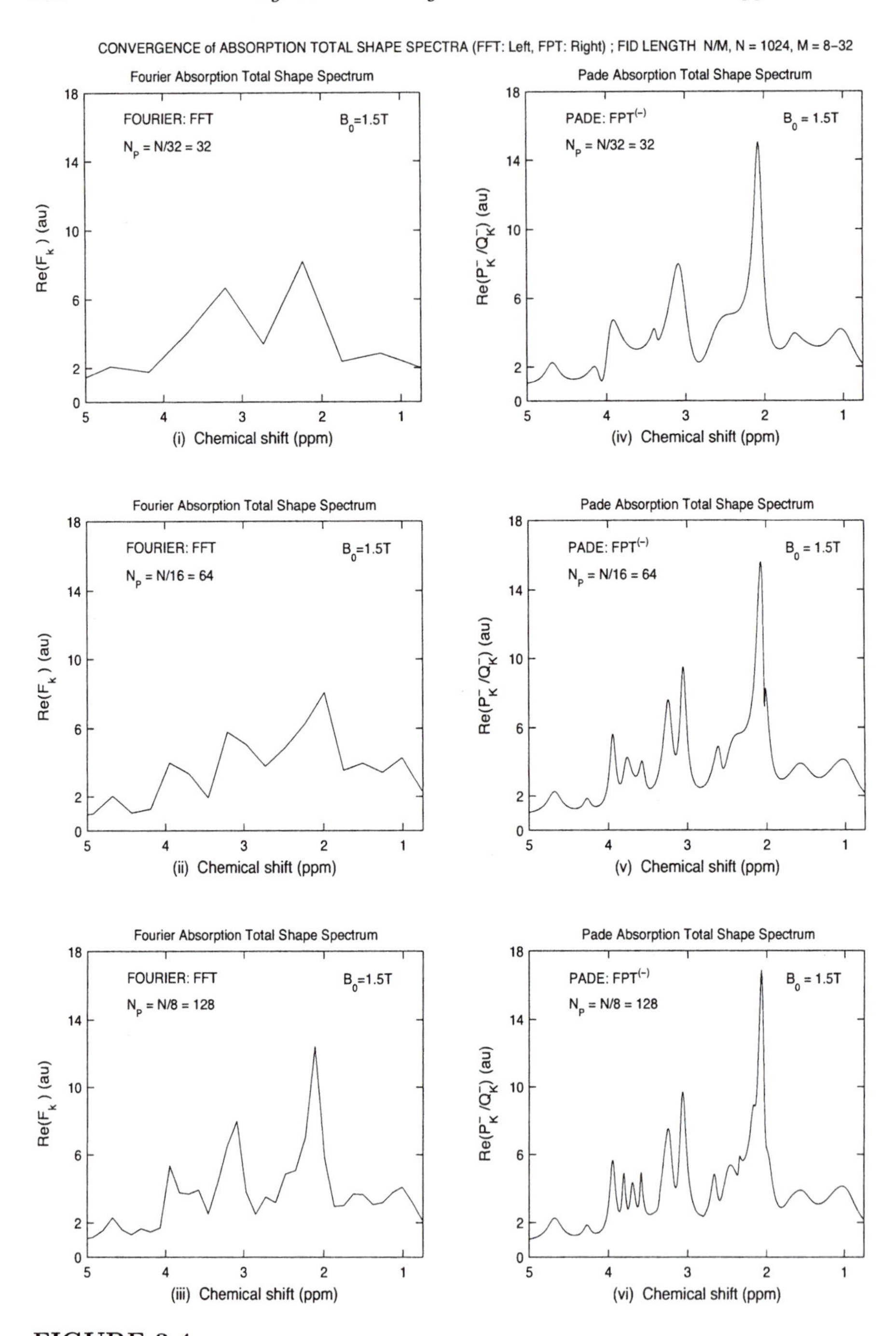

FIGURE 3.4

Convergence of absorption total shape spectra in the FFT (left) and $\text{FPT}^{(-)}$ (right) for signal lengths $N/32 = 32, N/16 = 64, N/8 = 128$.

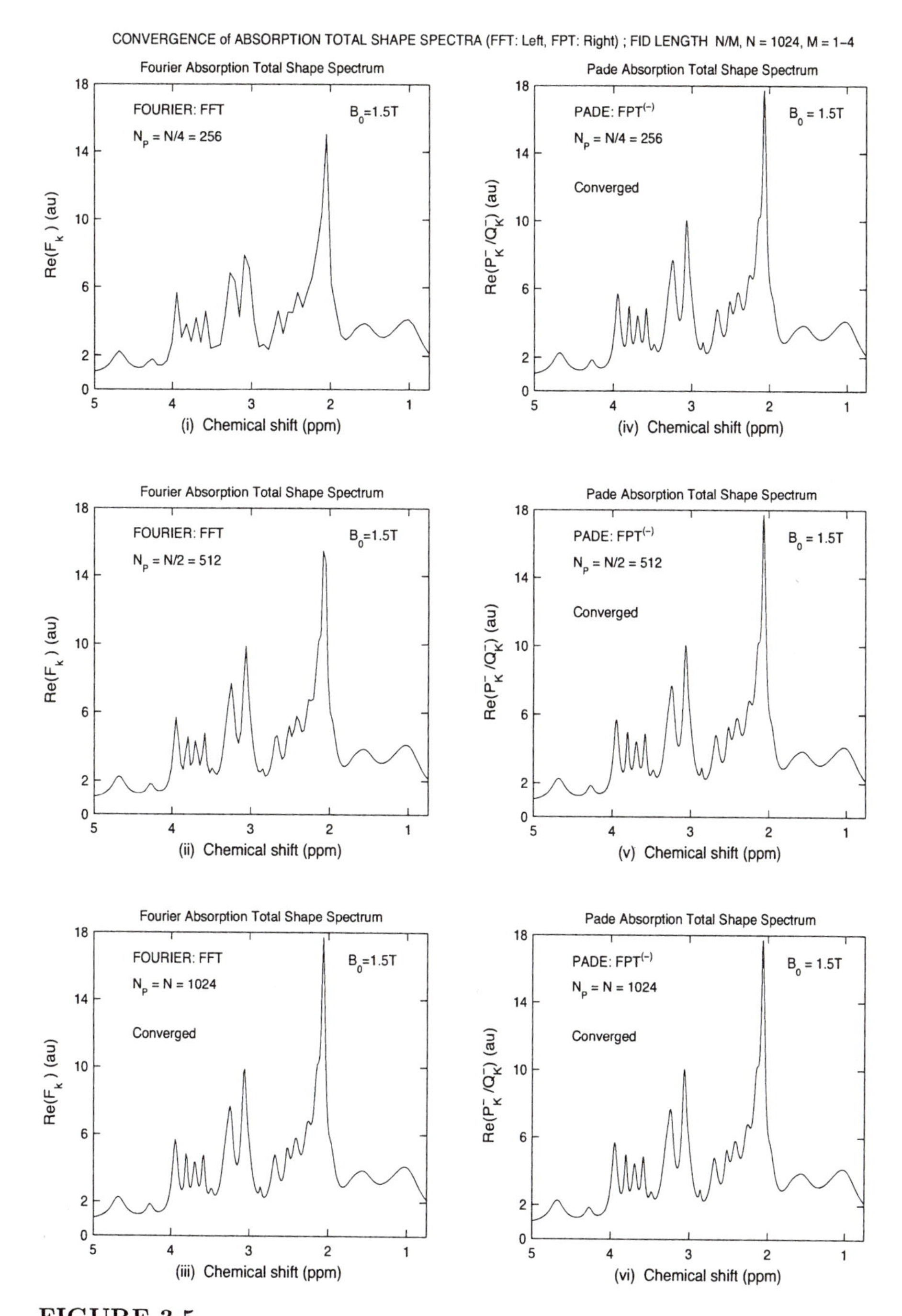

FIGURE 3.5
Convergence of absorption total shape spectra in the FFT (left) and $\text{FPT}^{(-)}$ (right) for signal lengths $N/4 = 256, N/2 = 512, N = 1024$.

are given by the FFT for the exact absorption total shape spectrum, (panel (iii) in Fig. 3.5). This compares poorly with the $\text{FPT}^{(-)}$ as seen on panel (iv) in Fig. 3.5, where full convergence to the exact input result is attained by using only a quarter $N/4 = 256$ of the full signal length N.

Within the $\text{FPT}^{(-)}$ the same conclusion also holds for one half $N/2$ and the full length N of the input FID as seen on panels (v) and (vi). Taken together, Figs. 3.4 and 3.5 provide a powerful demonstration that the $\text{FPT}^{(-)}$ outperforms the FFT with respect to the convergence rate for varying partial signal lengths $N/M\,(M = 2-32)$. This can be seen by referring to panels (i) and (ii) in Fig. 3.3. There it can be observed that the examined FID becomes negligible after exhausting the first $N/4 = 256$ data points, i.e., after $T/4 = 256$ ms. Therefore, an estimator with a resolution which is not restricted by $T = N\tau$ can be expected to yield accurate reconstructions by utilizing only the first quarter of the FID. It is this expectation which is fulfilled by the $\text{FPT}^{(-)}$, as shown in Fig. 3.5. In contrast, the resolving power $2\pi/T$ in the FFT is pre-fixed only by T. Thus, the entire FID is necessary to obtain a converged total shape spectrum.

The above conclusion regarding the $\text{FPT}^{(-)}$ also applies to the $\text{FPT}^{(+)}$. As shown in Ref. [19], at higher partial signal lengths, absorption total shape spectra generated by the $\text{FPT}^{(+)}$ are either very close or identical to the corresponding results of the $\text{FPT}^{(-)}$ from Figs. 3.4 and 3.5. This cross-validation between the two variants of the Padé-based reconstructions enhances the overall fidelity in the FPT.

3.3 Residual spectra and consecutive difference spectra

3.3.1 Residual or error absorption total shape spectra

Figure 3.6 displays the residual or error absorption total shape spectra in the $\text{FPT}^{(-)}$. These spectra are obtained by subtracting the absorption total shape spectra for the full signal length from those for the partial lengths, $\text{Re}(P_K^-/Q_K^-)[N] - \text{Re}(P_K^-/Q_K^-)[N/M]$. The label $[L]$ indicates that altogether, L signal points are used, where L is either N or N/M, with N being the full signal length ($N=1024$), and M is the truncation number ($M=2,4,8,16,32,64$). Here, the number $N/M < N$ represents the partial signal length, N_P.

The stable and rapid convergence within the $\text{FPT}^{(-)}$ that was observed previously in Figs. 3.4 and 3.5, is further reflected in the residual spectra from Fig. 3.6 as a systematic decline in the global error. Even at a quarter $N/4 = 256$ of the full signal length, the error spectra become zero throughout the entire frequency range. This continues to be the case for $N/M > 256$. Thus, the partial length $N/4$ could justifiably be used rather than N for computing these error spectra. Namely, the new differences

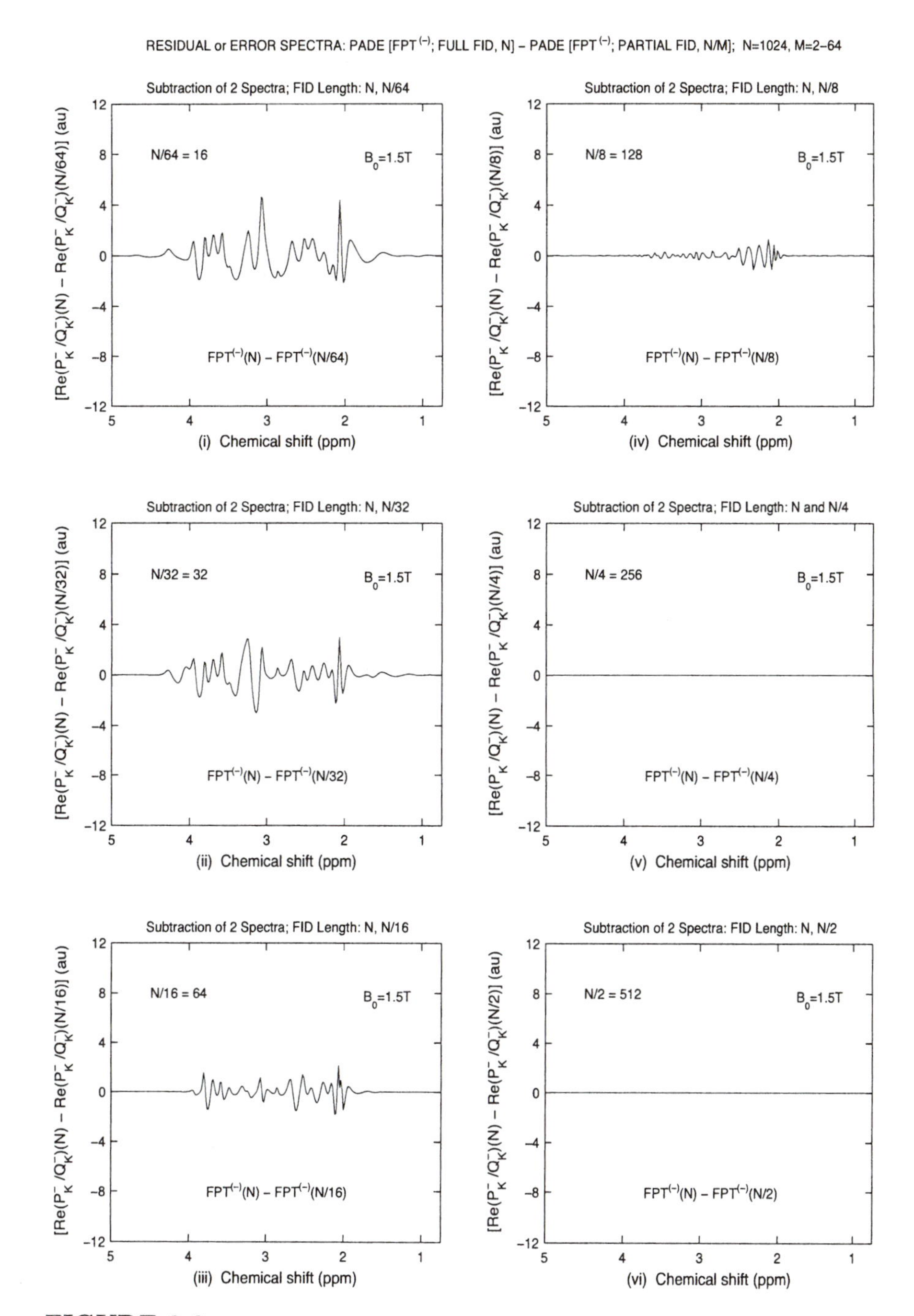

FIGURE 3.6

Residual spectra or error spectra for absorption total shape spectra in the $\mathrm{FPT}^{(-)}$ for signal lengths $N/64 = 16, N/32 = 32, N/16 = 64, N/8 = 128, N/4 = 256, N/2 = 512\,(N = 1024)$.

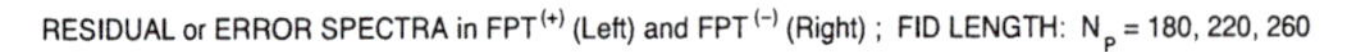

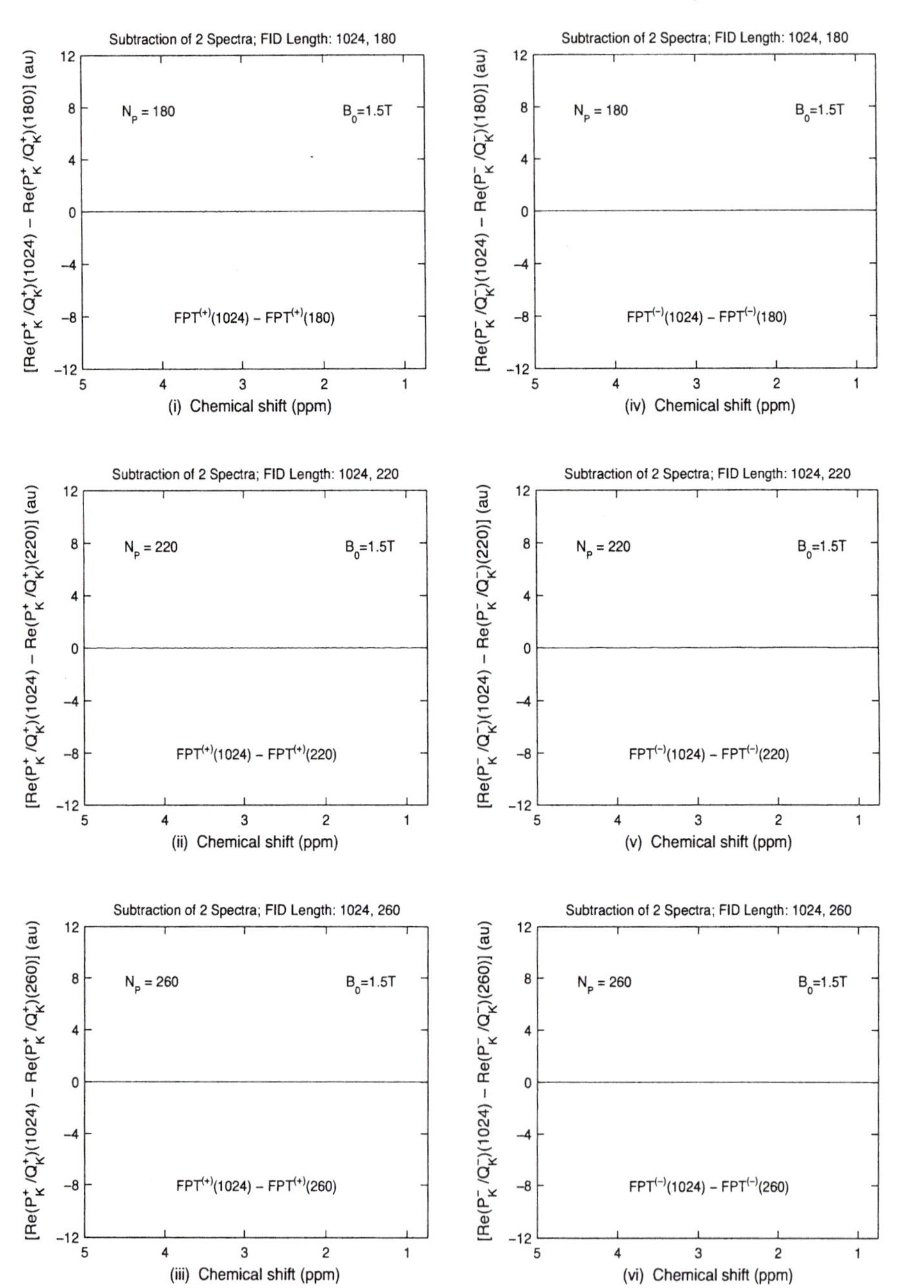

FIGURE 3.7

Residual spectra or error spectra for absorption total shape spectra in the $\text{FPT}^{(+)}$ (left) and $\text{FPT}^{(-)}$ (right) near full convergence for signal lengths $N_P = 180, 220, 260$.

$\mathrm{Re}(P_K^-/Q_K^-)[N/4] - \mathrm{Re}(P_K^-/Q_K^-)[N/M]$ for $M = 2, 8, 16, 32, 64$ would appear precisely the same as those on panels (i)-(iii), (iv) and (vi). Clearly, in this latter context, panel (v) is not mentioned, since in this sub-plot the error spectrum would be zero by definition, $\mathrm{Re}(P_K^-/Q_K^-)[N/4] - \mathrm{Re}(P_K^-/Q_K^-)[N/M] \equiv 0$ for $M = 4$. All told, the residual spectrum in the $\mathrm{FPT}^{(-)}$ converges fully by using at most the first quarter of the entire time signal, as expected from panel (iv) in Fig. 3.5. This finding also holds for the $\mathrm{FPT}^{(+)}$ (not shown), as could similarly be anticipated from Fig. 3.3, where exactly the same absorption total shape spectra are obtained in the $\mathrm{FPT}^{(+)}$ and $\mathrm{FPT}^{(-)}$ at $N/4$.

3.3.2 Residual or error absorption total shape spectra near full convergence

Figure 3.7 shows the residual absorption total shape spectra in the $\mathrm{FPT}^{(+)}$ (left column) and $\mathrm{FPT}^{(-)}$ (right column) computed from the difference of two absorption spectra $\mathrm{Re}(P_K^-/Q_K^-)[N] - \mathrm{Re}(P_K^-/Q_K^-)[N_P]$ where $N_P = 180, 220, 260$. This figure zooms into the range of the partial signal length near full convergence. It is clear that these residual or error spectra in both variants of the FPT are for all practical purposes equal to zero throughout the considered frequency range. This confirms full convergence of all the total shape spectra at $N_P = 180$ even though the peak $k = 11$ is absent, as seen earlier on panels (i) and (iv) of Table 3.3.

3.3.3 Consecutive difference spectra for absorption envelope spectra

On Fig. 3.8 we show the consecutive difference absorption envelope spectra obtained in the $\mathrm{FPT}^{(-)}$. These spectra are generated by subtracting the absorption envelope spectra at the two consecutive signal lengths via $\mathrm{Re}(P_K^-/Q_K^-)[N/M] - \mathrm{Re}(P_K^-/Q_K^-)[N/(2M)]$. Here, according to the notation from Fig. 3.6, the symbol $[L]$ implies that L signal points are used. However, in Fig. 3.8, the number L is either N/M or $N/(2M)$, where, as previously, N is the full signal length ($N = 1024$) and M is the truncation number ($M = 2, 4, 8, 16, 32, 64$).

These consecutive difference spectra are found in Fig. 3.8 to be zero at $N/4 = 256$. The same finding also extends to $N/M > 256$. Through this alternative error analysis, the result once again corroborates that the $\mathrm{FPT}^{(-)}$ has reached full convergence with only a quarter of the entire FID. This is also the case, as well, for the $\mathrm{FPT}^{(+)}$ (not shown).

The main reason for showing the consecutive difference spectra in Fig. 3.8, although the error spectra have already been analyzed in Fig. 3.6, is to display the finer details of the convergence rate on the local level by using adjacent values of the signal length. This is a helpful supplement to the estimates of the global error available from the associated error spectra $\mathrm{Re}(P_K^-/Q_K^-)[N] - \mathrm{Re}(P_K^-/Q_K^-)[N/M]$ from Fig. 3.6. Clearly, the latter formula is also a differ-

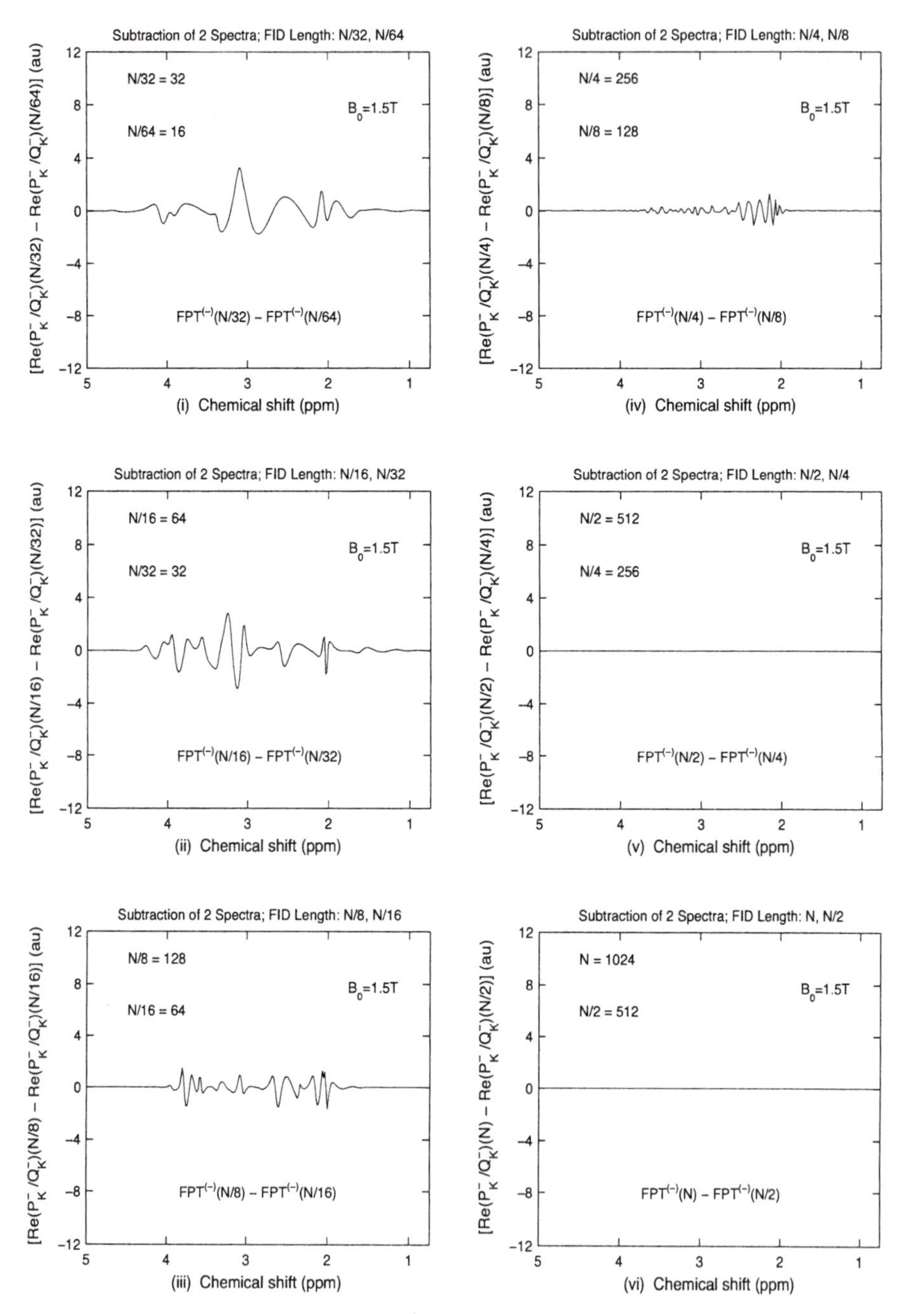

FIGURE 3.8

Consecutive difference spectra for absorption total shape spectra in the $\mathrm{FPT}^{(-)}$ for the adjacent signal lengths $N/32 - N/64$ (i), $N/16 - N/32$ (ii), $N/8 - N/16$ (iii), $N/4 - N/8$ (iv), $N/2 - N/4$ (v), $N - N/2$ (vi).

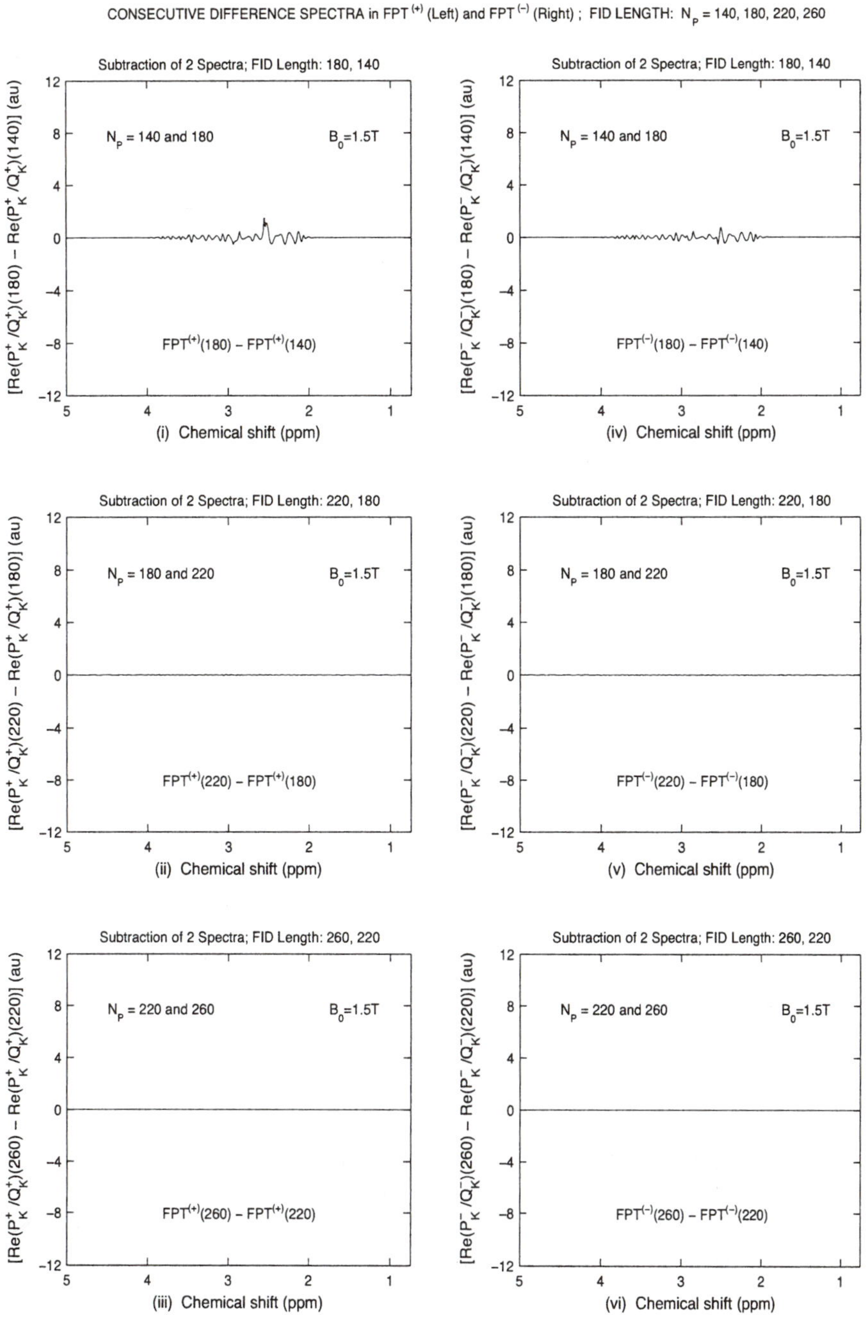

CONSECUTIVE DIFFERENCE SPECTRA in $\mathrm{FPT}^{(+)}$ (Left) and $\mathrm{FPT}^{(-)}$ (Right) ; FID LENGTH: N_P = 140, 180, 220, 260

Subtraction of 2 Spectra; FID Length: 180, 140
$[\mathrm{Re}(P_K^+/Q_K^+)(180) - \mathrm{Re}(P_K^+/Q_K^+)(140)]$ (au)
12 8 4 0 −4 −8 −12
N_P = 140 and 180
B_0=1.5T
$\mathrm{FPT}^{(+)}(180) - \mathrm{FPT}^{(+)}(140)$
5 4 3 2 1
(i) Chemical shift (ppm)

Subtraction of 2 Spectra; FID Length: 180, 140
$[\mathrm{Re}(P_K^-/Q_K^-)(180) - \mathrm{Re}(P_K^-/Q_K^-)(140)]$ (au)
12 8 4 0 −4 −8 −12
N_P = 140 and 180
B_0=1.5T
$\mathrm{FPT}^{(-)}(180) - \mathrm{FPT}^{(-)}(140)$
5 4 3 2 1
(iv) Chemical shift (ppm)

Subtraction of 2 Spectra; FID Length: 220, 180
$[\mathrm{Re}(P_K^+/Q_K^+)(220) - \mathrm{Re}(P_K^+/Q_K^+)(180)]$ (au)
12 8 4 0 −4 −8 −12
N_P = 180 and 220
B_0=1.5T
$\mathrm{FPT}^{(+)}(220) - \mathrm{FPT}^{(+)}(180)$
5 4 3 2 1
(ii) Chemical shift (ppm)

Subtraction of 2 Spectra; FID Length: 220, 180
$[\mathrm{Re}(P_K^-/Q_K^-)(220) - \mathrm{Re}(P_K^-/Q_K^-)(180)]$ (au)
12 8 4 0 −4 −8 −12
N_P = 180 and 220
B_0=1.5T
$\mathrm{FPT}^{(-)}(220) - \mathrm{FPT}^{(-)}(180)$
5 4 3 2 1
(v) Chemical shift (ppm)

Subtraction of 2 Spectra; FID Length: 260, 220
$[\mathrm{Re}(P_K^+/Q_K^+)(260) - \mathrm{Re}(P_K^+/Q_K^+)(220)]$ (au)
12 8 4 0 −4 −8 −12
N_P = 220 and 260
B_0=1.5T
$\mathrm{FPT}^{(+)}(260) - \mathrm{FPT}^{(+)}(220)$
5 4 3 2 1
(iii) Chemical shift (ppm)

Subtraction of 2 Spectra; FID Length: 260, 220
$[\mathrm{Re}(P_K^-/Q_K^-)(260) - \mathrm{Re}(P_K^-/Q_K^-)(220)]$ (au)
12 8 4 0 −4 −8 −12
N_P = 220 and 260
B_0=1.5T
$\mathrm{FPT}^{(-)}(260) - \mathrm{FPT}^{(-)}(220)$
5 4 3 2 1
(vi) Chemical shift (ppm)

FIGURE 3.9
Consecutive difference spectra for absorption total shape spectra in the $\mathrm{FPT}^{(+)}$ (left) and $\mathrm{FPT}^{(-)}$ (right) near full convergence for the adjacent signal lengths $N_P = 140, 180, 220, 260$.

ence spectrum for a varying partial length N/M which, however, only occurs in the second term, whereas the total signal length N is always fixed in the first term $\mathrm{Re}(P_K^-/Q_K^-)[N]$. Obviously, some of the details on panels (i) – (iii) on Figs. 3.6 and 3.8 are different. This is due to the lack of full convergence at $N/M < 256$. Even though convergence has not yet been reached at $N/8 = 128$ on panel (iv), the data for $\mathrm{Re}(P_K^-/Q_K^-)[N] - \mathrm{Re}(P_K^-/Q_K^-)[N/8]$ in Fig. 3.6 and $\mathrm{Re}(P_K^-/Q_K^-)[N/4] - \mathrm{Re}(P_K^-/Q_K^-)[N/8]$ in Fig. 3.8 are identical. As discussed, this is because the first terms in the differences $\mathrm{Re}(P_K^-/Q_K^-)[N]$ and $\mathrm{Re}(P_K^-/Q_K^-)[N/4]$ yield the same results.

3.3.4 Consecutive differences for absorption envelope spectra near full convergence

In Fig. 3.9 the consecutive difference spectra are shown in the $\mathrm{FPT}^{(+)}$ (left column) and $\mathrm{FPT}^{(-)}$ (right column). These were obtained by subtracting two absorption spectra $\mathrm{Re}(P_K^-/Q_K^-)[N_P] - \mathrm{Re}(P_K^-/Q_K^-)[N'_P]$, where $N_P = 180, 220, 260$ and $N'_P = 140, 180, 220$. This focuses upon a narrow region around full convergence. Panels (i) and (iv), prior to full convergence of the absorption total shape spectra, reveal some residual structures located in the frequency interval 2 ppm – 4 ppm. However, all such residuals have disappeared on the middle and lower panels resulting in practically zero-valued difference spectra.

3.4 Absorption component shape spectra of individual resonances

3.4.1 Absorption component spectra and metabolite maps

A comparison is given in Fig. 3.10 of the results for the $\mathrm{FPT}^{(+)}$ and $\mathrm{FPT}^{(-)}$ (left and right columns, respectively). The pertinent details with regard to panels (i) and (iv) as well as (ii) and (v) of Fig. 3.10 have already been presented in section 3.2 via Fig. 3.3. Nevertheless, further important information is presented on panels (ii) and (v) in Fig. 3.10 for the absorption total shape spectra by displaying the usual acronyms that locate the positions of the major MR-detectable metabolites associated with resonances stemming from FIDs encoded via MRS from a healthy human brain. Here, the same acronyms for several resonances (Cho, Glu, NAA) are seen at more than one chemical shift. This is a consequence of the so-called J-coupling [3]. On panels (iii) and (vi) in Fig. 3.10, the absorption component shape spectra are presented for each individual resonance. As explained previously, the sums of all of such component shape spectra yield the associated total shape spectra from panels (ii) and (v) in Fig. 3.10. Once again it is seen on panels (iii) and (vi) in Fig.

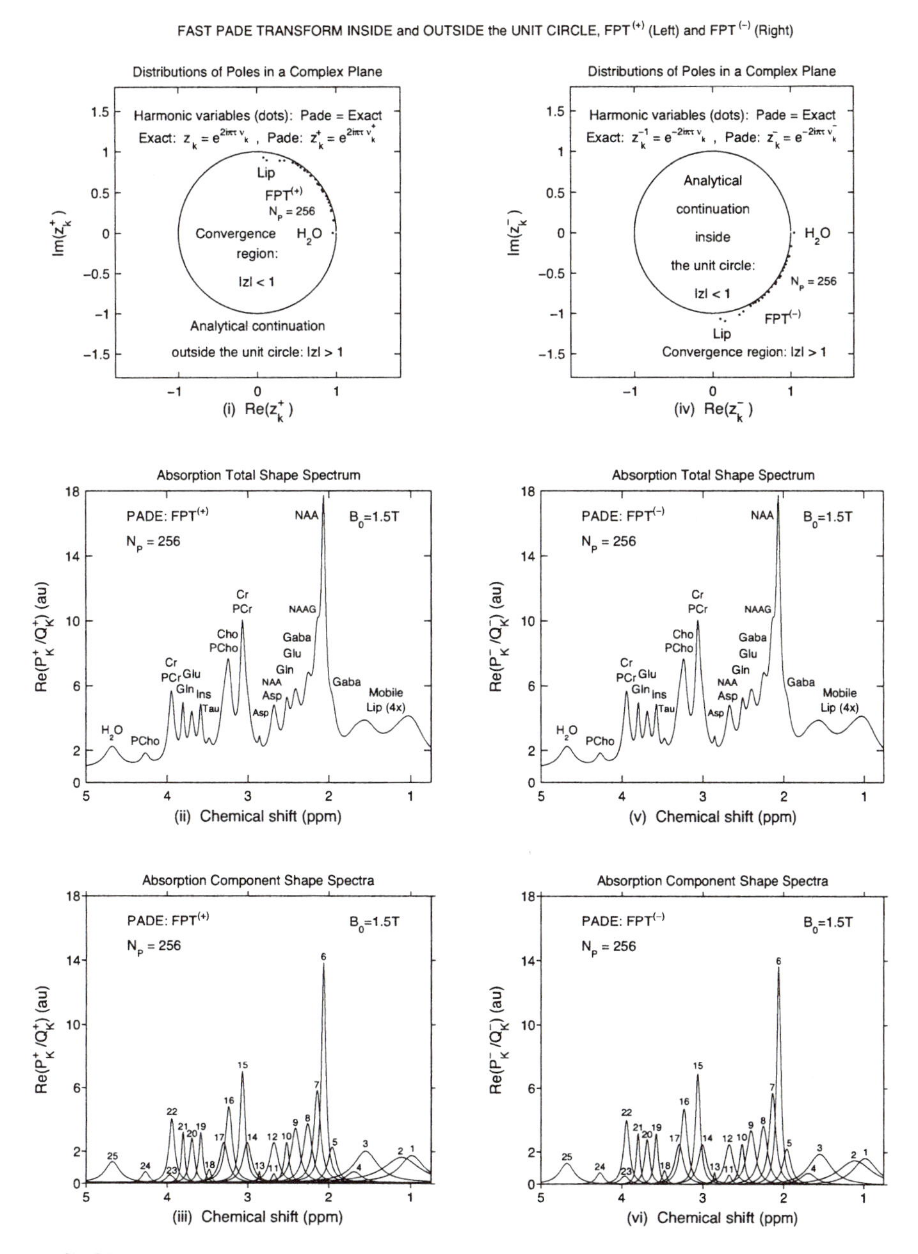

FIGURE 3.10

The initial convergence regions (top panels), absorption total shape spectra (middle panels) and absorption component shape spectra (bottom panels) in the $\mathrm{FPT}^{(+)}$ (left) and $\mathrm{FPT}^{(-)}$ (right). The middle panels display the usual acronyms for the main MR-detectable metabolites in the healthy human brain, whereas the bottom panels give the corresponding numbers of these metabolites according to Table 3.1.

3.10, that only a quarter $N/4 = 256$ of the full FID is necessary for both the $\text{FPT}^{(+)}$ and $\text{FPT}^{(-)}$ to fully resolve all the individual resonances, including the peaks that are isolated ($k = 8, 9, 10, 13, 18, 19, 20, 21, 24, 25$), overlapped ($k = 1, 2; k = 3, 4; k = 5, 6, 7; k = 14, 15; k = 16, 17$), tightly overlapped ($k = 22, 23$) as well as nearly degenerate ($k = 11, 12$). Furthermore, panels (iii) and (vi) of this figure show that the component shape spectra coincide in the $\text{FPT}^{(+)}$ and $\text{FPT}^{(-)}$. This observation is consistent with the conclusion from Fig. 3.3, where the corresponding envelope spectra on panels (iv) and (v) were seen to be the same in the $\text{FPT}^{(+)}$ and $\text{FPT}^{(-)}$. This is anticipated from the definition of the total shape spectrum in the FPT as the sum of the component shape spectra of all the constituent resonances. Obviously, this will be true only for unique reconstructions such as those achieved by means of the FPT.

In sharp contrast is the non-uniqueness of all fittings. In such reconstructions, a given absorption total shape spectrum is fitted by some pre-assigned component shape spectra that might very well differ markedly in nearly all the essential details from the corresponding exact counterparts.

3.4.2 Absorption component spectra and envelope spectra near full convergence

In Fig. 3.11 we see the absorption component shape spectra (left column) and total shape spectra (right column) from the $\text{FPT}^{(-)}$ computed near full convergence at 3 partial signal lengths $N_P = 180, 220, 260$. The three panels on the right column for the total shape spectra have all reached full convergence. However, on the left column for the corresponding component shape spectra, full convergence is achieved at $N_P = 220$, 260. On panel (i) for the component shape spectra at $N_P = 180$, peak $k = 11$ is absent, and peak $k = 12$ is over-estimated. Furthermore, the area of the 12th peak is over-estimated by the amount of the area of the absent 11th peak. As a consequence of this latter compensation, the total shape spectrum has not reflected that either shortcoming had occurred. Namely, the total shape spectrum on panel (iv) for $N_P = 180$ reached complete convergence even though peak $k = 11$ was missing and peak $k = 12$ was over-estimated. This full convergence is also seen in the corresponding zero-valued spectra for the residual $\text{Re}(P_K^-/Q_K^-)[1024] - \text{Re}(P_K^-/Q_K^-)[180]$ ($N = 1024, N_P = 180$) and consecutive difference $\text{Re}(P_K^-/Q_K^-)[220] - \text{Re}(P_K^-/Q_K^-)[180]$ ($N_P = 220, 180$) shown on panel (iv) in Figs. 3.7 and 3.12, respectively.

By way of recapitulation, the left column in Fig. 3.12 re-displays the Padé absorption component and total shape spectra, but this time superimposed on top of each other at $N_P = 180, 220, 260$. This is particularly illuminating when such shape spectra are juxtaposed to the corresponding three consecutive difference spectra on the right column in Fig. 3.12. Three consecutive difference spectra, built from the corresponding total shape spectra, are seen as identical on panels (iv), (v) and (vi) in Fig. 3.12, despite the lack of convergence of the

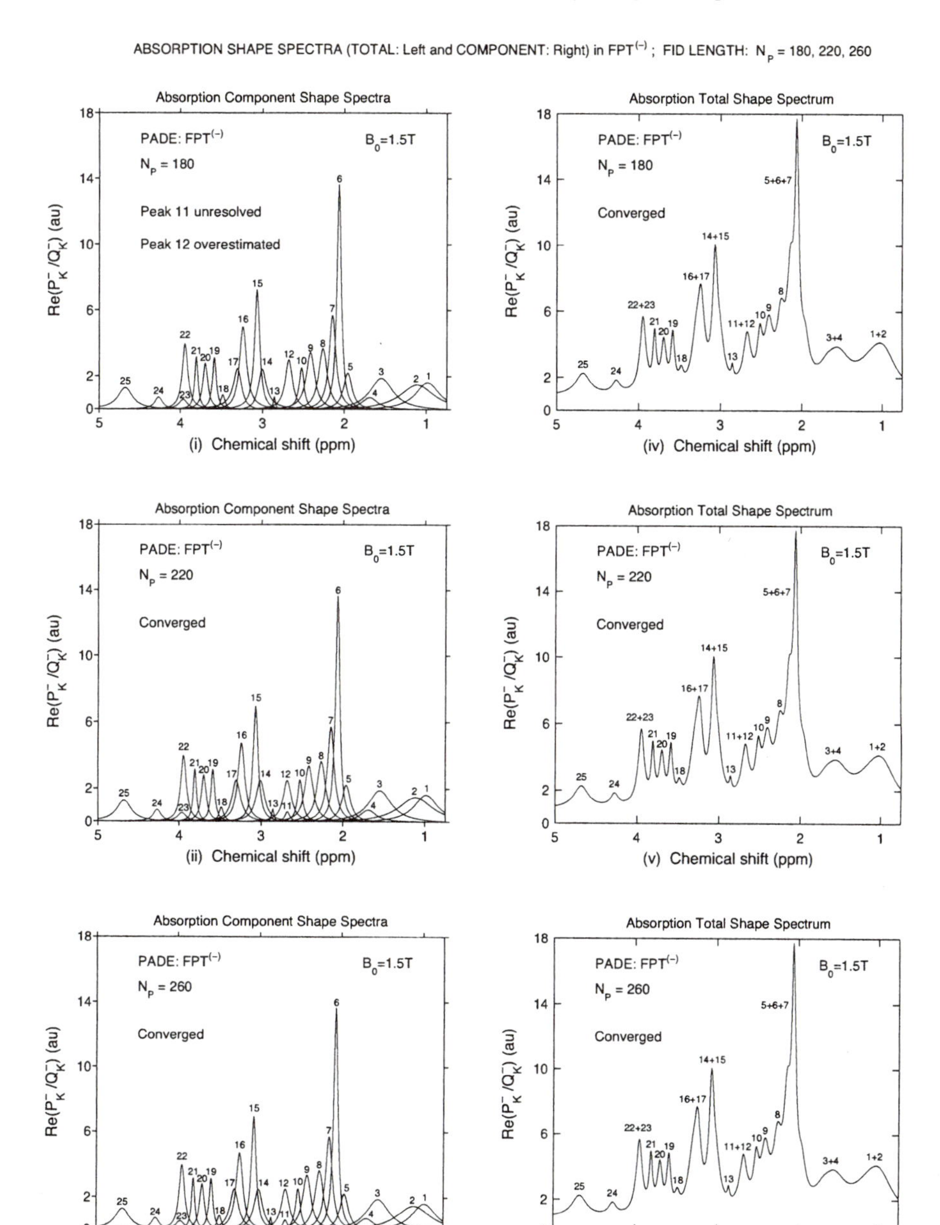

FIGURE 3.11

Absorption component shape spectra (left) and absorption total shape spectra (right) from the $\mathrm{FPT}^{(-)}$ near full convergence for signal lengths $N_P = 180, 220, 260$. On panel (iv) for $N_P = 180$, the total shape spectrum reached full convergence, despite that on panel (i) for the corresponding component shape spectra, the 11th peak is missing and the 12th peak is over-estimated.

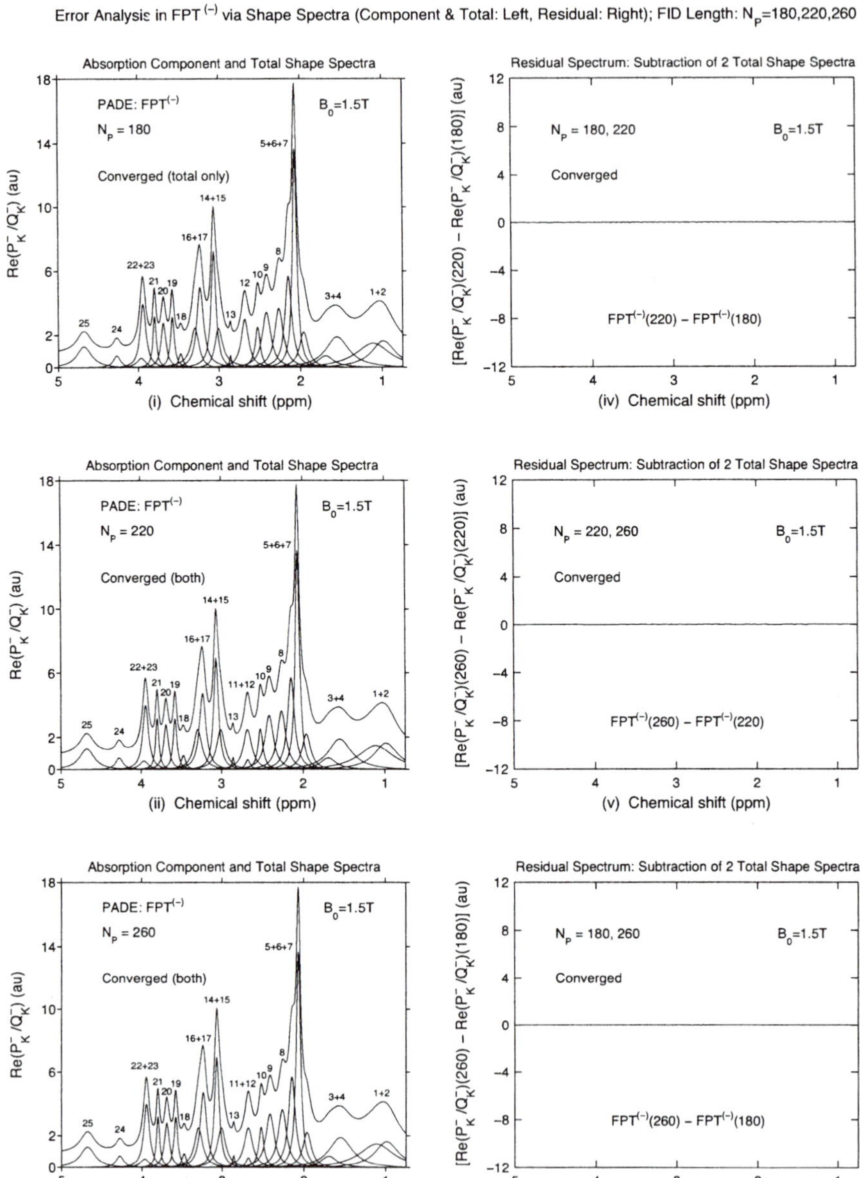

FIGURE 3.12

Left: absorption total shape spectra superimposed on top of the corresponding component shape spectra in the $\mathrm{FPT}^{(-)}$ near full convergence for signal lengths $N_P = 180, 220, 260$. Right: Consecutive difference spectra for absorption total shape spectra in the $\mathrm{FPT}^{(-)}$ for signal lengths $N_P = 180, 220, 260$.

component shape spectrum from panel (i) in the same figure.

Herein, we can therefore conclude that while obtaining the residual or error spectra at the level of background noise may be a necessary condition, this is not sufficient for judging the reliability of estimation. Therefore, it is recommended to pass beyond the point where full convergence of the total shape spectra has been reached for the first time (in this case above $N_P = 180$) in order to verify that anomalies as seen on panels (i) and (iv) of Fig. 3.11 do not occur in the final results. Such final results are obtained for $N_P = 220$ and 260 displayed on panels (ii) and (iii) for the components as well as panels (v) and (vi) for the envelopes. Clearly, monitoring the stability of the component spectra should be done in concert with inspection of the constancy of the reconstructed genuine spectral parameters.

3.5 Distributions of reconstructed spectral parameters in the complex plane

3.5.1 Distributions of spectral parameters in $\text{FPT}^{(+)}$

Figure 3.13 reveals further insights into the exact quantification within the $\text{FPT}^{(+)}$. As was previously the case in Figs. 3.3 and 3.10, all the obtained results are for $N/4 = 256$. The absorption total shape spectrum is shown on panel (iv) in Fig. 3.13, where the individual numbers of resonances are located near the related peaks. Thus each well-resolved isolated resonance is marked by the corresponding separate number, e.g., $k = 8, 9$, etc. Similarly, the overlapped, tightly overlapped and nearly degenerate resonances are labeled as the sum of the pertinent peak numbers, e.g., $k = 1 + 2$ or $k = 5 + 6 + 7$, etc. On panel (v) in Fig. 3.13 the absorption component shape spectra of the constituent resonances $k = 1 - 25$ are shown, and all the individual numbers are indicated for an easier comparison with panel (iv) on the same figure. Thereby, the hidden structures are well-delineated in these component shape spectra. These hidden resonances are those that are overlapped ($k = 1 + 2, 3 + 4, 5 + 6 + 7, 14 + 15, 16 + 17$), tightly overlapped ($k = 22 + 23$) and nearly degenerate ($k = 11 + 12$).

As previously, the resonance positions $\text{Re}(\nu_k^+)$ as chemical shifts are in descending order on the abscissa when proceeding from left to right. With respect to the ordinate, if only panel (vi) for the distribution of resonance frequencies in the corresponding complex plane is analyzed alone, without regard to the other panels in Fig. 3.13, there does not appear to be a need for inverting the imaginary frequency $\text{Im}(\nu_k^+)$. However, panel (vi) is most informative if it is seen not only independently of the other panels, but also in its relation to the rest of the panels in Fig. 3.13. In particular, panels (v)

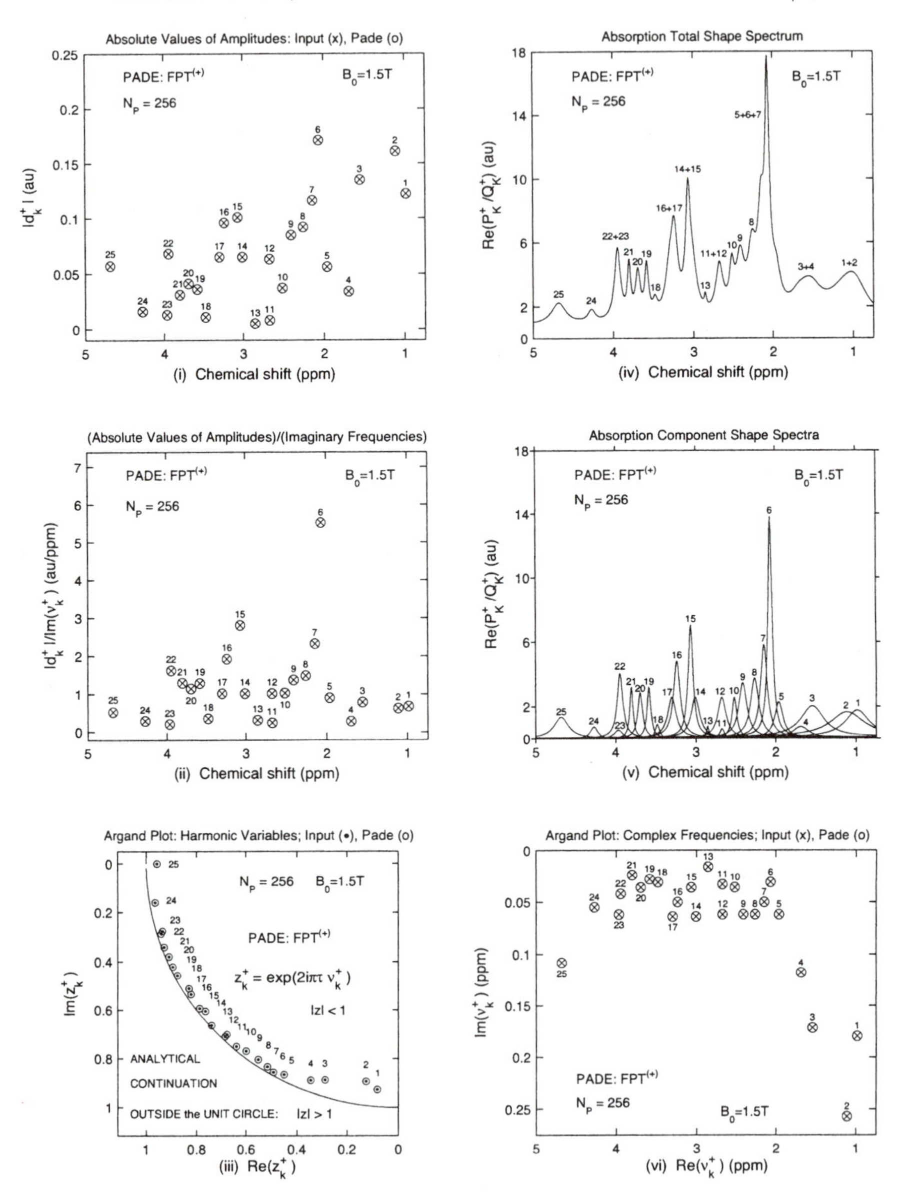

FIGURE 3.13

Configurations of the reconstructed spectral parameters in the $\mathrm{FPT}^{(+)}$. Panel (i): the absolute values $|d_k^+|$ of the amplitudes d_k^+ at the corresponding chemical shifts, $\mathrm{Re}(\nu_k^+)$. Panel (ii): the ratios $|d_k^+|/\mathrm{Im}(\nu_k^+)$ that are proportional to the peak heights. Panel (iii): distributions of poles via the harmonic variable z_k^+ in the complex z^+−plane. Panel (vi): distributions of the fundamental complex frequencies ν_k^+ in the complex ν^+−plane.

and (vi) should be reviewed together, as they are complementary.

For this reason, we chose to use the descending order for the values of $\mathrm{Im}(\nu_k^+)$ when proceeding from bottom to top of the ordinate axis. Thereby, the ordinate is also inverted similarly to the abscissa on panels (iii) and (vi) in Fig. 3.13. The convention used on panel (vi) provides an attractive layout for the configuration of the poles of complex frequencies according to the kind of resonances displayed on panel (v).

This is especially elucidating considering that panel (v) is placed immediately above panel (vi) on the same Fig. 3.13. Here, as seen panel (v), most resonances, e.g., $k = 5 - 24$ are rather narrow as implied by the relatively small values of $\mathrm{Im}(\nu_k^+)$. Thus, these imaginary frequencies are quite close to the real axis. As a result, these resonances are seen on panel (vi) in a group in the middle part of this sub-plot. In contrast, panel (v) in Fig. 3.13 shows wider resonances, e.g., $k = 1 - 4$ and $k = 25$ with larger values of $\mathrm{Im}(\nu_k^+)$. Thus, such imaginary frequencies are deeper in the complex plane and these resonances are quite distant from the real axis, as observed on the far left and the far right parts of panel (vi).

Besides panel (vi) in Fig. 3.13, graphic presentations of the reconstructed and the input data for the spectral parameters are also presented on panels (i) – (iii). Panel (i) in Fig. 3.13 depicts the distribution of the absolute values of the amplitudes at different chemical shifts. It follows from panel (i) that the quantities $|d_k^+|$ do not represent the heights of the absorption peaks from panels (iv) and (v). Instead, the absorption peak heights are directly proportional to the quotient $|d_k^+|/\mathrm{Im}(\nu_k^+)$, as per (2.186). Thus in Fig. 3.13, panel (ii) displays the distribution of the quotients of the absolute values of the amplitudes and the imaginary frequencies. It can be observed from panel (ii) that all the 25 ratios $|d_k^+|/\mathrm{Im}(\nu_k^+)$ are, in fact, proportional to the heights h_k of the corresponding peaks in the absorption component shape spectra from panel (v) in Fig. 3.13.

Panel (iii) from Fig. 3.13 shows, in the complex z^+–plane, the distribution of the Padé poles using the complex harmonic variable z_k^+. This is the zoomed version of panels (iii) or (i) from Figs. 3.3 or 3.10, respectively. The difference is in displaying only the first quadrant in Fig. 3.13, since the rest of the complex z^+–plane does not contain any genuine resonance. Moreover, panel (iii) from Figs. 3.3 and 3.13 differ in arrangements of the values of $\mathrm{Re}(z_k^+)$ and $\mathrm{Im}(z_k^+)$.

Unlike chemical shifts, there is no special reason in Fig. 3.3 or 3.10 to abide by the universal convention of the ascending order of the values for $\mathrm{Re}(z_k^+)$ when proceeding from left to right on the abscissa, and similarly for the ascending order of the values of $\mathrm{Im}(z_k^+)$ when passing from bottom to top on the ordinate. As such, this usual practice is seen on panel (iii) in Fig. 3.3 as well as on panels (i) and (iv) in Fig. 3.10. However, as to panel (iii) in Fig. 3.13, both axes $\mathrm{Re}(z_k^+)$ and $\mathrm{Im}(z_k^+)$ are reversed. This implies that the values on the abscissa and ordinate on panel (iii) in Fig. 3.13 are arranged in descending order when going from left to right on the abscissa and from bottom to top on the ordinate, respectively. This twofold inversion on panel (iii) in Fig. 3.13 is made to follow the arrangements of $\mathrm{Re}(\nu_k^+)$ and $\mathrm{Im}(\nu_k^+)$ that are both shown in descending order when proceeding from left to right on the abscissa or from bottom to top on the ordinate on panel (vi) on the same figure. This convenient layout reveals that all the Padé poles $k = 1 - 25$ are aligned one after the other from right to left regarding the abscissa. The same poles $k = 1 - 25$ are also packed together near the circumference ($|z| = 1$) of the unit circle in such a way that they follow each other according to their consecutive numbers from inside the unit circle, by being aligned upward with respect to the ordinate.

Panel (iii) in Fig. 3.13 shows that the poles contained in the harmonic variable z_k^+ are less scattered from each other relative to the associated distributions of the complex frequencies from panel (vi) in the same figure. The reason for this is in the exponential function of the complex frequency which is plotted on panel (iii), whereas the frequency itself is shown on panel (vi) in Fig. 3.13. It is seen on panel (iii) in Fig. 3.13 that all the genuine poles retrieved by the $\mathrm{FPT}^{(+)}$ are found inside the unit circle ($|z| < 1$), as expected. Notice that narrow resonances $k = 5 - 24$ are shown to be near the circumference ($|z| = 1$) of the unit circle. On the other hand, the wide resonances seen on panel (iii) in Fig. 3.13, such as $k = 1 - 4$ and $k = 25$, lie further from the borderline $|z| = 1$.

3.5.2 Distributions of spectral parameters in FPT$^{(-)}$

Figure 3.14 is, in many ways, similar to Fig. 3.13. The difference is that Fig. 3.14 displays the results of the $\mathrm{FPT}^{(-)}$. The interpretation of the results from the $\mathrm{FPT}^{(+)}$ as presented on panels (i), (ii) and (iv) – (vi) in Fig. 3.13 holds as well with respect to the corresponding findings from the $\mathrm{FPT}^{(-)}$ shown on panels (i), (ii) and (iv) – (vi) in Fig. 3.14.

This observation emerges from the fact that the $\mathrm{FPT}^{(+)}$ and $\mathrm{FPT}^{(-)}$ generate indistinguishable spectral parameters for the same $N/4 = 256$ signal points, as noted in sub-subsection 3.1.3. However, panel (iii) differs for Figs. 3.13 and 3.14, since the information presented here relates to the two complementary regions of the initial convergence, inside and outside the unit circle, for the $\mathrm{FPT}^{(+)}$ and $\mathrm{FPT}^{(-)}$, respectively.

In order to match the configuration from panel (iii) in Fig. 3.13, the Padé poles contained in the harmonic variable z_k^-, as displayed in the complex z^-– plane on panel (iii) in Fig. 3.14, are plotted with the values of $\mathrm{Im}(z_k^-)$ in ascending order when going from bottom to top of the ordinate. This is opposite to the ordering of $\mathrm{Im}(z_k^+)$ on panel (iii) in Fig. 3.13, as anticipated because $\mathrm{Im}(z_k^+) > 0$ and $\mathrm{Im}(z_k^-) < 0$.

Thus, in reconstructions by the $\mathrm{FPT}^{(+)}$ and $\mathrm{FPT}^{(-)}$, the harmonic variables z_k^+ and z_k^- are located in the first and the fourth quadrant of the complex z^+– and z^-–planes, respectively. This is apparent on panel (iii) of Fig. 3.3

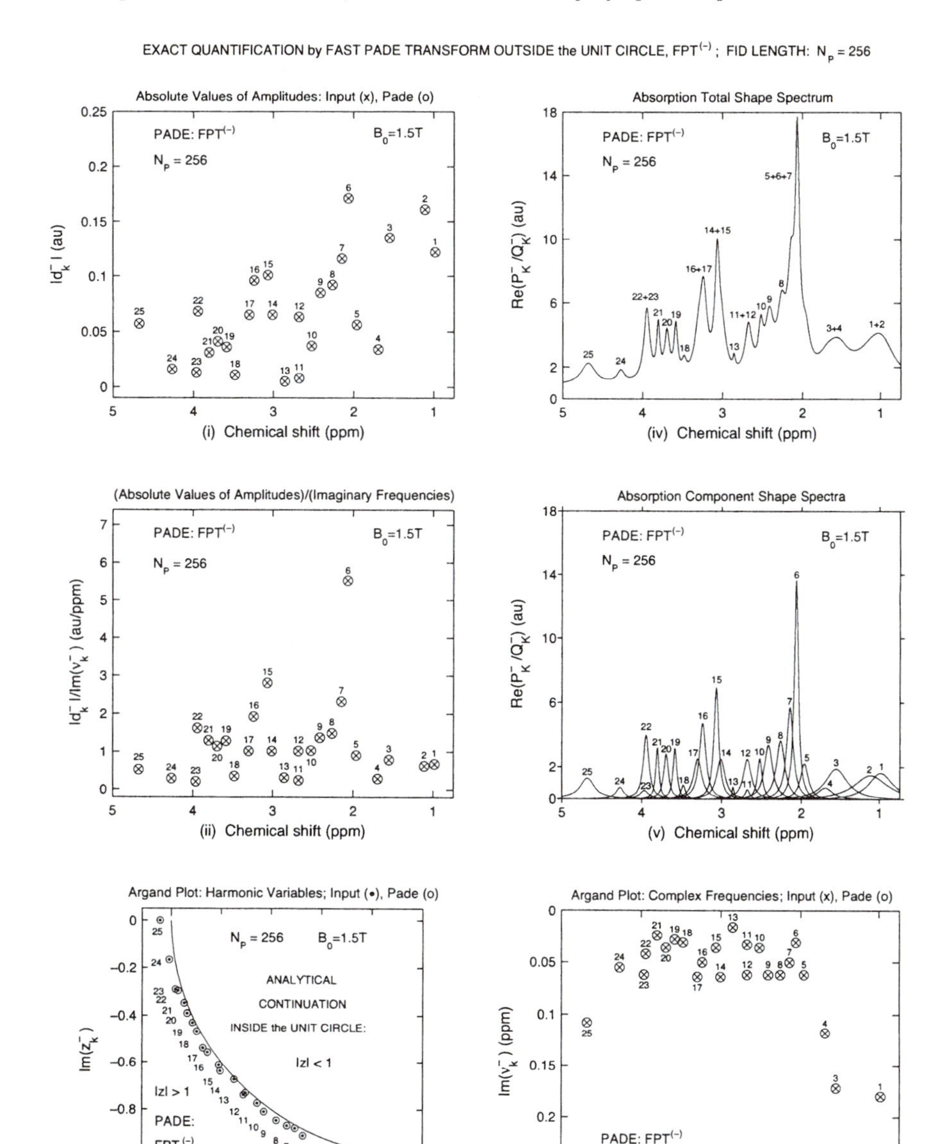

FIGURE 3.14

Configurations of the reconstructed spectral parameters in the $\mathrm{FPT}^{(-)}$. Panel (i): the absolute values $|d_k^-|$ of the amplitudes d_k^- at the corresponding chemical shifts, $\mathrm{Re}(\nu_k^+)$. Panel (ii): the ratios $|d_k^-|/\mathrm{Im}(\nu_k^-)$ that are proportional to the peak heights. Panel (iii): distributions of poles via the harmonic variable z_k^- in the complex z^-−plane. Panel (vi): distributions of the fundamental complex frequencies ν_k^- in the complex ν^-−plane.

or on panels (i) and (iv) in Fig. 3.10. Furthermore, on panel (iii) in Fig. 3.14 all the resonances reconstructed by means of the $\text{FPT}^{(-)}$ are observed to lie outside the unit circle ($|z| > 1$), as expected.

A careful inspection reveals that the kth heights $h_k^{\pm} \equiv |d_k^{\pm}|/\text{Im}(\nu_k^{\pm})$ shown on panels (ii) in Figs. 3.13 and Fig. 3.14 do not fully match the corresponding tops of the kth peaks in the component shape spectra $d_k^{\pm} z^{\pm 1}/(z - z_k^{\pm})$ from panel (v). This is explained by the fact that the heights $h_k^{\pm}$ from their definition in (2.186) are due to the line-shapes $d_k^{\pm}/(\omega - \omega_k^{\pm})$ rather than to the plotted spectra $d_k^{\pm} z^{\pm 1}/(z - z_k^{\pm})$. The former and the latter line-shapes are given in terms of the angular frequencies $\{\omega, \omega_k^{\pm}\}$ and harmonic variables $\{z^{\pm 1}, z_k^{\pm 1}\}$, respectively, where $z^{\pm 1} = \exp(\pm i\omega\tau)$ and $z_k^{\pm 1} = \exp(\pm i\omega_k\tau)$.

3.5.3 Convergence of fundamental frequencies in $\text{FPT}^{(-)}$

Figure 3.15 shows how the complex fundamental frequencies retrieved by the $\text{FPT}^{(-)}$ are configured in the complex frequency plane when utilizing six partial signal lengths ($N/32 = 32, N/16 = 64, N/8 = 128, N/4 = 256, N/2 = 512$) and the full FID ($N = 1024$). Consequently, panels (i), (ii) and (iii) show the frequency distribution of the 10, 14 and 20 resonances that are retrieved with $N/32 = 32, N/16 = 64$ and $N/8 = 128$, respectively.

It should be emphasized that the ordinate axis on panel (i) is enlarged with respect to all other panels on Fig. 3.15. This adjustment is due to a larger scatter of the two retrieved imaginary frequencies relative to the corresponding exact values. This scatter surpasses the window 0 – 0.275 ppm which is common to the remaining five ordinate axes on panels (ii) – (vi) in Fig. 3.15. Panels (iv), (v) and (vi) in Fig. 3.15 show the complex frequencies retrieved with $N/4 = 256, N/2 = 512$ and $N = 1024$, respectively. Note that on panel (iv) for $N/4 = 256$, the exact frequencies (x's) and the associated Padé-reconstructed values (circles) are in complete agreement with each other. Furthermore, the plot on panel (iv) coincides with the plots on panels (v) and (vi) for $N/2 = 512$ and $N = 1024$, respectively.

This graphic illustration of the full agreement between the complete set of exact $\{\nu_k\}$ and reconstructed $\{\nu_k^-\}$ frequencies, where the latter quantities are computed by the $\text{FPT}^{(-)}$ with the signal lengths $N/4 = 256, N/2 = 512$ and $N = 1024$, is anticipated from panels (iv) – (vi) in Table 3.2.

3.5.4 Distributions of fundamental frequencies in $\text{FPT}^{(\pm)}$ near full convergence

Figure 3.16 displays the distributions of the complex fundamental frequencies reconstructed by the $\text{FPT}^{(+)}$ (left column) and $\text{FPT}^{(-)}$ (right column) near full convergence at 3 partial signal lengths $N_P = 180, 220, 260$. On panels (i) and (iv) at $N_P = 180$, the frequency for peak $k = 11$ is missing from the reconstructed parameters in the $\text{FPT}^{(+)}$ and $\text{FPT}^{(-)}$. However, at $N_P = 220$

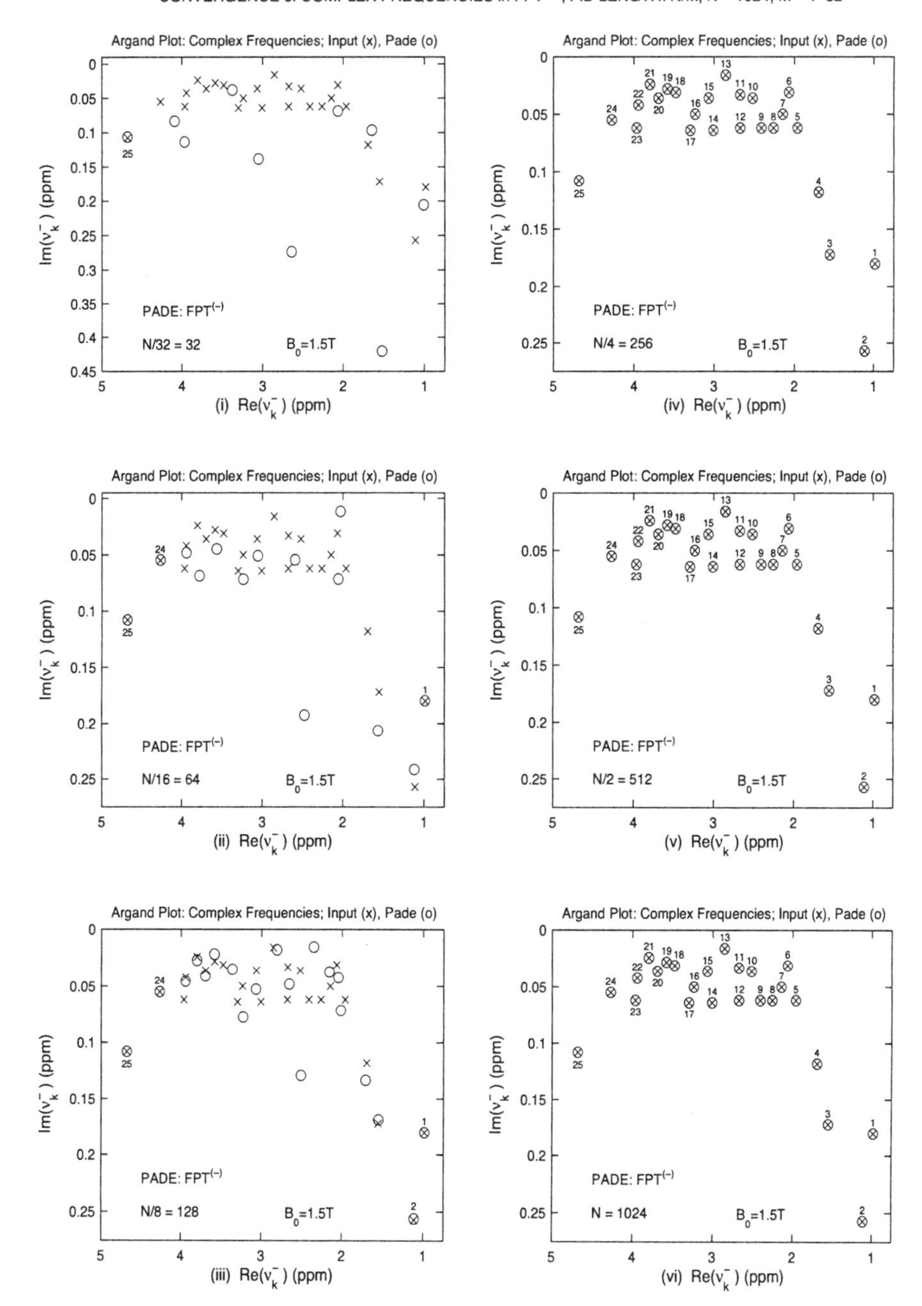

FIGURE 3.15

Convergence of the fundamental complex frequencies reconstructed by the $\mathrm{FPT}^{(-)}$ for signal lengths $N/32 = 32, N/16 = 64, N/8 = 128, N/4 = 256, N/2 = 512$ and $N = 1024$.

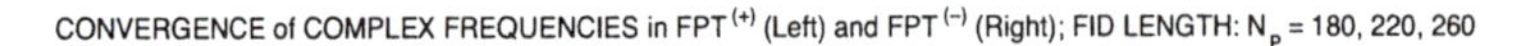

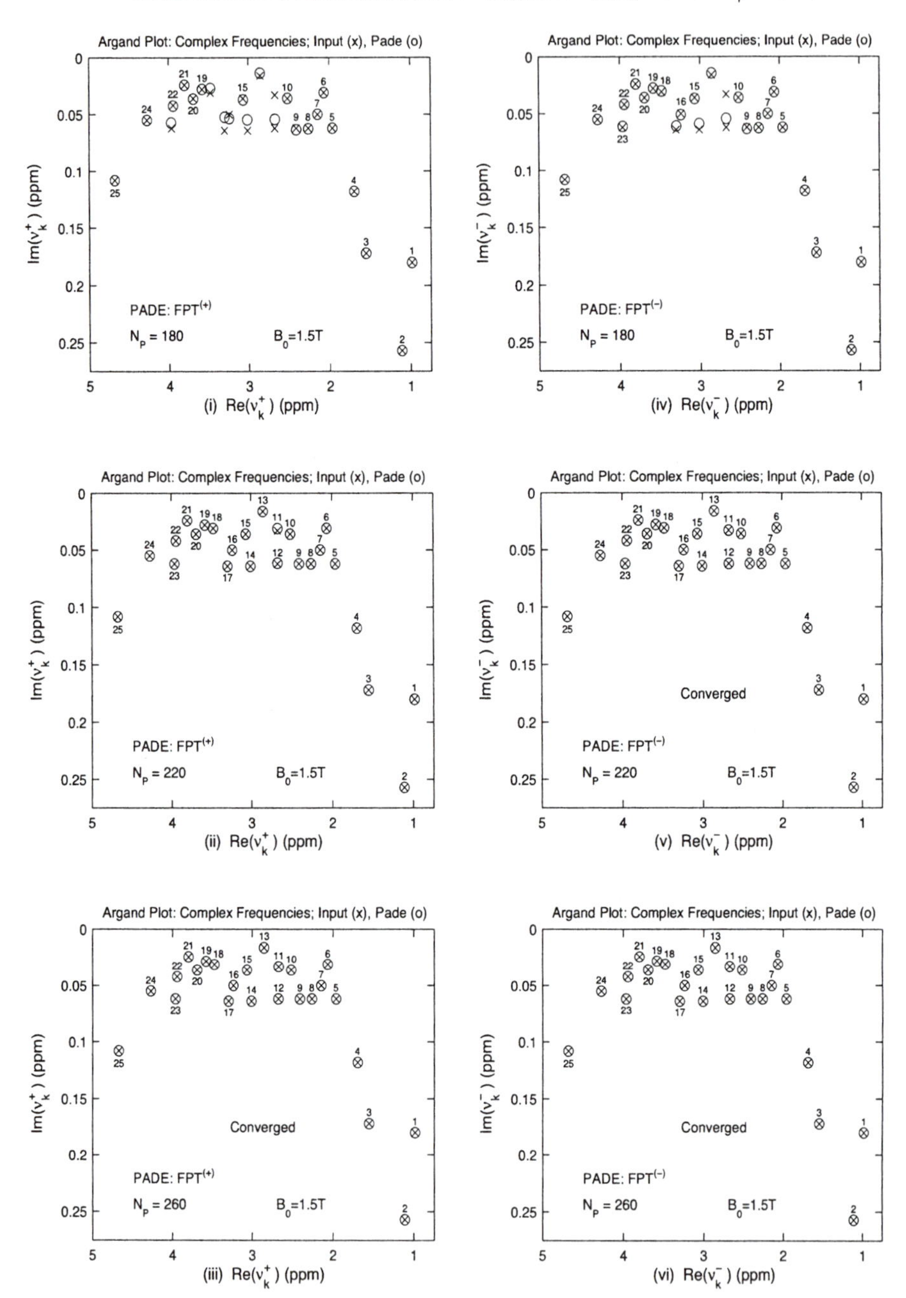

FIGURE 3.16

Convergence of the fundamental complex frequencies reconstructed by the $\mathrm{FPT}^{(+)}$ (left) and $\mathrm{FPT}^{(-)}$ (right) near full convergence for signal lengths $N_P = 180, 220, 260$.

the absent 11th peak is recovered by both variants of the FPT, as observed on panels (ii) and (v). Specifically, in the $\text{FPT}^{(+)}$ full convergence to all the exact digits is achieved at $N_P = 260$ in panel (iii). On the other hand, the $\text{FPT}^{(-)}$ completely converges at $N_P = 220$ and $N_P = 260$ in panels (v) and (vi). We have confirmed that stable estimation of the fundamental frequencies in the $\text{FPT}^{(\pm)}$ continues at $N_P > 260$.

3.5.5 Convergence of fundamental amplitudes in $\text{FPT}^{(-)}$

Figure 3.17 displays the distributions of the absolute values of amplitudes (related to the corresponding fundamental frequencies) that were reconstructed with the $\text{FPT}^{(-)}$. These distributions are computed as a function of chemical shifts by using the same sequence of signal lengths as in Fig. 3.15, i.e., $N/32 = 32, N/16 = 64, N/8 = 128, N/4 = 256, N/2 = 512$ and $N = 1024$. On panels (i), (ii) and (iii) in Fig. 3.17 we show the absolute values of the amplitude distributions of the 10, 14 and 20 resonances reconstructed for $N/32 = 32, N/16 = 64$ and $N/8 = 128$, respectively. The ordinate axes on the entire left column via panels (i) - (iii) in Fig. 3.17 are enlarged compared to panels (iv) - (vi) on the same figure. A similar enlargement was needed as well for Fig. 3.15 for the distribution of the retrieved complex frequencies, but only on panel (i) for the shortest partial signal length $N/32 = 32$. We then see that the reconstructed $|d_k^-|$'s in Fig. 3.17 are more scattered from the associated exact values than in the case of the corresponding complex frequencies on the left column in Fig. 3.15.

Thus, prior to convergence, there appears to be a greater sensitivity of the reconstructed absolute values of the amplitudes on panels (i) - (iii) in Fig. 3.17, relative to the corresponding complex frequencies from panels (i) - (iii) in Fig. 3.15, when both sets of spectral parameters are computed with the same partial signal length in the interval $N/M < N/4 = 256$. This is anticipated because the complex frequencies $\{\nu_k^-\}$ extracted by the $\text{FPT}^{(-)}$ are based only upon the roots of the denominator polynomial Q_K^-. However, the absolute values of the amplitudes $\{|d_k^-|\}$ rely upon both the numerator P_K^- and denominator Q_K^- polynomials, as per (2.183).

Clearly, any extra numerical inaccuracies due to the additional parameters to be obtained for the expansion coefficients $\{p_r^-\}$ in P_K^- yield more pronounced departures from the exact results for the amplitudes than for frequencies, prior to reaching full convergence. Also, the reconstructed amplitudes converge more slowly than the corresponding frequencies for the considered FID, because it is difficult to recover exactly zero values for all the phases $\{\phi_k^-\}$ to match the associated exact set $\{\phi_k\}$. This is reflected on panels (i)–(iii) in Fig. 3.17, because here the non-converged $|d_k^-|$'s contain an admixture of $\text{Im}(d_k^-) \neq 0$, i.e., $\phi_k^- \neq 0$. Thus, in these cases we have $|d_k^-| \neq d_k^-$.

In contrast, on panels (iv) – (vi) in Fig. 3.17, all the 25 absolute values of the amplitudes $|d_k^-|$ show the feature $\text{Im}(d_k^-) = 0$, or equivalently, $\phi_k^- = 0$.

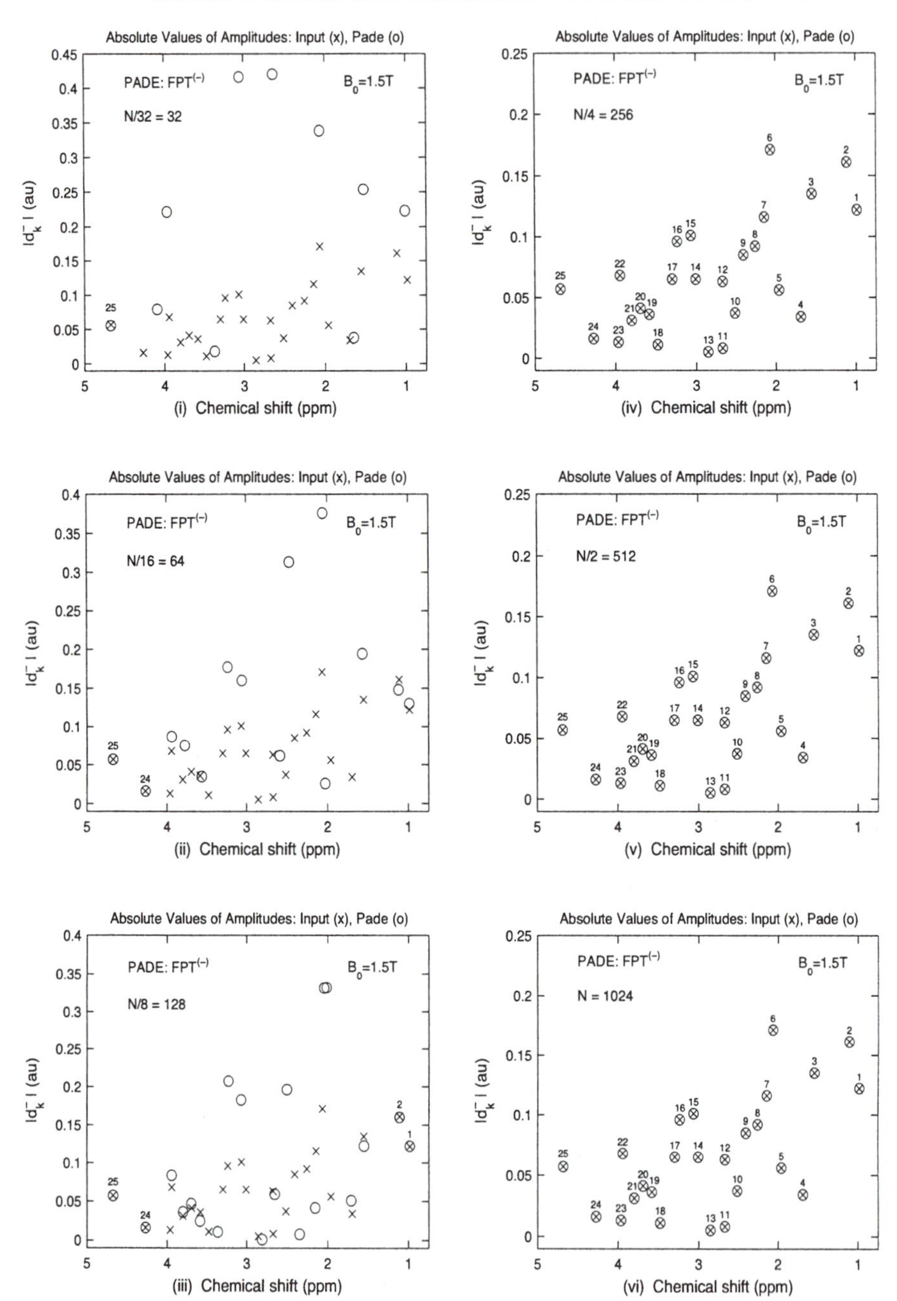

FIGURE 3.17
Convergence of the absolute values of the amplitudes reconstructed by the $\text{FPT}^{(-)}$ for signal lengths $N/32 = 32, N/16 = 64, N/8 = 128, N/4 = 256, N/2 = 512$ and $N = 1024$.

Thus, in the cases of panels (iv) – (vi) in Fig. 3.17, it follows that $|d_k^-| = d_k^-$ as in the corresponding exact values for $|d_k|$ from Table 3.1, where $\phi_k = 0$, so that $|d_k| = d_k \, (1 \le k \le K)$. In the $\text{FPT}^{(-)}$, any inaccuracies in the intermediate stage of computations should disappear for all the 25 reconstructed amplitudes once convergence has occurred. This expectation, based on theoretical grounds, is actually fulfilled already at $N/M \ge N/4 = 256$ as seen on panels (iv), (v) and (vi) on Fig. 3.17 where a quarter, one half and full signal length is used, respectively. This graphic visualization stems from the exact agreement between all the absolute values of the amplitudes in the input data (Table 3.1) and the corresponding values obtained by the $\text{FPT}^{(-)}$ with $N/4 = 256, N/2 = 512$ and $N = 1024$ (Table 3.2).

3.5.6 Distribution of fundamental amplitudes in $\text{FPT}^{(\pm)}$ near full convergence

Figure 3.18 shows the absolute values of the amplitudes reconstructed by the $\text{FPT}^{(+)}$ (left column) and $\text{FPT}^{(-)}$ (right column) near full convergence at three partial signal lengths $N_P = 180, 220, 260$. In the $\text{FPT}^{(+)}$ complete convergence to all the exact digits is attained at $N_P = 260$ in panel (iii). In contrast, the $\text{FPT}^{(-)}$ fully converges at $N_P = 220$ and $N_P = 260$ in panels (v) and (vi). Our computations show that stable estimation of the absolute values of the amplitudes in the $\text{FPT}^{(\pm)}$ continues at $N_P > 260$.

We emphasize that the most interesting features are actually shown prior to full convergence in the $\text{FPT}^{(-)}$. This can be seen on, e.g., panel (iv) from Figs. 3.16 and 3.18, particularly by referring to panels (i) and (iv) in Fig. 3.11 for the component and total shape spectra. It should be recalled that Fig. 3.11 shows that the converged total shape spectrum from panel (iv) is obtained without achieving convergence of the component spectra. This was due to a compensation effect through which the 11th peak was absent, and the 12th peak was over-estimated.

Panel (iv) in Figs. 3.16 and Fig. 3.18 provides the explanation for this compensation. Peak $k = 12$ on panels (iv) in Fig. 3.16 has a smaller imaginary frequency than the corresponding exact value, $\text{Im}(\nu_{12}^-) < \text{Im}(\nu_{12})$. In contrast, panel (iv) in Fig. 3.18 it is seen that $|d_{12}^-| > |d_{12}|$ so that $|d_{12}^-|/\text{Im}(\nu_{12}^-) > |d_{12}|/\text{Im}(\nu_{12})$. Here, $\{\nu_k, d_k\}$ and $\{\nu_k^-, d_k^-\}$ are the pairs of the input and $\text{FPT}^{(-)}$–reconstructed parameters, respectively.

According to (2.186), the ratio $|d_k^-|/\text{Im}(\nu_k^-)$ is proportional to the height h_k^- of the kth peak, i.e., $h_k^- = |d_k^-|/\text{Im}(2\pi\nu_k^-)$ and $h_k = |d_k|/\text{Im}(2\pi\nu_k)$. The height h_k itself is proportional to the kth peak area which is, in turn, proportional to the concentration of the kth resonance. Consequently, the over-estimation of peak $k = 12$ stems from the observed relation $h_{12}^- > h_{12}$. Moreover, a review of the numerical values of these spectral parameters from Table 3.3 demonstrates that the area of the 12th peak is over-estimated by the corresponding amount of the absent 11th peak. Thus, this compensation effect

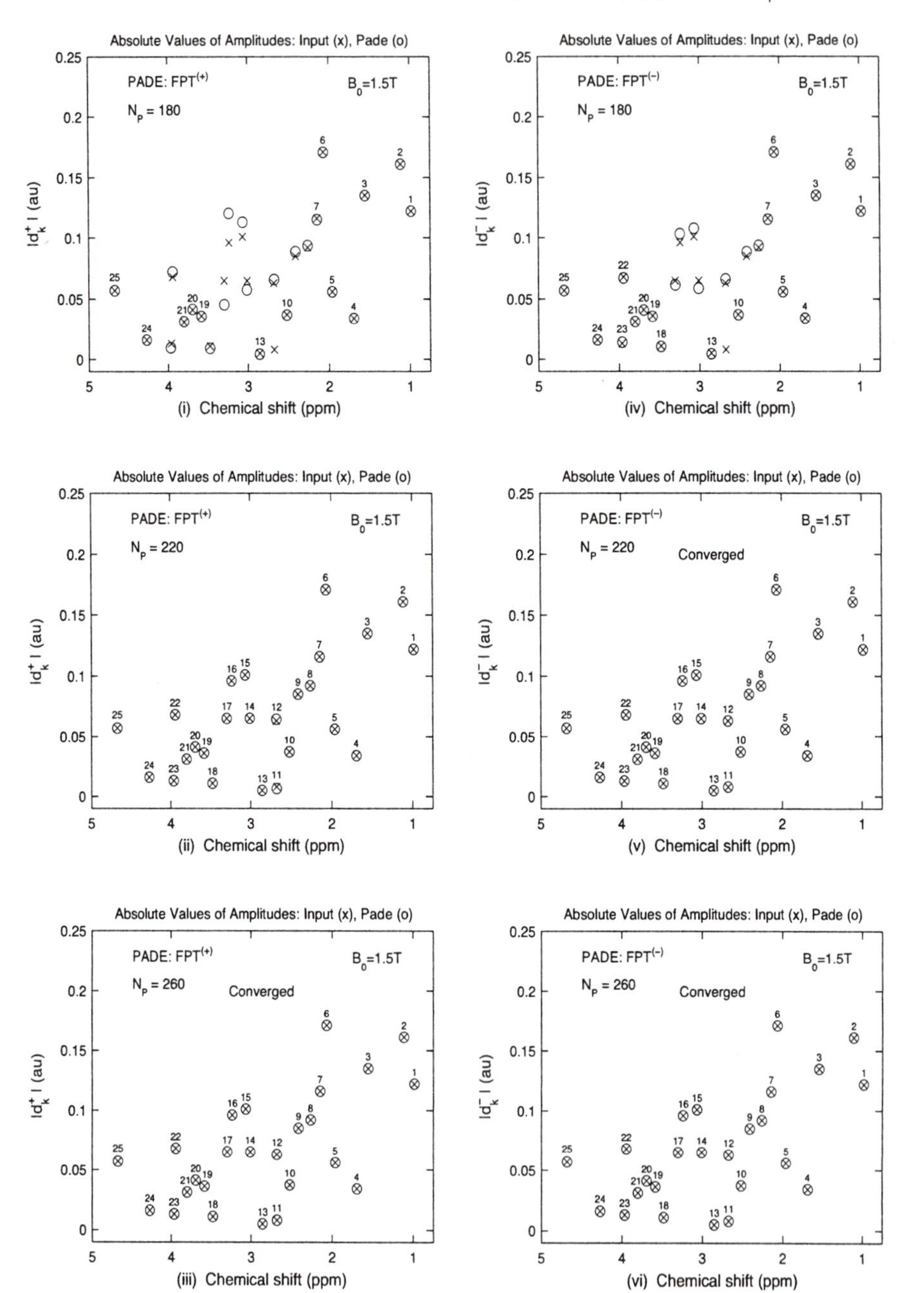

FIGURE 3.18
Convergence of the absolute values of the amplitudes reconstructed by the $\mathrm{FPT}^{(+)}$ (left) and $\mathrm{FPT}^{(-)}$ (right) near full convergence for signal lengths $N_P = 180, 220, 260$.

makes the total shape spectrum converge, on panel (iv) in Fig. 3.11, even in the absence of the peak $k = 11$ and the over-estimation of the peak $k = 12$. This also clearly shows that it is tenuous to use an envelope spectrum in attempts to describe the corresponding component spectra, as is customarily attempted in every fitting routine used in MRS and beyond.

3.6 Preview of illustrations for the concept of Froissart doublets

A critically important feature of the FPT is its ability to reconstruct exactly the total true number of physical harmonics in the given FID. Each such FID is a sum of damped complex exponentials with stationary and non-stationary (polynomial type) amplitudes, associated with non-degenerate (Lorentzian) and degenerate (non-Lorentzian) spectra. This type of FID is ubiquitous across the interdisciplinary research fields, including MRS [5]. In practice, determination of the exact number of resonances can be accomplished via Froissart doublets [44] or pole-zero cancellations [45]. The total number of genuine resonances is given by the degree K of the denominator polynomials $Q_K^{\pm}$. The only known information about this degree K is that it must obey the inequality $2K \leq N$.

Algebraically, the $2K$ unknown spectral parameters (frequencies and amplitudes) require at least $2K$ signal points from the whole set of N available FID entries, as mentioned. To determine K unequivocally, we compute a short sequence of the FPTs by varying the degree K' of the polynomials in the Padé spectra $\{P_{K'}^{\pm}/Q_{K'}^{\pm}\}$ until all the results stabilize/saturate. When this happens, e.g., at some $K' = K''$, we are sure that the true number K is obtained as $K = K''$. If we keep increasing the running order K' of the FPT beyond the stabilized value K, we would always obtain the same results for $K' = K + m$ and for K using any positive integer m according to (2.226). The mechanism by which this is achieved (i.e., the maintenance of the overall stability, including the constancy of the value of the true number of resonances) is provided by pole-zero cancellations or Froissart doublets [44].

By not knowing the exact number K in advance (as in encoding via, e.g., MRS), we would keep increasing the order $K' = K + m$, and this would lead to extra zeros from $P_{K+m}^{\pm}$ and $Q_{K+m}^{\pm}$. All the zeros of $P_{K+m}^{\pm}$ and $Q_{K+m}^{\pm}$ are the respective zeros and poles in the spectra $P_{K+m}^{\pm}/Q_{K+m}^{\pm}$ because these latter rational polynomials are meromorphic functions. These extra zeros and poles are spurious, as they cannot be found in the input FID, which is built from K true harmonics alone. However, such spurious poles and zeros in the spectra $P_{K+m}^{\pm}/Q_{K+m}^{\pm}$ for $m > 1$, beyond the stabilized number K of resonances, will automatically cancel each other, because of the special form of the Padé spectrum given by the polynomial quotients. Hence stability of the Padé estimation, as per (2.226).

This stabilization condition is the signature of the determination of the exact total number K of resonances. If the quantification problem is solved first, then the ensuing stability in (2.226) will be due to the constancy of all the spectral parameters that are reconstructed exactly by the FPT from the given FID. Of course, once all the parameters are reconstructed, it is not mandatory to search for saturation of the corresponding spectra in the FPT. In such a parametric signal processing within the FPT, the exact number of genuine resonances is determined by monitoring solely the constancy of all the found spectral parameters.

In the non-parametric version of the FPT, the exact number of genuine resonances is determined without spectral analysis, i.e., with no recourse to reconstruction of peak parameters. In such a case, the Padé spectra $P^{\pm}_{K'}/Q^{\pm}_{K'}$ are explicitly computed. This is done in a given frequency window or in the whole Nyquist range by systematically increasing the running order K'. When saturation of these computed spectra occurs at a certain value of K', the true number K of genuine resonances is extracted via $K' = K$. Here, the constancy of spectra is not achieved through Froissart doublets, because poles and zeros are not extracted at all. Rather, the polynomials $P^{\pm}_{K+m}$ and $Q^{\pm}_{K+m}$ *de facto* contain certain factored polynomials $S^{\pm}_{m}$, i.e., $P^{\pm}_{K+m} = \tilde{P}^{\pm}_{K}\bar{S}^{\pm}_{m}$ and $Q^{\pm}_{K+m} = \tilde{Q}^{\pm}_{K}\bar{S}^{\pm}_{m}$. Thus, the sought saturation of spectra occurs by cancellation of the common factors $\bar{S}^{\pm}_{m}(z^{\pm 1})$, in the Padé quotient $P^{\pm}_{K+m}/Q^{\pm}_{K+m} = \{\tilde{P}^{\pm}_{K}\bar{S}^{\pm}_{m}\}/\{\tilde{Q}^{\pm}_{K}\bar{S}^{\pm}_{m}\} = \tilde{P}^{\pm}_{K}/\tilde{Q}^{\pm}_{K}$.

Even in the non-parametric FPT, we may think of Froissart doublets being implicitly present in this cancellation of $\bar{S}^{\pm}_{m}$ as a whole. Of course, whether or not we solve the characteristic equations $\bar{S}^{\pm}_{m}(z^{\pm 1}) = 0$, polynomials $\bar{S}^{\pm}_{m}$ do inherently contain their roots $\bar{z}^{\pm 1}_{s}$ $(1 \le s \le m)$ such that their canonical forms $\bar{S}^{\pm}_{m} \sim \prod_{s=1}^{m}(z^{\pm 1} - \bar{z}^{\pm 1}_{s})$ implicitly exist. Therefore, given that $\bar{S}^{\pm}_{m}$ disappear altogether as the common factors in the quotients $P^{\pm}_{K+m}(z^{\pm 1})/Q^{\pm}_{K+m}(z^{\pm 1})$, this occurrence may be viewed as implicit cancellations of the hidden/latent joint terms $z^{\pm 1} - \bar{z}^{\pm 1}_{s}$ $(1 \le s \le m)$ in the numerators $P^{\pm}_{K+m}(z^{\pm 1})$ and denominators $Q^{\pm}_{K+m}(z^{\pm 1})$. Here, the same zeros $\{\bar{z}^{\pm 1}_{s}\}$ that satisfy the characteristic equation $\bar{S}^{\pm}_{m}(\bar{z}^{\pm 1}_{s}) = 0$, act simultaneously as zeros and poles depending whether they appear in the numerator or denominator of the quotients $\tilde{P}^{\pm}_{K}\bar{S}^{\pm}_{m}/\tilde{Q}^{\pm}_{K}\bar{S}^{\pm}_{m}$.

In this way, one may say that irrespective of whether poles or zeros $\{\bar{z}^{\pm 1}_{s}\}$ are explicitly available or not, Froissart doublets implicitly act by effectively reducing $P^{\pm}_{K+m}/Q^{\pm}_{K+m}$ to $\tilde{P}^{\pm}_{K}/\tilde{Q}^{\pm}_{K}$. Such a lowering of the degrees of the characteristic polynomials from $K+m$ to K, with parametric *and* non-parametric estimations within the $\text{FPT}^{(\pm)}$ promotes the concept of Froissart doublets to the status of an efficient method for reduction of the dimensionality of the problem. Note that the problem of dimensionality reduction in itself is a very important issue in the field of system theory, especially when dealing with large degrees of freedom [5]. Large systems are difficult to handle in any computation and, therefore, it is essential to reduce their original dimension without information loss. This is important even without the obvious concern for computational demands because capturing the essence of the investigated large system by an adequate extraction of a relatively small number of the main parametrizing characteristics allows simple and yet reliable descriptions. To achieve this goal, the parametric version of the FPT does not need a specially designed procedure, since Froissart doublets solve the dimensionality reduction problem *en route* while separating genuine from spurious information. This can be accomplished either in the whole Nyquist range or in one or more sub-intervals (windows) using the original time signal of the full length N. Alternatively, the Nyquist interval can be split into any number of sub-intervals and the original time signal could be beamspaced, i.e., decimated to a much shorter length $N_D \ll N$ with no information loss in any given window. Such a method, called decimated Padé approximant (DPA), was introduced by Belkić *et al.* [13] and explicitly combined with Froissart doublets by O'Sullivan *et al.* [63, 64].

Using the parametric variant of the FPT, both the pole-zero stabilization leading to the true K, and the exact reconstruction of all the fundamental frequencies, as well as the corresponding amplitudes are illustrated together in Fig. 3.19. Only a small number of all the obtained Froissart doublets appears in the shown frequency window in Fig. 3.19. The selected sub-interval 0 – 6 ppm is important because all the MR-detectable brain metabolites lie within this chemical shift domain of the full Nyquist range. Froissart doublets as spurious resonances are detected by the confluence of poles and zeros in the list of the Padé-reconstructed spectral parameters, as per (2.186). It is seen on panel (i) in Fig. 3.19 that the $\mathrm{FPT}^{(+)}$ disentangles the physical from unphysical resonances by the opposite signs of their imaginary frequencies.

In other words, the $\mathrm{FPT}^{(+)}$ provides the exact separation of the genuine from any spurious, i.e., noise-like content of the investigated time signal. By contrast, in the $\mathrm{FPT}^{(-)}$ depicted on panel (ii) in Fig. 3.19, genuine and spurious resonances are mixed together, since they all have the same positive sign of their imaginary frequencies. Nevertheless, the emergence of Froissart doublets also remains evidently clear in the $\mathrm{FPT}^{(-)}$ via coincidence of poles and zeros, with the ensuing unambiguous identification of spurious resonances. Precisely due to pole-zero coincidences, each Froissart doublet has zero-valued amplitudes, as seen on panel (iii) in Fig. 3.19, as per (2.227). This result, as another signature of Froissart doublets, represents a further check of consistency and fidelity of separation of genuine from spurious resonances and this is the basis of the SNS concept mentioned in chapter 1. Note that the full auxiliary lines on each sub-plot in Fig. 3.19 are drawn merely to transparently delineate the areas with Froissart doublets.

Once the Froissart doublets are identified and discarded from the whole set of the results, only the reconstructed parameters of the genuine resonances will remain in the output data. Crucially, however, the latter set of Padé-retrieved spectral parameters also contains the exact number K of true resonances as

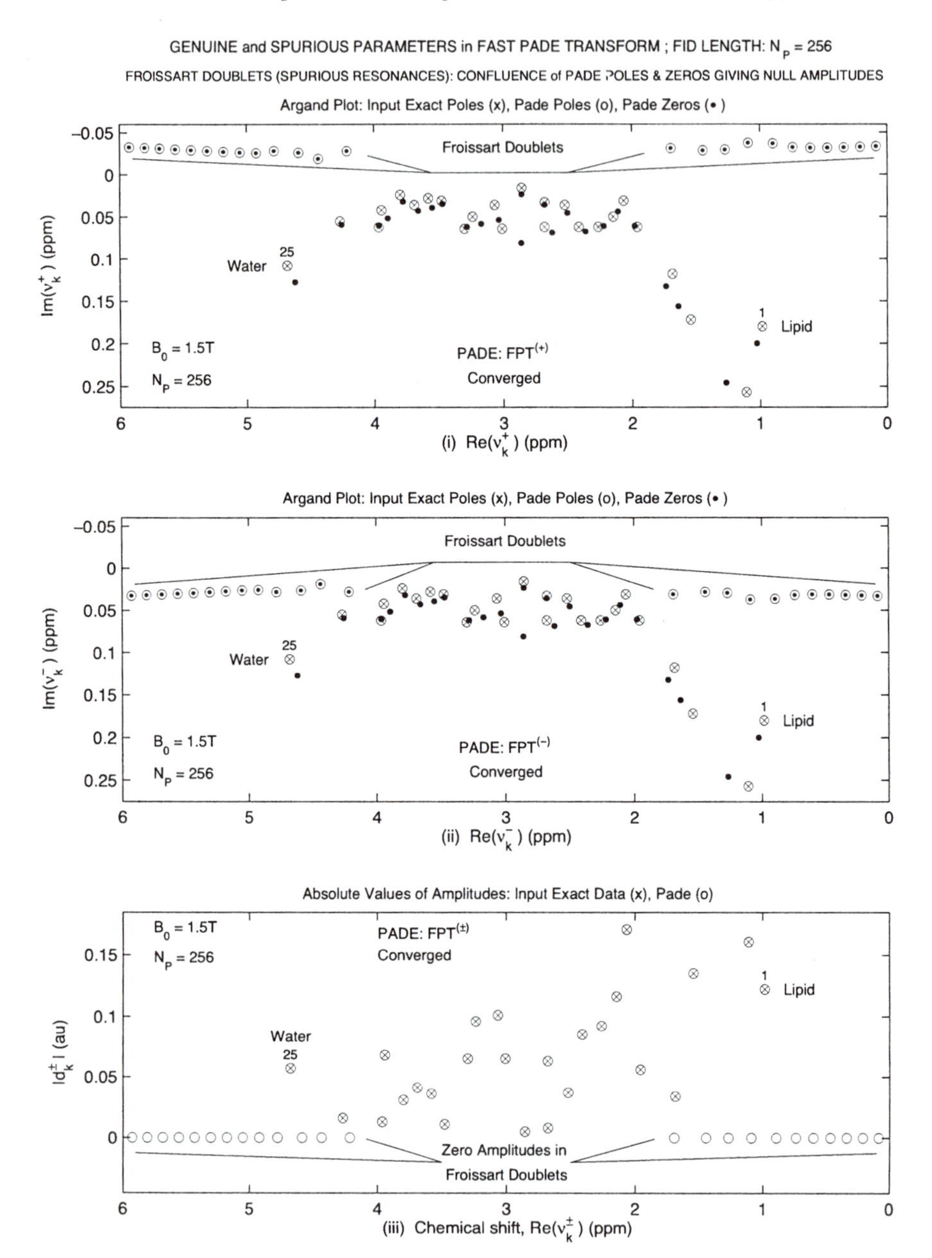

FIGURE 3.19
A subset of the whole set of Froissart doublets in the $\text{FPT}^{(\pm)}$ at a quarter of the FID length, $N/4 = 256\,(N = 1024)$. On panel (i), the $\text{FPT}^{(+)}$ achieves a total separation of genuine from spurious resonances that are mixed together in the $\text{FPT}^{(-)}$ on panel (ii). Panel (iii) shows genuine and spurious amplitudes in the $\text{FPT}^{(\pm)}$. The reconstructed converged amplitudes are identical in the $\text{FPT}^{(+)}$ and $\text{FPT}^{(-)}$. All the spurious amplitudes are zero-valued.

the difference between the total number of all the found resonances and the number of Froissart doublets. It can be seen in Fig. 3.19, that despite using the diagonal forms of the FPT$^{(\pm)}$ spectra, $P_K^{\pm}(z^{\pm 1})/Q_K^{\pm}(z^{\pm 1})$, there are only 24 genuine zeros as opposed to 25 genuine poles. The missing zero (located far away from the shown interval of the linear frequency $\nu = -i(\ln z)/(2\pi\tau)$ is the so-called "ghost" zero of the numerator polynomials $P_K^{\pm}(z^{\pm 1})$ and corresponds to the harmonic variables $z = 0$ in the FPT$^{(+)}$ or $z = \infty$ in the FPT$^{(-)}$. The original Maclaurin series $\sum_{n=0}^{\infty} c_n z^{-n}$ is not defined at the origin, since the value $z = 0$ of the harmonic variable $z = \exp(i\omega\tau)$ for $\mathrm{Im}(\omega) > 0$ is the singularity point[2].

Overall, Froissart doublets simultaneously achieve three important goals: (i) noise reduction, (ii) dimensionality reduction and (iii) stability enhancement. Stability against perturbations of the physical time signal under study is critical to the reliability of spectral analysis. The main contributor to instability of systems is its spurious information. Being inherently unstable, spuriousness is unambiguously identified by the twofold signature of Froissart doublets (pole-zero coincidences and zero-valued amplitudes) and, as such, discarded from the output data in the FPT. What is left is genuine information which is stable.

In practice, due to various reasons (finite arithmetics, computational round-off-errors, uncertainties in measured signal points, etc.), Froissart doublets do not elicit exact pole-zero matchings and, consequently, the corresponding amplitudes do not reduce to zero exactly, although they are very small. In such a case, the common factors in the Padé numerator and denominators, $P_K^{\pm}(z^{\pm 1})/Q_K^{\pm}(z^{\pm 1})$, will be the product of the quotient terms of the type $[z^{\pm 1} - (z_k^{\pm 1} + \tilde{\varepsilon}_k^{\pm})]/[z^{\pm 1} - (z_k^{\pm 1} + \varepsilon_k^{\pm})]$ where $|\varepsilon_k^{\pm}| \ll 1$ and $|\tilde{\varepsilon}_k^{\pm}| \ll 1$ as well as $\varepsilon_k^{\pm} \neq \varepsilon_k^{\pm}$. Nevertheless, the concept of Froissart doublets will be preserved in this case, as well, albeit in an approximate form, with the following rationale. The locations of poles $\{z_k^{\pm 1} + \varepsilon_k^{\pm}\}$ and zeros $\{z_k^{\pm 1} + \tilde{\varepsilon}_k^{\pm}\}$ will still be very close to each other due to $|\varepsilon_k^{\pm}| \ll 1$ and $|\tilde{\varepsilon}_k^{\pm}| \ll 1$. To find the qualitative behavior of the corresponding amplitudes $\{d_k^{\pm}\}$, it suffices to mention that the amplitudes are the residues of the Padé polynomial quotients. Here, the Cauchy notion of a residue can be conceived as the residual or remainder from a difference between the given pole and zero.

Thus, in general, $d_k^{\pm} \propto z_k^{\pm 1} - \tilde{z}_k^{\pm 1}$ where certain non-zero constants of proportionality need not be given for the present purpose (for more details on this, see chapter 6). When poles $z_k^{\pm 1}$ and zeros $\tilde{z}_k^{\pm 1}$ in these amplitudes are replaced by the mentioned terms $z_k^{\pm 1} + \varepsilon_k^{\pm}$ and $z_k^{\pm 1} + \tilde{\varepsilon}_k^{\pm}$, we have $d_k^{\pm} \propto (z_k^{\pm 1} + \varepsilon_k^{\pm}) - (z_k^{\pm 1} + \tilde{\varepsilon}_k^{\pm})$, so that $d_k^{\pm} \propto \varepsilon_k^{\pm} - \tilde{\varepsilon}_k^{\pm}$. It then follows from here, on the account of $|\varepsilon_k^{\pm}| \ll 1$ and $|\tilde{\varepsilon}_k^{\pm}| \ll 1$ that $|d_k^{\pm}| \ll 1$. Therefore, for approximate pole-zero coincidences, the ensuing amplitudes are indeed very small. Hence the

[2]Being located literally far away from both genuine and Froissart poles and zeros, one could use the term "outliers" (from statistical analysis of data) for "ghost" zeros and "ghost" poles (for more details, see chapter 5)

concept of the approximate SNS: appearance of quasi Froissart doublets and very small amplitudes, and this was set to explain. The smallness of Froissart amplitudes implies their sensitivity to even the slightest perturbations.

Therefore, these small amplitudes are the cause for great instability of spurious resonances, i.e., Froissart doublets. This is in sharp contrast to the stability of genuine resonances. Such a diametrically opposite behavior of physical and unphysical resonances greatly facilitates the task of distinguishing one from the other. In practical computations with, e.g., parametric signal processing, this is easily accomplished by merely monitoring the Padé table when passing from one Padé approximant $[K/K]$ to another $[(K+m)/(K+m)]\,(m = 1, 2, 3, ...)$. In so doing, as the order of the PA changes, we would observe that some resonances are robustly stable, whereas the others exhibit great instability. Then the former resonances are identified as genuine and the latter as spurious. A more detailed exposition on the theory and illustrations of Froissart doublets is given in chapters 5 and 6, respectively (for our previous studies, see Refs. [11, 24, 30, 31, 34]).

3.7 The importance of exact quantification for MRS

Single voxel *in vivo* MRS uses encoded time signals to extract biochemical information about concentrations of metabolites of the examined tissue. This can be critically important for cancer diagnostics, because malignant tumors, benign processes and healthy tissue usually differ markedly in their spectral information. These important differences are potentially detectable by MRS which can yield the initial information for the assignment of a particular disease to the identified spectral pattern of the scanned tissue.

In vitro MRS, which employs strong external magnetic fields, can resolve some fifty resonances. With this high resolution and excellent SNR, valuable diagnostic information is provided about tissue metabolites. However, such an excellent performance of *in vitro* MRS is not matched by the corresponding *in vivo* MRS on clinical scanners that operate at considerably weaker fields and with poorer SNR. Moreover, the usual information from *in vivo* MRS is limited to a relatively small number of metabolites that can be resolved and eventually quantified for the purpose of clinical diagnostics.

The main reason for this limitation is that signal processing for *in vivo* MRS relies predominantly upon the FFT. The diagnostic value of *in vivo* MRS could certainly be enhanced by employing high-resolution signal processors, such as the FPT. Especially because of its capability to resolve closely overlapped resonances, the FPT can be advantageously applied to clinical FIDs encoded by *in vivo* MRS at short echo times. Most frequently, FIDs encoded via MRS use long TE, because a larger number of short-lived tightly packed resonances create enormous problems for attempts at quantification by processing via fitting techniques. Furthermore, the powerful extrapolation characteristics of the FPT imply that long total acquisition times T of encoded FIDs, as required by the FFT, may not be needed. Such extrapolation advantages of the FPT also improve SNR relative to the FFT. Poor SNR has been cited in the literature as one of the major obstacles to the overall diagnostic performance of *in vivo* MRS [7, 116].

Originally, MRS was imported to medicine from basic laboratory research in physics and chemistry where it is known as NMR, which is one of the best spectroscopic methods available. However, the accompanying well-developed theoretical methods in the corresponding signal processing from physics and chemistry have not yet been widely imported to MRS. This situation should be improved on a more satisfying and deeper level in this extremely beneficial inter-disciplinary cross-fertilization.

Recall that the information which is sought from the tissue via MRS is acquired in two separate stages. In the first instance, the experimental measurement is used to encode a number of FIDs in a scanner[3]. Next, in the second step, analytical methods are used through mathematical processing to quantify the averaged FID. The results of parametric signal processing are the spectral parameters. These parameters are used to generate a frequency spectrum which has a number of peaks representing resonances at different chemical shifts. These peaks can be isolated, overlapped, tightly overlapped, nearly degenerate or degenerate (confluent). Each metabolite is linked to one or more resonances. The extracted spectral parameters determine the areas underneath these peaks in the frequency spectrum. The peak areas are proportional to the sought concentrations of metabolites. The retrieved spectral parameters for every individual resonance are the pairs of fundamental complex frequencies and the associated complex amplitudes. The peak position or chemical shift is the real part of the complex frequency. The peak width is proportional to the imaginary part of the complex frequency. The relaxation times of metabolites are equal to the inverse of the imaginary frequencies, as per (2.186). The peak height is proportional to the quotient of the absolute value of the amplitude and the associated imaginary frequency. The phase of the individual harmonic component of the FID is the phase of the complex-valued amplitude.

Thus, the reconstructed spectral parameters provide all the quantitative physical and biochemical information about metabolites of the scanned tissue, such as their concentrations, relaxation times, etc. The process of reconstruction of the pairs of complex-valued spectral parameters and the true number of resonances for the given FID represents an inverse nonlinear problem. In many research areas, this type of problem is encountered under a number of different

[3]Usually the number of encoded FIDs is of the order of 200. Such measured FIDs are afterwards averaged to improve SNR. The averaged FID is then subjected to signal processing.

names, e.g., spectral analysis, harmonic analysis or complex-valued moment problem (mathematics), harmonic inversion (physics, chemistry), quantification (medicine: MRS, MRSI), etc.

We underscore the key role played by signal processing in MRS. This is due to the fact that the most clinically relevant quantitative information about the examined tissue cannot be extracted from encoded FIDs without analytical methods. These mathematical methods are needed to properly interpret the measured time signals. Such a theoretical interpretation of the encoded data, through solving the underlying inverse problem, yields the sought information about metabolite relaxation times, concentrations and the like. All of this is accomplished via an adequate reconstruction of the frequency-dependent parameters for the spectral profiles of the unknown fundamental components that are ingrained in the encoded FID. Generally, an inverse problem consists of finding the causes from the observation or measurement of the effects of the investigated phenomenon. In fact, nearly all important measurements in science belong to the category of inverse problems. These inverse problems are extremely abundant in medicine. They are the essence of many powerful diagnostic tools without which modern hospitals and clinics would be basically non-functional. The most prominent examples of these versatile diagnostic modalities, that originate from basic research in physics and chemistry, are NMR, MRI, MRS, MRSI, CT, PET, SPECT, US, etc. The common feature of these powerful diagnostic tools is the impossibility for direct interpretations of the measurements. Acquisition of the needed information requires theory whose mathematical methods are capable of solving reconstruction or inverse problems that are inherent in all the mentioned diagnostic modalities.

Surprisingly, for too long a time now, signal processing in MRS has been dominated by the FFT. Such an occurrence is at variance with the quantification problem, which is the very goal of MRS. As noted, the FFT is a non-quantifying estimator of envelope spectra. This qualitative information is usually complemented in MRS by fitting either encoded FIDs or Fourier spectra. These fittings employ some free, LS-adjusted parameters in attempts to quantify the given MRS data. However, none of the fitting algorithms from MRS and elsewhere can provide the unequivocal solution for the mentioned inverse problem. Namely, fittings cannot yield the unique estimates for the sought spectral parameters that are the positions, widths, heights and phases of the peak profiles, nor about the total number of resonances. As seen herein, this is the case because a given total shape spectrum can be fitted by several different sets of component shape spectra. Each set contains a number of metabolites. However, any fixed resonance can be of vastly different spectral characteristics when passing from one to another set of component shape spectra. Thus, these fittings are non-unique in MRS and in other research fields. This typically leads to two difficulties in practice, such as over-fitting and under-fitting. Over-fitting retrieves some non-existent, i.e., spurious metabolites. Under-fitting fails to reconstruct some of the existing, i.e., genuine metabolites. In both cases, a fitting bias exists, entailing guessing the number of the sought physical resonances. This could severely hamper diagnostics that are supposed to be aided by signal processing. For example, in brain tumor diagnostics via MRS, as reviewed in chapter 8, several inconclusive findings have been reported with contradictions depending on whether some typical metabolites were included or excluded from the basis set of *in vitro* spectra used in the LCModel for fitting the corresponding *in vivo* data [144, 145].

All fitting algorithms available in MRS use the customary LS adjustments. This is accomplished in either the time or the frequency domain. In the frequency domain, fittings attempt to adjust the envelope spectrum by a chosen model in order to reconstruct the most important component shape spectra of individual resonances and their spectral parameters. As stated earlier, the LS technique is a minimization of the LS residual, which represents the squared difference between the input and the modeled information. Any selected model in fittings is based upon variations of the free parameters until eventually a minimum of the LS residual is found. Invariably, a detected minimum is a local extremum rather than the global minimum. If interpreted properly, this would have grave consequences especially for non-linear fitting (e.g., VARPRO, AMARES, LCModel, etc.) which yields several local minimae with comparable χ^2 and LS residuals for the given Fourier envelope spectrum, thus giving different predictions of spectral parameters with no way to tell which of the set of estimates is correct. The parameters to be estimated are often constrained by some conditions to facilitate fittings. Similarly, in the time domain, the studied time signal is fitted in MRS via adjustable parameters to match the damped harmonic oscillations in the encoded raw data. Regardless of the studied domain (time or frequency), fitting prescriptions from MRS invariably use non-orthogonal expansion functions. In most cases these basis sets are damped complex exponentials, Gaussians and/or their combinations as implemented in VARPRO, AMARES, etc. Alternatively, rather than modeling individual resonances, fittings in MRS also use the entire model spectra with some pre-selected metabolites that play the role of a non-orthogonal basis set expansion as done in the LCModel. As mentioned, in all such fittings, non-orthogonality of expansion sets implies that any alteration in the weighted sum of the LS residuals caused by changes of one or more adjustable parameters in the selected mathematical model, could be compensated to a large extent by completely independent variations of the other free parameters [78]. This arbitrariness becomes particularly problematic when the number of adjustable parameters in fittings is not small. Such a mathematical ill-conditioning leads to non-uniqueness of all the LS fitting algorithms currently used in MRS and elsewhere.

In contradistinction, the FPT yields unique reconstructions. In the FPT, the complex frequencies are unequivocally determined by solving the characteristic equation, which as the so-called secular equation is equivalent to the Schrödinger equation [13]. This leaves no room whatsoever for freedom in 'varying' the amplitudes, since they are unambiguously fixed by their definitions as the Cauchy residues (2.183) or (2.192) of the associated Green functions. There is also the uniqueness proof for these amplitudes. This proof demonstrates that, for the same set of retrieved complex frequencies, all the various ways of computing the associated amplitudes must yield identical results [5].

The fundamental frequencies for the genuine resonances reconstructed via the HLSVD and FPT under the same conditions should therefore coincide with each other. Consequently, despite the fact that in the HLSVD, the amplitudes are computed by solving a system of linear equations, rather than through the analytical expression of the FPT, the results from both methods must be the same. This should serve as a check of the numerical solutions for the amplitudes in the HLSVD. Obviously, the procedure in the FPT is optimal for the amplitudes of resonances encountered in MRS, since these Padé residues are available in the closed, analytical formula (2.183) or (2.192).

As stated in chapter 1, it has frequently been contended in MRS that the so-called prior information from fittings should be used in non-fitting algorithms, as well. Moreover, it is claimed that the overall quality of quantification is pre-determined by some prior information. However, the alleged 'biochemical' prior information used in MRS is most often a pre-assigned relationship, e.g., fixed phases or amplitude ratios or quotients of peak areas of some resonances (or imposed connections among widths of several peaks in a selected part of the spectrum or throughout the Nyquist interval, etc.), as done in VARPRO, AMARES and LCModel. These unsubstantiated claims imply that such prior information should be viewed as a requirement for successful quantifications in MRS. This is incorrect, since the introduction of this type of prior information actually represents an attempt to mitigate the effect of subjectivity of fittings. These attempts via certain imposed constraints germane only to fitting are, in fact, aimed at reducing the error from under- or over-estimation of subjectively pre-selected metabolites in an expansion set. Thus, for example, under-fitting would result in an over-estimation of the actual values of the spectral parameters (amplitudes) because of the lack of sufficient number of resonances in an expansion sets. Such an outcome is due to some attempts to minimize the LS residuals. This is particularly unwelcome in medical diagnostics via MRS, since such minimizations could easily over-estimate the peak areas and this, in turn, would undermine the estimates for the related concentrations. The fitted peak areas are affected by attempts to minimize the error from the missing genuine resonances in a chosen basis set by over-weighting the relative contribution from the included resonances.

These outlined fundamental deficiencies of fitting approaches to quantifications in MRS also indicate the need for better estimations of spectral parameters. These non-fitting strategies should not contain any free parameters and should provide unique solutions for restoration of the unknown harmonic components from FIDs in MRS.

The FPT is well-suited to fulfill this task of quantification in MRS. This can be appreciated by noting that the FPT is the exact filter for all the

FIDs built from attenuated complex exponentials with stationary and non-stationary amplitudes that yield Lorentzian and non-Lorentzian spectra in MRS. The term 'filter' denotes an operation applied to the time signal with the result given by a baseline constant c_{∞}. This constant is the limiting stationary value of the time signal from which all the harmonics (transients) have been filtered out in a transformation whose outcome is equivalent to taking the limit of infinite times ($n\tau \longrightarrow \infty$) in the considered FID, i.e., c_n. This exact filter is the Shanks transform, i.e., the ST which is, as noted, the FPT applied directly in the time domain [5]. The implicit time limiting process $t \longrightarrow \infty$, which goes hand in hand with the ST, well illustrates the extrapolation features of the time-domain version of the FPT. The ST is defined by (2.198) as a quotient of two Hankel determinants. In the case of higher dimensions, these determinants are not practical for extensive numerical computations. Nevertheless, the ST is salvaged by the Wynn $\varepsilon-$algorithm (2.201) which computes the ratios of the two Hankel determinant recursively. This recursion is actually another computational algorithm in the FPT. In order to confirm that these statements about filtering of harmonics from an FID are plausible, it should be recalled that there is a simpler and more familiar accelerator called the Aitken extrapolation, or equivalently, the Δ^2-process [5, 77]. If an FID has only one harmonic ($K = 1$), then the Aitken transformation is the exact filter in the above-defined sense. A direct generalization of the Aitken transform to an FID with any higher number of harmonics ($K > 1$) represents the ST as the multi-exponential version of the single-exponential time signal.

The Padé methodology is not just another spectral analyzer added to the other already existing ones. In addition to its practical advantages of uniqueness and exactness for solving the quantification problem in MRS, the most fundamental characteristic which distinguishes the Padé methodology from the conventional signal processors is that it is much more than a powerful computational framework. Namely, the Padé methodology is the cornerstone of the very formulation of the Schwinger variational principles, the Heisenberg scattering matrix, the Green function, to name only a few of the leading strategies in the most successful physics theory – quantum mechanics [5].

Overall, in this chapter, we have focused upon the possibility of obtaining the exact solution of the quantification problem for noiseless synthesized FIDs from theoretical modelings within MRS. We have employed the two equivalent variants of the fast Padé transform, i.e., the $\mathrm{FPT}^{(+)}$ and $\mathrm{FPT}^{(-)}$ defined with their initial convergence regions that reside inside and outside the unit circle, respectively. Quantification in MRS entails mathematically ill-conditioned harmonic inversion. As stated, other terminologies used to describe this problem include: spectral analysis, reconstruction, spectral decomposition, etc. In this problem, the FID is given, but the damped fundamental harmonics from which it is built are not known. By means of the FPT, we succeeded to exactly reconstruct the spectral parameters that are the complex frequencies and the associated complex amplitudes. These extracted quantities can be interpreted to yield the physical and biochemical information about the metabolites in the tissue from which the FID has been encoded.

We have shown that the quantification problem in MRS can be solved exactly with the FPT applied to a typical synthesized noiseless FID. As noted, the values of the complex frequencies and amplitudes for the 25 component resonances are akin to time signals encoded *in vivo* on clinical MR scanners at B_0 =1.5T using short TE of 20 ms at bandwidth 1000 Hz and the signal length 1024 [88]. Similarity between the presently synthesized FID and time signals encoded via *in vivo* MRS was further ensured by selecting spectral parameters that would generate isolated, overlapped, tightly overlapped as well as almost degenerate resonances.

It should be recalled that chemical shifts are the resonating frequencies of the tissue protons. These are much smaller than the Larmor precessing frequency of proton magnetization vectors around the axis of the external static magnetic field of strength B_0. It is this very small correction to the Larmor frequency that is the basis of the entire MRS. Without these small corrections, all the scanned protons would resonate at a fixed identical Larmor frequency and, therefore, the spectrum would be comprised of just a single peak. In fact, this would be the case if all the protons under investigation were free. However, protons are bound in various molecular compounds in the tissue. The electronic clouds of various molecules are able to produce different shieldings of proton magnetic fields. Due to such varying shieldings, protons resonate with the external field at frequencies that are slightly shifted relative to the Larmor frequency. This shielding depends on the proton chemical surrounding. Nevertheless, especially for the human brain metabolites investigated via proton MRS, the range of such shifting is quite small. Namely, for the human brain investigated via proton MRS, most of the chemical shifts associated with the main part of a customary spectrum are found between 0 ppm and 5 ppm. This range is just a fraction of the entire Nyquist interval. The resonances selected for our synthesized FID correspond to a linear combination of damped complex exponentials with stationary amplitudes.

As discussed, the FPT can also handle degenerate resonances that lead to non-Lorentzian spectra. In such a case, the corresponding FID becomes a sum of damped complex exponentials with non-stationary (or time-dependent) amplitudes. Parametric as well as non-parametric analyses can be performed by the FPT. As described, the former is achieved by explicitly solving the quantification problem. Non-parametric analysis is performed as a transform, with frequency-by-frequency computation of the response function. This is done without necessarily reconstructing spectral parameters themselves.

The two different versions of the FPT via the $\text{FPT}^{(+)}$ and $\text{FPT}^{(-)}$ unify the PA and PzT, where z is the frequency-dependent complex harmonic variable. These two versions of the FPT were created with the same truncated input Maclaurin series expansion for the exact spectrum as expressed by the exact Green function of the system under examination. The expansion coefficients in this Maclaurin development are the time signal points, that are equivalent to the quantum-mechanical auto-correlation functions. The $\text{FPT}^{(+)}$ and $\text{FPT}^{(-)}$

are both formulated as the unique polynomial quotients in their respective independent variables z and $1/z$. Thus, as mentioned, the initial convergence regions of the $\mathrm{FPT}^{(+)}$ and $\mathrm{FPT}^{(-)}$ are inside ($|z| < 1$) and outside ($|z| > 1$) the unit circle, respectively. As rational analytical functions, both polynomial quotients expand their initial validity to the entire complex plane of the harmonic variables according to the Cauchy principle of analytical continuation. Thereby, the $\mathrm{FPT}^{(+)}$ and $\mathrm{FPT}^{(-)}$ can be computed everywhere with the exception of the singular points given by the poles that represent the locations of fundamental frequencies from the examined FID.

Herein, the original Maclaurin series is defined in the variable $1/z$. Consequently, this exact Green function expansion is convergent for $|z| > 1$ and divergent for $|z| < 1$, respectively. Thus, the $\mathrm{FPT}^{(+)}$ analytically continues a divergent series forcing it to converge for $|z| < 1$. In contrast, the $\mathrm{FPT}^{(-)}$ has the relatively easier task of accelerating the convergence rate of an already convergent series for $|z| > 1$.

It should be pointed out, however, that analytical continuation entails more than transforming an initial series in terms of $1/z$ into another series in powers of z. Analytical continuation also denotes that a function originally defined, e.g., only for real frequencies, can be mapped into another more general function of complex frequencies. Both the $\mathrm{FPT}^{(+)}$ and $\mathrm{FPT}^{(-)}$ accomplish this task. Therefore, in fact, they both perform analytical continuation: $\mathrm{FPT}^{(+)}$ for $|z| > 1$ and $\mathrm{FPT}^{(-)}$ for $|z| < 1$. Insofar as the defining polynomial quotients from the $\mathrm{FPT}^{(+)}$ and $\mathrm{FPT}^{(-)}$ are expanded in their own Maclaurin series, they coincide with the original Maclaurin expansion at a fixed order depending on the degrees of the numerator and denominator polynomials. In other words, all the converged results from the $\mathrm{FPT}^{(+)}$ and $\mathrm{FPT}^{(-)}$ are expected to be identical. This is the basis of the uniqueness of the FPT.

This internal cross-validation of the FPT is confirmed with numerical illustrations by carrying out the exact reconstructions of all the input spectral parameters. Furthermore, as discussed, the FPT accomplishes this task by using a quarter of the full MR time signal.

For a given MR time signal, the spectrum in the FPT can be formulated, for example, in the diagonal form as the quotient of two unique polynomials P_K/Q_K. Then, the sought exact number of resonances due to the fundamental harmonics from the given FID can be determined unequivocally in the FPT by the degree K of the denominator polynomial Q_K. In sharp contrast, as said earlier, all the other estimators used in MRS and especially fitting algorithms can only guess the true number K. We have demonstrated that the FPT takes a completely different approach by viewing the exact number K of resonances as yet another spectral parameter to be recovered together with the complex frequencies and amplitudes in the process of solving the quantification problem. We should herein emphasize that the entire strategy of the FPT is completely orthogonal to the usual practice with attempted quantifications in MRS. Therefore, the Padé methodology actually represents a paradigm shift in the field of signal processing.

As discussed, the FPT does not need a total shape spectrum such as the one from the FFT or any other processor in order to reconstruct the spectral parameters. On the contrary, the FPT begins by extracting the complex frequencies and the corresponding amplitudes directly from the input raw data. The spectral analysis could actually finish at that very point. Of course, when all the spectral parameters are faithfully reconstructed, the shape spectra in any mode can be subsequently constructed in a straightforward manner through the available explicit formulae.

The computational work needed for solving the quantification problem via the FPT is reasonably straightforward. The complex resonance frequencies are reconstructed by determining the roots of the denominator polynomial Q_K. The optimal method for rooting a polynomial of arbitrarily large order is to solve the equivalent eigenvalue problem of the Hessenberg matrix, which is extraordinarily sparse (the polynomial coefficients on the first row, unity on the main diagonal and zeros elsewhere) [13, 77]. The Padé denominator polynomial Q_K is actually the characteristic or secular polynomial from quantum mechanics and linear algebra in numerical analysis. After determining the fundamental frequencies, the corresponding unique amplitudes are taken from the analytical expression for the Cauchy residue of the quotient P_K/Q_K. This is in contrast to the procedures needed in, e.g., the HLSVD in which a system of linear equations must be solved.

We emphasize here that these features are convenient, so that the FPT quite naturally arrives at the quantifications in MRS. A robust algorithmic implementation of the FPT, as done in the present illustrations using exclusively the standard MATLAB routines, requires unprecedentedly minimal computational effort. Furthermore, the results are numerically exact in the face of the typical finite precision arithmetic with its attendant round-off errors. Herein we explicitly showed that by using less than a quarter of the full time signal length, the $\text{FPT}^{(-)}$ can reconstruct exactly all the spectral parameters of the investigated 25 resonances. This includes isolated, overlapped, tightly overlapped as well as nearly degenerate peaks.

It is of interest to note that the Fourier analysis itself actually corroborates the success of Padé analysis. It should be recalled that from the Fourier analysis, the exact one-sided Fourier integral from zero to infinity over a time signal from MRS, modeled by a linear combination of damped complex exponentials with constant amplitudes, is expressed exactly by the corresponding sum of Lorentzians. This, in turn, is the frequency spectrum. Moreover, continuing within the Fourier analytical framework, this latter sum is expressed explicitly in the representation of the Heaviside partial fractions. When the sum over these partial fractions is performed, the result is invariably the ratio of two unique polynomials, and this is how the standard Padé approximant is obtained. However, the Fourier analysis stops short before this latter statement becomes explicit. Nevertheless, with a deeper insight it can be seen that the exact spectrum for these FIDs is a rational function given by a polynomial quotient. Thus, for this class of functions that describe spectra as encountered in MRS, the Padé approximant is, in point of fact, the exact theory. This consideration in itself further confirms that this processor should be considered optimal for MRS.

Thus, even the Fourier analysis naturally leads us to the Padé-based MRS. Furthermore, these observations are completely consistent with an independent formulation of the FPT. Namely, the Maclaurin series or the $z-$transform, the Green function and the Heaviside partial fraction representation containing implicitly the Padé polynomial quotient, are all incorporated into the FPT from the outset, and there is actually no reference to the Fourier integral or Fourier analysis. Moreover, while these expansions for the exact spectrum might provisionally be the starting point of the FPT, the end result of the Padé spectral analysis is the set of fundamental frequencies and amplitudes as the reconstructed attenuated harmonic constituents of the FID from the time domain. Exactly the same results are gleaned if the analysis begins with the FID rather than its spectrum. Thus, the process of estimation in the FPT is performed by handling the time and frequency domain on the same footing. Both time and frequency domain analysis generate exactly the same characteristic equation or the eigen-problem, whose solutions are the spectral parameters. Therefore, for the full quantitative information about the system, it is sufficient to specify these parameters as the fundamental frequencies and the corresponding amplitudes without necessarily indicating the way in which they might be used. Such spectral parameters can subsequently be employed to generate spectra in various modes, to retrieve the FID or to extrapolate it to unmeasured values of the time signal, etc.

A system will be deemed robust if it is capable of maintaining stability in the face of external excitations or perturbations. In numerous systems across a wide range inter-disciplinary fields, including basic sciences and engineering, optimal stability is modeled by rational response functions via polynomial quotients as predicted by the Padé analysis. Thus, instead of defining these rational polynomials via their frequency representations of the Heaviside partial fractions, it is sufficient to simply give the defining parameters of these functions as the fundamental frequencies and amplitudes. This is reminiscent of an alternative definition of an ordinary single polynomial, not necessarily through its development in powers of the independent variable. Rather, this is done via the set of all its genuine parameters as the expansion coefficients. It should also be pointed out that these general concepts concerning Padé-based MRS render the interface between the time and frequency domain analysis rather loose. But this is as it should be. In contrast, from the very beginning, fittings are defined either the time or frequency domain. These alternatives depend upon whether the FID or its spectrum is used to adjust the free, fitting parameters via some *ad hoc* models [79]–[83].

As mentioned, quantification in MRS is mathematically ill-conditioned. This is due to the lack of a continuous dependence of the sought solution upon the input data. The intrinsic difficulty of this problem, irrespective of the methods used, can yield large variations in the reconstructed spectral parameters even for rather small uncertainties in the encoded time signal. These uncertainties can arise from a number of sources, including experimental errors, statistical effects manifested as noise, etc. In addition, the methods of spectral analysis can contribute to further uncertainties via computational round-off errors, algorithmic instabilities, etc. In attempting to deal with all these types of corruption of the analyzed data, one typically confronts major obstacles. This is certainly the case also for newly designed methods such as the FPT which is herein applied to the specific area of MRS.

The prudent approach to such a challenging problem would, therefore, be to systematically evaluate the validity and the overall performance of a given processor using realistically synthesized time signals. These would be noise-free as well as noise-corrupted. Such a systematic strategy is necessary to gain confidence while establishing the validity of the selected estimator when the solution of the quantification problem is known. This gives a much better chance for reliability of estimations for FIDs that are measured experimentally. For the theoretically-generated time signals, the fundamental frequencies and amplitudes are known from the onset. Thereby, the evaluation can be focused upon testing the accuracy, robustness and reliability of the estimator. Synthesized noiseless FIDs merit special investigation as done herein, since every step in the analysis is completely controllable, starting with the input data until all the information is reconstructed. Thus, we hereby demonstrate that the FPT exhibits an unprecedentedly high performance for each critical step in the exact solution of the quantification problem in MRS.

From the high-resolution spectral analysis presented herein for noiseless synthesized FIDs, and considering the results obtained together with their importance for furthering the field of MRS, in particular for applications to clinical oncology, the following central question is asked. Is it realistic to utilize the remarkable reliability of the FPT, according to all the predictions of the quantum-mechanical spectral analysis and signal processing, to carry out the most accurate and robust quantification also for noise-corrupted synthesized as well as encoded *in vivo* MRS time signals? The answer in the affirmative to this key question can be surmised from our investigations on the level of estimation of envelope spectra computed by the FPT using FIDs encoded by *in vivo* MRS [8, 9]. Although in these investigations, quantification problems have explicitly been solved by the FPT prior to construction of such spectra, the concrete detailed information was not reported, especially about the sensitivity of all the reconstructed fundamental frequencies and amplitudes to the presence of noise in the encoded FIDs. This should be done in a systematic manner for both synthesized noise-corrupted and encoded FIDs. We have, in fact, performed this demanding task in chapter 6 by extending strictly the Padé methodology to noise-polluted FIDs. As a preview, we stress that the role of Froissart doublets or pole-zero cancellation for distinguishing physical from non-physical harmonics, as demonstrated herein for noiseless FIDs, is of key importance for identifying the genuine resonances in noise-corrupted FIDs and clearly differentiating them from spurious content.

4

Harmonic transients in time signals

A large number of parametric estimators [146]–[205] can be used for tackling spectral analysis of time signals. In so doing, long-time practice across interdisciplinary fields firmly established that the most adequate strategy for this purpose is the powerful methodology of rational approximations. Moreover, this same practice has also found that the most frequently employed rational function is the Padé approximant to which many of the existing methods [146]–[205] can either be reduced or become equivalent. This is a good motivation for devoting this chapter to the theory of rational approximations with the highlight on the leading role of the Padé approximant in this field.

4.1 Rational response function to generic external perturbations

From the outset, the quantification problem in MRS is the same as the harmonic inversion from quantum theory of resonances and spectroscopy [6, 92]. Thus, the general quantum-mechanical relaxation formalism can advantageously be used to solve the quantification problem in MRS. To this end, we shall use the quantum-mechanical parametric signal processing via the frequency-dependent Green function $G(\omega)$. With this strategy, upon its algorithmic implementation through the FPT, it becomes feasible to reconstruct exactly all the spectral parameters [10, 11, 24, 34]. The FPT provides the frequency spectrum as the unique ratio of two polynomials

$$\mathrm{G}_{L,K}(\omega) \propto \frac{P_L(\omega)}{Q_K(\omega)}. \tag{4.1}$$

The two subscripts L and K are written in the label for the Green function $\mathrm{G}_{L,K}$ to indicate the degrees of the numerator and denominator polynomial, respectively. As discussed, for the given degrees L and K, these polynomial quotients can yield the off-diagonal ($L \neq K$) and the diagonal ($L = K$) variant of the FPT. Here, the particular case $L = K - 1$ of the off-diagonal FPT is called the para-diagonal fast Padé transform. In practice, the most frequently used versions are the para-diagonal and the diagonal FPT because of their pronounced stability. This stability is the main reason for wide applications of the Padé approximant via continued fractions for computation of virtually all elementary and special functions. As mentioned earlier, the diagonal $[K/K]_f(z)$ and para-diagonal $[(K \pm 1)/K]_f(z)$ Padé approximant to a given function $f(z)$ are equivalent to the continued fractions [5].

In MRS, virtually all spectra have rolling backgrounds that can be naturally described by the off-diagonal FPT. This is due to the fact that the quotient P_L/Q_K for $L > K$ can always be expressed as a sum of a polynomial B_{L-K} and the diagonal FPT which is the remaining quotient A_K/Q_K (see also chapter 2, section 2.6)

$$\mathrm{G}_{L,K}(\omega) \propto \frac{P_L(\omega)}{Q_K(\omega)} = B_{L-K}(\omega) + \frac{A_K(\omega)}{Q_K(\omega)}. \tag{4.2}$$

The polynomial B_{L-K} describes the background, and the new diagonal rational function A_K/Q_K is responsible for the polar structures of the spectrum, i.e., its resonances. Crucially, the particular mathematical form (4.1) of the quantum-mechanical rational response function to the given external perturbations or excitations is dictated by the intrinsically quantum origin of time signals from MRS. The same form is also prescribed by the resolvent structure of the Green operator, which generates the entire dynamics of the investigated general system [5].

The specific rational form of a generic response function $R(\omega)$ in many research fields is also implied by a standard connection between the input $I(\omega)$ and the output $O(\omega)$ data in linear systems

$$\text{Output} = \text{Response} \times \text{Input} \qquad\qquad O(\omega) = R(\omega)I(\omega). \tag{4.3}$$

Therefore, for a general system, $R(\omega)$ ought to be a rational algebraic function as a ratio of the output and input data

$$\text{Response} = \frac{\text{Output}}{\text{Input}} \qquad\qquad R(\omega) = \frac{O(\omega)}{I(\omega)}. \tag{4.4}$$

The simplest form for the response function is given by a quotient of two polynomials $O_L(\omega)$ and $I_K(\omega)$. In such a case, $R(\omega)$ as denoted by $R_{L,K}(\omega)$ becomes

$$R_{L,K}(\omega) \propto \frac{O_L(\omega)}{I_K(\omega)} \tag{4.5}$$

and this is precisely the FPT from (4.1). Nevertheless, simplicity is not the sole reason which guides us to a rational polynomial for the response function. More importantly, the rational functional form is dictated by the physics of the response mechanism of any studied system upon exposure to external perturbations. This occurs because a system excited by an external field is automatically set into an oscillatory type of motion for which a differential equation of a definite order represents the appropriate mathematical description. A like differential equation also describes the internal state of the system prior to any external perturbation. Algebraically, these differential or difference equations are equivalent to the corresponding characteristic or secular equations. This follows from the fact that the degree of a given characteristic polynomial is equal to the order of the corresponding differential equation. Thus, solving a differential equation is equivalent to obtaining all the roots of the associated characteristic polynomial. Physically, the roots of the characteristic polynomial $Q_K(\omega)$ represent the fundamental frequencies $\{\omega_k\}$ $(1 \leq k \leq K)$ of the intrinsic oscillations of the investigated system. Integer K is the total number of fundamental damped harmonics which are given in terms of the complex frequencies $\{\omega_k\}$ and the corresponding complex amplitudes $\{d_k\}$.

Thus, under the most general circumstances in virtually all research fields, including MRS, the adequate physical considerations invariably lead to a polynomial quotient for the response function, or a frequency spectrum, as prescribed precisely by the FPT. This is anticipated, since the formalism of the Green function is fully equivalent to the Schrödinger equation through which quantum physics becomes applicable to any system, including living organisms. One of the justifications for this statement is the proven equivalence of generic time signals with quantum-mechanical auto-correlation functions [5]. Due to this fact, every data analysis based upon time signals is automatically within the realm of quantum mechanics. Since such signals are abundantly present in living organisms, there is no need for any further proof that quantum mechanics is applicable to life sciences. While direct solutions of the Schrödinger equations are possible to find only for the simplest systems of a very limited practical utility, the Green functions can be computed for any physical system regardless of its complexity as long as, e.g., time signal points are available. We have noted that such time signals that are precisely of the same structure as auto-correlation functions from quantum physics are encountered in MRS and in many other fields [5]. Therefore, the quantum-mechanical formalism of the Green functions implemented via, e.g., the FPT can be used to spectrally analyze these time signals that are either generated theoretically as auto-correlation functions or measured experimentally as free induction decay curves.

4.2 The exact solution for the general harmonic inversion problem

The adequacy of modeling any investigated phenomenon chiefly depends upon the correct description of the underlying physical process. The succinct outlines from the preceding section indicate that the FPT is the method of choice for solving the quantification problem in MRS and in other fields that use time signals. Both physics and mathematics support this fact. Physics, because time signals in MRS originate from attenuated harmonic oscillators that are described by quantum-mechanical auto-correlation functions. In fact, for such time signals, the FPT is the exact filter in the sense of being capable of filtering out all the harmonics without invoking any approximation [5]. Mathematics, because the exact spectrum for these time signals is given by a polynomial quotient. Moreover, if a function to be described is itself a ratio of two polynomials, as is actually the case with the exact Green function, the fast Padé transform represents, by construction, the exact theory. Hence optimality of the FPT for signal processing.

We have presented two equivalent Padé variants of the Green functions, $G^{(+)}$ and $G^{(-)}$, corresponding to the incoming and the outgoing boundary conditions, inside and outside the unit circle in the complex plane of the harmonic variable z. Recall that these variants of the FPT are denoted by $\mathrm{FPT}^{(+)}$ and $\mathrm{FPT}^{(-)}$, respectively. The $\mathrm{FPT}^{(+)}$ and $\mathrm{FPT}^{(-)}$ are identical to the causal and anti-causal Padé-z transforms studied extensively in mathematical statistics and the engineering literature on signal processing [5]. In chapter 3, we applied the $\mathrm{FPT}^{(+)}$ and $\mathrm{FPT}^{(-)}$ to time signals from MRS to explicitly demonstrate that both versions of the fast Padé transform can solve exactly a standard quantification problem from MRS.

It was seen that the FPT is a very efficient signal processor, since it belongs to the category of fast algorithms as does the fast Fourier transform. This becomes possible with the FPT by employing the Euclid algorithm implemented with continued fractions [86, 87]. In this latter algorithm, the FPT necessitates $N(\log_2 N)^2$ multiplications, that are comparable to the conventional $N\log_2 N$ multiplications within the FFT.

4.3 General time series

In examinations of time evolution of a given system, one usually begins from an initial state Φ_0 prepared at the instant $t_0 = 0$, which is associated with $n = 0$. Integer n counts the discrete time $t \equiv t_n = n\tau \, (n = 0, 1, 2, ...)$ where τ is the sampling time or dwell time. One of the important properties of any time signal is its total duration $T = N\tau$ which is in encoding called the total acquisition time. In the Schrödinger picture of quantum mechanics, for a given state Φ_0, the vector Φ_n is deduced from $\Phi_n = \hat{\mathrm{U}}\Phi_0$. Here, $\hat{\mathrm{U}} = \hat{\mathrm{U}}(\tau)$ is the evolution operator $\hat{\mathrm{U}} = \exp{(-i\hat{\Omega})\tau}$ where $\hat{\Omega}$ represents the system's dynamical non-Hermitean operator, which is the Hamiltonian $\hat{\mathrm{H}}$ in quantum mechanics. Moreover, the time signal $\{c_n\} \, (0 \leq n \leq N-1)$ of length N is equivalent to the quantum-mechanical auto-correlation function $c_n = (\Phi_0|\hat{\mathrm{U}}^n|\Phi_0)$. As customary with dissipative dynamics, the symmetric scalar product is employed $(f|g) = (g|f)$ without complex conjugation of either function. With the special choice $t_0 = 0$, the first element of the data or Hankel matrix $\mathbf{H}_n(c_0) \equiv \{c_{i+j}\}$ is the signal point $c_0 = (\Phi_0|\Phi_0) \neq 0$. The general element $c_{i,j} = c_{i+j}$ of the matrix $\mathbf{H}_n(c_0)$ is the overlap between the two Schrödinger or Krylov states $c_{i+j} = (\Phi_j|\Phi_i) = (\Phi_0|\hat{\mathrm{U}}^{i+j}|\Phi_0) \equiv \mathrm{U}_{i,j}^{(0)}$. The matrix which gives the sought spectrum is not $\mathbf{H}_n(c_0)$, but rather the evolution matrix $\mathbf{H}_n(c_1)$ with the general elements $c_{i+j+1} = (\Phi_j|\hat{\mathrm{U}}|\Phi_i) \equiv \mathrm{U}_{i,j}^{(1)}$.

The matrices $\mathbf{H}_n(c_0)$ and $\mathbf{H}_n(c_1)$ are the particular cases of the general delayed Hankel matrix $\mathbf{H}_n(c_s) = \{c_{n+m+s}\} \equiv \mathbf{U}_n^{(s)}$ of the form

$$\mathbf{H}_n(c_s) = \mathbf{U}_n^{(s)} \begin{pmatrix} c_s & c_{s+1} & c_{s+2} & \cdots & c_{s+n-1} \\ c_{s+1} & c_{s+2} & c_{s+3} & \cdots & c_{s+n} \\ c_{s+2} & c_{s+3} & c_{s+4} & \cdots & c_{s+n+1} \\ \vdots & \vdots & \vdots & \ddots & \vdots \\ c_{s+n-1} & c_{s+n} & c_{s+n+1} & \cdots & c_{s+2n-2} \end{pmatrix} \tag{4.6}$$

where $\mathbf{U}_n^{(s)} = \{\mathrm{U}_{i,j}^{(s)}\}_{i,j=0}^{n-1}$. Here, the signal point c_s appearing in the small parentheses in $\mathbf{H}_n(c_s)$ represents the leading element from the first row and first column in the matrix (4.6). The corresponding delayed Hankel determinant $\mathrm{H}_n(c_s)) \equiv \det \mathbf{H}_n(c_s)$ is defined by (2.200). In the special cases, $s = 0$ and $s = 1$ the Hankel matrices $\mathbf{H}_n(c_0)$ and $\mathbf{H}_n(c_1)$ coincide with the overlap matrix $\mathbf{S}_n = \mathbf{U}_n^{(0)}$ and the evolution or relaxation matrix $\mathbf{U}_n = \mathbf{U}_n^{(1)}$ introduced in terms of the Schrödinger basis set $\{|\Phi_n)\}$.

In practice, many applications necessitate the usage of the non-zero initial time, $t_0 \neq 0$. In this chapter in addition to $t_0 = 0$, we will also consider the delayed time signals with the non-zero initial time $t_s = s\tau (0 \leq s < N-1)$. The corresponding mathematical model for the delayed time signal $\{c_{n+s}\}$ $(0 \leq n \leq N-1$ and $0 \leq s < N-1)$ is given by a sum of attenuated complex exponentials

$$c_{n+s} = \sum_{k=1}^{K} d_k \mathrm{e}^{-i(n+s)\omega_k \tau} = \sum_{k=1}^{K} d_k^{(s)} u_k^n \tag{4.7}$$

$$d_k^{(s)} = d_k u_k^s \qquad d_k \equiv d_k^{(0)} \qquad u = \mathrm{e}^{-i\omega\tau} \qquad u_k = \mathrm{e}^{-i\omega_k\tau} \tag{4.8}$$

where ω_k and $d_k^{(s)}$ are the fundamental angular frequencies and amplitudes, respectively. This should be compared to the non-delayed time signal

$$c_n = \sum_{k=1}^{K} d_k \mathrm{e}^{-in\omega_k\tau} = \sum_{k=1}^{K} d_k u_k^n. \tag{4.9}$$

Evidently, spectral analysis of non-delayed $\{c_n\}$ and delayed time signals $\{c_{n+s}\}$ $(s \neq 0)$ can be accomplished on the same footing.

4.4 The response or the Green function

A non-zero initial time $t_s \neq 0$ in place of $t_0 = 0$ can also be employed in spectral methods that rely upon the Green function, as in the state-space formulation of the FPT. Here, the main quantity is the delayed Green operator which represents the delayed resolvent

$$\hat{\mathrm{R}}^{(s)}(u) \equiv (\hat{1}u - \hat{\mathrm{U}})^{-1}\hat{\mathrm{U}}^s = \sum_{n=0}^{\infty} \hat{\mathrm{U}}^{n+s}(\tau)u^{-n-1}. \tag{4.10}$$

The associated delayed Green function $G^{(s)}(u)$ is defined by the standard matrix element $(\Phi_0|\hat{\mathrm{R}}^{(s)}(u)|\Phi_0)$ which together with (4.10) yields

$$G^{(s)}(u) = (\Phi_0|\hat{\mathrm{R}}^{(s)}(u)|\Phi_0) = \sum_{n=0}^{\infty} c_{n+s}u^{-n-1}. \tag{4.11}$$

This result is a direct generalization of the customary non-delayed Green function

$$G(u) = \sum_{n=0}^{\infty} c_n u^{-n-1} \tag{4.12}$$

where $G^{(0)}(u) \equiv G(u)$. In signal processing, the delayed counterpart of the standard Green function can also be obtained by splitting the original infinite time interval $[0, \infty]$ into two parts, as in $[0, \infty] = [0, s-1] + [s, \infty]$. This maps (4.12) to

$$\begin{aligned} G(u) &= \sum_{n=0}^{s-1} c_n u^{-n-1} + u^{-s} \sum_{n=0}^{\infty} c_{n+s}u^{-n-1} \\ &= \sum_{n=0}^{s-1} c_n u^{-n-1} + u^{-s} G^{(s)}(u) \end{aligned} \tag{4.13}$$

where $\sum_{n=0}^{-1} \equiv 0$. This procedure yields a relationship between the two exact Green functions $G^{(s)}(u)$ and $G^{(0)}(u)$ that are related to the case with and without the delay, i.e., $s \neq 0$ and $s = 0$, respectively. It is seen from (4.12) that the delayed spectrum $G^{(s)}(u)$ is multiplied by the term u^{-s} which compensates for the time evolution of the system from $t_0 = 0$ to $t_s \neq 0$.

By substituting (4.7) into the rhs of (4.10), we have

$$\sum_{n=0}^{\infty} c_{n+s}u^{-n-1} = \sum_{n=0}^{\infty} \left\{ \sum_{k=1}^{K} d_k^{(s)} u_k^n \right\} u^{-n-1}$$
$$\sum_{k=1}^{K} d_k^{(s)} u^{-1} \left\{ \sum_{n=0}^{\infty} (u_k/u)^n \right\} = \sum_{k=1}^{K} d_k^{(s)} u^{-1}(1 - u_k/u)^{-1}$$

where the geometric series $\sum_{n=0}^{\infty} x^n = 1/(1-x)$ is used, so that

$$\sum_{n=0}^{\infty} c_{n+s} u^{-n-1} = \sum_{k=1}^{K} \frac{d_k^{(s)}}{u - u_k}. \tag{4.14}$$

This expression represents the most important property of the mathematical model (4.7). Here, the lhs of (4.14) is recognized as the *exact* delayed spectrum which corresponds to the delayed signal (4.7). In theory, the exactness stems from the presence of an infinite sum, i.e., a series over signal points. We see that the model (4.7) is indeed powerful, since it does not use any approximation to reduce the *infinite* sum over the signal points $\sum_{n=0}^{\infty} c_{n+s} u^{-n-1}$ to the finite Heaviside partial fractions, $\sum_{k=1}^{K} d_k^{(s)}/(u-u_k)$, with the retrieved spectral parameters $\{u_k, d_k^{(s)}\}$. Naturally, in practice, only a finite number of signal points is conventionally available and, therefore, the exact spectrum $\sum_{n=0}^{\infty} c_{n+s} u^{-n-1}$ cannot be obtained. However, even in this case, the finite geometric sum $\sum_{n=0}^{N-1} x^n = (1-x^N)/(1-x)$ can be used to derive the following exact result for the truncated spectrum

$$\sum_{n=0}^{N-1} c_{n+s} u^{-n-1} = \sum_{k=1}^{K} d_k^{(s)} \frac{1-(u_k/u)^N}{u - u_k}. \tag{4.15}$$

This formula tends to (4.14) as $N \longrightarrow \infty$ if $|u_k/u| < 1\,(1 \le k \le K)$. In the response functions (4.14) and (4.15) associated with the infinite ($N = \infty$) and finite ($N < \infty$) signal length, respectively, the exact delayed spectrum for the model (4.7) is represented by rational functions. Specifically, the two rational functions from the rhs of (4.14) and (4.15) are polynomial quotients that are seen as the Padé approximants for the input sums $\sum_{n=0}^{\infty} c_{n+s} u^{-n-1}$ and $\sum_{n=0}^{N-1} c_{n+s} u^{-n-1}$, respectively. Hence, for any processing method employed to recover the spectral parameters $\{u_k, d_k^{(s)}\}$ from the input delayed time signal of the form (4.7), the ensuing spectrum invariably coincides with the Padé approximant through the unique quotient of two polynomials for the given power series expansion such as the Green function (4.12). This shows that among all the signal processors, the FPT indeed appears as optimal for parametric estimations of spectra based upon the time signal (4.7).

4.5 The key prior knowledge: Internal structure of time signals

As stated, by construction, the Padé approximant becomes the exact theory if the input function is a rational function given as a quotient of two polynomials, i.e., a rational polynomial. Of course, the exactness overrides the required optimality and, therefore, the Padé approximant has no competitor for investigating those input functions that are themselves defined as rational polynomials. In signal processing, the input function is *not* a quotient of two polynomials, but rather a single series $\sum_n c_{n+s}u^{-n-1}$ for which the exact processor does not exist. Nevertheless, this situation can be salvaged, and an exact theory could still be formulated for the spectrum $\sum_n c_{n+s}u^{-n-1}$ by invoking the appropriate prior information, such as the existence of a harmonic structure of the signal via (4.7). In this case, with no additional approximation, the single sum $\sum_{n=0}^{\infty} c_{n+s}u^{-n-1}$ from the original input data coincides exactly with the rational polynomial (4.14) or (4.15), for which the Padé approximation is again the exact theory.

Regarding the general issue of prior information, the minimal knowledge needed in advance could be the assumption that the time signal *possesses a structure*. For instance, by merely plotting a given signal, it might become possible for a trained eye to qualitatively discern certain oscillatory patterns with a harmonic-type structure. These structures often become more pronounced by considering the associated derivative of the time signal. Alternatively, the Fourier shape spectrum in the frequency domain would give a more definitive indication of an underlying structure via a clearer emergence of a number of peaks. Here, the otherwise generic structure would become more specific through the appearance of the resonant nature of the spectrum.

Despite such a favorable circumstance, the Fourier method does not exploit this key finding from its own analysis, since only the envelope spectrum is obtained as the final result of signal processing. Such an occurrence is reflected in the so-called 'Fourier uncertainty principle'. This principle states that for a fixed total acquisition time T, or equivalently, a given full signal length N at a fixed bandwidth, the corresponding Fourier spectrum cannot predict an angular frequency resolution better than the Fourier bound $2\pi/T$. The obvious drawback of this principle is in the implication that, regardless of their intrinsic nature, all the signals of the identical acquisition time T will possess the same frequency resolution $2\pi/T$. In other words, all Fourier spectra stemming from time signals of the same length N will have the same resolution. As such, in advance of both measurement and processing, the Fourier method pre-assigns the frequencies at which all the spectra should exist and these are the frequencies from the Fourier grid $2\pi k/T$, $(k = 0, 1, \ldots, N-1)$. Such a severe limitation of the Fourier analysis rules out the chance for interpolation or extrapolation. However, without an interpolation property, the Fourier method is restricted to the pre-determined minimal separation $\omega_{\text{min}} = 2\pi/T$ between any two adjacent frequencies. Further, without an extrapolation feature, the Fourier processing has no predictive power. In lieu of extrapolation, the Fourier method employs zero filling or zero padding beyond N or uses signal's periodic extensions $c_{n+N} = c_n$. However, in most circumstance encountered in practice, time signals are non-periodic and, therefore, no extrapolation or any other new information can be obtained by periodic extensions, as mentioned in chapter 1.

An attempt to simultaneously circumvent the mentioned fundamental restrictions of the Fourier method could be made by first acknowledging the fact that each signal has its own inner structure. Subsequently, it would be necessary to unfold the hidden structure and try to parametrize it. Finally, a quantitative analysis should be carried out to determine the spectral features by retrieving the resonance parameters $\{u_k, d_k^{(s)}\}$ from the input time signal. Such a strategy would lead to parametric estimations of spectra as a major advantage relative to a shape processing by the Fourier analysis. Specifically, the fact that a given time signal possesses its intrinsic structure through, e.g., a number of constituent harmonics $\{d_k^{(s)} u_k^n\}$, offers a definite chance for resolution improvement beyond the Fourier bound $2\pi/T$. As discussed, in parametric estimations, the total acquisition time T of the signal is not critical for definition of the spectral resolution. As such, parametric processing of spectra can lower the Fourier bound $2\pi/T$, thus yielding improved resolution beyond the prescription of the Fourier uncertainty principle. Moreover, the Fourier uncertainty principle within parametric processing is replaced by a milder limitation imposed by 'the informational principle'. According to this latter principle, no more information could possibly be obtained from a spectrum in the frequency domain than what has already been encoded originally in the time domain. Such a conservation of information translates into an algebraic condition, which requires a minimum of $2K$ signal points to retrieve all the spectral parameters $\{u_k, d_k^{(s)}\}\,(1 \le k \le K)$. This condition is rooted in the requirement that the underlying system of linear equations must be at least determined by the number of equations equal to the number of unknown parameters. In practice, all experimentally measured time signals are corrupted with noise and this obstacle is partially counter-balanced by solving the associated over-determined system in which the number of signal points exceeds the number $2K$ of the sought parameters $\{u_k, d_k^{(s)}\}$.

By definition, the delayed signal $\{c_{n+s}\}$ originates from a physical system which has already evolved from $t_0 = 0$ to $t_s \ne 0$ before beginning to count the time. It could also happen that in some experimental measurements, e.g., via ICRMS, the first few hundred signal points might be of such poor quality that they must be discarded. In such cases, the spectral analysis deals with the delayed Hankel matrix $\mathbf{H}_n(c_s) = \{c_{i+j+s}\}$ where $c_s = (\Phi_0|\hat{\mathrm{U}}^s|\Phi_0) \ne 0$ appears as the first element. Within the FFT, this effect of a delay in the considered signal is taken into account merely by skipping the first s points $\{c_r\}$ $(0 \le r \le s-1)$ from the whole signal $\{c_n\}$ $(0 \le n \le N-1)$. However, this is not good, since the resulting FFT spectra are of unacceptably poor quality, which cannot be amended due to information loss in some of the skipped original data. Hence, it is essential to have a signal processor which can properly analyze data matrices $\{c_{i+j+s}\}$ associated with the evolution of the examined system from a non-zero initial time $t_s = s\tau \ne 0$.

In many inter-disciplinary applications using signal processing methods, the problem of spectral analysis of data records with delayed time series is known to be very important. Of course, theoretical developments always favor the analysis of the general delayed Hankel matrix $\mathbf{H}_n(c_s)$, since the needed associated counterparts $\mathbf{H}_n(c_0)$ and $\mathbf{H}_n(c_1)$ appear as two special cases taken at $s = 0$ and $s = 1$, respectively. Naturally, a general approach using $\mathbf{H}_n(c_s)$ has a chance to yield certain advantages over a particular procedure given by the evolution matrix $\mathbf{H}_n(c_1)$, particularly in the domain of generating more fruitful algorithms. This will be shown to be the case in the present chapter. Specifically, it will be demonstrated that the sole introduction of a non-zero initial time $t_s \neq 0$ has far-reaching consequences for spectral analysis, even when one actually intends to keep the whole signal $\{c_n\}$ $(0 \leq n \leq N-1)$ of total length N. For instance, processing the evolution matrix $\mathbf{H}_n(c_1)$ usually proceeds through matrix diagonalizations, or equivalently, rooting the corresponding characteristic equation to find the spectral parameters $\{u_k, d_k\}$ of the signal (4.9). By contrast, both of these customary procedures can be alleviated altogether by using $\mathbf{H}_n(c_s)$ for a fixed integer $s > 0$ rather than setting $s = 1$, as in the evolution matrix $\mathbf{H}_n(c_1)$. Then the sought spectral parameters $\{u_k, d_k\}$ could be obtained non-conventionally from convergence of the coefficients of the corresponding 'delayed' continued fractions as the values of number s are systematically augmented [5].

In contradistinction to the FFT, the space methods, such as the FPT, from the Schrödinger picture of quantum mechanics can spectrally analyze Hankel matrices with an arbitrary initial time t_s. These signal processors employ the evolution operator $\hat{\mathrm{U}}$ which generates the state Φ_s of the considered system at the delayed moment t_s by the transformation. $\Phi_s = \hat{\mathrm{U}}^s\Phi_0$. In this expression, the time delay is described by starting the analysis from Φ_s rather than from Φ_0 which refers to $s = 0$. Of course, evolution of the system in the time interval $[t_0, t_s]$ of non-zero length has to be taken into account. The needed correction is easily made in the state space methods through multiplication of the vector Φ_0 by the operator $\hat{\mathrm{U}}^{-s} = \exp(is\hat{\Omega}\tau)$. This cancels out the evolution effect accumulated in the state vector in the time interval $t \in [0, t_s]$. Thus, when the first s signal points are skipped, all the matrix elements in the state-space version of the FPT must be altered by the counteracting operator $\exp(is\hat{\Omega}\tau)$. As such, in these signal processors, taking the instant $t_s \neq 0$ in place of $t_0 = 0$ for the initial time engenders no difficulty whatsoever. Moreover, the state-space methods accomplish spectral analysis by diagonalizing the Hankel matrix $\mathbf{H}_n(c_s)$ for a fixed integer s in the same fashion as for $\mathbf{H}_n(c_0)$ and $\mathbf{H}_n(c_1)$. The same is true for the FPT which replaces diagonalization by rooting the characteristic polynomials. Importantly, relative to $\mathbf{H}_n(c_1)$, it is advantageous to diagonalize the delayed evolution matrix $\mathbf{H}_n(c_s) = \mathbf{U}_n^{(s)} = \{\mathrm{U}_{i,j}^{(s)}\}$, where the general element $\mathrm{U}_{i,j}^{(s)}$ is due to the sth power of $\hat{\mathrm{U}}$ from $\mathrm{U}_{i,j}^{(s)} = (\Phi_j|\hat{\mathrm{U}}^s|\Phi_i)$. The advantage is in the key possibility of identifying spurious roots. To accomplish this task in practice, one usually diagonalizes the matrix $\mathbf{U}_n^{(s)}$ for $s = 1$ and $s = 2$ in order to compare the obtained eigen-frequencies. The same frequencies for different values of s

are retained as physical/genuine, whereas those frequencies that are altered when passing from $s = 1$ to $s = 2$ are rejected as unphysical/spurious. Similarly, in the non-state-space variant of the FPT, one uses different powers s of the evolution operator $\hat{\mathrm{U}}$ that are implicit in the para-diagonal elements $[(n+s-1)/n]_G(u)$ of the Padé table corresponding to the delayed counterpart $G^{(s)}(u)$ of the Green function $G(u)$ from (4.12). In this case, one selects several values of s to discriminate between physical and spurious eigen-roots of the denominator polynomials that are the said characteristic polynomials. The eigen-roots that are stable/unstable for different s are considered as physical/unphysical, respectively. These practical advantages lend support to taking the initial times t_s different from the conventional one, $t_0 = 0$.

4.6 The Rutishauser quotient-difference recursive algorithm

The Padé approximant can alternatively be written in the form of continued fractions, i.e., the CFs, as a staircase with descending quotients. There exist several equivalent symbolic notations for a given CF and two of them read as

$$\cfrac{A_1}{B_1 + \cfrac{A_2}{B_2 + \cfrac{A_3}{B_3 + \ddots}}} \equiv \frac{A_1}{B_1}{+}\frac{A_2}{B_2}{+}\frac{A_3}{B_3}{+}\cdots \tag{4.16}$$

where the plus signs on the rhs of (4.16) are dropped, i.e., lowered to point at a 'step-down' process in forming the CF. The lhs of (4.16) represents quite a natural way of writing the staircase-shaped continued fraction. However, for frequent usage, the rhs of (4.16) is obviously more economical since it takes less space. The rhs of (4.16) could be alternatively written by employing the ordinary plus signs via $A_1/(B_1 + A_2/(B_2 + A_3/(B_3 + \cdots)))$.

The infinite and mth order delayed CF [5] related to the time series (4.11) are introduced as

$$G^{\mathrm{CF}(s)}(u) = \frac{a_1^{(s)}}{u} - \frac{a_2^{(s)}}{1} - \frac{a_3^{(s)}}{u} - \cdots - \frac{a_{2r}^{(s)}}{1} - \frac{a_{2r+1}^{(s)}}{u} - \cdots \tag{4.17}$$

$$G_m^{\mathrm{CF}(s)}(u) = \frac{a_1^{(s)}}{u} - \frac{a_2^{(s)}}{1} - \frac{a_3^{(s)}}{u} - \cdots - \frac{a_{2m}^{(s)}}{1} - \frac{a_{2m+1}^{(s)}}{u} \tag{4.18}$$

where $a_n^{(s)}$ are the expansion coefficients. Further, we can define the even (e) and odd (o) contracted continued fractions (CCF) of order n that contain $2n$ and $2n+1$ terms from the corresponding CFs

$$G_{\mathrm{e},n}^{\mathrm{CCF}(s)}(u) = G_{2n}^{\mathrm{CF}(s)}(u) \qquad G_{\mathrm{o},n}^{\mathrm{CCF}(s)}(u) = G_{2n+1}^{\mathrm{CF}(s)}(u) \tag{4.19}$$

respectively. Each element of the set $\{a_n^{(s)}\}$ can be obtained from the equality between the expansion coefficients of the series of the rhs of (4.17) developed in powers of u^{-1} and the signal points $\{c_{n+s}\}$ from (4.11). Such a procedure is similar to the one from Ref. [17] for non-delayed time signals and Green functions when $s = 0$. Thus, it suffices here to give only some of the main results that will be needed in the analysis which follows. For example, the expressions for the general CF expansion coefficients are given by

$$a_{2n}^{(s)} = \frac{\mathrm{H}_n(c_{s+1})\mathrm{H}_{n-1}(c_s)}{\mathrm{H}_{n-1}(c_{s+1})\mathrm{H}_n(c_s)} \qquad a_{2n+1}^{(s)} = \frac{\mathrm{H}_{n-1}(c_{s+1})\mathrm{H}_{n+1}(c_s)}{\mathrm{H}_n(c_s)\mathrm{H}_n(c_{s+1})} \tag{4.20}$$

$$\alpha_n^{(s)} = a_{2n+1}^{(s)} + a_{2n+2}^{(s)} \qquad [\beta_n^{(s)}]^2 = a_{2n}^{(s)}a_{2n+1}^{(s)} \qquad (n \geq 1) \tag{4.21}$$

where the Hankel determinant $\mathrm{H}_n(c_s)$ is given in (2.200). The parameters $\alpha_n^{(s)}$ and $[\beta_n^{(s)}]^2$ are the Lanczos coupling constants in the nearest neighbor approximation [5]. Using (4.20) and (4.21), a recursive algorithm can also be established for computations of all the coefficients $\{a_n^{(s)}\}$. Thus, with the help of the alternative notation

$$a_{2n}^{(s)} \equiv q_n^{(s)} \qquad a_{2n+1}^{(s)} \equiv e_n^{(s)} \tag{4.22}$$

the product of $q_n^{(s)}$ with $e_n^{(s)}$ becomes

$$q_n^{(s)}e_n^{(s)} = \frac{\mathrm{H}_{n-1}(c_s)\mathrm{H}_{n+1}(c_s)}{\mathrm{H}_n^2(c_s)}. \tag{4.23}$$

Similarly, the product of $q_{n+1}^{(s)}$ and $e_n^{(s)}$ reads as

$$q_{n+1}^{(s)}e_n^{(s)} = q_n^{(s+1)}e_n^{(s+1)} \qquad s \geq 0. \tag{4.24}$$

Further, employing the following identity of the Hankel determinants

$$[\mathrm{H}_n(c_s)]^2 = \mathrm{H}_n(c_{s-1})\mathrm{H}_n(c_{s+1}) - \mathrm{H}_{n+1}(c_{s-1})\mathrm{H}_{n-1}(c_{s+1}) \tag{4.25}$$

we can find the sum of $q_n^{(s+1)}$ and $e_{n-1}^{(s+1)}$ in the form

$$q_n^{(s)} + e_n^{(s)} = q_n^{(s+1)} + e_{n-1}^{(s+1)}. \tag{4.26}$$

The relationships (4.24) and (4.26) for the delayed continued fraction coefficients represent the Rutishauser quotient-difference (QD) algorithm [156]

$$\left.\begin{array}{c} e_n^{(s)} = e_{n-1}^{(s+1)} + q_n^{(s+1)} - q_n^{(s)} \\[2ex] q_{n+1}^{(s)} = q_n^{(s+1)}\dfrac{e_n^{(s+1)}}{e_n^{(s)}} \\[2ex] e_0^{(s)} = 0 \qquad (s \geq 1) \qquad q_1^{(s)} = \dfrac{c_{s+1}}{c_s} \qquad (s \geq 0) \end{array}\right\}. \tag{4.27}$$

In the recursions from (4.27), the vectors $q_n^{(s)}$ and $e_n^{(s)}$ are obtained by interchangeably forming their quotients and differences. This is the basis for the name 'quotient-difference' for the Rutishauser algorithm. The QD algorithm is one of the most extensively employed computational tools throughout numerical analysis. Transparently, the vectors $q_n^{(s)}$ and $e_n^{(s)}$ can be stored in a two-dimensional table as a double array of a lozenge form via

$$\begin{array}{cccccccc}
 & q_1^{(0)} & & & & & & \\
e_0^{(1)} & & e_1^{(0)} & & & & & \\
 & q_1^{(1)} & & q_2^{(0)} & & & & \\
e_0^{(2)} & & e_1^{(1)} & & e_2^{(0)} & & & \\
 & q_1^{(2)} & & q_2^{(1)} & & \ddots & & \\
e_0^{(3)} & & e_1^{(2)} & & e_2^{(1)} & & & \\
\vdots & q_1^{(3)} & & q_2^{(2)} & & \ddots & & \\
 & \vdots & e_1^{(3)} & & e_2^{(2)} & & & \\
 & & \vdots & \vdots & \vdots & \ddots & &
\end{array} \tag{4.28}$$

where the first column is filled with zeros $e_0^{(0)} = 0$. It follows from the table (4.28) that the subscript (n) and superscript (s) denote a column and a counter-diagonal, respectively. Moreover, the columns of the arrays $q_n^{(s)}$ and $e_n^{(s)}$ are interleaved. Here, the starting values are $e_0^{(s)} = 0$ $(s = 1, 2, \ldots)$ and $q_1^{(s)} = c_{s+1}/c_s$ $(s = 0, 1, 2, \ldots)$. Subsequent arrays $q_n^{(s)}$ and $e_n^{(s)}$ are obtained by two intertwined recursions of quantities that are located at the vertices (corners) of the lozenge in the table (4.28). The column containing only the vectors $e_n^{(s)}$ are generated through the differences $e_n^{(s)} = q_n^{(s+1)} - q_n^{(s)} + e_{n-1}^{(s+1)}$ $(n = 1, 2, \ldots$ and $s = 0, 1, 2, \ldots)$. Similarly, the columns with the arrays $q_n^{(s)}$ are constructed by means of the quotients $q_n^{(s)} = q_{n-1}^{(s+1)} e_{n-1}^{(s+1)} / e_{n-1}^{(s)}$ $(n = 2, 3, \ldots$ and $s = 0, 1, 2, \ldots)$. This is the complete procedure by which the QD algorithm creates one column at a time by alternatively generating the quotients and differences of the q- and e-quantities through the recursion relations from (4.27).

4.7 The Gordon product-difference recursive algorithm

The well-known Lanczos algorithm [5, 148] has numerical difficulties including the loss of orthogonality of the elements of the Lanczos basis set $\{\psi_n\}$. This often seriously deteriorates the accuracy of the coupling parameters that are computed during the construction of the state vectors $\{\psi_n\}$. Because the

Lanczos coupling constants are exceptionally important for spectral analysis, it is necessary to look for more stable algorithms than the Lanczos recursion for state vectors $\{\psi_n\}$, but that still rely upon the signal points $\{c_n\}$ or auto-correlation functions as the only input data. Given that the major task of the Lanczos algorithm consists of obtaining the couplings, it is recommended to try to bypass the construction of state vectors $\{\psi_n\}$ whose orthogonality could be destroyed in the course of the recursion. To this end, one can resort to at least two recursive algorithms that fulfill the mentioned requirements by using only the signal points and by simultaneously alleviating altogether generation of the Lanczos state vectors $\{\psi_n\}$. These are the Rutishauser QD algorithm [156] and the Gordon [158] product-difference (PD) algorithm. Both algorithms can compute the entire set of the Lanczos coupling constants $\{\alpha_n^{(s)}, \beta_n^{(s)}\}$ for arbitrarily large values of n. This fact is important to emphasize particularly because of a statement from *Numerical Recipes* [77] claiming that computing the Lanczos coupling parameters generated by, e.g., the power moments $\{\mu_n\}$ (equivalent to $\{c_n\}$) must be viewed as useless due to their mathematical ill-conditioning. This claim does not hold true for the power moments generated by the QD [156] and PD [158] algorithms.

The most general prescription for the PD algorithm is facilitated by introducing an auxiliary matrix $\boldsymbol{\lambda}^{(s)} = \{\lambda_{n,m}^{(s)}\}$ with zero-valued elements below the main counter-diagonal as

$$\boldsymbol{\lambda}^{(s)} = \begin{pmatrix} \lambda_{1,1}^{(s)} & \lambda_{1,2}^{(s)} & \lambda_{1,3}^{(s)} & \cdots & \lambda_{1,n-2}^{(s)} & \lambda_{1,n-1}^{(s)} & \lambda_{1,n}^{(s)} \\ \lambda_{2,1}^{(s)} & \lambda_{2,2}^{(s)} & \lambda_{2,3}^{(s)} & \cdots & \lambda_{2,n-2}^{(s)} & \lambda_{2,n-1}^{(s)} & 0 \\ \lambda_{3,1}^{(s)} & \lambda_{3,2}^{(s)} & \lambda_{3,3}^{(s)} & \cdots & \lambda_{3,n-2}^{(s)} & 0 & 0 \\ \vdots & \vdots & \vdots & \ddots & \vdots & \vdots & \vdots \\ \lambda_{n-2,1}^{(s)} & \lambda_{n-2,2}^{(s)} & \lambda_{n-2,3}^{(s)} & \cdots & 0 & 0 & 0 \\ \lambda_{n-1,1}^{(s)} & \lambda_{n-1,2}^{(s)} & 0 & \cdots & 0 & 0 & 0 \\ \lambda_{n,1}^{(s)} & 0 & 0 & \cdots & 0 & 0 & 0 \end{pmatrix}. \tag{4.29}$$

Here, the first column of this matrix is filled with zeros, except the element $\lambda_{1,1}^{(s)}$ which is set to unity. The second column contains the signal points with the alternating sign via

$$\boldsymbol{\lambda}^{(s)} = \begin{pmatrix} 1 & c_s & c_{s+1} & \cdots & \lambda_{1,n-2}^{(s)} & \lambda_{1,n-1}^{(s)} & \lambda_{1,n}^{(s)} \\ 0 & -c_{s+1} & -c_{s+2} & \cdots & \lambda_{2,n-2}^{(s)} & \lambda_{2,n-1}^{(s)} & 0 \\ 0 & c_{s+2} & c_{s+3} & \cdots & \lambda_{3,n-2}^{(s)} & 0 & 0 \\ \vdots & \vdots & \vdots & \ddots & \vdots & \vdots & \vdots \\ 0 & (-1)^{n-1}c_{n+s-3} & (-1)^{n-1}c_{n+s-2} & \cdots & 0 & 0 & 0 \\ 0 & (-1)^{n}c_{n+s-2} & 0 & \cdots & 0 & 0 & 0 \\ 0 & 0 & 0 & \cdots & 0 & 0 & 0 \end{pmatrix}. \tag{4.30}$$

The general matrix element $\lambda_{n,m}^{(s)}$ can be introduced by the 2×2 determinant

$$\lambda_{n,m}^{(s)} \equiv - \begin{vmatrix} \lambda_{1,m-2}^{(s)} & \lambda_{1,m-1}^{(s)} \\ \lambda_{n+1,m-2}^{(s)} & \lambda_{n+1,m-1}^{(s)} \end{vmatrix} . \tag{4.31}$$

The alternative form of this expression is the following recursion

$$\lambda_{n,m}^{(s)} = \lambda_{1,m-1}^{(s)} \lambda_{n+1,m-2}^{(s)} - \lambda_{1,m-2}^{(s)} \lambda_{n+1,m-1}^{(s)} \tag{4.32}$$

which is initialized by the elements

$$\lambda_{n,1}^{(s)} = \delta_{n,1} \qquad \lambda_{n,2}^{(s)} = (-1)^{n+1} c_{n+s-1} \qquad \lambda_{n,3}^{(s)} = (-1)^{n+1} c_{n+s} \tag{4.33}$$

where $\delta_{n,1}$ is the Kronecker symbol. Upon generation of the arrays $\{\lambda_{i,j}^{(s)}\}$, one can compute all the coefficients $\{a_n^{(s)}\}$ of the delayed continued fractions (4.17) by means of the formula

$$a_n^{(s)} = \frac{\lambda_{1,n+1}^{(s)}}{\lambda_{1,n-1}^{(s)} \lambda_{1,n}^{(s)}} \qquad (n = 1, 2, 3, \ldots). \tag{4.34}$$

Inserting (4.34) into (4.22), we arrive at

$$q_n^{(s)} = \frac{\lambda_{1,2n+1}^{(s)}}{\lambda_{1,2n-1}^{(s)} \lambda_{1,2n}^{(s)}} \qquad e_n^{(s)} = \frac{\lambda_{1,2n+2}^{(s)}}{\lambda_{1,2n}^{(s)} \lambda_{1,2n+1}^{(s)}}. \tag{4.35}$$

The Lanczos coupling parameters $\{\alpha_n^{(s)}, [\beta_n^{(s)}]^2\}$ are deduced by substituting the string $\{a_n^{(s)}\}$ into (4.21). The dependence of the pair $\{\alpha_n^{(s)}, [\beta_n^{(s)}]^2\}$ upon the matrix elements $\{\lambda_{1,n}^{(s)}\}$ can be made explicit by means of (4.21) and (4.34) that yields the expression

$$\alpha_n^{(s)} = \frac{[\lambda_{1,2n+2}^{(s)}]^2 + \lambda_{1,2n}^{(s)} \lambda_{1,2n+3}^{(s)}}{\lambda_{1,2n}^{(s)} \lambda_{1,2n+1}^{(s)} \lambda_{1,2n+2}^{(s)}} \qquad [\beta_n^{(s)}]^2 = \frac{\lambda_{1,2n+2}^{(s)}}{\lambda_{1,2n-1}^{(s)} [\lambda_{1,2n}^{(s)}]^2}. \tag{4.36}$$

Thus, the recursion (4.31) of the vectors $\{\lambda_{n,m}^{(s)}\}$ invokes only their products and differences, but no divisions. This gives the name for the 'product-difference' algorithm. Originally, the PD algorithm for non-delayed signals or moments ($s = 0$) was proposed by Gordon [158]. The generalization of the PD algorithm to delayed time signals or moments $\{c_{n+s}\} = \{\mu_{n+s}\}$ was given by Belkić [5, 17]. It follows from (4.34) that the PD algorithm performs the division only once at the end of the computations while obtaining the delayed CF coefficients $\{a_n^{(s)}\}$. Because of this advantageous feature, the PD algorithm is error-free for those signal points $\{c_{n+s}\}$ that are integers. In practice, integer data matrices $\{c_{n+s}\}$ are measured experimentally in MRS, NMR, ICRMS, etc. Likewise, the same infinite precision, i.e., without round-off errors is also possible in the PD algorithm for auto-correlation functions $\{c_n\}$ or power moments $\{\mu_{n+s}\}$ expressed as rational numbers. For many circumstances in physics, such as systems exposed to external fields (Zeeman effect, harmonic oscillator, etc.), the role of signal points is taken by expansion coefficients that are obtained exactly as rational numbers from quantum-mechanical perturbation theory. In such a case, one would operate directly with rational numbers using symbolic language programming, e.g., MAPLE [185]–[187]. Regarding efficiency, the computational complexity of the PD algorithm for the CF coefficients $\{a_m^{(s)}\}$ $(1 \leq m \leq n)$ is of the order of n^2 multiplications. This should be compared with a direct computation of the Hankel determinant $\mathrm{H}_n(c_s)$ of the dimension n from the definition (4.20) for $\{a_n^{(s)}\}$ where, e.g., the Cramer rule would require the formidable $n!$ multiplications that would prohibit any meaningful application for large n.

As seen in section 4.6, the QD algorithm (4.27) for the auxiliary double array $\{q_n^{(s)}, e_n^{(s)}\}$ performs divisions in each iteration. In a finite-precision arithmetic, this could produce round-off errors that might cause the QD algorithm to break down for non-integer signal points. When the input data $\{c_{n+s}\}$ are non-zero integers, divisions would yield rational numbers in the course of the QD recursion. This also would be innocuous with error-free results if infinite precision arithmetic using rational numbers is employed by means of, e.g., MAPLE [185]–[187].

Remarkably, it is possible to judiciously combine the vectors $\{\lambda_{1,m}^{(s)}\}$ to recursively generate Hankel determinants of arbitrary orders as

$$\mathrm{H}_n(c_s) = \frac{\lambda_{1,2n}^{(s)}}{\lambda_{2n-2}^{(s)}}\mathrm{H}_{n-1}(c_s) \qquad\qquad \lambda_n^{(s)} = \prod_{m=1}^{n} \lambda_{1,m}^{(s)}. \tag{4.37}$$

The recursion (4.37) can explicitly be solved by iterations and the result is given by

$$\mathrm{H}_n(c_s) = c_s \prod_{m=2}^{n} \frac{\lambda_{1,2m}^{(s)}}{\lambda_{2m-2}^{(s)}} \qquad\qquad \mathrm{H}_1(c_s) = c_s \qquad (n = 2, 3, \ldots). \tag{4.38}$$

With this expression, the result of the general-order Hankel determinant $\mathrm{H}_n(c_s)$ can efficiently be obtained from the recursively pre-computed string $\{\lambda_{1,m}^{(s)}\}$. A computationally very attractive property of the formula (4.38) is that it effectively performs only the computation of the simplest 2×2 determinant (4.31). If the data $\{c_{n+s}\}$ are integers, the corresponding Hankel determinant $\mathrm{H}_n(c_s)$ will also be an integer number, say $N_n^{(s)}$. In this case, the result (4.38) for $\mathrm{H}_n(c_s)$ would be a rational number. However, by definition (4.37), the numerator in (4.38) reads as $\prod_{m=2}^{n} \lambda_{1,2m}^{(s)} = N_n^{(s)} \prod_{m=2}^{n} \lambda_{2m-2}^{(s)}$ so that $\mathrm{H}_n(c_s) = N_n^{(s)}$, as it ought to be. To preserve such a feature in computations, integer algebra should be used in which the integer numbers $\{\lambda_{i,j}^{(s)}\}$

are kept in their composite intermediate forms without performing the final multiplications in (4.38). This would permit the exact cancellation of the denominator $\prod_{m=2}^{n} \lambda_{2m-2}^{(s)}$ by the associated portion of the numerator in $\mathrm{H}_n(c_s)$ from (4.38) to give the exact result in the integer form $\mathrm{H}_n(c_s) = N_n^{(s)}$.

The extension of the PD algorithm from its original non-delayed versions of Gordon [158] to the delayed version of Belkić [5] is especially advantageous with respect to the eigen-values $\{u_k\}$. When only the non-delayed CF coefficients $\{a_n\} \equiv \{a_n^{(0)}\}$ are available as in Ref. [158], the eigen-values $\{u_k\}$ are generated either by solving the eigen-problem for the Jacobi matrix or by rooting the corresponding characteristic polynomial $Q_K(u) = 0$ [5]. On the other hand, the delayed CF coefficients $\{a_n^{(s)}\}$ can avoid altogether these two latter conventional procedures and offer an alternative way of computing the eigen-values of data matrices from the following limiting procedure

$$u_k = \lim_{s\longrightarrow\infty} a_{2k}^{(s)} = \lim_{s\longrightarrow\infty} \frac{\lambda_{1,2k+1}^{(s)}}{\lambda_{1,2k-1}^{(s)}\lambda_{1,2k}^{(s)}}. \tag{4.39}$$

For checking purposes, it is also useful to apply the same limit $s \longrightarrow \infty$ to the string $\{a_{2n+1}^{(s)}\}$ and this gives

$$\lim_{s\longrightarrow\infty} a_{2k+1}^{(s)} = \lim_{s\longrightarrow\infty} \frac{\lambda_{1,2k+2}^{(s)}}{\lambda_{1,2k}^{(s)}\lambda_{1,2k+1}^{(s)}} = 0. \tag{4.40}$$

To verify the accuracy of the results for the eigen-values $\{u_k\}$ computed by means of the relation $u_k = \lim_{s\to\infty} a_{2k}^{(s)}$ from (4.39) within the delayed PD algorithm we can use the analytical expression for $a_{2k}^{(s)}$. The exact closed formula for the general delayed CF coefficients $a_n^{(s)}$ has been obtained by Belkić [5, 17] as

$$a_{n+1}^{(s)} = \frac{c_{n+s} - \sigma_n^{(s)}\pi_{n-1}^{(s)} - \lambda_n^{(s)}\pi_{n-4}^{(s)}}{\pi_n^{(s)}} \tag{4.41}$$

$$\pi_n^{(s)} = \prod_{i=1}^{n} a_i^{(s)} \qquad\qquad \sigma_n^{(s)} = \left[\sum_{j=2}^{n} a_j^{(s)}\right]^2 \tag{4.42}$$

$$\lambda_n^{(s)} = \sum_{j=\left[\frac{n-1}{2}\right]}^{n-3} a_j^{(s)}[\xi_j^{(s)}]^2 \qquad\qquad \xi_j^{(s)} = \sum_{k=2}^{j+1} a_k^{(s)} \sum_{\ell=2}^{k+1} a_\ell^{(s)} \tag{4.43}$$

$$\pi_n^{(s)} \equiv 0 \quad (n \le 0) \qquad\qquad \sigma_n^{(s)} \equiv 0 \quad (n \le 3) \tag{4.44}$$

where the symbol $[n/2]$ denotes the integer part of $n/2$. Once the CF coefficients $\{a_n^{(s)}\}$ have been generated, all the input time signal points $\{c_{n+s}\}$ could be reconstructed exactly using the explicit expression

$$c_{n+s} = \pi_{n+1}^{(s)} + \sigma_n^{(s)}\pi_{n-1}^{(s)} + \lambda_n^{(s)}\pi_{n-4}^{(s)}. \tag{4.45}$$

Moreover, having obtained the exact delayed CF coefficients $\{a_n^{(s)}\}$ from (4.41), one could deduce the exact delayed continued fractions of a fixed order as the explicit polynomial quotient, which is the Padé approximant. Thus, the even-order delayed CF is the Padé approximant of the form

$$G_{2n}^{\mathrm{CF}(s)}(u) = a_1^{(s)} \frac{\tilde{P}_n^{\mathrm{CF}(s)}(u)}{\tilde{Q}_n^{\mathrm{CF}(s)}(u)}. \tag{4.46}$$

Similarly, the delayed odd-order CF, which is labeled as $G_{2n-1}^{\mathrm{CF}(s)}(u)$, is extracted from the even-order CF by putting $a_{2n}^{(s)} \equiv 0$

$$G_{2n-1}^{\mathrm{CF}(s)}(u) \equiv \{G_{2n}^{\mathrm{CF}(s)}(u)\}_{a_{2n}^{(s)}=0} \qquad (n = 1, 2, 3, \ldots). \tag{4.47}$$

Here, the polynomials $\tilde{P}_n^{\mathrm{CF}(s)}(u)$ and $\tilde{Q}_n^{\mathrm{CF}(s)}(u)$ from (4.46) can be introduced by their general power series representations

$$\tilde{P}_n^{\mathrm{CF}(s)}(u) = \sum_{r=0}^{n-1} \tilde{p}_{n,n-r}^{(s)} u^r \qquad \tilde{Q}_n^{\mathrm{CF}(s)}(u) = \sum_{r=0}^{n} \tilde{q}_{n,n-r}^{(s)} u^r. \tag{4.48}$$

The polynomial expansion coefficients $\tilde{p}_{n,n-r}^{(s)}$ and $\tilde{q}_{n,n-r}^{(s)}$ are available as the analytical expressions that were derived in protracted calculations by Belkić [5, 17]

$$\tilde{p}_{n,m}^{(s)} = (-1)^{m-1} \underbrace{\sum_{r_1=3}^{2(n-m+2)} a_{r_1}^{(s)} \sum_{r_2=r_1+2}^{2(n-m+3)} a_{r_2}^{(s)} \cdots \sum_{r_{m-1}=r_{m-2}+2}^{2n} a_{r_{m-1}}^{(s)}}_{m-1 \text{ summations}} \tag{4.49}$$

$$\tilde{q}_{n,m}^{(s)} = (-1)^{m} \underbrace{\sum_{r_1=2}^{2(n-m+1)} a_{r_1}^{(s)} \sum_{r_2=r_1+2}^{2(n-m+2)} a_{r_2}^{(s)} \sum_{r_3=r_2+2}^{2(n-m+3)} a_{r_3}^{(s)} \cdots \sum_{r_m=r_{m-1}+2}^{2n} a_{r_m}^{(s)}}_{m \text{ summations}} \tag{4.50}$$

where $n \geq m$. With these and the other listed closed expressions, the FPT is established as the only parametric estimator by means of which signal processing can entirely be carried out from the explicit analytical formulae. This overrides the mathematical ill-conditioning of the alternative numerical algorithms.

4.8 The Lanczos algorithm for continued fractions

Employing (4.22), the infinite and the mth order delayed continued fractions $G^{\mathrm{CF}(s)}(u)$ and $G_m^{\mathrm{CF}(s)}(u)$ from (4.17) and (4.18) become, respectively

$$G^{\mathrm{CF}(s)}(u) = \frac{c_s}{u}\frac{}{-}\frac{q_1^{(s)}}{1}\frac{}{-}\frac{e_1^{(s)}}{u}\frac{}{-}\cdots\frac{}{-}\frac{q_r^{(s)}}{1}\frac{}{-}\frac{e_r^{(s)}}{u}\frac{}{-}\cdots \tag{4.51}$$

$$G_m^{\mathrm{CF}(s)}(u) = \frac{c_s}{u}\frac{}{-}\frac{q_1^{(s)}}{1}\frac{}{-}\frac{e_1^{(s)}}{u}\frac{}{-}\cdots\frac{}{-}\frac{q_m^{(s)}}{1}\frac{}{-}\frac{e_m^{(s)}}{u}. \tag{4.52}$$

Similarly, the infinite and the mth order of the even part of the associated Lanczos continued fractions (LCF) can be introduced as

$$\begin{aligned} G_{\mathrm{e}}^{\mathrm{LCF}(s)}(u) &= \frac{c_s}{u-q_1^{(s)}-e_0^{(s)}}\frac{}{-}\frac{q_1^{(s)}e_1^{(s)}}{u-q_2^{(s)}-e_1^{(s)}}\frac{}{-}\cdots\frac{}{-}\frac{q_r^{(s)}e_r^{(s)}}{u-q_{r+1}^{(s)}-e_r^{(s)}}\frac{}{-}\cdots \\ &= \frac{c_s}{u-\alpha_0^{(s)}}\frac{}{-}\frac{[\beta_1^{(s)}]^2}{u-\alpha_1^{(s)}}\frac{}{-}\cdots\frac{}{-}\frac{[\beta_r^{(s)}]^2}{u-\alpha_r^{(s)}}\frac{}{-}\cdots \end{aligned} \tag{4.53}$$

$$\begin{aligned} G_{\mathrm{e},m}^{\mathrm{LCF}(s)}(u) &= \frac{c_s}{u-q_1^{(s)}-e_0^{(s)}}\frac{}{-}\frac{q_1^{(s)}e_1^{(s)}}{u-q_2^{(s)}-e_1^{(s)}}\frac{}{-}\cdots\frac{}{-}\frac{q_m^{(s)}e_m^{(s)}}{u-q_{m+1}^{(s)}-e_m^{(s)}} \\ &= \frac{c_s}{u-\alpha_0^{(s)}}\frac{}{-}\frac{[\beta_1^{(s)}]^2}{u-\alpha_1^{(s)}}\frac{}{-}\cdots\frac{}{-}\frac{[\beta_m^{(s)}]^2}{u-\alpha_m^{(s)}}. \end{aligned} \tag{4.54}$$

A detailed derivation [5] shows that there is a general relationship between $G_{\mathrm{e},m}^{\mathrm{LCF}(s)}(u)$ and $G_m^{\mathrm{CF}(s)}(u)$ which by reference to (4.19) can be written as

$$G_{\mathrm{e},n}^{\mathrm{LCF}(s)}(u) = G_{\mathrm{e},n}^{\mathrm{CCF}(s)}(u) = G_{2n}^{\mathrm{CF}(s)}(u). \tag{4.55}$$

There exists also the infinite-order and the mth order odd part of (4.51) that are labeled by $G_{\mathrm{o}}^{\mathrm{LCF}(s)}(u)$ and $G_{\mathrm{o},m}^{\mathrm{LCF}(s)}(u)$, respectively

$$\begin{aligned} G_{\mathrm{o}}^{\mathrm{LCF}(s)}(u) &= \frac{c_s}{u} \\ &\times\left[1+\frac{q_1^{(s)}}{u-q_1^{(s)}-e_1^{(s)}}\frac{}{-}\frac{q_2^{(s)}e_1^{(s)}}{u-q_2^{(s)}-e_2^{(s)}}\frac{}{-}\cdots\frac{}{-}\frac{q_r^{(s)}e_{r-1}^{(s)}}{u-q_r^{(s)}-e_r^{(s)}}\frac{}{-}\cdots\right] \end{aligned} \tag{4.56}$$

$$\begin{aligned} G_{\mathrm{o},m}^{\mathrm{LCF}(s)}(u) &= \frac{1}{u} \\ &\times\left[c_s+\frac{c_{s+1}}{u-q_1^{(s)}-e_1^{(s)}}\frac{}{-}\frac{q_2^{(s)}e_1^{(s)}}{u-q_2^{(s)}-e_2^{(s)}}\frac{}{-}\cdots\frac{}{-}\frac{q_m^{(s)}e_{m-1}^{(s)}}{u-q_m^{(s)}-e_m^{(s)}}\right] \\ &= \frac{1}{u}\left\{c_s+\frac{c_{s+1}}{u-\alpha_0^{(s+1)}}\frac{}{-}\frac{[\beta_1^{(s+1)}]^2}{u-\alpha_1^{(s+1)}}\frac{}{-}\cdots\frac{}{-}\frac{[\beta_{m-1}^{(s+1)}]^2}{u-\alpha_{m-1}^{(s+1)}}\right\} \end{aligned} \tag{4.57}$$

where $m > 0\,(m = 1, 2, 3, \ldots)$. If (4.56) is compared with the identity

$$\sum_{n=0}^{\infty} c_{n+s} u^{-n-1} = \frac{c_s}{u} + \frac{1}{u} \sum_{n=0}^{\infty} c_{n+s+1} u^{-n-1} \tag{4.58}$$

the following relation is obtained

$$\begin{aligned}
\sum_{n=0}^{\infty} & c_{n+s+1} u^{-n-1} \\
&= c_s \left[\frac{q_1^{(s)}}{u - q_1^{(s)} - e_1^{(s)}} - \frac{q_2^{(s)} e_1^{(s)}}{u - q_2^{(s)} - e_2^{(s)}} - \cdots - \frac{q_{r+1}^{(s)} e_r^{(s)}}{u - q_{r+1}^{(s)} - e_{r+1}^{(s)}} - \cdots \right] \\
&= c_s \left\{ \frac{q_1^{(s)}}{u - \gamma_1^{(s)}} - \frac{[\delta_1^{(s)}]^2}{u - \gamma_2^{(s)}} - \cdots - \frac{[\delta_r^{(s)}]^2}{u - \gamma_{r+1}^{(s)}} - \cdots \right\}
\end{aligned} \tag{4.59}$$

where

$$\gamma_n^{(s)} = q_n^{(s)} + e_n^{(s)} \qquad [\delta_n^{(s)}]^2 = q_{n+1}^{(s)} e_n^{(s)}. \tag{4.60}$$

On the other hand, by employing (4.21) and (4.27) we arrive at

$$q_n^{(s)} + e_n^{(s)} = q_n^{(s+1)} + e_{n-1}^{(s+1)} = \alpha_{n-1}^{(s+1)} \tag{4.61}$$

$$q_{n+1}^{(s)} e_n^{(s)} = q_n^{(s+1)} e_n^{(s+1)} = [\beta_n^{(s+1)}]^2 \tag{4.62}$$

$$\gamma_n^{(s)} = \alpha_{n-1}^{(s+1)} \qquad [\delta_n^{(s)}]^2 = [\beta_n^{(s+1)}]^2 \tag{4.63}$$

so that

$$\begin{aligned}
\sum_{n=0}^{\infty} & c_{n+s+1} u^{-n-1} \\
&= \frac{c_{s+1}}{u - q_1^{(s)} - e_1^{(s)}} - \frac{q_2^{(s)} e_1^{(s)}}{u - q_2^{(s)} - e_2^{(s)}} - \cdots - \frac{q_{r+1}^{(s)} e_r^{(s)}}{u - q_{r+1}^{(s)} - e_{r+1}^{(s)}} - \cdots \\
&= \frac{c_{s+1}}{u - \alpha_0^{(s+1)}} - \frac{[\beta_1^{(s+1)}]^2}{u - \alpha_1^{(s+1)}} - \cdots - \frac{[\beta_r^{(s+1)}]^2}{u - \alpha_r^{(s+1)}} - \cdots.
\end{aligned} \tag{4.64}$$

Here, the second line of (4.64) is identical to the second line of (4.53), provided that $s+1$ is used instead of s, as it ought to be. As such, (4.64) is, in fact, the proof that (4.56) is correct. Returning now to (4.57) for the odd part of $G_n^{\mathrm{CCF}(s)}(u)$

$$G_{\mathrm{o},n}^{\mathrm{LCF}(s)}(u) = G_{\mathrm{o},n}^{\mathrm{CCF}(s)}(u) = G_{2n+1}^{\mathrm{CF}(s)}(u). \tag{4.65}$$

Hence, we see from (4.55) and (4.65) that the even and odd parts of the delayed Lanczos approximants $G_{\mathrm{e},n}^{\mathrm{LCF}(s)}(u)$ and $G_{\mathrm{o},n}^{\mathrm{LCF}(s)}(u)$ of order n $(n = 1, 2, 3, \ldots)$ coincide with the delayed continued fractions $G_{2n}^{\mathrm{CF}(s)}(u)$ and $G_{2n+1}^{\mathrm{CF}(s)}(u)$ of orders $2n$ and $2n+1$, respectively.

4.9 The Padé-Lanczos approximant

Here, we define the delayed Padé-Lanczos approximant (PLA) [5, 17] by

$$G_{L,K}^{\mathrm{PLA}(s)}(u) = \frac{c_s}{\beta_1^{(s)}}\frac{P_L^{(s)}(u)}{Q_K^{(s)}(u)}. \tag{4.66}$$

As usual, the special case $L = K$ of (4.66) yields the para-diagonal variant denoted as[1]

$$G_{n,n}^{\mathrm{PLA}(s)}(u) \equiv G_n^{\mathrm{PLA}(s)}(u) \qquad G_n^{\mathrm{PLA}(s)}(u) = \frac{c_s}{\beta_1^{(s)}}\frac{P_n^{(s)}(u)}{Q_n^{(s)}(u)}. \tag{4.67}$$

Functions $Q_n^{(s)}(u)$ and $P_n^{(s)}(u)$ are the delayed Lanczos polynomials of the first and second kind, respectively. They can be introduced through their recursion relations

$$\left.\begin{array}{c} \beta_{n+1}^{(s)}P_{n+1}^{(s)}(u) = [u-\alpha_n^{(s)}]P_n^{(s)}(u) - \beta_n^{(s)}P_{n-1}^{(s)}(u) \\ P_0^{(s)}(u) = 0 \qquad\qquad P_1^{(s)}(u) = 1 \end{array}\right\} \tag{4.68}$$

$$\left.\begin{array}{c} \beta_{n+1}^{(s)}Q_{n+1}^{(s)}(u) = [u-\alpha_n^{(s)}]Q_n^{(s)}(u) - \beta_n^{(s)}Q_{n-1}^{(s)}(u) \\ Q_{-1}^{(s)}(u) = 0 \qquad\qquad Q_0^{(s)}(u) = 1 \end{array}\right\}. \tag{4.69}$$

Alternatively, the polynomials $P_n^{(s)}(u)$ and $Q_n^{(s)}(u)$ could be defined via the corresponding power series representations

$$P_n^{(s)}(u) = \sum_{r=0}^{n-1} p_{n,n-r}^{(s)}u^r \qquad Q_n^{(s)}(u) = \sum_{r=0}^{n} q_{n,n-r}^{(s)}u^r. \tag{4.70}$$

The polynomial expansion coefficients $p_{n,n-r}^{(s)}$ and $q_{n,n-r}^{(s)}$ can be generated by means of the following recursions

$$\left.\begin{array}{c} \beta_{n+1}^{(s)}p_{n+1,n+1-r}^{(s)} = p_{n,n+1-r}^{(s)} - \alpha_n^{(s)}p_{n,n-r}^{(s)} - \beta_n^{(s)}p_{n-1,n-1-r}^{(s)} \\ p_{0,0}^{(s)} = 0 \qquad\qquad p_{1,1}^{(s)} = 1 \end{array}\right\} \tag{4.71}$$

$$\left.\begin{array}{c} \beta_{n+1}^{(s)}q_{n+1,n+1-r}^{(s)} = q_{n,n+1-r}^{(s)} - \alpha_n^{(s)}q_{n,n-r}^{(s)} - \beta_n^{(s)}q_{n-1,n-1-r}^{(s)} \\ q_{0,0}^{(s)} = 1 \qquad\qquad q_{1,1}^{(s)} = -\alpha_0^{(s)}/\beta_1^{(s)} \end{array}\right\} \tag{4.72}$$

$$p_{n,-1}^{(s)} = 0 \qquad q_{n,-1}^{(s)} = 0 \qquad p_{n,m}^{(s)} = 0 \qquad q_{n,m}^{(s)} = 0 \qquad (m > n). \tag{4.73}$$

---

[1] Exceptionally, in the Lanczos algorithm, the numerator $P_L^{(s)}$ and denominator $Q_K^{(s)}$ polynomials in the quotient $P_L^{(s)}/Q_K^{(s)}$ are of degree $L-1$, and K, respectively. For this reason, the case $L = K$, i.e., $P_K^{(s)}/Q_K^{(s)}$, does not actually represent the diagonal, but rather the para-diagonal Padé approximant, $[(K-1)/K]$ [5].

Formally, the recursions (4.68) and (4.69) for $P_n^{(s)}(u)$ and $Q_n^{(s)}(u)$ are the same, with the exception of the different initializations.

The derivation from Ref. [5, 17] shows that the Green functions $G_{\mathrm{e},n}^{\mathrm{LCF}(s)}(u)$ and $G_{\mathrm{o},n}^{\mathrm{LCF}(s)}(u)$ are linked to the delayed Padé-Lanczos approximant (4.66) by the expression

$$G_n^{\mathrm{PLA}(s)}(u) = G_{\mathrm{e},n}^{\mathrm{LCF}(s)}(u) \qquad (n = 1, 2, 3, \ldots). \tag{4.74}$$

Therefore, the delayed Padé-Lanczos approximant $G_n^{\mathrm{PLA}(s)}(u)$ and the even part of the delayed Lanczos continued fraction $G_{\mathrm{e},n}^{\mathrm{LCF}(s)}(u)$ coincide with each other for any order n.

Of course, it is important to investigate whether the odd part of the delayed LCF, i.e., $G_{\mathrm{o},n}^{\mathrm{LCF}(s)}(u)$, could also be located within the Padé-Lanczos general table for $G_{n,m}^{\mathrm{PLA}(s)}(u)$. To this end, by considering, e.g., the diagonal case $(L = K + 1)$ in (4.66)

$$G_{n+1,n}^{\mathrm{PLA}(s)}(u) \equiv \tilde{G}_n^{\mathrm{PLA}(s)}(u) = \frac{c_s}{\beta_1^{(s)}} \frac{P_{n+1}^{(s)}(u)}{Q_n^{(s)}(u)} \tag{4.75}$$

it can be shown that [5, 17]

$$\tilde{G}_n^{\mathrm{PLA}(s)}(u) \neq G_{\mathrm{o},n}^{\mathrm{LCF}(s)}(u) \qquad (n = 1, 2, \ldots). \tag{4.76}$$

Moreover, in general, there are no integers n and m for which $G_{n,m}^{\mathrm{PLA}(s)}(u)$ can be equal to $G_{\mathrm{o},n}^{\mathrm{LCF}(s)}(u)$ [5, 17]. This occurs because the denominator in $G_{\mathrm{o},n}^{\mathrm{LCF}(s)}(u)$ is a polynomial $\gamma_0 u + \gamma_1 u^2 + \cdots + \gamma_n u^{n+1}$ without the constant free term $\propto u^0 \equiv 1$. The additional term u in the denominator of $G_{\mathrm{o},n}^{\mathrm{LCF}(s)}(u)$ relative to $G_{\mathrm{e},n}^{\mathrm{LCF}(s)}(u)$ implies that $G_{o,n}^{\mathrm{LCF}(s)}(u)$ might originate from the Padé approximant in the variable u^{-1} instead of u. This is proven in the next section.

4.10 The fast Padé transform FPT$^{(-)}$ outside the unit circle

In this section, the analysis begins with the exact delayed Green function (4.11). The expansion (4.11) is the Maclaurin series in powers of the variable $u^{-1} \equiv z = \exp(i\omega\tau)$ and, hence, convergent for $|u| > 1$, i.e., outside the unit circle in the u-complex plane. It will prove convenient to rewrite $G^{(s)}$ in terms of an auxiliary function $\mathcal{G}^{(s)}$ as follows

$$G^{(s)}(u) = \sum_{n=0}^{\infty} c_{n+s} u^{-n-1} = u^{-1} \mathcal{G}^{(s)}(u^{-1}) \tag{4.77}$$

$$\mathcal{G}^{(s)}(u^{-1}) = \sum_{n=0}^{\infty} c_{n+s} u^{-n} = \lim_{N\to\infty} \mathcal{G}_N^{(s)}(u^{-1}) \tag{4.78}$$

$$\mathcal{G}_N^{(s)}(u^{-1}) = \sum_{n=0}^{N-1} c_{n+s} u^{-n} \tag{4.79}$$

$$G_N^{(s)}(u) \equiv \sum_{n=0}^{N-1} c_{n+s} u^{-n-1} = u^{-1}\mathcal{G}_N^{(s)}(u^{-1}). \tag{4.80}$$

With respect to the finite-rank Green function $\mathcal{G}_N^{(s)}(u^{-1})$, we define the delayed diagonal Padé approximant, $\mathcal{G}_K^{\mathrm{PA}(s)-}(u^{-1})$, or equivalently, the delayed diagonal fast Padé transform $\mathcal{G}_K^{\mathrm{FPT}(s)-}(u^{-1})$

$$\mathcal{G}_K^{\mathrm{PA}(s)-}(u^{-1}) \equiv \mathcal{G}_K^{\mathrm{FPT}(s)-}(u^{-1}) \tag{4.81}$$

as the following polynomial quotient

$$\mathrm{FPT}^{(s)-}: \qquad \mathcal{G}_K^{\mathrm{FPT}(s)-}(u^{-1}) = \frac{A_K^{(s)-}(u^{-1})}{B_K^{(s)-}(u^{-1})}. \tag{4.82}$$

In this expression, the numerator and denominator polynomials $A_K^{(s)-}(u^{-1})$ and $B_K^{(s)-}(u^{-1})$ are given in the same variable u^{-1} similarly to the function $\mathcal{G}_K^{(s)}(u^{-1})$. The polynomials $A_K^{(s)-}(u^{-1})$ and $B_K^{(s)-}(u^{-1})$ have the same degree K and can be written as sums of powers of variable u^{-1} via

$$A_K^{(s)-}(u^{-1}) = \sum_{r=0}^{K} a_r^{(s)-} u^{-r} \qquad B_K^{(s)-}(u^{-1}) = \sum_{r=0}^{K} b_r^{(s)-} u^{-r}. \tag{4.83}$$

The associated delayed diagonal fast Padé transform relative to $G_N^{(s)}(u)$ reads

$$G_K^{\mathrm{FPT}(s)-}(u^{-1}) \equiv u^{-1}\mathcal{G}_K^{\mathrm{FPT}(s)-}(u^{-1}) = u^{-1}\frac{A_K^{(s)-}(u^{-1})}{B_K^{(s)-}(u^{-1})}. \tag{4.84}$$

The unknown expansion coefficients $a_r^{(s)-}$ and $b_r^{(s)-}$ from (4.83) can be found by setting the equality $\mathcal{G}_N^{(s)}(u^{-1}) = \mathcal{G}_K^{\mathrm{FPT}(s)-}(u^{-1})$

$$\mathcal{G}_N^{(s)}(u) \equiv \sum_{n=0}^{N-1} c_{n+s} u^{-n} \approx \frac{A_K^{(s)-}(u^{-1})}{B_K^{(s)-}(u^{-1})}. \tag{4.85}$$

Multiplication of (4.85) by $B_K^{(s)-}(u^{-1})$ yields the equations

$$\left.\begin{aligned} B_K^{(s)-}(u^{-1})\sum_{n=0}^{N-1} c_{n+s} u^{-n} &= A_K^{(s)-}(u^{-1}) \\ \left[\sum_{r=0}^{K} b_r^{(s)-} u^{-r}\right]\left[\sum_{n=0}^{N-1} c_{n+s} u^{-n}\right] &= \sum_{r=0}^{K} a_r^{(s)-} u^{-r} \end{aligned}\right\}. \tag{4.86}$$

Further, by carrying out the multiplication of the two sums on the lhs of (4.86) and comparing the resulting coefficients of the like powers of u^{-1} with their counterparts from the rhs of (4.86), we obtain

$$a_k^{(s)-} = \sum_{r=0}^{k} b_r^{(s)-} c_{k-r+s} \qquad (0 \le k \le K) \tag{4.87}$$

$$c_k = -\sum_{r=1}^{K} b_r^{(s)-} c_{k-r}. \tag{4.88}$$

Introducing the integer M via

$$M = N - 1 - K - s \tag{4.89}$$

we rewrite (4.88) in the form

$$c_{K+s+m} + \sum_{r=1}^{K} b_r^{(s)-} c_{K+s+m-r} = 0 \qquad 0 \le m \le M. \tag{4.90}$$

This represents an implicit system of linear equations for the unknown coefficients $b_r^{(s)-}$. Moreover, the same system (4.90) can be made explicit by letting the integer m vary from 1 to M, so that

$$\left.\begin{array}{c} c_{K+s+1}b_0^{(s)-} + c_{K+s}b_1^{(s)-} + \cdots + c_{s+1}b_K^{(s)-} = 0 \\ c_{K+s+2}b_0^{(s)-} + c_{K+s+1}b_1^{(s)-} + \cdots + c_{s+2}b_K^{(s)-} = 0 \\ \vdots \\ c_{M+K+s}b_0^{(s)-} + c_{M+K+s-1}b_1^{(s)-} + \cdots + c_{M+s}b_K^{(s)-} = 0 \end{array}\right\}. \tag{4.91}$$

We see that (4.87) will be a system of linear equations if the suffix k is varied from 0 to K and, therefore

$$\left.\begin{array}{rl} a_0^{(s)-} &= c_s b_0^{(s)-} \\ a_1^{(s)-} &= c_{s+1}b_0^{(s)-} + c_s b_1^{(s)-} \\ &\vdots \\ a_K^{(s)-} &= c_{K+s}b_0^{(s)-} + c_{K+s-1}b_1^{(s)-} + \cdots + c_s b_K^{(s)-} \end{array}\right\}. \tag{4.92}$$

The equivalent matrix forms of the systems (4.91) and (4.92) are

$$\begin{pmatrix} c_{K+s} & c_{K-1+s} & \cdots & c_{s+1} \\ c_{K+s+1} & c_{K+s} & \cdots & c_{s+2} \\ \vdots & \vdots & \ddots & \vdots \\ c_{K+s+M-1} & c_{K+s+M-2} & \cdots & c_{s+M} \end{pmatrix} \begin{pmatrix} b_1^{(s)-} \\ b_2^{(s)-} \\ \vdots \\ b_K^{(s)-} \end{pmatrix} = -b_0^{(s)-} \begin{pmatrix} c_{K+s+1} \\ c_{K+s+2} \\ \vdots \\ c_{K+s+M} \end{pmatrix} \tag{4.93}$$

$$\begin{pmatrix} a_0^{(s)-} \\ a_1^{(s)-} \\ \vdots \\ a_K^{(s)-} \end{pmatrix} = \begin{pmatrix} c_s & 0 & \cdots & 0 \\ c_{s+1} & c_s & \cdots & 0 \\ \vdots & \vdots & \ddots & \vdots \\ c_{K+s} & c_{K+s-1} & \cdots & c_s \end{pmatrix} \begin{pmatrix} b_0^{(s)-} \\ b_1^{(s)-} \\ \vdots \\ b_K^{(s)-} \end{pmatrix}. \tag{4.94}$$

Then, the explicit calculation from Refs. [5, 17] yields

$$\begin{aligned} \mathcal{G}_n^{\mathrm{FPT}(s)-}(u^{-1}) &= \mathcal{G}_{\mathrm{o},n}^{\mathrm{LCF}(s)}(u) \\ &= \mathcal{G}_{\mathrm{o},n}^{\mathrm{CCF}(s)}(u) = \mathcal{G}_{2n+1}^{\mathrm{CF}(s)}(u) \qquad (n = 1, 2, 3\ldots) \end{aligned} \tag{4.95}$$

so that

$$\begin{aligned} G_n^{\mathrm{FPT}(s)-}(u^{-1}) &= G_{\mathrm{o},n}^{\mathrm{LCF}(s)}(u) \\ &= G_{\mathrm{o},n}^{\mathrm{CCF}(s)}(u) = G_{2n+1}^{\mathrm{CF}(s)}(u) \qquad (n = 1, 2, 3\ldots). \end{aligned} \tag{4.96}$$

These results show that the delayed diagonal fast Padé transform, $\mathrm{FPT}^{(s)-}$, with the convergence region outside the unit circle ($|u| > 1$) is the same as the odd part of the delayed Lanczos continued fraction of order n. Furthermore, $G_n^{\mathrm{FPT}(s)-}(u^{-1})$ and the original truncated Green function (4.79) are convergent for $|u| > 1$ for $N \to \infty$. Thus, outside the unit circle, the delayed diagonal fast Padé transform $G_n^{\mathrm{FPT}(s)-}(u^{-1})$ represents an accelerator of an already convergent series which is the Green function (4.77).

4.11 The fast Padé transform $\mathrm{FPT}^{(+)}$ inside the unit circle

There exists another version of the delayed diagonal Padé approximant for the same function $\mathcal{G}_N^{(s)}(u^{-1})$ from (4.79). This version can be obtained from (4.14) which we recast as

$$\mathcal{G}^{(s)}(u) = \sum_{n=0}^{\infty} c_{n+s} u^{-n} = \sum_{k=1}^{K} \frac{d_k^{(s)} u}{u - u_k}. \tag{4.97}$$

The summation over k is an implicit ratio of two polynomials in the variable u and, therefore, it represents the Padé approximant. A particular property of (4.97) is that the numerator polynomial has no free term, i.e., a constant independent of u. Hence, the sought variant of the rational function should be taken as a ratio of two polynomials in u. This version of the delayed diagonal Padé approximant, or equivalently, delayed diagonal fast Padé transform will be denoted by $G_K^{\mathrm{PA}(s)+}(u)$ or $G_K^{\mathrm{FPT}(s)+}(u)$, respectively

$$\mathcal{G}_K^{\mathrm{PA}(s)+}(u) \equiv \mathcal{G}_K^{\mathrm{FPT}(s)+}(u). \tag{4.98}$$

Of course, the original function (to be represented by $\mathcal{G}_K^{\rm FPT(s)+}(u)$ in the form of a rational polynomial) is the same Green function $\mathcal{G}_N^{(s)}(u^{-1})$ as it was in (4.81)

$$\text{FPT}^{(s)+}: \qquad \mathcal{G}_K^{\rm FPT(s)+}(u) = \frac{A_K^{(s)+}(u)}{B_K^{(s)+}(u)}. \tag{4.99}$$

The polynomials $A_K^{(s)+}(u)$ and $B_K^{(s)+}(u)$ of the same degree K are given by

$$A_K^{(s)+}(u) = \sum_{r=1}^{K} a_r^{(s)+} u^r \qquad\qquad B_K^{(s)+}(u) = \sum_{r=0}^{K} b_r^{(s)+} u^r. \tag{4.100}$$

As in the sum over k in (4.97), the variable of the numerator and denominator polynomials from (4.99) is set to be u. This is to be contrasted to variable u^{-1}, which appears in the original sum (4.80).

The corresponding delayed diagonal fast Padé transform with respect to the truncated Green function $\mathcal{G}_N^{(s)}(u^{-1})$ from (4.79) reads as

$$G_K^{\rm FPT(s)+}(u) = u^{-1}\mathcal{G}_K^{\rm FPT(s)+}(u) = u^{-1}\frac{A_K^{(s)+}(u)}{B_K^{(s)+}(u)}. \tag{4.101}$$

The numerator polynomial $A_K^{(s)+}(u)$ from (4.99) or (4.101) is seen not to have the free term, $a_0^{(s)+} = 0$, i.e., the sum over r begins with $r = 1$, such that the first term is given by $a_1^{(s)+}u$. Otherwise, the convergence range of $G_K^{\rm FPT(s)+}(u)$ is located inside the unit circle ($|u| < 1$) where the original sum $\mathcal{G}^{(s)}(u^{-1})$ from (4.78) is divergent. We can extract the polynomials $A_K^{(s)+}(u)$ and $B_K^{(s)+}(u)$ from the condition

$$\mathcal{G}_N^{(s)}(u) \equiv \sum_{n=0}^{N-1} c_{n+s} u^{-n} = \frac{A_K^{(s)+}(u)}{B_K^{(s)+}(u)}. \tag{4.102}$$

Specifically, multiplying (4.102) by $B_K^{(s)+}(u)$ we arrive at

$$\left.\begin{aligned} B_K^{(s)+}(u)\sum_{n=0}^{N-1} c_{n+s}u^{-n} = A_K^{(s)+}(u) \\ \left[\sum_{r=0}^{K} b_r^{(s)+}u^r\right]\left[\sum_{n=0}^{N-1} c_{n+s}u^{-n}\right] = \sum_{r=1}^{K} a_r^{(s)+}u^r \end{aligned}\right\}. \tag{4.103}$$

Here, performing the indicated multiplications of the two sums and equating the coefficients of the same powers of the expansion variable yields

$$b_0^{(s)+}c_{n+s} + \sum_{r=1}^{K} b_r^{(s)+}c_{n+s+r} = 0 \qquad\qquad (n = 1, 2, \ldots, M). \tag{4.104}$$

Employing (4.89), we can express (4.104) in the explicit form as

$$\left.\begin{array}{llllll} c_s b_0^{(s)+} & + c_{s+1} b_1^{(s)+} & + c_{s+2} b_2^{(s)+} & + \cdots & + c_{s+K} b_K^{(s)+} & = 0 \\ c_{s+1} b_0^{(s)+} & + c_{s+2} b_1^{(s)+} & + c_{s+3} b_2^{(s)+} & + \cdots & + c_{s+K+1} b_K^{(s)+} & = 0 \\ & & \vdots & & & \\ c_{M+s} b_0^{(s)+} & + c_{M+s+1} b_1^{(s)+} & + c_{M+s+2} b_2^{(s)+} & + \cdots & + c_{M+s+K} b_K^{(s)+} & = 0 \end{array}\right\}. \tag{4.105}$$

These equations have their matrix representations given by

$$\begin{pmatrix} c_{s+1} & c_{s+2} & c_{s+3} & \cdots & c_{s+K} \\ c_{s+2} & c_{s+3} & c_{s+4} & \cdots & c_{s+K+1} \\ c_{s+3} & c_{s+4} & c_{s+5} & \cdots & c_{s+K+2} \\ \vdots & \vdots & \vdots & \ddots & \vdots \\ c_{s+M+1} & c_{s+M+2} & c_{s+M+3} & \cdots & c_{s+K+M} \end{pmatrix} \begin{pmatrix} b_1^{(s)+} \\ b_2^{(s)+} \\ b_3^{(s)+} \\ \vdots \\ b_K^{(s)+} \end{pmatrix} = - \begin{pmatrix} c_s b_0^{(s)+} \\ c_{s+1} b_0^{(s)+} \\ c_{s+2} b_0^{(s)+} \\ \vdots \\ c_{s+M} b_0^{(s)+} \end{pmatrix}. \tag{4.106}$$

The expansion coefficients $\{a_r^{(s)+}\}$ of the numerator polynomial $A_K^{(s)+}(u)$ can be obtained from the inhomogeneous part of the positive powers of the expansion variable from (4.103)

$$\left.\begin{array}{rrrrl} c_s b_1^{(s)+} + c_{s+1} b_2^{(s)+} & + c_{s+3} b_3^{(s)+} & + \cdots & + c_{s+K-1} b_K^{(s)+} & = a_1^{(s)+} \\ c_s b_2^{(s)+} & + c_{s+1} b_3^{(s)+} & + \cdots & + c_{s+K-2} b_K^{(s)+} & = a_2^{(s)+} \\ & & \ddots & \vdots & \quad \vdots \\ & & & c_s b_K^{(s)+} & = a_K^{(s)+} \end{array}\right\} \tag{4.107}$$

or through the equivalent matrix form

$$\begin{pmatrix} a_1^{(s)+} \\ a_2^{(s)+} \\ a_3^{(s)+} \\ \vdots \\ a_K^{(s)+} \end{pmatrix} = \begin{pmatrix} c_s & c_{s+1} & c_{s+2} & \cdots & c_{s+K-1} \\ 0 & c_s & c_{s+1} & \cdots & c_{s+K-2} \\ 0 & 0 & c_s & \cdots & c_{s+K-3} \\ \vdots & \vdots & \vdots & \ddots & \vdots \\ 0 & 0 & 0 & \cdots & c_s \end{pmatrix} \begin{pmatrix} b_1^{(s)+} \\ b_2^{(s)+} \\ b_3^{(s)+} \\ \vdots \\ b_K^{(s)+} \end{pmatrix}. \tag{4.108}$$

A detailed derivation from Ref. [5, 17] gives the following general relation

$$\begin{aligned} \mathcal{G}_n^{\mathrm{FPT}(s)+}(u) &= \mathcal{G}_{\mathrm{e},n}^{\mathrm{LCF}(s)}(u) \\ &= \mathcal{G}_{\mathrm{e},n}^{\mathrm{CCF}(s)}(u) = \mathcal{G}_{2n}^{\mathrm{CF}(s)}(u) \qquad (n = 1, 2, 3 \ldots) \end{aligned} \tag{4.109}$$

and, thus

$$G_n^{\mathrm{FPT}(s)+}(u) = G_{\mathrm{e},n}^{\mathrm{LCF}(s)}(u)$$
$$= G_{\mathrm{e},n}^{\mathrm{CCF}(s)}(u) = G_{2n}^{\mathrm{CF}(s)}(u) \qquad (n = 1, 2, 3 \ldots). \tag{4.110}$$

We see from this result that the delayed diagonal fast Padé transform, $\mathrm{FPT}^{(s)+}$, with the convergence region inside the unit circle ($|u| < 1$) coincides with the even part of the delayed Lanczos continued fraction for any order n. Furthermore, in this method, the ensuing Padé-Green function $G_n^{\mathrm{FPT}(s)+}(u)$ is convergent for $|u| < 1$, as opposed to the original truncated Green function (4.79) which is divergent in the same region, i.e., inside the unit circle. Therefore, inside the unit circle, the delayed diagonal fast Padé transform $G_n^{\mathrm{FPT}(s)+}(u)$ employs the Cauchy concept of analytical continuation to induce or force convergence into the originally divergent series, which is the Green function (4.77).

Overall, we can conclude that the introduction of $G_n^{\mathrm{FPT}(s)\pm}(u^{\pm 1})$ is extremely helpful in proving that the same LCF, i.e., $G_n^{\mathrm{LCF}(s)}(u)$ contains implicitly both $G_n^{\mathrm{FPT}(s)-}(u^{-1})$ (as an accelerator of monotonically converging series) and $G_n^{\mathrm{FPT}(s)+}(u)$ (as an analytical continuator of divergent series/sequences). Specifically, $G_n^{\mathrm{FPT}(s)-}(u^{-1})$ and $G_n^{\mathrm{FPT}(s)+}(u)$ are equal to the odd and even part of $G_n^{\mathrm{LCF}(s)}(u)$, i.e., $G_n^{\mathrm{FPT}(s)-}(u^{-1}) = G_{\mathrm{o},n}^{\mathrm{LCF}(s)}(u)$ and $G_n^{\mathrm{FPT}(s)+}(u) = G_{\mathrm{e},n}^{\mathrm{LCF}(s)}(u)$, respectively. This means that the $\mathrm{FPT}^{(s)-}$ and $\mathrm{FPT}^{(s)+}$ contain twice as many terms as the original CF, i.e, $G_n^{\mathrm{FPT}(s)-}(u^{-1}) = G_{2n+1}^{\mathrm{CF}(s)}(u)$ and $G_n^{\mathrm{FPT}(s)+}(u) = G_{2n}^{\mathrm{CF}(s)}(u)$. Hence, the $\mathrm{FPT}^{(s)-}$ and $\mathrm{FPT}^{(s)+}$ are equal to the odd and even contracted continued fractions, $G_n^{\mathrm{FPT}(s)-}(u^{-1}) = G_{\mathrm{o},n}^{\mathrm{CCF}(s)}(u)$ and $G_n^{\mathrm{FPT}(s)+}(u) = G_{\mathrm{e},n}^{\mathrm{CCF}(s)}(u)$, as per (4.96) and (4.110), respectively. When such equivalences are established, it appears as optimal to employ only the quantities $G_{\mathrm{e},n}^{\mathrm{LCF}(s)}(u)$ and $G_{\mathrm{o},n}^{\mathrm{LCF}(s)}(u)$, as the two equivalent expressions for $G_n^{\mathrm{FPT}(s)+}(u) = G_{\mathrm{e},2n}^{\mathrm{CF}(s)}(u)$ and $G_n^{\mathrm{FPT}(s)-}(u^{-1}) = G_{\mathrm{o},2n+1}^{\mathrm{CF}(s)}(u)$ for any fixed $s\,(s = 0, 1, 2, ...)$, in order to generate the two sets of observables in the $\mathrm{FPT}^{(s)+}$ and $\mathrm{FPT}^{(s)-}$ that converge inside $|u| < 1$ and outside $|u| > 1$ the unit circle, respectively. Crucially, the Padé-Green functions $G_n^{\mathrm{FPT}(s)+}(u)$ and $G_n^{\mathrm{FPT}(s)-}(u^{-1})$ are, respectively, the lower and upper bounds of the computed observables (spectra, eigen-frequencies, density of states, etc.). For instance, $\mathrm{Re}(G_n^{\mathrm{FPT}(s)+}(u)) = \mathrm{Re}(G_{\mathrm{e},n}^{\mathrm{LCF}(s)}(u))$ and $\mathrm{Re}(G_n^{\mathrm{FPT}(s)-}(u^{-1})) = \mathrm{Re}(G_{\mathrm{o},n}^{\mathrm{LCF}(s)}(u))$ represent the lower and the upper limits of the envelope of the absorption total shape spectrum for a given time signal $\{c_{n+s}\}$ $(0 \le n \le N-1)$. Likewise, the eigen-frequencies and residues $\{\omega_k^{(s)+}, d_k^{(s)+}\}$ and $\{\omega_k^{(s)-}, d_k^{(s)-}\}$ $(k = 1, 2, ..., K)$, that are obtained using the two variants of the FPT, i.e., $G_K^{\mathrm{FPT}(s)+}(u) = G_{\mathrm{e},K}^{\mathrm{LCF}(s)}(u)$ and $G_K^{\mathrm{FPT}(s)-}(u^{-1}) = G_{\mathrm{o},K}^{\mathrm{LCF}(s)}(u)$ are, respectively, the lower and upper limits of the true, i.e., exact values $\{\omega_k, d_k\}$.

5

Signal-noise separation via Froissart doublets

The FPT converges upon reaching constancy or stabilization of the whole set of the retrieved frequencies and amplitudes of all the physical resonances. Furthermore, such a stabilization is a true signature of the reconstruction of the exact number of resonances. Namely, once the stage at which full convergence is achieved, with any further augmentation of the partial signal length N_P towards the full signal length N, all the fundamental frequencies and amplitudes will "stay put", i.e., they shall still remain constant [11, 34]. Critically, we shall see in the illustrations in chapter 6 that this stability is maintained within machine accuracy at any value N_P between the step when convergence has first been reached up to the full signal length N. Moreover, during reconstruction of the exact input data, the convergence stage itself is reached with an exponential convergence rate [188]. This accomplishment establishes the FPT as a spectral analyzer which is capable of yielding an exponentially accurate approximation for time signals from MRS and other related fields [5]. Both the stability of all the spectral parameters for every genuine resonance and the constancy of the estimate of the true number of resonances can be established by the concept of Froissart doublets [44] or pole-zero cancellation [45]. As discussed, by this phenomenon, all the additional poles and zeros of the Padé spectrum $P^{\pm}_{K+m}/Q^{\pm}_{K+m}$ for $m > 1$, beyond the stabilized number K of resonances, will automatically cancel each other via (2.186). With such a feature, the FPT becomes safe-guarded against contamination of the final results by spurious resonances, because all the poles due to unphysical resonances originating from the denominator polynomial become identical to the associated zeros of the numerator polynomial. Evidently, this yields pole-zero cancellation by way of the polynomial quotient in the FPT from (2.186). Advantageously, pole-zero cancellations can be exploited to separate spurious from genuine contents of the signal. Because these unphysical poles and zeros always appear as pairs in the FPT, they are considered as doublets. Moreover, they are known as Froissart doublets after Froissart [44] who discovered by empirical means this remarkable phenomenon, which is unique to the Padé approximant precisely because of modeling via the polynomial quotient.

By its nature, noise represents spurious information which corrupts the genuine part of the signal. However, pole-zero cancellations could be used to discriminate between noise and the physical information in the investigated time signal. This is the most important application of Froissart doublets in MRS [11, 24, 30, 31, 34] and also in other similar fields where the FPT could be used [5]. In this way, spurious content (noise and noise-like pollution) which is usually viewed as a burden in every data can, in fact, be advantageously used to tease out the genuine information from the output results of the analysis.

5.1 Critical importance of poles and zeros in generic spectra

As stated, a Froissart doublet [44] is a spectral pair consisting of a pole and a zero that coincide with each other. Hence, for an investigation of Froissart doublets, one needs zeros and poles of the complex Padé spectra $P_K^{\pm}(z^{\pm 1})/Q_K^{\pm}(z^{\pm 1})$ in the FPT$^{(\pm)}$. Such zeros and poles are defined as the solutions of the characteristic equations for the polynomials in the numerator

$$P_K^{\pm}(z^{\pm 1}) = 0 \tag{5.1}$$

and denominator

$$Q_K^{\pm}(z^{\pm 1}) = 0. \tag{5.2}$$

Thus far, the solutions of the numerator and denominator characteristic equations from (5.1) and (5.2) were denoted by $\tilde{z}_k^{\pm 1}$ in (2.190) and $z_k^{\pm 1}$ in (2.181). However, from now on we shall use the alternative notations $z_{k,P}^{\pm 1} \equiv z_{k,P}^{\pm}$ and $z_{k,P}^{\pm 1} \equiv z_{k,Q}^{\pm}$ for the zeros of polynomials $P_K^{\pm}(z^{\pm 1})$ and $Q_K^{\pm}(z^{\pm 1})$, respectively. Here, the second subscript P in $z_{k,P}^{\pm}$ and likewise Q in $z_{k,Q}^{\pm}$ is introduced to indicate that these eigen-roots $z_{k,P}^{\pm} = \exp(\pm i\omega_{k,P}^{\pm}\tau)$ and $z_{k,Q}^{\pm} = \exp(\pm i\omega_{k,Q}^{\pm}\tau)$ satisfy their respective characteristic equations $P_K^{\pm}(z_{k,P}^{\pm}) = 0$ and $Q_K^{\pm}(z_{k,Q}^{\pm}) = 0$, according to (5.1) and (5.2).

5.2 Spectral representations via Padé poles and zeros as pFPT$^{(\pm)}$ and zFPT$^{(\pm)}$

By studying pole-zero cancellations, we are naturally led to the introduction of the two pairs of complementary representations of the FPT$^{(\pm)}$. One of these pairs is zFPT$^{(\pm)}$ named the 'zeros of the FPT$^{(\pm)}$', whereas the other pair is pFPT$^{(\pm)}$ called the 'poles of the FPT$^{(\pm)}$'. The zFPT$^{(\pm)}$ and pFPT$^{(\pm)}$ can autonomously give spectra by employing exclusively either the zeros $\{z_{k,P}^{\pm}\}$

through the following representation

$$\text{Spectra in zFPT}^{(\pm)} \propto g_{K,P}^{\pm} \prod_{r=1}^{K} (z^{\pm 1} - z_{r,P}^{\pm}) \tag{5.3}$$

or the poles $\{z_{k,Q}^{\pm}\}$ via

$$\text{Spectra in pFPT}^{(\pm)} \propto g_{K,Q}^{\pm} \prod_{s=1}^{K} (z^{\pm 1} - z_{s,Q}^{\pm}) \tag{5.4}$$

where $g_{K,P}^{\pm}$ and $g_{K,Q}^{\pm}$ represent the so-called gain factors. Clearly, if zeros $\{z_{k,P}^{\pm}\}$ and poles $\{z_{k,Q}^{\pm}\}$ are simultaneously employed within shape spectra and/or in quantification, the usual composite representations $\text{FPT}^{(\pm)}$ are obtained through the union of the two new constituent representations, the $\text{zFPT}^{(\pm)}$ and $\text{pFPT}^{(\pm)}$.

Alternatively, the $\text{zFPT}^{(\pm)}$ and $\text{pFPT}^{(\pm)}$ can be investigated via the canonical forms of the polynomials $P_K^{\pm}(z^{\pm 1})$

$$P_K^{\pm}(z^{\pm 1}) = p_K^{\pm} \prod_{r=1}^{K} (z^{\pm 1} - z_{r,P}^{\pm}) \tag{5.5}$$

and $Q_K^{\pm}(z^{\pm 1})$

$$Q_K^{\pm}(z^{\pm 1}) = q_K^{\pm} \prod_{s=1}^{K} (z^{\pm 1} - z_{s,Q}^{\pm}). \tag{5.6}$$

With these formulae it becomes possible to write the corresponding expressions for the general derivatives of the polynomials $P_K^{\pm}(z^{\pm 1})$ and $Q_K^{\pm}(z^{\pm 1})$. For instance, the first derivatives of $Q_K^{\pm}(z^{\pm 1})$, which will be needed in this chapter at $z^{\pm 1} = z_{k,Q}^{\pm}$, can be obtained by the following explicit expressions

$$\begin{aligned} Q_K^{\pm\prime}(z_{k,Q}^{\pm}) &\equiv \left\{ \frac{\mathrm{d}}{\mathrm{d}z^{\pm 1}} Q_K^{\pm}(z^{\pm 1}) \right\}_{z^{\pm 1} = z_{k,Q}^{\pm}} \\ &= q_K^{\pm} \prod_{s=1, s\neq k}^{K} (z_{k,Q}^{\pm} - z_{s,Q}^{\pm}). \end{aligned} \tag{5.7}$$

Here, it is seen that for simple poles, defined as the non-confluent zeros $z_{k,Q}^{\pm} \neq z_{k',Q}^{\pm}$ $(k' \neq k)$ of the denominator polynomial $Q_K^{\pm}(z^{\pm 1})$, the first derivative $Q_K^{\pm\prime}(z_{k,Q}^{\pm})$ is always different from zero

$$Q_K^{\pm\prime}(z_{k,Q}^{\pm}) \neq 0. \tag{5.8}$$

5.3 Padé canonical spectra

Substituting (5.5) and (5.7) into (5.9) gives the canonical forms of the polynomial quotients from the FPT, i.e., $P_K^{\pm}(z^{\pm 1})/Q_K^{\pm}(z^{\pm 1})$

$$\frac{P_K^{\pm}(z^{\pm 1})}{Q_K^{\pm}(z^{\pm 1})} = \frac{p_K^{\pm}}{q_K^{\pm}} \frac{\prod_{r=1}^{K}(z^{\pm 1} - z_{r,P}^{\pm})}{\prod_{s=1}^{K}(z^{\pm 1} - z_{s,Q}^{\pm})}. \tag{5.9}$$

The spectra from (5.9) can also be expressed by a single product symbol as

$$\frac{P_K^{\pm}(z^{\pm 1})}{Q_K^{\pm}(z^{\pm 1})} = \frac{p_K^{\pm}}{q_K^{\pm}} \prod_{k=1}^{K} \frac{(z^{\pm 1} - z_{k,P}^{\pm})}{(z^{\pm 1} - z_{k,Q}^{\pm})}. \tag{5.10}$$

Physically, the degree K of the denominator polynomials in the $\text{FPT}^{(\pm)}$ represents the total number K_T of poles, $K_T \equiv K$. Otherwise, the number K_T is defined as the sum of the numbers of the genuine (K_G) and spurious (K_S) poles, $K_T = K_G + K_S$. Genuine poles are, in fact, the signal poles that represent the true, physical content of the investigated FID. Spurious or extraneous poles are the non-physical ingredient of the input FID and, as such, ought to be dropped from the final results of the spectral analysis. By definition, a noiseless input FID possesses no spurious information. Nevertheless, spuriousness can still appear in the spectral analysis of a noiseless FID in all signal processors. Among the major origins of this computationally generated noise (without counting the obvious round-off errors) is under-estimation or over-estimation of the otherwise unknown, exact number K_G. Generally, spurious poles predominantly consist of Froissart doublets via the couples of the coinciding Froissart zeros and poles

$$z_{k,P}^{\pm} = z_{k,Q}^{\pm} \qquad\qquad k \in \mathcal{K}_F. \tag{5.11}$$

Here, the collection $\mathcal{K}_F$ represents the set of counting indices k for Froissart poles, whose number is denoted by K_F. There also exist some extraneous isolated poles (called "ghost" poles) that do not have the matching zeros to form the corresponding pairs. Moreover, there could be some extraneous isolated zeros (named "ghost" zeros) that do not have the associated matching poles. For example, a "ghost" zero in the $\text{FPT}^{(+)}$ is the point $z = 0$ which is one of the K zeros of $P_K^+(z)$. In numerical computations by means of the $\text{FPT}^{(+)}$, one would easily detect the trivial zero $z = 0$ for any order K. However, in signal processing, this zero should be discarded from the outset, since the point $z = 0$ is excluded from the domain of definition of the original

Maclaurin expansion (4.79), from which the $\text{FPT}^{(+)}$ is derived. Obviously, this latter point is associated precisely to $z^{-1} = \infty$ in the $\text{FPT}^{(-)}$, since in this version of the FPT, the harmonic expansion variable is z^{-1}. Thus, the "ghost" zero $z^{-1} = \infty$ is one of the K zeros of $P_K^-(z^{-1})$ in the $\text{FPT}^{(-)}$ and, therefore, it should be ignored for the same reason used to eliminate the point $z = 0$ from the $\text{FPT}^{(+)}$. Naturally, in numerical computations, the point $z^{-1} = \infty$ cannot be reached exactly. Nevertheless, one of the zeros from the reconstructed set $\{z_{k,P}^-\}$ must have a very large real and imaginary part, and this will become more pronounced by increasing the degree K of $P_K^-(z^{-1})$ in the $\text{FPT}^{(-)}$. The two "ghost" zeros, $z = 0$ and $z^{-1} = \infty$ in the $\text{FPT}^{(+)}$ and $\text{FPT}^{(-)}$, respectively, have been found in our computations, exactly as per the outlined description. Moreover, the computations from chapters 3 and 6 with noise-free and noise-corrupted FIDs did not detect any "ghost" poles and, therefore, it follows

$$K_T = K_G + K_F. \tag{5.12}$$

Note that although the numerator polynomials $P_K^\pm(z^{\pm 1})$ are of degree K, the total number of their zeros will be $K-1$ instead of K because of the excluded "ghost" zero in each polynomial. This was already discussed in chapter 3 and will be addressed again in chapter 6 in more detailed illustrations.

5.4 Signal-noise separation with exclusive reliance upon resonant frequencies

As explained, the sets of all the poles $\{z_{k,Q}^\pm\}$ consist of the two disjoint subsets of the genuine and Froissart poles

$$\{z_{k,Q}^\pm\}_{k\in\mathcal{K}_T} = \{z_{k,Q}^\pm\}_{k\in\mathcal{K}_G} \oplus \{z_{k,Q}^\pm\}_{k\in\mathcal{K}_F}. \tag{5.13}$$

Here, $\mathcal{K}_G$ is the set of the counting indices k for the genuine poles and $\mathcal{K}_T$ is the set of all the values of k. Since there are no common elements in the two subsets $\{z_{k,Q}^\pm\}_{k\in\mathcal{K}_G}$ and $\{z_{k,Q}^\pm\}_{k\in\mathcal{K}_F}$, their sums from (5.13) represent the so-called direct sums as labeled by the usual symbol $\oplus$ for disjoint sets. Therefore, it is sufficient to count the number K_F of Froissart doublets to obtain the exact number K_G of the genuine poles through $K_G = K_T - K_F$ as in (5.12). In practice, as soon as the $\text{FPT}^{(\pm)}$ are found to fully converge, one should make a straightforward binning of all the reconstructed poles $\{z_{k,P}^\pm\}_{k\in\mathcal{K}_T}$ into two sets $\{z_{k,Q}^\pm\}_{k\in\mathcal{K}_F}$ and $\{z_{k,Q}^\pm\}_{k\in\mathcal{K}_G}$ depending on whether or not $z_{k,Q}^\pm = z_{k,P}^\pm$, i.e., whether or not (5.11) is fulfilled. This grouping allows an unambiguous recovery of the true number K_G of the genuine poles associated with the exact numerical values of the corresponding harmonic variables

$$\{z_{k,Q}^{\pm}\}_{k\in\mathcal{K}_T} \supseteq \begin{cases} \{z_{k,Q}^{\pm}\}_{k\in\mathcal{K}_G} & z_{k,Q}^{\pm} \neq z_{k,P}^{\pm} \quad : \quad \text{genuine poles} \\ \{z_{k,Q}^{\pm}\}_{k\in\mathcal{K}_F} & z_{k,Q}^{\pm} = z_{k,P}^{\pm} \quad : \quad \text{Froissart poles.} \end{cases} \tag{5.14}$$

The associated genuine fundamental frequencies $\{\nu_{k,Q}^{\pm}\}_{k\in\mathcal{K}_G}$ are extracted from the whole set $\{\nu_{k,Q}^{\pm}\}_{k\in\mathcal{K}_T}$ via

$$\{\nu_{k,Q}^{\pm}\}_{k\in\mathcal{K}_T} \supseteq \begin{cases} \{\nu_{k,Q}^{\pm}\}_{k\in\mathcal{K}_G} & \nu_{k,Q}^{\pm} \neq \nu_{k,P}^{\pm} \quad : \quad \text{genuine poles} \\ \{\nu_{k,Q}^{\pm}\}_{k\in\mathcal{K}_F} & \nu_{k,Q}^{\pm} = \nu_{k,P}^{\pm} \quad : \quad \text{Froissart poles.} \end{cases} \tag{5.15}$$

In these expressions, the following definitions of $\nu_{k,P}^{\pm}$ and $\nu_{k,Q}^{\pm}$ are employed in terms of $z_{k,P}^{\pm}$ and $z_{k,Q}^{\pm}$, respectively

$$\nu_{k,P}^{\pm} = \mp\frac{i}{2\pi\tau}\ln(z_{k,P}^{\pm}) \qquad \nu_{k,Q}^{\pm} = \mp\frac{i}{2\pi\tau}\ln(z_{k,Q}^{\pm}). \tag{5.16}$$

This disentangling of the genuine from spurious (noise and noise-like) information is the signature of the concept of signal-noise-separation, i.e., the SNS. As noted, even noiseless FIDs have spurious information which is noise-like, since it behaves similarly to noise. Namely, Froissart doublets are found in spectral analysis by the FPT for any noise-free time signal and they accumulate at the circumference ($|z| = 1$) of the unit circle in the plane of the complex harmonic variable. The borderline $|z| = 1$ of the essential singularity points represents the natural limit (in the Weierstrass sense) with the maximal probability of 1 (except at most for a nearly zero-area set [58, 59, 61]) for finding the locations of all the poles of the response function for time signals generated from random numbers [44, 52, 58, 59].

In this context, it is important to recall that in the FFT, all frequencies from the Fourier grid $\omega_k^{\rm FFT} \equiv 2\pi k/T\,(0 \le k \le N-1)$ are purely real and, therefore, located exactly at the circumference $|z| = 1$ of the unit circle. As such, the Fourier "fundamental" frequencies $\omega_k^{\rm FFT}$ are entirely embedded in noise throughout the circumference $|z| = 1$ of the unit circle with no procedure available for separation of physical signal from noise within the Fourier analysis.

Overall, as stated before, due to its linearity, the FFT imports noise as intact from the time to the frequency domain. Even worse, this situation is further aggravated in the complex plane of the harmonic variable z by the distribution of the corresponding Fourier grid points $z_k^{\rm FFT} = \exp{(i\omega_k^{\rm FFT}\tau)}$. Namely, all the elements of the Fourier set $\{z_k^{\rm FFT}\}\,(0 \le k \le N-1)$ exhibit the feature $|z_k^{\rm FFT}| = 1$ which places them precisely at the unit distance from the origin of the unit circle. Consequently, all the $z_k^{\rm FFT}$'s are maximally mixed with noise, which predominantly populates the circumference $|z| = 1$

of the same unit circle. This illustrates that the FFT is least suitable for: (i) processing FIDs with poor SNR, and for (ii) solving the quantification problem via post-processing total shape spectra from the FFT in an attempt to extract peak parameters, as done in the usual fitting algorithms in MRS, e.g., VARPRO, AMARES, LCModel, etc.

5.5 Model reduction problem via Padé canonical spectra

Employing (5.12), the canonical representations from (5.9) can be rewritten in the following forms

$$\frac{P^{\pm}_{K_T}(z^{\pm 1})}{Q^{\pm}_{K_T}(z^{\pm 1})} = \frac{p^{\pm}_K}{q^{\pm}_K}\frac{\prod_{r=1}^{K_T}(z^{\pm 1}-z^{\pm}_{r,P})}{\prod_{s=1}^{K_T}(z^{\pm 1}-z^{\pm}_{s,Q})}$$
$$= \frac{p^{\pm}_K}{q^{\pm}_K}\frac{\prod_{r=1}^{K_G}(z^{\pm 1}-z^{\pm}_{r,P})}{\prod_{s=1}^{K_G}(z^{\pm 1}-z^{\pm}_{s,Q})}\left\{\frac{\prod_{r'=K_G+1}^{K_G+K_F}(z^{\pm 1}-z^{\pm}_{r',P})}{\prod_{s'=K_G+1}^{K_G+K_F}(z^{\pm 1}-z^{\pm}_{s',Q})}\right\}_{r',s'\in\mathcal{K}_F}. \tag{5.17}$$

The canonical quotient in the curly brackets of (5.17) is equal to unity. This stems from the exact cancellation of the numerator and denominator polynomials, as a result of the coincidence of the invoked poles and zeros via (5.11) in Froissart doublets. Thus, we can simplify (5.17) as

$$\frac{P^{\pm}_{K_T}(z^{\pm 1})}{Q^{\pm}_{K_T}(z^{\pm 1})} = \frac{p^{\pm}_K}{q^{\pm}_K}\frac{\prod_{r=1}^{K_G+K_F}(z^{\pm 1}-z^{\pm}_{r,P})}{\prod_{s=1}^{K_G+K_F}(z^{\pm 1}-z^{\pm}_{s,Q})} = \frac{p^{\pm}_K}{q^{\pm}_K}\frac{\prod_{r=1}^{K_G}(z^{\pm 1}-z^{\pm}_{r,P})}{\prod_{s=1}^{K_G}(z^{\pm 1}-z^{\pm}_{s,Q})}. \tag{5.18}$$

Upon convergence in the $\mathrm{FPT}^{(\pm)}$, it follows

$$p^{\pm}_{K_G+K_F} = p^{\pm}_{K_G} \tag{5.19}$$

$$q^{\pm}_{K_G+K_F} = q^{\pm}_{K_G}. \tag{5.20}$$

Using (5.19) and (5.20), we can reduce (5.18) to the sought relationship

$$\frac{P^{\pm}_{K_G+K_F}(z^{\pm 1})}{Q^{\pm}_{K_G+K_F}(z^{\pm 1})} = \frac{P^{\pm}_{K_G}(z^{\pm 1})}{Q^{\pm}_{K_G}(z^{\pm 1})} \qquad (K_F = 1, 2, 3, ...). \tag{5.21}$$

This is the proof of the earlier stated result from (2.186). In an alternative derivation, we could avoid dealing with the equalities in (5.19) and (5.20) by defining $P_K^{\pm}$ and $Q_K^{\pm}$ as the monic polynomials. Recall that a polynomial is defined as monic when its coefficient, as a multiplier of the highest power of the expansion variable, is equal to unity. As such, $P_K^{\pm}$ and $Q_K^{\pm}$ can be monic polynomials when all their expansion coefficients are divided by $p_K^{\pm}$ and $q_K^{\pm}$, respectively.

As they stand, the expressions from (5.21) directly transcend Froissart doublets via pole-zero cancellations. This cancellation, in turn, decreases the order of the FPT from $K = K_T$ to $K_T - K_F = K_G$. Hence, the concept of Froissart doublets via pole-zero cancellation represents an efficient procedure for reduction of the order of the model for Padé-based quantification in MRS and elsewhere.

In fact, pole-zero cancellations diminish the dimensionality of the interim problem, which would be of the order $K_T = K_F + K_G$ without the elimination of the K_F Froissart doublets. Thus, upon discarding the K_F Froissart doublets, we are left with the order K_G, which is then necessarily the exact order of the original problem. From the physics viewpoint, this means that the recovered order K_G is indeed the exact number of genuine poles. This is the way the true number of genuine resonances is unambiguously reconstructed from the input FID by applying the $\mathrm{FPT}^{(\pm)}$.

5.6 Denoising Froissart filter

The critical point for retrieving the true number K_G is the virtue of the $\mathrm{FPT}^{(\pm)}$ to unequivocally differentiate between the genuine (Padé) and spurious (Froissart) poles and zeros. This achievement is based on pole-zero cancellations that effectively filter out all the spurious, i.e., Froissart poles from the full solution of the quantification problem. Such a filtering leaves us with the genuine poles alone, and this is vital to the ultimate solution of the quantification problem. Hence, it is appropriate to coin the term the denoising Froissart filter, i.e., the DFF, for the outlined procedure.

One of the direct meanings of the term 'denoising' is the 'noise reduction', where Froissart doublets are considered as noise because of their spuriousness. Such a terminology can be used for noise-corrupted as well as noise-free input FIDs. Of course, in both these cases, the exact number K_G is unknown in advance of spectral analysis. This implies that any estimate $K' \neq K_G$ would inevitably produce a non-zero difference FID(input) − FID(reconstructed by using K'). Such a difference is spurious and, thus, acts implicitly as noise for both noiseless and noisy input FIDs.

5.7 Signal-noise separation with exclusive reliance upon resonant amplitudes

The derived proof of determining the exact number of genuine resonances is based exclusively on the retrieved signal poles $z_{k,Q}^{\pm}$, i.e., the quantities that include only the complex frequencies, with no information about the associated complex amplitudes $d_k^{\pm}$. However, it is likewise important to know whether the genuine and spurious resonances could also be distinguished by their amplitudes. To investigate this subject, it would be advantageous to have closed, analytical expressions for the amplitudes $d_k^{\pm}$ corresponding to the signal poles $z_{k,Q}^{\pm}$. Recall that the amplitudes $d_k^{\pm}$ are defined as the Cauchy residues of the rational polynomial $P_K^{\pm}(z^{\pm 1})/Q_K^{\pm}(z^{\pm 1})$. Such residues for the simple poles of $Q_K^{\pm}(z^{\pm 1})$ are defined by the expressions

$$d_k^{\pm} = \lim_{z^{\pm 1} \to z_{k,Q}^{\pm}} \left\{ (z^{\pm 1} - z_{k,Q}^{\pm}) \frac{P_K^{\pm}(z^{\pm 1})}{Q_K^{\pm}(z^{\pm 1})} \right\}. \tag{5.22}$$

With the help of the canonical form (5.11), we can perform the limit in (5.22) with the results

$$d_k^{\pm} = \frac{p_K^{\pm}}{q_K^{\pm}} \times \lim_{z^{\pm 1} \to z_{k,Q}^{\pm}} \left\{ \frac{[z^{\pm 1} - z_{k,Q}^{\pm}] \prod_{r=1}^{K} (z^{\pm 1} - z_{r,P}^{\pm})}{A \cdots (z^{\pm 1} - z_{k-1,Q}^{\pm})[z^{\pm 1} - z_{k,Q}^{\pm}](z^{\pm 1} - z_{k+1,Q}^{\pm}) \cdots B} \right\}$$

where $A = (z^{\pm 1} - z_{1,Q}^{\pm})$ and $B = (z^{\pm 1} - z_{K_T,Q}^{\pm})$. In these formulae for $d_k^{\pm}$, we can cancel out the common terms $[z^{\pm 1} - z_{k,Q}^{\pm}]$ in the square brackets from the numerator and denominator. Therefore, in the remaining expressions, $z^{\pm 1}$ can be replaced by $z_{k,Q}^{\pm}$ so that

$$d_k^{\pm} = \frac{p_K^{\pm}}{q_K^{\pm}} \frac{\prod_{r=1}^{K} (z_{k,Q}^{\pm} - z_{r,P}^{\pm})}{\prod_{s=1, s \neq k}^{K} (z_{k,Q}^{\pm} - z_{s,Q}^{\pm})} \qquad k \in \mathcal{K}_T. \tag{5.23}$$

More concisely, and similarly to (5.10), we can rewrite (5.23) as

$$d_k^{\pm} = \frac{p_K^{\pm}}{q_K^{\pm}} \prod_{k'=1}^{K} \frac{(z_{k,Q}^{\pm} - z_{k',P}^{\pm})}{(z_{k,Q}^{\pm} - z_{k',Q}^{\pm})_{k' \neq k}}. \tag{5.24}$$

We see that the denominator in (5.23) and (5.24) is never zero in the case of simple poles, similarly to (5.7) and (5.8). Moreover, the numerator in (5.23) represents the canonical form of $P_K^{\pm}(z_{k,Q}^{\pm})$ by reference to (5.5). Similarly, the denominator in (5.23) is the canonical form of the first derivative with respect to $z^{\pm 1}$ of $Q_K^{\pm}(z)$ taken at $z^{\pm 1} = z_{k,Q}^{\pm}$ according to (5.7). This yields the following equivalent formulae for $d_k^{\pm}$

$$d_k^{\pm} = \frac{P_K^{\pm}(z_{k,Q}^{\pm})}{Q_K^{\pm\prime}(z_{k,Q}^{\pm})} \tag{5.25}$$

where we always have $Q_K^{\pm\prime}(z_{k,Q}^{\pm}) \neq 0$, as in (5.8). The formulae from (5.25) could also be established by using definition (5.22) without referring to any special representation of the pertinent polynomials. With this goal, we employ the characteristic equation $Q_K^{\pm}(z_{k,Q}^{\pm}) = 0$ from (5.2) and the definition of the first derivative via $Q_K^{\pm\prime}(z_{k,Q}^{\pm}) = \lim_{z^{\pm 1} \to z_{k,Q}^{\pm}} [Q_K^{\pm}(z^{\pm 1}) - Q_K^{\pm}(z_{k,Q}^{\pm})](z^{\pm 1} - z_{k,Q}^{\pm})$. This precisely reproduces (5.25) as follows

$$\begin{aligned}
d_k^{\pm} &= \lim_{z^{\pm 1} \to z_{k,Q}^{\pm}} \left\{ (z^{\pm 1} - z_{k,Q}^{\pm}) \frac{P_K^{\pm}(z^{\pm 1})}{Q_K^{\pm}(z^{\pm 1})} \right\} \\
&= P_K^{\pm}(z_{k,Q}^{\pm}) \left\{ \lim_{z^{\pm 1} - z_{k,Q}^{\pm}} \frac{Q_K^{\pm}(z^{\pm 1}) - Q_K^{\pm}(z_{k,Q}^{\pm})}{z^{\pm 1} - z_{k,Q}^{\pm}} \right\}^{-1} \\
&= \frac{P_K^{\pm}(z_{k,Q}^{\pm})}{Q_K^{\pm\prime}(z_{k,Q}^{\pm})} \qquad \text{(QED)}.
\end{aligned}$$

To find out whether the amplitudes $d_k^{\pm}$ can be used to distinguish between the genuine and spurious resonances, we employ the same prescription as in (5.17) to rewrite (5.23) as

$$\begin{aligned}
d_k^{\pm} &= \frac{p_K^{\pm}}{q_K^{\pm}} \frac{\prod_{r=1}^{K_T} (z_{k,Q}^{\pm} - z_{r,P}^{\pm})}{\prod_{s=1, s\neq k}^{K_T} (z_{k,Q}^{\pm} - z_{s,Q}^{\pm})} \qquad (5.26) \\
&= \frac{p_K^{\pm}}{q_K^{\pm}} \frac{\prod_{r=1}^{K_G} (z_{k,Q}^{\pm} - z_{r,P}^{\pm})}{\prod_{s=1, s\neq k}^{K_G} (z_{k,Q}^{\pm} - z_{s,Q}^{\pm})} \left\{ \frac{\prod_{r'=K_G+1}^{K_G+K_F} (z_{k,Q}^{\pm} - z_{r',P}^{\pm})}{\prod_{s'=K_G+1, s'\neq k}^{K_G+K_F} (z_{k,Q}^{\pm} - z_{s',Q}^{\pm})} \right\}_{r',s' \in \mathcal{K}_F} \qquad k \in \mathcal{K}_T.
\end{aligned}$$

Since the two sets $\mathcal{K}_G$ and $\mathcal{K}_F$ are disjoint, we know that if $k \in \mathcal{K}_G$ (respectively, $k \in \mathcal{K}_F$), then the amplitudes $d_k^{\pm}$ on the lhs of (5.26) will correspond to the genuine (respectively, spurious) resonances. Hence, for genuine resonances, i.e., $k \in \mathcal{K}_G$ (which automatically implies that $k \notin \mathcal{K}_F$, i.e., $k \neq k'$ for $k' \in \mathcal{K}_F$), the genuine amplitudes can be obtained from (5.26) via

$$d_k^{\pm} = \frac{p_K^{\pm}}{q_K^{\pm}} \frac{\prod_{r=1}^{K_T}(z_{k,Q}^{\pm} - z_{r,P}^{\pm})}{\prod_{s=1,s\neq k}^{K_T}(z_{k,Q}^{\pm} - z_{s,Q}^{\pm})} \tag{5.27}$$

$$= \frac{p_K^{\pm}}{q_K^{\pm}} \frac{\prod_{r=1}^{K_G}(z_{k,Q}^{\pm} - z_{r,P}^{\pm})}{\prod_{s=1,s\neq k}^{K_G}(z_{k,Q}^{\pm} - z_{s,Q}^{\pm})} \left\{ \frac{\prod_{r'=K_G+1,r'\neq k}^{K_G+K_F}(z_{k,Q}^{\pm} - z_{r',P}^{\pm})}{\prod_{s'=K_G+1,s'\neq k}^{K_G+K_F}(z_{k,Q}^{\pm} - z_{s',Q}^{\pm})} \right\}_{r',s'\in\mathcal{K}_F} \quad k \in \mathcal{K}_G.$$

In (5.27), the rational polynomial in the curly brackets is equal to unity, because of the coincidence of the constituent polynomials in the numerators, $\prod_{r'=K_G+1,r'\neq k}^{K_G+K_F}(z_{k,Q}^{\pm} - z_{r',P}^{\pm})$ and denominators, $\prod_{s'=K_G+1,s'\neq k}^{K_G+K_F}(z_{k,Q}^{\pm} - z_{s',Q}^{\pm})$. This maps (5.27) into the form

$$d_k^{\pm} = \frac{p_K^{\pm}}{q_K^{\pm}} \frac{\prod_{r=1}^{K_G}(z_{k,Q}^{\pm} - z_{r,P}^{\pm})}{\prod_{s=1,s\neq k}^{K_G}(z_{k,Q}^{\pm} - z_{s,Q}^{\pm})} \qquad k \in \mathcal{K}_G \qquad \text{(genuine)}. \tag{5.28}$$

Similarly, for $k \in \mathcal{K}_F$, the Froissart or spurious amplitudes are identified from (5.26) as

$$d_k^{\pm} = \frac{p_K^{\pm}}{q_K^{\pm}} \frac{\prod_{r=1}^{K_T}(z_{k,Q}^{\pm} - z_{r,P}^{\pm})}{\prod_{s=1,s\neq k}^{K_T}(z_{k,Q}^{\pm} - z_{s,Q}^{\pm})} \tag{5.29}$$

$$= \frac{p_K^{\pm}}{q_K^{\pm}} \frac{\prod_{r=1}^{K_G}(z_{k,Q}^{\pm} - z_{r,P}^{\pm})}{\prod_{s=1,s\neq k}^{K_G}(z_{k,Q}^{\pm} - z_{s,Q}^{\pm})} \left\{ \frac{\prod_{r'=K_G+1,r'=k}^{K_G+K_F}(z_{k,Q}^{\pm} - z_{r',P}^{\pm})}{\prod_{s'=K_G+1,s'\neq k}^{K_G+K_F}(z_{k,Q}^{\pm} - z_{s',Q}^{\pm})} \right\}_{r',s'\in\mathcal{K}_F} \quad k \in \mathcal{K}_F.$$

The polynomial quotients in the curly brackets from (5.29) are equal to zero. This is due to the fact that the corresponding numerator polynomials

$\prod_{r'=K_G+1,r'=k}^{K_G+K_F}(z_{k,Q}^{\pm} - z_{r',P}^{\pm})$ are equal to zero, because the following null-factors are always present in the product: $[(z_{k,Q}^{\pm} - z_{r',P}^{\pm})]_{r'=k} = (z_{k,Q}^{\pm} - z_{k,P}^{\pm}) = 0$ for $z_{k,Q}^{\pm} = z_{k,P}^{\pm}$ where $k \in \mathcal{K}_F$, according to the definition (5.11) of Froissart doublets as pole-zero cancellations. Therefore, *all* the Froissart amplitudes are found to be zero-valued

$$d_k^{\pm} = 0 \quad \text{(spurious)} \qquad k \in \mathcal{K}_F \qquad \Longleftrightarrow \qquad z_{k,Q}^{\pm} = z_{k,P}^{\pm}. \tag{5.30}$$

This also follows from the alternative formulae $d_k^{\pm} = P_K^{\pm}(z_{k,Q}^{\pm})/Q_K^{\pm\prime}(z_{k,Q}^{\pm})$ in (5.25) via (5.1) with $P_K^{\pm}(z_{k,Q}^{\pm}) = 0$ for $z_{k,P}^{\pm} = z_{k,Q}^{\pm}$ and (5.8) where $Q_K^{\pm\prime}(z_{k,Q}^{\pm}) \neq 0$. Using (5.13), we can decompose the whole sets $\{d_k^{\pm}\}$ into two disjoint sets comprised of the genuine and spurious amplitudes

$$\{d_k^{\pm}\}_{k\in\mathcal{K}_T} = \{d_k^{\pm}\}_{k\in\mathcal{K}_G} \oplus \{d_k^{\pm}\}_{k\in\mathcal{K}_F}. \tag{5.31}$$

The members of these two subsets from (5.31) are given by (5.28) and (5.30). This can be recapitulated via

$$d_k^{\pm} = \begin{cases} \dfrac{p_K^{\pm}}{q_K^{\pm}} \dfrac{\prod_{r=1}^{K_G}(z_{k,Q}^{\pm} - z_{r,P}^{\pm})}{\prod_{s=1,s\neq k}^{K_G}(z_{k,Q}^{\pm} - z_{s,Q}^{\pm})} & k \in \mathcal{K}_G \quad : \quad \text{genuine amplitudes} \\[2ex] 0 & k \in \mathcal{K}_F \quad : \quad \text{Froissart amplitudes.} \end{cases} \tag{5.32}$$

To give another interpretation of the remarkable result (5.30), it is instructive to return to the canonical formula (5.24) for the residues $d_k^{\pm}$. In these expressions for $d_k^{\pm}$, we shall single out the critical terms $z_{k,Q}^{\pm} - z_{k,P}^{\pm}$ by factoring them in front of the product symbol in (5.24), thus leaving certain complex-valued non-zero remainders denoted by $R_k^{\pm} \neq 0$

$$d_k^{\pm} = \left(z_{k,Q}^{\pm} - z_{k,P}^{\pm}\right) R_k^{\pm} \tag{5.33}$$

$$R_k^{\pm} = \frac{p_K^{\pm}}{q_K^{\pm}} \prod_{k'=1}^{K} \left(\frac{z_{k,Q}^{\pm} - z_{k',P}^{\pm}}{z_{k,Q}^{\pm} - z_{k',Q}^{\pm}}\right)_{k'\neq k} \qquad R_k^{\pm} \neq 0\,. \tag{5.34}$$

The formula (5.33) states that the kth residues $d_k^{\pm}$ are proportional to the *distance* between the poles $z_{k,Q}^{\pm}$ and zeros $z_{k,P}^{\pm}$ with the non-zero complex constants of proportionality being $R_k^{\pm}$. This is a plausible geometric interpretation of the residues in the complex plane of the harmonic variable z. Since the remainders $R_k^{\pm}$ are always non-zero, they can act merely as modulators of the complex-valued distance or metric $\sim (z_{k,Q}^{\pm} - z_{k,P}^{\pm})$. In other words, if the residues $d_k^{\pm}$ are indeed found to be zero-valued, this could only be due to the zero distance between the corresponding poles and zeros $z_{k,Q}^{\pm} - z_{k,P}^{\pm} = 0$, as per (5.30). With such a graphic picture, it becomes crystal clear why Froissart doublets must have zero amplitudes. Such an occurrence was already encountered in chapter 2 via expression (2.227) and illustrated in chapter 3 through a preview via Fig. 3.19.

5.8 Padé partial fraction spectra

When the spectral parameters $\{z_{k,Q}^{\pm}, d_k^{\pm}\}$ are reconstructed by the $\mathrm{FPT}^{(\pm)}$ following the explained procedure, yet another form for the Padé complex mode spectra $P_K^{\pm}(z^{\pm 1})/Q_K^{\pm}(z^{\pm 1})$ can be obtained. These spectra are the Heaviside or Padé partial fractions that have the following representations for the diagonal versions of the $\mathrm{FPT}^{(\pm)}$ where the numerator and denominator polynomials have the same degree K

$$\frac{P_K^{\pm}(z^{\pm 1})}{Q_K^{\pm}(z^{\pm 1})} = b_0^{\pm} + \sum_{k=1}^{K} \frac{d_k^{\pm} z^{\pm 1}}{z^{\pm 1} - z_{k,Q}^{\pm}}. \tag{5.35}$$

The factored terms $b_0^{\pm}$ in (5.35) are called baseline constants that describe the corresponding flat backgrounds

$$b_0^{\pm} \equiv \frac{p_0^{\pm}}{q_0^{\pm}}. \tag{5.36}$$

It is possible to invert the frequency spectra from (5.35) by the inverse fast Padé transforms (IFPT). These inversions yield the corresponding time signals

$$c_n^{\pm} = b_0^{\pm}\delta(n) + \sum_{k=1}^{K} d_k^{\pm} z_{k,Q}^{\pm n} = b_0^{\pm}\delta(n) + \sum_{k=1}^{K} d_k^{\pm} \mathrm{e}^{\pm 2i\pi n\tau\nu_{k,Q}^{\pm}} \tag{5.37}$$

where the nth power of $z_{k,Q}^{\pm}$ is denoted by $z_{k,Q}^{\pm n} \equiv (z_{k,Q}^{\pm})^n$. The symbol $\delta(n)$ is the usual discrete unit impulse (or discrete unit sample, or discrete unit-step time signal). The meaning of this quantity is the same as in the Kronecker $\delta-$symbol $\delta_{n,0}$, i.e., $\delta(n) = 1$ for $n = 0$ and $\delta(n) = 0$ for $n \neq 0$. For this reason, $\delta(n)$ is alternatively called the Kronecker discrete time sequence. In practice, care must be exercised not to interpret $\delta(n)$ as a sampled version of the associated continuous Dirac delta function $\delta(t)$. The latter function $\delta(t)$ *cannot* be sampled, because it is infinite at time $t = 0$ [5, 84].

Within the $\mathrm{FPT}^{(+)}$, the formulae (5.35) and (5.37) for the spectra and the corresponding reconstructed time signal, respectively, can be simplified by using the property

$$b_0^+ = 0 \tag{5.38}$$

which originates from $p_0^+ \equiv 0$ according to the derivation from Ref. [5, 34]. Thus, using (5.35) and (5.37) we have

$$\left.\begin{aligned} \frac{P_K^+(z)}{Q_K^+(z)} &= \sum_{k=1}^{K} \frac{d_k^+ z}{z - z_{k,Q}^+} \\ c_n^+ &= \sum_{k=1}^{K} d_k^+ z_{k,Q}^{+n} = \sum_{k=1}^{K} d_k^+ \mathrm{e}^{2i\pi n\tau\nu_{k,Q}^+} \end{aligned}\right\} \tag{5.39}$$

$$\left.\begin{aligned} \frac{P_K^-(z)}{Q_K^-(z)} &= \frac{p_0^-}{q_0^-} + \sum_{k=1}^{K} \frac{d_k^- z^{-1}}{z^{-1} - z_{k,Q}^-} \\ c_n^- &= \frac{p_0^-}{q_0^-} + \sum_{k=1}^{K} d_k^- z_{k,Q}^{-n} = \frac{p_0^-}{q_0^-} + \sum_{k=1}^{K} d_k^- \mathrm{e}^{-2i\pi n\tau\nu_{k,Q}^-} \end{aligned}\right\}. \tag{5.40}$$

5.9 Model reduction problem via Heaviside or Padé partial fraction spectra

In an alternative derivation, we can show that the model order reduction can also be accomplished by using the formulae (5.35) and (5.37) for the Heaviside or Padé partial fractions

$$\begin{aligned} \frac{P_{K_T}^\pm(z^{\pm 1})}{Q_{K_T}^\pm(z^{\pm 1})} &= b_0^\pm + \sum_{k=1}^{K_G+K_F} \frac{d_k^\pm z^{\pm 1}}{z^{\pm 1} - z_{k,Q}^\pm} \\ &= b_0^\pm + \sum_{k=1}^{K_G} \frac{d_k^\pm z^{\pm 1}}{z^{\pm 1} - z_{k,Q}^\pm} + \left\{ \sum_{k=K_G+1}^{K_G+K_F} \frac{d_k^\pm z^{\pm 1}}{z^{\pm 1} - z_{k,Q}^\pm} \right\}_{k\in\mathcal{K}_F} \end{aligned} \tag{5.41}$$

as well as for the time signals

$$c_n^\pm = b_0^\pm \delta(n) + \sum_{k=1}^{K_T} d_k^\pm z_{k,Q}^{\pm n} = b_0^\pm \delta(n) + \sum_{k=K_G+1}^{K_G+K_F} d_k^\pm z_{k,Q}^{\pm n}$$

$$\therefore \qquad c_n^\pm = b_0^\pm \delta(n) + \sum_{k=1}^{K_G} d_k^\pm z_{k,Q}^{\pm n} + \left\{ \sum_{k=K_G+1}^{K_G+K_F} d_k^\pm z_{k,Q}^{\pm n} \right\}_{k\in\mathcal{K}_F}. \tag{5.42}$$

Here, the summations in the curly brackets from (5.41) and (5.42) are equal to zero, because of the vanishing Froissart amplitudes $\{d_k^{\pm}\}$ $(k \in \mathcal{K}_F)$, according to (5.32). This implies

$$\frac{P^{\pm}_{K_G+K_F}(z^{\pm 1})}{Q^{\pm}_{K_G+K_F}(z^{\pm 1})} = b_0^{\pm} + \sum_{k=1}^{K_G+K_F} \frac{d_k^{\pm} z^{\pm 1}}{z^{\pm 1} - z^{\pm}_{k,Q}}$$
$$= b_0^{\pm} + \sum_{k=1}^{K_G} \frac{d_k^{\pm} z^{\pm 1}}{z^{\pm 1} - z^{\pm}_{k,Q}} = \frac{P^{\pm}_{K_G}(z^{\pm 1})}{Q^{\pm}_{K_G}(z^{\pm 1})}$$

$$\therefore \qquad \frac{P^{\pm}_{K_G+K_F}(z^{\pm 1})}{Q^{\pm}_{K_G+K_F}(z^{\pm 1})} = \frac{P^{\pm}_{K_G}(z^{\pm 1})}{Q^{\pm}_{K_G}(z^{\pm 1})} \tag{5.43}$$

$$c_n^{\pm} = b_0^{\pm}\delta(n) + \sum_{k=1}^{K_G+K_F} d_k^{\pm} z^{\pm n}_{k,Q}$$
$$= b_0^{\pm}\delta(n) + \sum_{k=1}^{K_G} d_k^{\pm} z^{\pm n}_{k,Q}$$
$$= b_0^{\pm}\delta(n) + \sum_{k=1}^{K_G} d_k^{\pm} \mathrm{e}^{\pm 2i\pi n\tau\nu^{\pm}_{k,Q}}$$

$$\therefore \qquad c_n^{\pm} = b_0^{\pm}\delta(n) + \sum_{k=1}^{K_G+K_F} d_k^{\pm} z^{\pm n}_{k,Q} = b_0^{\pm}\delta(n) + \sum_{k=1}^{K_G} d_k^{\pm} z^{\pm n}_{k,Q}. \tag{5.44}$$

The model order reduction in the $\mathrm{FPT}^{(+)}$ can be performed by using either (5.39) or substituting (5.38) into (5.43) and (5.44) with the results

$$\frac{P^{+}_{K_G+K_F}(z)}{Q^{+}_{K_G+K_F}(z)} = \sum_{k=1}^{K_G+K_F} \frac{d_k^{+} z}{z - z^{+}_{k,Q}} = \sum_{k=1}^{K_G} \frac{d_k^{+} z}{z - z^{+}_{k,Q}} = \frac{P^{+}_{K_G}(z)}{Q^{+}_{K_G}(z)} \tag{5.45}$$

$$c_n^{+} = \sum_{k=1}^{K_G+K_F} d_k^{+} z^{+n}_{k,Q} = \sum_{k=1}^{K_G} d_k^{+} z^{+n}_{k,Q} = \sum_{k=1}^{K_G} d_k^{+} \mathrm{e}^{2i\pi n\tau\nu^{+}_{k,Q}}. \tag{5.46}$$

5.10 Disentangling genuine from spurious resonances

Once both the FPT$^{(+)}$ and FPT$^{(-)}$ have attained convergence, they yield the same spectral parameters for the genuine resonances

$$\left.\begin{array}{c} z_{k,P}^{+} = z_{k,P}^{-} \quad z_{k,Q}^{+} = z_{k,Q}^{-} \\[2ex] \nu_{k,P}^{+} = \nu_{k,P}^{-} \quad \nu_{k,Q}^{+} = \nu_{k,Q}^{-} \\[2ex] d_k^{+} = d_k^{-} \end{array}\right\} \quad k \in \mathcal{K}_G \quad : \quad \text{genuine} \quad \text{resonances.} \tag{5.47}$$

By contrast, the remaining spectral parameters for all Froissart resonances never converge because of their spuriousness. Even the slightest alteration in the signal length can change the distributions of the parameters of spurious resonances in the complex planes. Furthermore, Froissart harmonic variables and Froissart frequencies differ when passing from the FPT$^{(+)}$ to FPT$^{(-)}$

$$\left.\begin{array}{c} z_{k,P}^{+} \neq z_{k,P}^{-} \quad z_{k,Q}^{+} \neq z_{k,Q}^{-} \\[2ex] \nu_{k,P}^{+} \neq \nu_{k,P}^{-} \quad \nu_{k,Q}^{+} \neq \nu_{k,Q}^{-} \\[2ex] d_k^{+} = 0 = d_k^{-} \end{array}\right\} \quad k \in \mathcal{K}_F \quad : \quad \text{Froissart} \quad \text{resonances.} \tag{5.48}$$

According to (5.48), both sets of Froissart amplitudes $\{d_k^{\pm}\}$ $(k \in \mathcal{K}_F)$ in the FPT$^{(+)}$ and FPT$^{(-)}$ are equal to zero, in agreement with (5.32).

In practice, the most efficient manner to spot Froissart doublets while inspecting tables containing the output list of all the spectral parameters reconstructed by the FPT$^{(\pm)}$ for noise-free (noise-corrupted) input FIDs is to look for zero or near zero amplitudes, $d_k^{\pm} = 0$ or $d_k^{\pm} \approx 0$, respectively. An even simpler way is to inspect the corresponding panels in the Argand plots of frequency distributions in the complex $\nu^{\pm}$−planes mounted on top of the absolute values of amplitudes as a function of chemical shifts. This would lead to an easy observation of Froissart doublets through zero or near zero amplitudes associated with pole-zero coincidences or near-coincidences in the case of noiseless or noisy time signals, respectively. We have already encountered a part of such a graphic inspection in Fig. 3.19 for a noise-free input FID. A related monitoring of Froissart doublets for the corresponding noise-corrupted FID will be presented in chapter 6.

Robust disentangling of genuine from spurious resonances is a part of the SNS concept, which also underlies the determination of the exact or optimally accurate number K of resonances for noiseless or noise-polluted FIDs, respectively. In section 2.8 of chapter 2, several procedures were analyzed for finding the true number K. One of the discussed procedures is based upon Froissart doublets for determining the exact number K of genuine resonances, and this will be illustrated in chapter 6 for both noiseless and noisy time signals.

6

Machine accurate quantification and illustrated signal-noise separation

6.1 Formulation of the most stringent test for quantification in MRS

Time signals with their fundamental harmonics are spectrally analyzed while solving the quantification problem. The solution of this problem permits reconstruction of all the physical complex-valued frequencies and the associated amplitudes. These spectral parameters are the sole constituents of the fundamental harmonics of the investigated time signal. Every single harmonic in its role of a transient possesses its resonant frequency, relaxation time, intensity and phase. Such elements completely determine the underlying normal-mode damped oscillations in the generic signal. These four real-valued parameters can be directly deduced from the retrieved complex frequencies and amplitudes to give the peak areas of the associated resonance profiles in the corresponding spectrum. For clinical diagnostics via MRS, the most relevant quantities are metabolite concentrations that are proportional to the reconstructed peak areas. Therefore, metabolite concentrations of the examined tissue can be extracted from the spectrally analyzed time signal. Hence the clinical importance of the quantification problem in MRS.

In this chapter, we shall tackle the most difficult numerical aspects in solving the quantification problem in MRS. The primary goal of chapters 4 and 5 was to establish the methods of rational approximations and their algorithms that would be capable of performing a rigorous computation within finite arithmetics to retrieve exactly all the input spectral parameters of each resonance from synthesized noiseless time signals. Specifically, the methodologies expounded in the two preceding chapters would be considered as fully adequate if they could produce machine accurate output from machine accurate input data. A favorable outcome from this evidently most stringent test would demonstrate that the estimator selected for this reconstruction problem is extraordinarily accurate, stable and robust even against round-off errors from computations.

Conversely, any estimator which does not meet these highest standards imposed onto spectral analysis with ideal input time signals would hardly have any systematic reliability for processing insufficiently accurate FIDs corrupted with noise as encoded via MRS. Additionally, and most importantly, in the illustrations which follow, we will also spectrally analyze simulated time signals embedded in random Gaussian distributed noise in order to see whether it would still be possible to unambiguously reconstruct all the genuine resonances, including the weakest ones as well as those that are tightly overlapped and practically degenerate. Thereby, it will be demonstrated that the FPT is also a highly reliable method for quantifying noise-polluted time signals reminiscent of those encoded via MRS in clinical diagnostics.

Nevertheless, this comprehensiveness in the envisaged testing would still be incomplete without being supplemented by the critical demand for unequivocal separation of genuine from spurious information in the output data. Such a demanding list of prerequisites for adequacy in signal processing might be approached by many methods [146]–[205], but with varying and usually insufficient success in parametric estimations. Based upon our thorough analysis from chapter 3 with accuracy pre-determined by 4-digit input data, and relying upon the conclusions on a number of our earlier studies [5, 10, 11, 24, 34], we shall opt for the fast Padé transform to accomplish the formulated task of paramount significance for quantification in MRS.

From a theoretical viewpoint, all the sought complex frequencies and amplitudes can unambiguously and exactly be retrieved from a given input time signal by the FPT. In its role of the response or system function, given by the unique quotient of two polynomials, the FPT is the most frequently applied estimator in various research fields. In principle, the FPT can achieve the so-called spectral convergence, which represents the exponential convergence rate as a function of the signal length for a fixed bandwidth. This unique property equips the FPT with unprecedented high-resolution capabilities that are, in fact, theoretically unrestricted. In the present chapter, we will illustrate these features by the exact reconstruction (within machine accuracy) of all the spectral parameters from an input time signal containing 25 harmonics that are complex, attenuated exponentials. To make the already difficult task even more challenging, we shall include two practically degenerate resonances with chemical shifts differing by a remarkably small fraction of only 10^{-11} ppm. We will show that by employing merely a quarter of the full signal length, the FPT is capable of exactly reconstructing all the input spectral parameters defined with 12 digits of accuracy. Hence, achieving 12-digit accuracy in reconstructions by using 12-digit accurate input data proves that the FPT is robust even against round-off-errors.

In particular, we shall demonstrate that when the FPT is near the convergence region, an unparalleled "phase transition" takes place, because literally two additional signal points are sufficient to attain the full 12 digit accuracy with the exponentially fast rate of convergence. Such an achievement represents the crucial proof-of-principle for the high-resolution power of the FPT for machine accurate input time signals. Finally, we shall demonstrate that the FPT can unambiguously separate genuine from spurious resonances using both noise-free and noise-corrupted FIDs. All of the conclusions reached in this chapter about the performance of the FPT in MRS are also valid in, e.g, analytical chemistry when using ICRMS or NMR spectroscopy, and in other applied sciences and technologies with signal processing for data analysis [5].

6.2 The key factors for high resolution in quantification

Following the outlined task, we shall apply the FPT to provide machine accurate quantification, and illustrations of signal-noise separation. Machine accuracy is sought to benchmark the FPT on the noise-free FIDs. Passing this most stringent test is far from being academic or formal. Quite the contrary, those estimators that cannot achieve a comparable accuracy as the FPT for fully controlled noiseless FIDs have no guarantee whatsoever for reliable performance in the case of uncontrolled FIDs such as those which are encoded *in vivo* by means of MRS. The maximal sought accuracy which is possible for noise-corrupted synthesized or encoded FIDs is 3-4 stable digits in the reconstructed concentrations of the genuine, clinically relevant metabolites. This level of requested accuracy can readily be achieved by the FPT for synthesized noise-corrupted FIDs as well as encoded *in vivo* time signals.

Before proceeding further, we must enumerate the key factors that influence the overall performance of estimators in general parametric reconstructions, especially when aiming at machine accuracy. First of all, the resolving power and convergence rate of a given estimator depend on certain obvious characteristics of time signals, such as SNR and the total acquisition time T. In addition, however, there are more subtle aspects of signal processing that can be of critical importance with respect to accuracy and robustness of the signal processor under examination. Among these key features are the configurations of poles and zeros in the complex plane, the smallest distance among poles on the one hand and zeros on the other, particularly when such distances are compared to the level of noise, the density of signal poles and zeros in the chosen portion or throughout the Nyquist range, inter-separations among poles and zeros, their distance from the real frequency axis and the smallest imaginary frequencies (the longest lifetimes of transients) in the spectrum.

As thoroughly presented and analyzed in chapter 3, Argand plots provide a remarkably convenient mathematical tool for scrutinizing the enumerated subtle features of signal processing. These plots display the imaginary part as a function of the corresponding real part of a given complex-valued quantity, such as the harmonic variables $z_k^{\pm}$, the fundamental frequencies $\nu_k^{\pm}$ and the corresponding amplitudes $d_k^{\pm}$. It is also useful to depict the dependence of the absolute values of the amplitudes, $|d_k^{\pm}|$, or peak heights $|d_k^{\pm}|/\mathrm{Im}(\nu_k^{\pm})$. As discussed in chapter 3, these are proportional to the concentrations of the associated resonances/metabolites for $\mathrm{Im}(\nu_k^{\pm}) > 0$. Such highly illustrative types of graphs will also be plotted in the present chapter, with important implications related to the concept of Froissart doublets [44] through pole-zero coincidences [45] using noiseless and noise-corrupted time signals that are theoretically generated (synthesized, simulated).

6.3 The goals and plan for presentation of results

The results from this chapter on quantification by the $\mathrm{FPT}^{(\pm)}$ for synthesized time signals are given in the two subsequent sections. In the first section 6.4, we deal with machine accurate input and output data for a noiseless FID. Machine accuracy is defined here to be the precision of 12 digits. This should not be considered as a limitation, since other higher accuracies could also be considered with equal success. The second section 6.5 is devoted to graphic presentations of all the findings. Here, we use a noise-free as well as a noise-corrupted simulated time signal. Accuracy of the spectral parameters of both of these FIDs is restricted to 4-digits to mimic the realistic, MR-encoded data in clinical scanners that operate at $B_0 = 1.5\mathrm{T}$ [88]. In section 6.5, 12-digit accuracy could also be used for the input parameters to generate a noiseless FID, but it would be pointless to add noise to this latter time signal to obtain the corresponding noisy FID. The reason being that the added noise itself would affect, say the 4th digit. Therefore, it is more reasonable to consistently use 4-digit accuracy for both the input and output data when dealing with noise-corrupted FIDs. Moreover, for an unambiguous interpretation of any departure of distributions of noisy reconstructed spectral parameters in complex planes from their noiseless counterparts, it is necessary that the noise-free input data themselves from section 6.5 are also given with 4-digit accuracy. Further, for the same reason of avoiding potential influence of other uncontrollable factors that could lead to ambiguities in our comparative analyses of the results from the noiseless and noisy FIDs, we choose the noiseless time signal $\{c_n\}$ in section 6.5 to be identical to the noiseless part of the noisy FID, which is of the form $\{c_n + r_n\}$, where r_n is a given random noise.

The reason for not performing comparisons of the results of quantification for a noiseless c'_n and noisy $\{c_n + r_n\}$ with $\{c'_n\} \neq \{c_n\}$ is that even very small differences between $\{c'_n\}$ and $\{c_n\}$ could lead to different distributions of poles and zeros of the response functions in complex planes. This would impinge upon a clear-cut interpretation because any detected displacements of these poles and zeros for the noisy $\{c_n + r_n\}$ relative to the noiseless case $\{c'_n\}$ for $\{c'_n\} \neq \{c_n\}$ would stem from the two inseparable origins. One origin could be a small difference (say, in the 4th digit) between $\{c'_n\}$ and $\{c_n\}$, i.e., $|c'_n - c_n| \sim 10^{-4}$. The other origin could be due to noise r_n itself in $\{c_n + r_n\}$

when $\{r_n\}$ is of a level which might also affect the 4th digit, $|r_n| \sim 10^{-4}$. The main emphasis in section 6.5 is on separation of stable (genuine) from unstable (spurious) resonances. Spurious spectral structures are extremely sensitive to even the slightest changes in the input data. Therefore, for the outlined reasons of being able to clearly see what is caused by noise alone, when comparing the results of quantifications in the noiseless and noisy cases in section 6.5, we shall use time signals $\{c_n\}$ and $\{c_n + r_n\}$, respectively, with the common part $\{c_n\}$ built from the same 4-digit accurate input parameters.

Importantly, the envisaged comparisons will always be performed for the same total (N) as well as partial (N_P) signal lengths and the identical sampling time τ (i.e., the same acquisition time $T = N\tau$) for the noiseless $\{c_n\}$ and noisy $\{c_n + r_n\}$ data, respectively. To cohere with chapter 3, the common noiseless $\{c_n\}$ from section 6.5 is built from the same 4-digit accurate input parameters given in Table 3.1. Furthermore, for both the 12- and 4-digit accuracies from sections 6.4 and 6.5, the deterministic time signals $\{c_n\}$ $(0 \le n \le N-1)$ have the full signal length $N = 1024$ and are sampled at the dwell time τ =1 ms. This same N and τ were also employed for the noiseless FID from chapter 3. To such input data $\{c_n\}$ $(0 \le n \le N-1)$, random numbers as stochastic signal points $\{r_n\}$ $(0 \le n \le N-1)$ are added and used in section 6.5. As in chapter 3, each deterministic time signal point c_n is taken to be comprised of 25 genuine fundamental harmonics $\{z_k\}$. More precisely, these c_n's are built from the corresponding sum of 25 complex damped exponentials $\{z_k\}$ with constant amplitudes $\{d_k\}$. Once sampling of such a deterministic time signal is completed, the only known quantities that remain available are τ, N and this tabulated FID. In other words, the previously given input data set $\{z_k, d_k\}$ and the number K of the elements of this set are to be considered as unknown. This is where spectral analysis comes into play. Specifically, the process of solving the quantification problem for this FID amounts to reconstruction of the exact 100 complex-valued spectral parameters $\{z_k, d_k\}$ $(1 \le k \le 25)$, after the true number $K = 25$ has been unequivocally retrieved. To re-emphasize, these deterministic time signal points $\{c_n\}$ for the noise-free FID, that are repeated in their noisy counterparts $\{c_n + r_n\}$, are sampled from the formula $c_n = \sum_{k=1}^{K} d_k z_k^n$ $(0 \le n \le N-1)$, as per (3.1), with the exact data for the input values of the fundamental spectral parameters $\{\nu_k, d_k\}$. Here, the elements $\{z_k\}$ are the signal poles, $z_k = \mathrm{e}^{i\omega_k\tau}$ where $\omega_k = 2\pi\nu_k$ and $\mathrm{Im}(\omega_k) > 0$ $(1 \le k \le K)$.

In the 12-digit input data from section 6.4, an unprecedented challenge is placed on the subsequent usage of the Padé-guided reconstruction by introducing the minuscule difference of only 1×10^{-11} ppm in the last digit of chemical shifts between the 11th and 12th fundamental harmonics. Further, yet another level of remarkable challenge is designed for the Padé-optimized quantification by entering all the 25 phases ϕ_k of the input amplitudes d_k $(1 \le k \le 25)$ as precisely 12-digit zeros that are extremely difficult to reconstruct with the input accuracy. Likewise, along the lines of chapter 3, in the present section 6.5

dealing with the 4-digit input data, the 11th and 12th chemical shifts differ in the 4th decimal place only by 1 unit, i.e., 1×10^{-4} ppm, and all the phases of d_k are defined as 4-digit zeros. Thus, in both the 4- and 12-digit input data considered in the present chapter, we have $|d_k| = d_k$ due to zero-valued phases $\phi_k \equiv \text{Arg}(d_k)$ for $1 \le k \le 25$. The chemical shift splittings between the 11th and 12th fundamental harmonics in both accuracies from sections 6.4 and 6.5 mimic the near degeneracy. The FPT can indeed handle the exact degeneracy, too, as we verified in a check computation by entering the 12 identical digits for the 11th and 12th chemical shifts. This feature of the FPT remains valid for any other pairs or larger groups of resonances. Moreover, not only chemical shifts, but the other three spectral parameters can also be selected to coincide among the chosen transients. Such a capability of the FPT to treat non-degenerate, near-degenerate and degenerate harmonics in quantification problems on the same footing is very important, since MR-encoded spectra are abundant with tightly overlapping and nearly degenerate peaks.

As stated, the data are presented in two sections 6.4 and 6.5 each of which is split into two sub-sections. All the tables deal only with genuine spectral parameters, whereas every figure contains genuine and spurious reconstructed data. The tables for noiseless cases are in section 6.4 with sub-sections 6.4.1 for the input data and 6.4.2 for the output data in the $\text{FPT}^{(-)}$. The graphs for noiseless and noisy data are in section 6.5 with sub-sections 6.5.1 and 6.5.2 for reconstructions by the $\text{FPT}^{(\pm)}$ at full convergence ($N_P = N/4 = 256$) and near convergence ($N_P = 180, 220, 260$) of genuine resonances. As always, when convergence is addressed, this refers to data for genuine resonances, since spectral parameters for spurious resonances never converge.

Section 6.4 with tabular data contains the 12-digit accurate input data in sub-section 6.4.1 given in Table 6.1. The output data reconstructed by the $\text{FPT}^{(-)}$ are presented in sub-section 6.4.2 via Table 6.2 for two partial signal lengths of the non-FFT type ($N_P \neq 2^m$, m positive integers), e.g., $N_P = 180$ (prior to convergence with accuracy 2 – 7 digits) and $N_P = 220$ (full convergence with 12 digit accuracy throughout). Also reported in sub-section 6.4.2 is Table 6.3 for 12-digit accurate reconstruction of all the spectral parameters with the FFT-type signal lengths ($N_P = 2^m$, m positive integers) using the specifications $N_P = 2^8 = 256$ (a quarter of the full FID) and $N = 2^{10} = 1024$ (the whole FID). The output data for $N_P = 180$ and $N_P = 220$ are given to prove the exponential convergence rate of the $\text{FPT}^{(-)}$ around the convergence point which occurs first at $N_P = 210$ [6, 35]. This zooming would not be possible if the $\text{FPT}^{(-)}$ were limited to the FFT-type structured number 2^m (m positive integers) for signal lengths. The FFT-type partial ($N_P = 2^8 = 256$) and full ($N = 2^{10} = 1024$) signal lengths are used to show the constancy of machine accurate Padé reconstructions.

Section 6.5 presents Figs. 6.1–6.14 that are multifaceted in conveying six different spectral aspects using the input and output data:

• (i) two time signals, a noise-free and a noise-corrupted input FID, to test the FPT in two clear-cut situations, with and without external perturbations,

• (ii) two variants of the fast Padé transform for cross-validations of quantifications via the $\text{FPT}^{(+)}$ and $\text{FPT}^{(-)}$ for the output data,

• (iii) four partial signal lengths in both versions of the FPT to monitor the convergence rate for the fixed bandwidth,

• (iv) structured and unstructured partial N_P to exhibit the advantages of non-Fourier partial signal lengths; the former is of the FFT type, $N_P = N/4 = 2^8 = 256$ at convergence, and the latter of the non-FFT type $N_P \neq 2^m$ (m positive integers) via $N_P = 180, 220$ and 260 near convergence,

• (v) two types of graphs to take advantage of different distributions of the same quantities: Argand plots (real versus imaginary numbers) and physical quantities as a function of chemical shifts,

• (vi) two coordinate systems, polar (Euler) and rectangular (Descartes), in complex planes to illuminate the complementary interpretations of harmonic variables (signal poles and zeros) and fundamental frequencies.

Admittedly, Figs. 6.1–6.14 from section 6.5 will appear to be quite involved. This is due to their said multi-level informational content, intended to be conveyed in a way which is both thorough and concise. To simultaneously achieve thoroughness and conciseness, as well as to make these figures as self-explanatory as possible, we equipped them with the following utilities:

• the minimum pertinent text and formulae, e.g., for spectral parameters such as harmonic variables, linear frequencies and amplitudes that are sought via reconstructions in quantifications,

• the canonical representations of the $\text{FPT}^{(\pm)}$ with signal poles and zeros,

• a selection of the main features for the two sub-representations of the fast Padé transform via $\text{pFPT}^{(\pm)}$ ("poles of the FPT") and $\text{zFPT}^{(\pm)}$ ("zeros of the FPT"), with the understanding that the complete FPT representation involves all poles and all zeros,

• the characteristic equations and the formulae for the amplitudes in the $\text{FPT}^{(\pm)}$ to recall the source of the reconstructed spectral parameters,

• the expressions for the signatures of Froissart doublets (pole-zero confluences and the accompanying zero amplitudes),

• the way in which the exact number K_{G} of genuine resonances is reconstructed, as the difference between the total number K of all found resonances and the number K_{F} of Froissart doublets, $K_{\text{G}} = K - K_{\text{F}}$.

Despite such elaborated texts and formulae appearing in these graphs either through the main titles, sub-titles on sub-plots or within a number of panels, the overall visual and informational display is nevertheless free from clutter[1]. These additions to Figs. 6.1–6.14 are meant to be both a facilitator (for an easier follow-up of all the graphic illustrations) and, most importantly, a reminder to the corresponding theoretical analyses from, e.g., chapter 5 for an in-depth connection to the physical/mathematical meaning of the displayed

[1]The figures in chapter 3 were also aided by the appropriate texts, titles, sub-titles and, occasionally, formulae, all of which were aimed at helping the reader to grasp the main messages from those graphic illustrations.

quantities. In this way, the mathematical physics from the preceding theoretical expositions becomes fully ingrained into Figs. 6.1–6.14, that are followed by their pertinent practical interpretations in section 6.6, thus establishing one whole – comprehensive data analysis of the exact solution of the quantification problem in MRS. To adhere to the main theme of the present book, these specific illustrations are for biomedical FIDs, but the entire powerful and versatile Padé-Froissart methodology remains valid for a multitude of other applications across inter-disciplinary research dealing with time signals in the form of sums of damped complex exponentials [5, 6].

6.4 Numerical presentation of the spectral parameters

6.4.1 Input spectral parameters with 12-digit accuracy

Table 6.1 gives the 12-digit input data for the quantification problem under examination. As mentioned, these data are the complex fundamental frequencies and the corresponding amplitudes from a synthesized noiseless FID, whose associated spectrum is comprised of a total of 25 resonances, some of which are individual although tightly packed peaks, while others are closely-overlapped or nearly degenerate. As in chapter 3, the first 4 digits of all the 12-digit values of the spectral parameters were chosen to closely correspond to the typical frequencies and amplitudes found in proton MR time signals as encoded *in vivo* from a healthy human brain at 1.5T [88].

Furthermore, all the input parameters from Table 6.1 are set to be in exact 4-digit mutual agreement with the corresponding data from Table 3.1 in chapter 3. The remaining 8 digits in the 12-digit data are chosen in an arbitrary manner without any special ordering/rule, except for care being taken to avoid the situations where rounding in Table 6.1 would preclude the imposed exact agreement with the 4-digit input data from Table 3.1. The columns in Table 6.1 of the input fundamental transients are headed by labels n_k°, $\mathrm{Re}(\nu_k)\,(\mathrm{ppm})$, $\mathrm{Im}(\nu_k)\,(\mathrm{ppm})$, $|d_k|\,(\mathrm{au})$ and M_k that represent the running number, real and imaginary frequencies (both in ppm), absolute values (moduli) of amplitudes (in arbitrary units) and the metabolite assignments, respectively. For convenience, we shall interchangeably use the notations k, # and n_k° for the running number of resonances. Of particular note are the crossings of the 2nd column with the 11th and 12th rows where the two chemical shifts $\mathrm{Re}(\nu_{11}) = 2.67602157683$ ppm and $\mathrm{Re}(\nu_{12}) = 2.67602157684$ ppm are separated by an extraordinarily small splitting $\mathrm{Re}(\nu_{12}) - \mathrm{Re}(\nu_{11}) = 1 \times 10^{-11}$ ppm. This maximally sharpens the corresponding earlier splitting $\mathrm{Re}(\nu_{12}) - \mathrm{Re}(\nu_{11}) = 1 \times 10^{-4}$ ppm for the 4-digit data in Table 3.1 from chapter 3. Within the first common 4 digits, all the other remaining tight separations among several metabolites stay unaltered relative to Table 3.1

TABLE 6.1
Twelve digit numerical values for all the input spectral parameters: the real $\mathrm{Re}(\nu_k)$ and the imaginary $\mathrm{Im}(\nu_k)$ part of complex-valued frequencies ν_k, and the absolute values $|d_k|$ of amplitudes d_k of 25 damped complex exponentials from the synthesized time signal similar to a short echo time (~20 ms) encoded FID via MRS at the magnetic field strength B_0=1.5 T from a healthy human brain as in Ref. [88]. Every phase $\{\phi_k\}$ of the amplitudes is equal to zero, such that each d_k is purely real, $d_k = |d_k| \exp(i\phi_k) = |d_k|$.

INPUT DATA for ALL SPECTRAL PARAMETERS of a SYNTHESIZED TIME SIGNAL or FID

n_k^o	$\mathrm{Re}(\nu_k)$ (ppm)	$\mathrm{Im}(\nu_k)$ (ppm)	$|d_k|$ (au)	M_k
1	0.98503147283	0.18003158592	0.12202458746	Lip
2	1.11201283942	0.25701463975	0.16101129387	Lip
3	1.54804437561	0.17202346512	0.13501532472	Lip
4	1.68902546965	0.11804159816	0.03403258327	Lip
5	1.95903149379	0.06201541256	0.05603432458	Gaba
6	2.06504523854	0.03103281453	0.17101383971	NAA
7	2.14502148923	0.05002435671	0.11602423465	NAAG
8	2.26103238547	0.06201591724	0.09203547326	Gaba
9	2.41101584725	0.06204158493	0.08502124847	Glu
10	2.51902465396	0.03602852174	0.03703732654	Gln
11	2.67602157683	0.03303465795	0.00801253419	Asp
12	2.67602157684	0.06201513458	0.06302642348	NAA
13	2.85501349581	0.01602478536	0.00502416737	Asp
14	3.00903254937	0.06404152367	0.06501248591	Cr
15	3.06702138733	0.03603245678	0.10103549132	PCr
16	3.23901528474	0.05001584372	0.09602363474	Cho
17	3.30102372562	0.06402158967	0.06501467513	PCho
18	3.48101278659	0.03104627543	0.01103173534	Tau
19	3.58402513273	0.02803513684	0.03602382512	m–Ins
20	3.69401319485	0.03601275683	0.04103493165	Glu
21	3.80303438676	0.02402174915	0.03103265173	Gln
22	3.94401324818	0.04204712356	0.06801528345	Cr
23	3.96502574264	0.06201328748	0.01302356831	PCr
24	4.27103483491	0.05503491376	0.01602147862	PCho
25	4.68000000000	0.10802134657	0.05703576583	Water

and, as such, need not be specially emphasized here.

6.4.2 Exponential convergence rates of Padé reconstructions of spectral parameters with 12-digit accuracy

The results of the $\text{FPT}^{(-)}$ for machine accurate exact reconstructions of the input spectral parameters are given in Tables 6.2 and 6.3. In Table 6.2, it can be seen that a varying level of accuracy is attained in the retrieved spectral parameters from the $\text{FPT}^{(-)}$ near full convergence at 2 partial signal lengths $N_P = 180, 220$. As stated, neither length is compliant with the FFT-type lengths, i.e., we have $N_P \neq 2^m$ (m positive integers). On panel (i) at $N_P = 180$, before full convergence is achieved, some 2 – 7 exactly reconstructed digits can be seen. Note that the 11th resonance is not detected here. Its absence is marked by the sign "–" in the first column at the corresponding vacant location ($n_k^\circ = 11$). However, at $N_P = 220$ an enormous increase in accuracy is observed on panel (ii), with all the 12 input digits exactly reconstructed for each spectral parameter of 25 resonances. Thus, with only 220 signal points out of 1024 entries from the full FID, the $\text{FPT}^{(-)}$ resolves unequivocally the two near degenerate frequencies separated from each other by an unprecedented chemical shift of merely 10^{-11} ppm. This shows that the $\text{FPT}^{(-)}$ has an exponential convergence rate (the spectral convergence) [188] to the exact numerical values within machine accuracy of all the reconstructed fundamental frequencies and the associated amplitudes. Moreover, these 12-digit output results for the 12-digit input data prove an indeed great robustness of the $\text{FPT}^{(-)}$ even against round-off-errors.

In Table 6.3, we examine the accuracy when the partial signal length N_P is selected via a composite number 2^m (m positive integers), as customarily used in the FFT. Here, on panels (i) and (ii), we show the results of the $\text{FPT}^{(-)}$ at a quarter $N/4 = 2^8 = 256$, and the full signal length $N = 2^{10} = 1024$. It can be seen that these two panels yield exactly the same 12-digit accurate results throughout. We have also confirmed this to hold true for one half of the full signal length, $N/2 = 2^9 = 512$ [6, 35]. As such, this constancy of genuine spectral parameters persists at any signal length after retrieving the true number of resonances. By contrast, spurious spectral parameters (not tabulated) do not ever converge.

Taken together, the findings from Tables 6.2 and 6.3 demonstrate that the $\text{FPT}^{(-)}$ continues to be stable by preserving 12-digit accuracy after the point at which full convergence is reached first at $N_P = 210$ [6, 35]. In other words, the addition of more signal points does not alter the stabilized results in any way. Such an outstanding feature is of major importance with regard to the stability of the $\text{FPT}^{(-)}$ in 12-digit accurate quantification within MRS. This is an extension of the previous conclusion on the stable reconstructions by the $\text{FPT}^{(-)}$ reached in chapter 3 with 4-digit accuracy, which is closer to the data precision encountered with MR-encoded *in vivo* FIDs.

TABLE 6.2
Two to twelve digit accuracy for the complex frequencies and amplitudes reconstructed by the $\mathrm{FPT}^{(-)}$ at partial signal lengths $N_P = 180$ and 220.

PROOF–OF–PRINCIPLE ACCURACY of $\mathrm{FPT}^{(-)}$ for QUANTIFICATION ; FID LENGTH: N_P = 180, 220

Accuracy of $\mathrm{FPT}^{(-)}$ for every parameter of each resonance: 2 – 7 Exact Digits (ED_k^-)

n_k^o	$\mathrm{Re}(\nu_k^-)$ (ppm)	ED_k^-	$\mathrm{Im}(\nu_k^-)$ (ppm)	ED_k^-	$\lvert d_k^-\rvert$ (au)	ED_k^-
1	0.9850414	5	0.1800362	5	0.1220540	4
2	1.1120321	5	0.2569838	4	0.1609574	4
3	1.5480312	5	0.1720063	5	0.1349446	4
4	1.6890837	4	0.1180436	6	0.0340359	5
5	1.9589616	4	0.0620059	5	0.0559247	4
6	2.0650203	5	0.0310142	5	0.1707245	4
7	2.1452638	4	0.0498990	3	0.1153469	3
8	2.2613748	4	0.0625487	3	0.0938680	3
9	2.4107665	4	0.0632107	3	0.0883779	3
10	2.5178270	3	0.0358322	4	0.0366481	4
–	–	–	–	–	–	–
12	2.6757002	4	0.0542097	2	0.0662198	2
13	2.8555390	3	0.0145410	3	0.0045151	4
14	3.0111655	3	0.0589646	3	0.0597872	2
15	3.0660983	3	0.0367101	3	0.1072950	2
16	3.2403252	3	0.0510424	3	0.1037047	2
17	3.2994935	3	0.0606944	3	0.0608127	2
18	3.4799781	3	0.0301731	3	0.0105106	4
19	3.5841372	4	0.0277386	4	0.0354354	3
20	3.6942574	4	0.0359717	4	0.0408940	4
21	3.8030809	4	0.0240682	4	0.0311048	4
22	3.9440858	4	0.0418930	4	0.0671907	3
23	3.9640522	2	0.0609769	3	0.0140023	3
24	4.2710342	6	0.0550358	6	0.0160220	6
25	4.6800001	7	0.1080210	7	0.0570354	6

(i) Partial FID Length: N_P = 180

Accuracy of $\mathrm{FPT}^{(-)}$ for every parameter of each resonance: 12 Exact Digits (ED_k^-)

n_k^o	$\mathrm{Re}(\nu_k^-)$ (ppm)	ED_k^-	$\mathrm{Im}(\nu_k^-)$ (ppm)	ED_k^-	$\lvert d_k^-\rvert$ (au)	ED_k^-
1	0.98503147283	12	0.18003158592	12	0.12202458746	12
2	1.11201283942	12	0.25701463975	12	0.16101129387	12
3	1.54804437561	12	0.17202346512	12	0.13501532472	12
4	1.68902546965	12	0.11804159816	12	0.03403258327	12
5	1.95903149379	12	0.06201541256	12	0.05603432458	12
6	2.06504523854	12	0.03103281453	12	0.17101383971	12
7	2.14502148923	12	0.05002435671	12	0.11602423465	12
8	2.26103238547	12	0.06201591724	12	0.09203547326	12
9	2.41101584725	12	0.06204158493	12	0.08502124847	12
10	2.51902465396	12	0.03602852174	12	0.03703732654	12
11	2.67602157683	12	0.03303465795	12	0.00801253419	12
12	2.67602157684	12	0.06201513458	12	0.06302642348	12
13	2.85501349581	12	0.01602478536	12	0.00502416737	12
14	3.00903254937	12	0.06404152367	12	0.06501248591	12
15	3.06702138733	12	0.03603245678	12	0.10103549132	12
16	3.23901528474	12	0.05001584372	12	0.09602363474	12
17	3.30102372562	12	0.06402158967	12	0.06501467513	12
18	3.48101278659	12	0.03104627543	12	0.01103173534	12
19	3.58402513273	12	0.02803513684	12	0.03602382512	12
20	3.69401319485	12	0.03601275683	12	0.04103493165	12
21	3.80303438676	12	0.02402174915	12	0.03103265173	12
22	3.94401324818	12	0.04204712356	12	0.06801528345	12
23	3.96502574264	12	0.06201328748	12	0.01302356831	12
24	4.27103483491	12	0.05503491376	12	0.01602147862	12
25	4.68000000000	12	0.10802134657	12	0.05703576583	12

(ii) Partial FID Length: N_P = 220

TABLE 6.3
Twelve digit accuracy for all the 25 complex frequencies and amplitudes reconstructed by the $\text{FPT}^{(-)}$ at signal lengths $N/4 = 256$ and $N = 1024$.

MACHINE ACCURACY of $\text{FPT}^{(-)}$ for QUANTIFICATION ; FID LENGTH: N/4 = 256, N = 1024

Accuracy of $\text{FPT}^{(-)}$ for every parameter of each resonance: 12 Exact Digits (ED_k^-)

n_k^o	$\text{Re}(\nu_k^-)$ (ppm)	ED_k^-	$\text{Im}(\nu_k^-)$ (ppm)	ED_k^-	$\lvert d_k^- \rvert$ (au)	ED_k^-
1	0.98503147283	12	0.18003158592	12	0.12202458746	12
2	1.11201283942	12	0.25701463975	12	0.16101129387	12
3	1.54804437561	12	0.17202346512	12	0.13501532472	12
4	1.68902546965	12	0.11804159816	12	0.03403258327	12
5	1.95903149379	12	0.06201541256	12	0.05603432458	12
6	2.06504523854	12	0.03103281453	12	0.17101383971	12
7	2.14502148923	12	0.05002435671	12	0.11602423465	12
8	2.26103238547	12	0.06201591724	12	0.09203547326	12
9	2.41101584725	12	0.06204158493	12	0.08502124847	12
10	2.51902465396	12	0.03602852174	12	0.03703732654	12
11	2.67602157683	12	0.03303465795	12	0.00801253419	12
12	2.67602157684	12	0.06201513458	12	0.06302642348	12
13	2.85501349581	12	0.01602478536	12	0.00502416737	12
14	3.00903254937	12	0.06404152367	12	0.06501248591	12
15	3.06702138733	12	0.03603245678	12	0.10103549132	12
16	3.23901528474	12	0.05001584372	12	0.09602363474	12
17	3.30102372562	12	0.06402158967	12	0.06501467513	12
18	3.48101278659	12	0.03104627543	12	0.01103173534	12
19	3.58402513273	12	0.02803513684	12	0.03602382512	12
20	3.69401319485	12	0.03601275683	12	0.04103493165	12
21	3.80303438676	12	0.02402174915	12	0.03103265173	12
22	3.94401324818	12	0.04204712356	12	0.06801528345	12
23	3.96502574264	12	0.06201328748	12	0.01302356831	12
24	4.27103483491	12	0.05503491376	12	0.01602147862	12
25	4.68000000000	12	0.10802134657	12	0.05703576583	12

(i) Partial FID Length: $N_P = N/4 = 256$

Accuracy of $\text{FPT}^{(-)}$ for every parameter of each resonance: 12 Exact Digits (ED_k^-)

n_k^o	$\text{Re}(\nu_k^-)$ (ppm)	ED_k^-	$\text{Im}(\nu_k^-)$ (ppm)	ED_k^-	$\lvert d_k^- \rvert$ (au)	ED_k^-
1	0.98503147283	12	0.18003158592	12	0.12202458746	12
2	1.11201283942	12	0.25701463975	12	0.16101129387	12
3	1.54804437561	12	0.17202346512	12	0.13501532472	12
4	1.68902546965	12	0.11804159816	12	0.03403258327	12
5	1.95903149379	12	0.06201541256	12	0.05603432458	12
6	2.06504523854	12	0.03103281453	12	0.17101383971	12
7	2.14502148923	12	0.05002435671	12	0.11602423465	12
8	2.26103238547	12	0.06201591724	12	0.09203547326	12
9	2.41101584725	12	0.06204158493	12	0.08502124847	12
10	2.51902465396	12	0.03602852174	12	0.03703732654	12
11	2.67602157683	12	0.03303465795	12	0.00801253419	12
12	2.67602157684	12	0.06201513458	12	0.06302642348	12
13	2.85501349581	12	0.01602478536	12	0.00502416737	12
14	3.00903254937	12	0.06404152367	12	0.06501248591	12
15	3.06702138733	12	0.03603245678	12	0.10103549132	12
16	3.23901528474	12	0.05001584372	12	0.09602363474	12
17	3.30102372562	12	0.06402158967	12	0.06501467513	12
18	3.48101278659	12	0.03104627543	12	0.01103173534	12
19	3.58402513273	12	0.02803513684	12	0.03602382512	12
20	3.69401319485	12	0.03601275683	12	0.04103493165	12
21	3.80303438676	12	0.02402174915	12	0.03103265173	12
22	3.94401324818	12	0.04204712356	12	0.06801528345	12
23	3.96502574264	12	0.06201328748	12	0.01302356831	12
24	4.27103483491	12	0.05503491376	12	0.01602147862	12
25	4.68000000000	12	0.10802134657	12	0.05703576583	12

(ii) Full FID Length: N = 1024

6.5 Signal-noise separation via Froissart doublets with pole-zero coincidences

Figures 6.1–6.14 reveal how the concept of Froissart doublets within the $\mathrm{FPT}^{(+)}$ and $\mathrm{FPT}^{(-)}$ can be successfully applied to synthesized noise-free and noise-corrupted FIDs. Here, as mentioned, for noise-free $\{c_n\}$ and noise-corrupted $\{c_n + r_n\}$ time signals, the 4-digit input data for all the 100 spectral parameters taken directly from Table 3.1 are used to carry out spectral analysis in this section. The specifics of the present model for noise run as follows. We add random numbers $\{r_n\}$ to the noiseless time signal $\{c_n\}$ to generate the noisy input FID data $\{c_n + r_n\}\,(0 \le n \le N-1)$. More precisely, this additive noise r_n is a set $\{r_n\}\,(0 \le n \le N-1)$ of N random Gauss-distributed zero mean numbers (orthogonal in the real and imaginary parts) with the standard deviation $\sigma = \lambda \times \mathrm{RMS}$. Here, λ is the selected noise level and the acronym RMS stands for root-mean-square (or equivalently, the quadratic mean) of the noiseless FID. For the given noiseless set $\{|c_n|\}$ generated with the 4-digit spectral parameters from Table 3.1, RMS is defined by the arithmetic mean (average) value $\mathrm{RMS} = (\sum_{n=0}^{N} |c_n|^2/N)^{1/2}$. According to our noise model, adding $\lambda\,\%$ noise $\{r_n\}$ to noiseless data $\{c_n\}$ of $\mathrm{RMS}_{\text{noise-free}}$ would produce noisy data $\{c_n + r_n\}$ whose $\mathrm{RMS}_{\text{noise-corrupted}}$ is $\lambda\,\%$ of $\mathrm{RMS}_{\text{noise-free}}$, so that $\mathrm{RMS}_{\text{noise-corrupted}} = \lambda\,\mathrm{RMS}_{\text{noise-free}}$. Here λ is a fixed number expressed in percent. For example, adding 10% noise would yield a new RMS (noisy), which is 10% of the old RMS (noiseless), $\sigma = 0.01\,\mathrm{RMS}_{\text{noise-free}}$. In the present computations, we shall fix the noise level λ to be a constant number equal to 0.00289, so that $\sigma = 0.00289\,\mathrm{RMS}$ where, as stated, $\mathrm{RMS} \equiv \mathrm{RMS}_{\text{noise-free}}$. The value 0.00289 in the standard deviation σ of noise is chosen to approximately match 1.5% of the height of the weakest resonance ($n_k^\circ = 13$) in the spectrum. This noise level is sufficient to illustrate the main principles of Froissart doublets. The FPT can also successfully handle FIDs with much higher noise levels for synthesized and encoded data, as shown in Refs. [6, 9, 35].

6.5.1 Converged Padé genuine resonances and lack of convergence of Froissart doublets in $\mathrm{FPT}^{(\pm)}$ with a quarter of full signal length

In practice, for the same level of accuracy in quantification, the $\mathrm{FPT}^{(+)}$ needs more signal points than the $\mathrm{FPT}^{(-)}$. However, in this sub-section, to simplify our comparative analyses of these two variants of the Padé methodology, we shall choose the same partial signal length to show the fully converged reconstructions in the $\mathrm{FPT}^{(\pm)}$ for the FFT-structured signal lengths (integer powers of 2). This convention need not be followed when displaying certain intermediate graphs prior to complete convergence in the $\text{FPT}^{(\pm)}$. For the latter cases, to be able to zoom into a narrow interval around the value of N_P for which convergence has been reached, we need to choose partial signal lengths that are not equal to integer powers of 2 (i.e., not the structured numbers from the FFT). This will be encountered in sub-section 6.5.2.

Sub-section 6.5.1 deals only with *converged* reconstructions in the $\text{FPT}^{(\pm)}$ as illustrated at the same N_P, which is a quarter of the full signal length, $N_P = N/4 = 256$. Figures 6.1–6.4 and 6.5–6.8 present the results from the $\text{FPT}^{(+)}$ and $\text{FPT}^{(-)}$, respectively. In the $\text{FPT}^{(+)}$, the noiseless data are given in Figs. 6.1 and 6.2, whereas the noisy cases are shown in Figs. 6.3 and 6.4. Likewise, in the $\text{FPT}^{(-)}$, the noiseless cases are displayed in Figs. 6.5 and 6.6, whereas the noisy data are plotted in Figs. 6.7 and 6.8.

Figures 6.1 and 6.3 [$\text{FPT}^{(+)}$] as well as Figs. 6.5 and 6.7 [$\text{FPT}^{(-)}$] display the Argand plots for all the harmonic variables via the distributions of the reconstructed poles $z_{k,Q}^{\pm}$ and zeros $z_{k,P}^{\pm}$ with respect to the unit circle in Euler polar coordinates. Similarly, Figs. 6.2 and 6.4 [$\text{FPT}^{(+)}$] as well as Figs. 6.6 and 6.8 [$\text{FPT}^{(-)}$] on panel (i) depict the Argand plots of linear frequencies of all the retrieved poles $\nu_{k,Q}^{\pm}$ and zeros $\nu_{k,P}^{\pm}$ in Descartes rectangular coordinates. Shown on panel (ii) in Figs. 6.2 and 6.4 [$\text{FPT}^{(+)}$] as well as in Figs. 6.6 and 6.8 [$\text{FPT}^{(-)}$] are the absolute values $|d_k^{\pm}|$ of the recovered amplitudes $d_k^{\pm}$. The input data $\{\nu_k, |d_k|\}$ are given in Figs. 6.2, 6.4, 6.6 and 6.8.

In Figs. 6.1–6.8, Froissart doublets from the $\text{FPT}^{(\pm)}$ are seen as being distributed along circles (or arcs) and lines (or wings) in the polar and rectangular coordinates, respectively. Away from the relatively narrow range of genuine spectral parameters, the distributions of Froissart doublets are configured in quite a regular way for the noise-free time signals in Figs. 6.1 and 6.2 for the $\text{FPT}^{(+)}$, as well as in Figs. 6.5 and 6.6 for the $\text{FPT}^{(-)}$. The reconstructed data for the noise-corrupted FID show a totally different pattern for constellations of Froissart doublets which are very irregularly configured in Figs. 6.3 and 6.4 for the $\text{FPT}^{(+)}$, as well as in Figs. 6.7 and 6.8 for the $\text{FPT}^{(-)}$. Such sharply distinct behaviors of Froissart doublets generated from noise-free and noise-corrupted FIDs are due a marked instability of spurious spectral structures to even the slightest changes in the input time signals. In the noisy time signal $\{c_n + r_n\}$, random perturbations $\{r_n\}$ of the corresponding noiseless counterpart $\{c_n\}$ cause random ruptures or bifurcations in the distributions of Froissart doublets along deformed circular arcs and chains for harmonic variables and linear frequencies, respectively. However, the most important observation here is that for both noiseless and noisy FIDs, pole-zero coincidences are seen to occur in a remarkably systematic manner. This permits a perfectly clear distinction of all spurious from true resonances for unperturbed as well as perturbed FIDs. From this unequivocal differentiation between noise-free and noise-corrupted MR time signals, all the correct values for the genuine spectral parameters can be exactly reconstructed, including the fundamental frequencies, the corresponding amplitudes and the original

EXACT RECONSTRUCTION of the TRUE NUMBER of GENUINE HARMONICS by FPT $^{(+)}$: NOISELESS FID

POLE – ZERO CANONICAL FORM of FPT $^{(+)}$: $P_K^+(z)/Q_K^+(z) = (p_K^+/q_K^+)\,\Pi_{k=1}^{K}(z-z_{k,P}^+)/(z-z_{k,Q}^+)$

zFPT $^{(+)}$ and pFPT $^{(+)}$; TWO COMPLEMENTARY SPECTRAL REPRESENTATIONS of FPT $^{(+)}$:

zFPT $^{(+)}$ (•) : ZEROS of FPT $^{(+)}$; ARGAND PLOT for ALL ZEROS $z_{k,P}^+$ in FPT $^{(+)}$ via $P_K^+(z) = 0$

pFPT $^{(+)}$ (o) : POLES of FPT $^{(+)}$; ARGAND PLOT for ALL POLES $z_{k,Q}^+$ in FPT $^{(+)}$ via $Q_K^+(z) = 0$

FROISSART DOUBLETS (POLE – ZERO CANCELLATIONS): $z_{k,P}^+ = z_{k,Q}^+$ (confluence of '•' & 'o')

GENUINE HARMONICS in the FIRST QUADRANT, INSIDE the UNIT CIRCLE C (|z| < 1), RANGING from # 1 (Lipid) to # 25 (Water)

FROISSART DOUBLETS OUTSIDE C : GENUINE & SPURIOUS HARMONICS SEPARATED in 2 DISJOINT REGIONS, |z| < 1 & |z| > 1

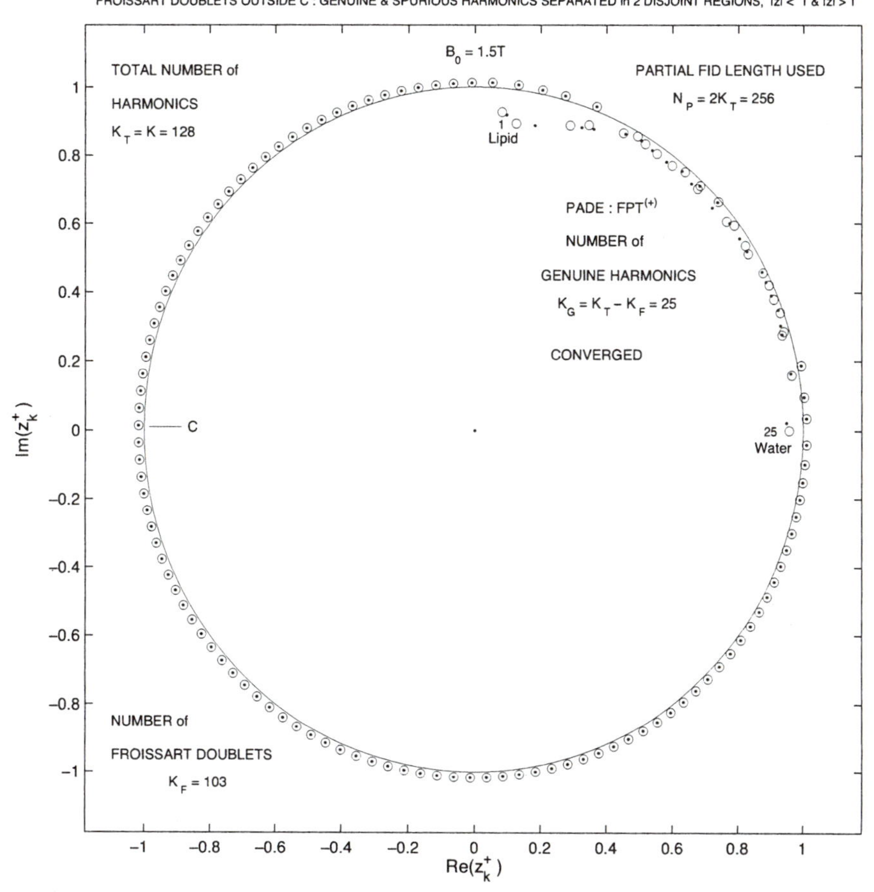

FIGURE 6.1

Froissart doublets for unequivocal determination of the exact number K_G of the genuine harmonics from the total number $K_T \equiv K$ of all the harmonics reconstructed by the $\mathrm{FPT}^{(+)}$ for the noiseless FID with input data from Table 3.1. The $\mathrm{FPT}^{(+)}$ separates the genuine from spurious harmonics in the two non-overlapping regions, inside and outside the unit circle C, respectively. The zero in the center of C is not physical; rather it is the so-called ghost zero as also discussed in chapter 3.

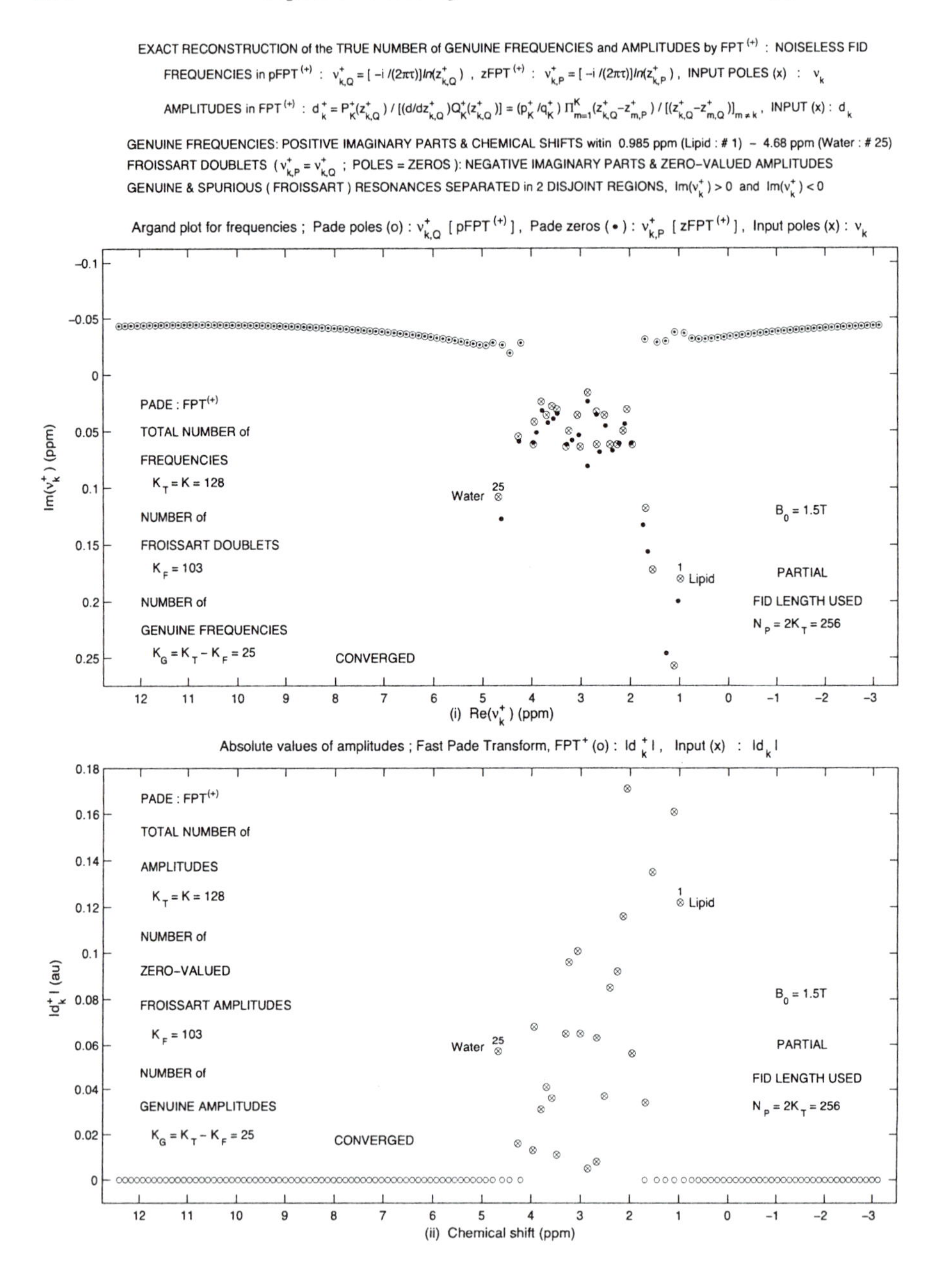

FIGURE 6.2

Froissart doublets for unequivocal determination of the exact number K_G of the genuine frequencies and amplitudes from the total number $K_T \equiv K$ of the spectral parameters reconstructed by the $\text{FPT}^{(+)}$ for the noiseless FID with input data from Table 3.1. In panel (i), the $\text{FPT}^{(+)}$ separates the genuine from spurious frequencies in the two non-overlapping regions, $\text{Im}(\nu^+) > 0$ and $\text{Im}(\nu^+) < 0$, respectively. In panel (ii), all the spurious (Froissart) amplitudes are unambiguously identified by their zero values.

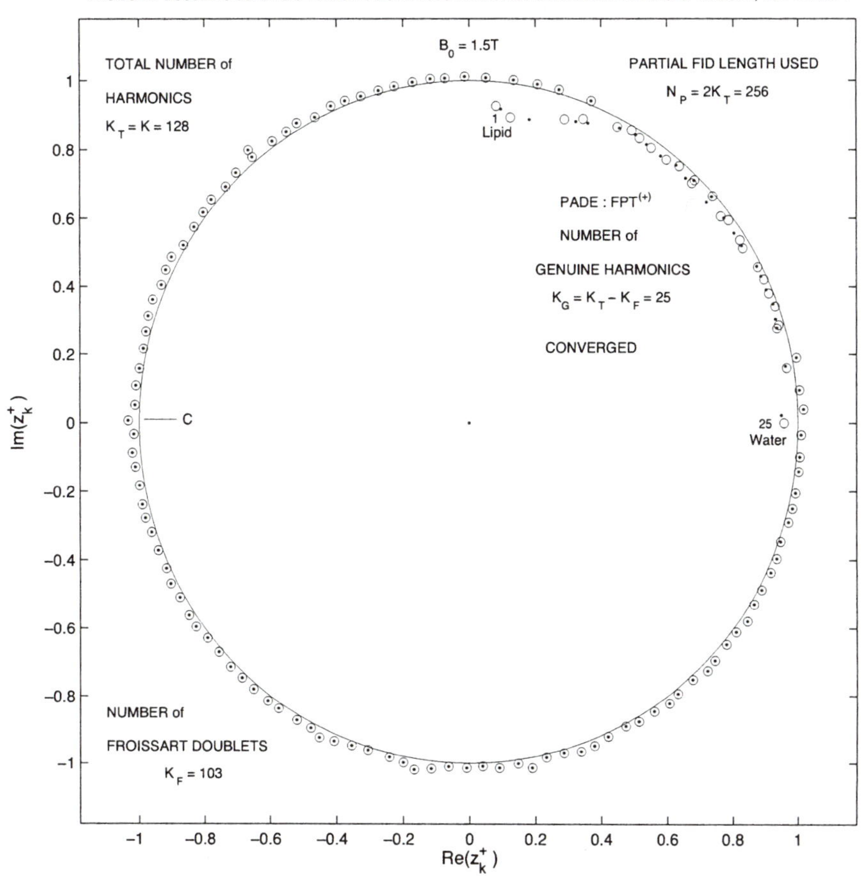

FIGURE 6.3

Froissart doublets for unequivocal determination of the exact number K_G of the genuine harmonics from the total number $K_T \equiv K$ of all the harmonics reconstructed by the $\mathrm{FPT}^{(+)}$ for the FID with input data from Table 3.1 corrupted with random noise. The $\mathrm{FPT}^{(+)}$ separates the genuine from spurious harmonics in the two non-overlapping regions, inside and outside the unit circle C, respectively.

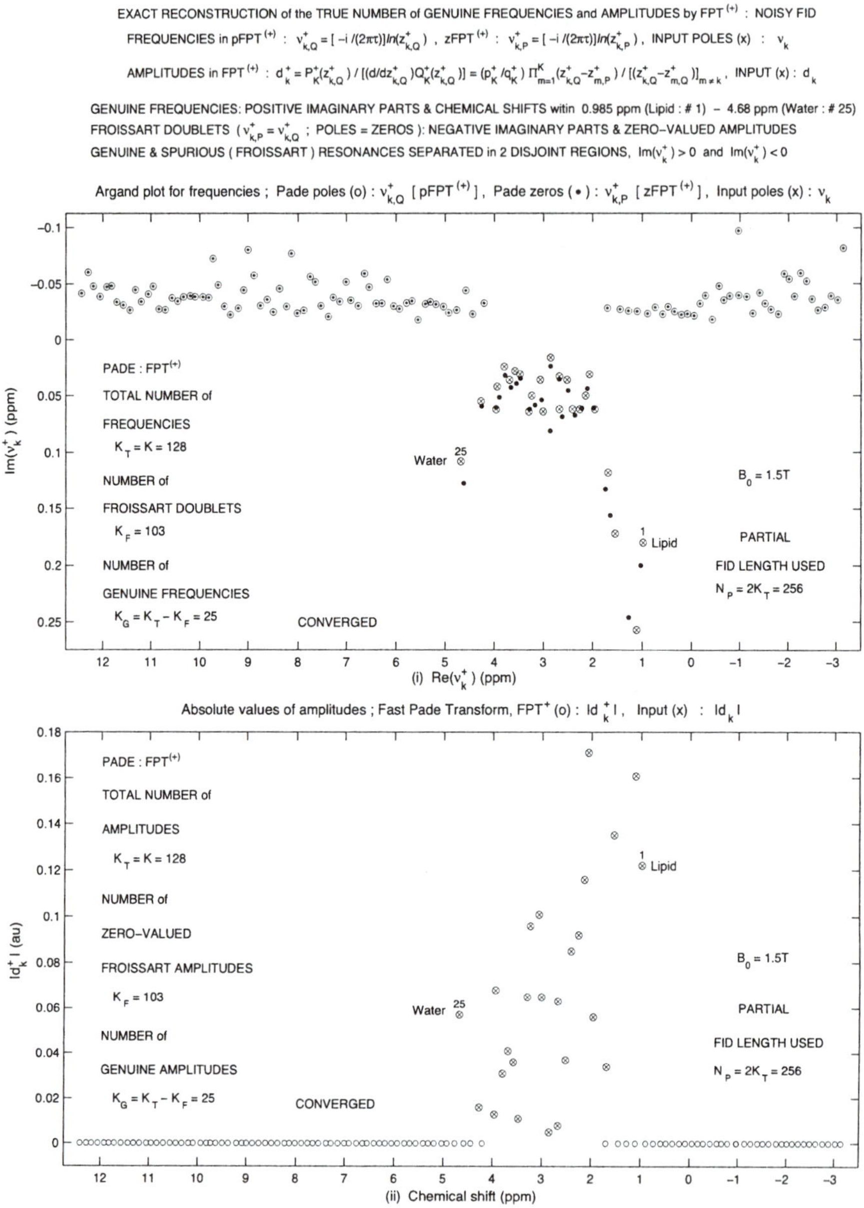

EXACT RECONSTRUCTION of the TRUE NUMBER of GENUINE FREQUENCIES and AMPLITUDES by $\mathrm{FPT}^{(+)}$: NOISY FID

FREQUENCIES in $\mathrm{pFPT}^{(+)}$: $\nu^+_{k,Q} = [-i/(2\pi\tau)]ln(z^+_{k,Q})$, $\mathrm{zFPT}^{(+)}$: $\nu^+_{k,P} = [-i/(2\pi\tau)]ln(z^+_{k,P})$, INPUT POLES (x) : ν_k

AMPLITUDES in $\mathrm{FPT}^{(+)}$: $d^+_k = P^+_K(z^+_{k,Q}) / [(d/dz^+_{k,Q})Q^+_K(z^+_{k,Q})] = (p^+_K/q^+_K)\,\Pi^K_{m=1}(z^+_{k,Q}-z^+_{m,P}) / [(z^+_{k,Q}-z^+_{m,Q})]_{m \neq k}$, INPUT (x) : d_k

GENUINE FREQUENCIES: POSITIVE IMAGINARY PARTS & CHEMICAL SHIFTS witin 0.985 ppm (Lipid : # 1) – 4.68 ppm (Water : # 25)
FROISSART DOUBLETS ($\nu^+_{k,P} = \nu^+_{k,Q}$; POLES = ZEROS): NEGATIVE IMAGINARY PARTS & ZERO-VALUED AMPLITUDES
GENUINE & SPURIOUS (FROISSART) RESONANCES SEPARATED in 2 DISJOINT REGIONS, $\mathrm{Im}(\nu^+_k) > 0$ and $\mathrm{Im}(\nu^+_k) < 0$

FIGURE 6.4
Froissart doublets for unequivocal determination of the exact number K_G of the genuine frequencies and amplitudes from the total number $K_T \equiv K$ of the spectral parameters retrieved by the $\mathrm{FPT}^{(+)}$ for the FID with input data from Table 3.1 corrupted with random noise. In panel (i), the $\mathrm{FPT}^{(+)}$ separates the genuine from spurious frequencies in two non-overlapping regions $\mathrm{Im}(\nu^+) > 0$ and $\mathrm{Im}(\nu^+) < 0$, respectively. In panel (ii), all spurious (Froissart) amplitudes are unambiguously identified by their zero values.

number of physical resonances as explicitly shown in Figs. 6.1–6.8. It can be seen that, upon stabilization via convergence of genuine resonances, the number of spurious resonances is systematically about four times larger than that of genuine resonances. Another common and interesting feature observed in Figs. 6.1–6.8 is that genuine and spurious resonances seem to repel each other as if they were some opposite electrostatic charges. The greater the distance between these resonances, the less interaction between them. This can best be seen for reconstructions from the noiseless FID as illustrated on panel (i) in Fig. 6.2 in the $\text{FPT}^{(+)}$, as well as in Fig. 6.6 in the $\text{FPT}^{(-)}$. Here, the distant wings of Froissart doublets appear quite unperturbed, but stronger disturbances of these spurious structures begin to set in near the range populated by genuine resonances, [0.985, 4.68] ppm. At larger signal lengths, such as $N/2 = 512$ and $N/2 = 1024$ [6, 35], many Froissart doublets penetrate into the region right above the genuine resonances. Still none of these spurious harmonics ever enter inside the unit circle, where all the harmonics of genuine resonances are located, as per the $\text{FPT}^{(+)}$. As such, in this variant of the FPT, the borderline $|z| = 1$ of the unit circle C acts as a cut which Froissart doublets cannot cross at all. This is a common occurrence at any partial length, including $N/4 = 256$ concerning Figs. 6.1 and 6.3.

Specifically, pole-zero confluences in Froissart doublets are seen using Euler polar coordinates for harmonic variables in Figs. 6.1 (noise-free) and 6.3 (noise-corrupted) for the $\text{FPT}^{(+)}$. This is also the case in Figs. 6.5 (noise-free) and 6.7 (noise-corrupted) for the $\text{FPT}^{(-)}$. Useful complementary illustrations for the formation of Froissart pairs through pole-zero coincidences can also be observed by employing Descartes rectangular coordinates for linear frequencies on panel (i) in Figs. 6.2 (noise-free) and 6.4 (noise-corrupted) for the $\text{FPT}^{(+)}$. The corresponding configurations from the $\text{FPT}^{(-)}$ are shown on panel (i) in Figs. 6.6 (noise-free) and 6.8 (noise-corrupted). On panel (ii) in Figs. 6.2 (noise-free) and 6.4 (noise-corrupted), displayed are Froissart zero-valued amplitudes for the $\text{FPT}^{(+)}$. The related constellations of the amplitudes for the $\text{FPT}^{(-)}$ are presented on panel (ii) in Figs. 6.6 (noise-free) and 6.8 (noise-corrupted). This twofold signature (pole-zero confluences and the accompanying zero amplitudes) represents the unequivocal manner by which the $\text{FPT}^{(\pm)}$ can disentangle genuine from spurious resonances. As mentioned in chapters 2, 3 and 5, one of the easiest visual means to immediately spot all Froissart doublets is to search for their zero amplitudes, as on panel (ii) for both considered time signals: noise-free in Figs. 6.2 [$\text{FPT}^{(+)}$] and 6.6 [$\text{FPT}^{(-)}$], as well as noise-corrupted in Figs. 6.4 [$\text{FPT}^{(+)}$] and 6.8 [$\text{FPT}^{(-)}$]. A small sub-interval of the whole Nyquist interval for linear frequencies reconstructed by the $\text{FPT}^{(\pm)}$ using the same noiseless FID as in Figs. 6.2 and 6.6 was presented and discussed earlier in chapter 3 via Fig. 3.19, as a preview of the more detailed illustrations and analyses from the current chapter.

As Figs. 6.5–6.8 clearly show, the $\text{FPT}^{(-)}$ mixes the genuine and spurious resonances in the same region $|z| > 1$ and $\text{Im}(\nu^-) > 0$. Still, the clear pattern of Froissart doublets for harmonic variables, linear frequencies and amplitudes

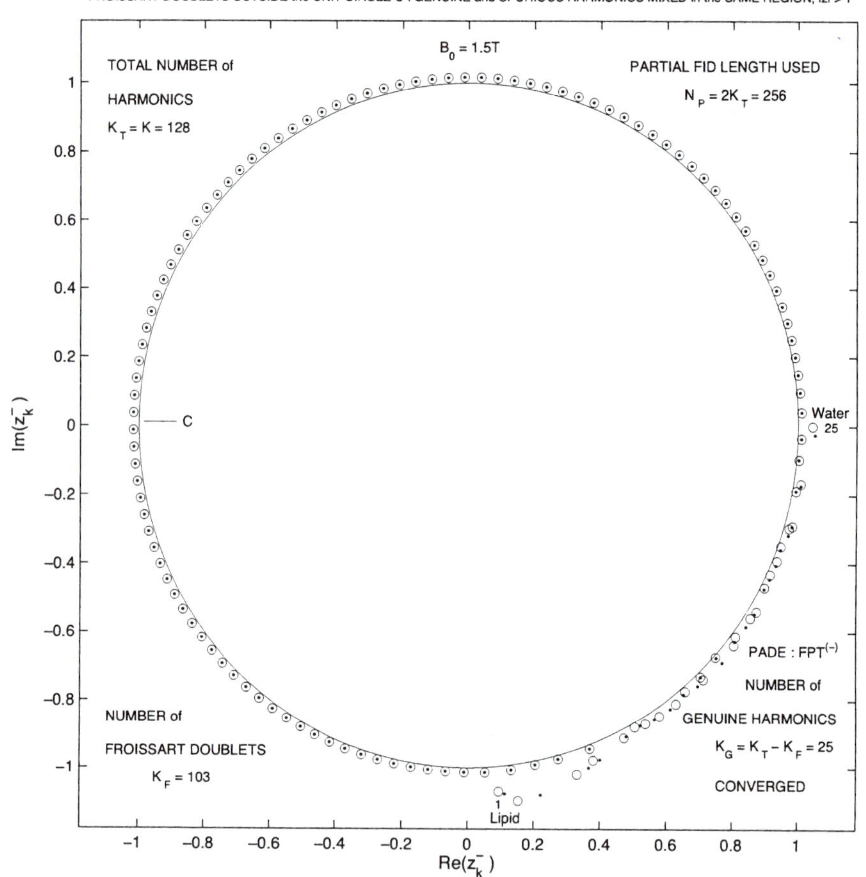

FIGURE 6.5
Froissart doublets for unequivocal determination of the exact number K_G of the genuine harmonics from the total number $K_T \equiv K$ of all the harmonics reconstructed by the $\mathrm{FPT}^{(-)}$ for the noiseless FID with input data from Table 3.1. The $\mathrm{FPT}^{(-)}$ mixes together the genuine and spurious harmonics in the same region outside the unit circle C. The ghost zero $z = 0$ from the $\mathrm{FPT}^{(+)}$ does not appear any longer at the center of C in the $\mathrm{FPT}^{(-)}$. Rather, it is moved to infinity (in theory, or to very large distance, in practice), since the $\mathrm{FPT}^{(-)}$ employs the inverse variable z^{-1}.

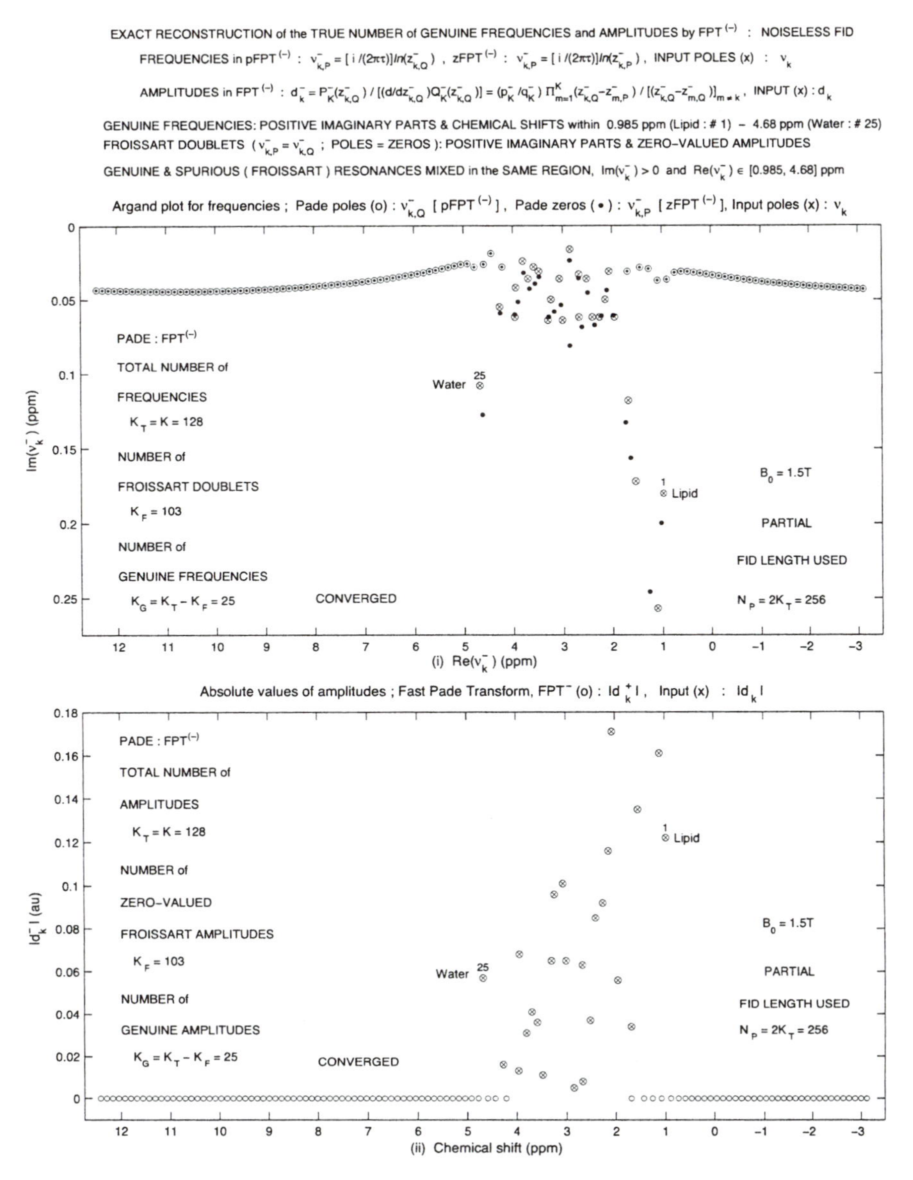

FIGURE 6.6
Froissart doublets for unequivocal determination of the exact number K_G of the genuine frequencies and amplitudes from the total number $K_T \equiv K$ of the spectral parameters reconstructed by the $\text{FPT}^{(-)}$ for the noiseless FID with input data from Table 3.1. In panel (i), the $\text{FPT}^{(-)}$ mixes the genuine and spurious frequencies in the same range, $\text{Im}(\nu^-) > 0$. In panel (ii), all the spurious (Froissart) amplitudes are unambiguously identified by their zero values.

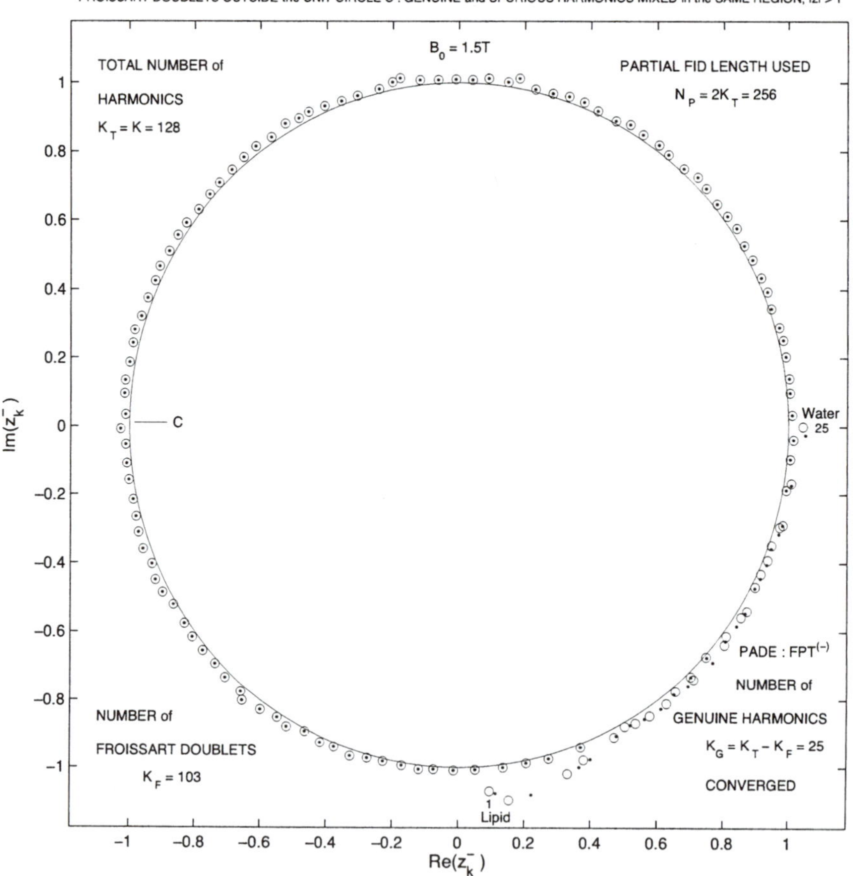

FIGURE 6.7
Froissart doublets for unequivocal determination of the exact number K_G of the genuine harmonics from the total number $K_T \equiv K$ of all the harmonics reconstructed by the $\text{FPT}^{(-)}$ for the FID with input data from Table 3.1 corrupted with random noise. The $\text{FPT}^{(-)}$ mixes together the genuine and spurious harmonics in the same region outside the unit circle C.

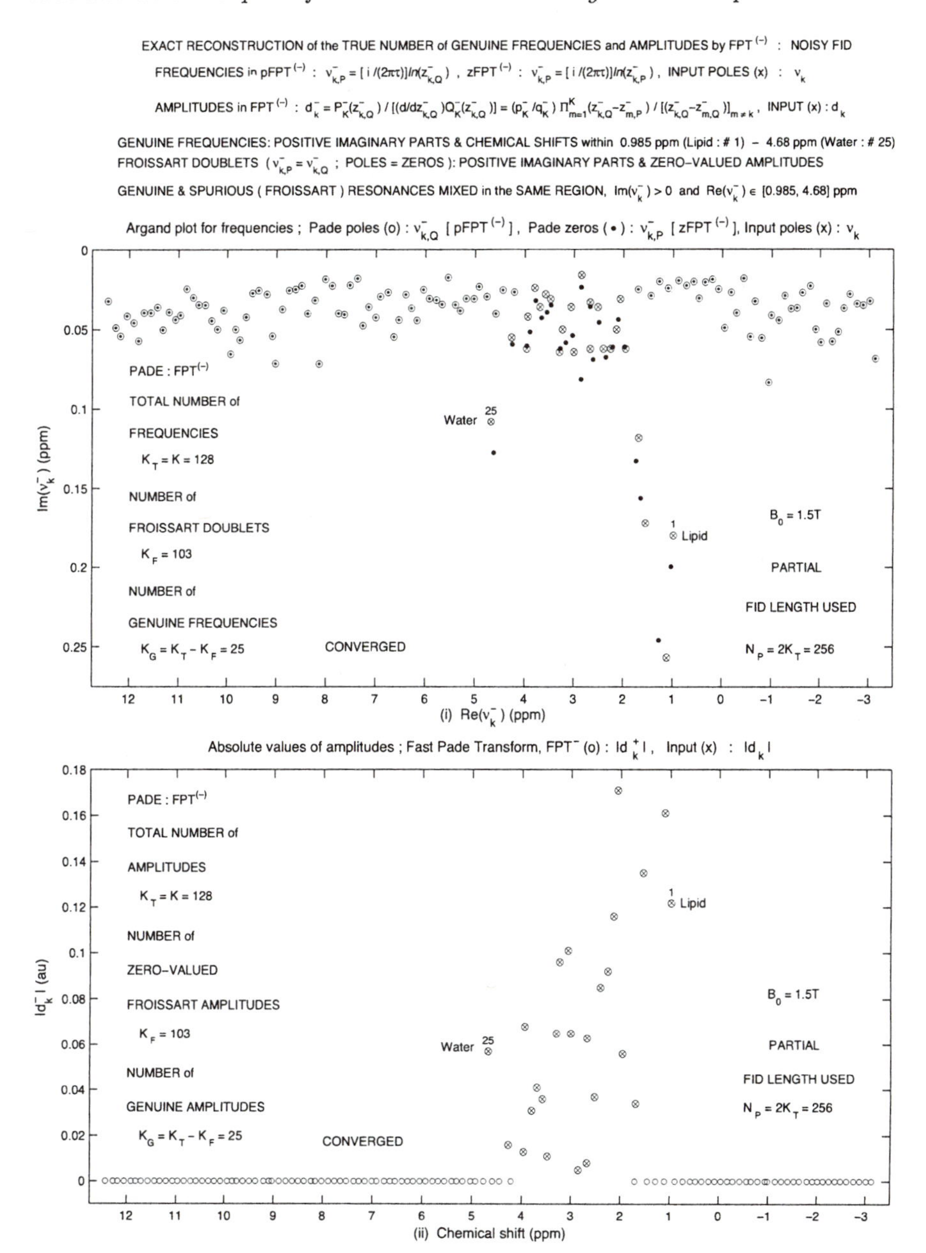

FIGURE 6.8

Froissart doublets for unequivocal determination of the exact number K_G of genuine frequencies and amplitudes from the total number $K_T \equiv K$ of spectral parameters retrieved by the $\text{FPT}^{(-)}$ for the FID with input data from Table 3.1 corrupted with random noise. In panel (i), the $\text{FPT}^{(-)}$ mixes genuine and spurious frequencies in the same range, $\text{Im}(\nu^-) > 0$. In panel (ii), all spurious (Froissart) amplitudes are unambiguously identified by their zero values.

allows the exact solution of the quantification problem by the $\text{FPT}^{(-)}$. On the other hand, the $\text{FPT}^{(+)}$ can be seen in Figs. 6.1–6.4 to sharply separate all the physical from every non-physical resonance in the two disjoint regions, inside $|z| < 1$ and outside $|z| > 1$ the unit circle for harmonic variables, as well as for frequencies via $\text{Im}(\nu^+) > 0$ and $\text{Im}(\nu^+) < 0$. This represents an unprecedented separation of genuine from the spurious (noisy or noise-like) informational contents of the investigated FIDs by using the $\text{FPT}^{(+)}$. Such a signal-noise separation is expected to play a major role in optimally reliable spectral analysis, not only for quantifications in MRS, but also in other areas of signal processing in many different branches of basic as well as applied sciences, including research and implementations in engineering, technology and industry [5, 6].

6.5.2 Zooming near convergence for Padé genuine resonances and instability of non-converged configurations of Froissart doublets in $\text{FPT}^{(\pm)}$

As discussed, there is yet another grouping of the figures from this chapter depending on the employed partial signal length N_P. Thus, the analysis from sub-section 6.5.1 is entirely devoted to quantification by the $\text{FPT}^{(\pm)}$ at a quarter of the whole signal length ($N_P = N/4 = 256$). At this value of the partial length N_P, full convergence is obtained in the $\text{FPT}^{(\pm)}$ for all the genuine resonances. However, it is also important to give a graphic display of the configurations of spectral parameters in complex planes prior to convergence[2]. For this reason, in the present sub-section, we shall zoom into a relatively small partial length interval ($N_P = 180, 220, 260$) around a quarter of the full signal length from the preceding analyses. Here, in dealing with these three non-FFT type partial signal lengths, we shall restrict our discussion to the noiseless FID. This is deemed sufficient in the present graphic illustrations of the main trend in changes of the mentioned configurations during the process of convergence towards the stabilized results for genuine resonances. Simultaneously, this should also suffice to illuminate the corresponding typical patterns of movements of Froissart doublets in complex planes with increased order of the diagonal $\text{FPT}^{(\pm)}$, i.e., with the augmented common degree of numerator and denominator Padé polynomials.

Thus, the output data obtained using the partial signal lengths $N_P = 180-260$ of the same noiseless FID from sub-section 6.5.1 are given in Figs. 6.9–6.11 for the $\text{FPT}^{(+)}$ and in Figs. 6.12–6.14 for the $\text{FPT}^{(-)}$. In Fig. 6.9 for the $\text{FPT}^{(+)}$, shown are the Argand plots in the Euler polar coordinates for harmonic variables $\{z_k^+\}$ at $N_P = 180$ and 260. Figure 6.12 deals with similar graphs for $\{z_k^-\}$ at $N_P = 180$ and 220 in the $\text{FPT}^{(-)}$. Linear frequencies $\{\nu_k^+\}$

[2]Recall that a similar line of thought was also present in sub-section 6.4.2 of section 6.4 for the tabular data.

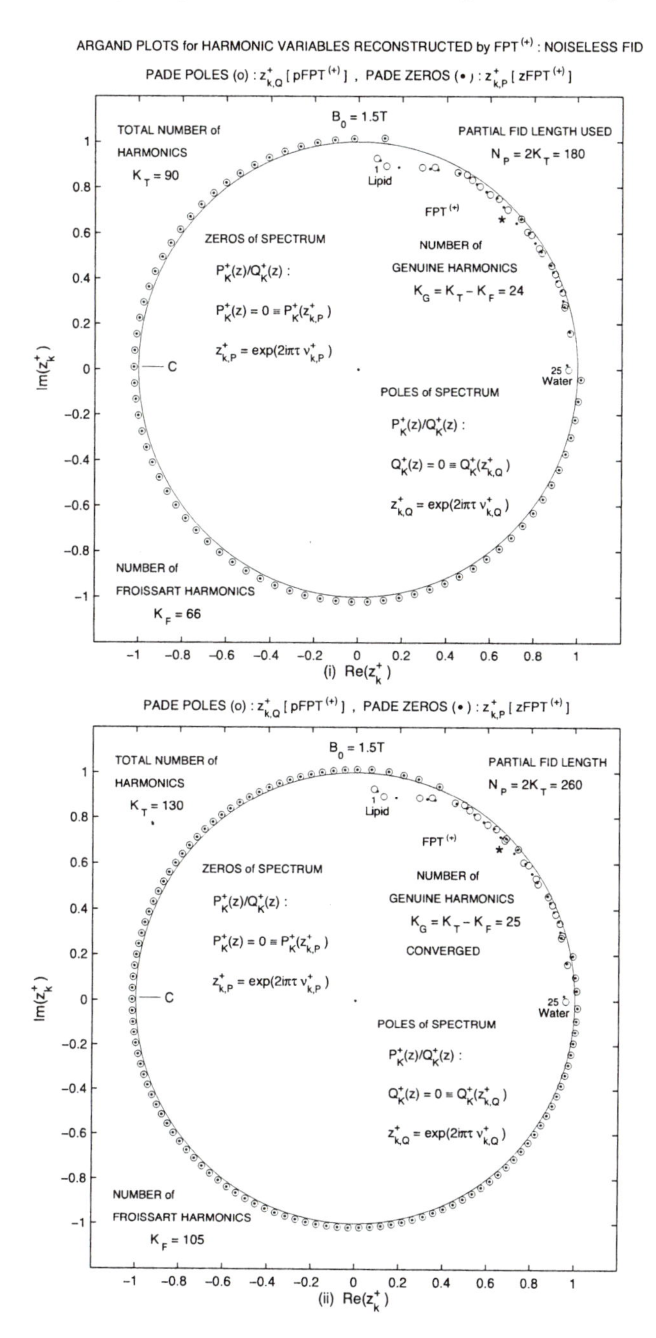

FIGURE 6.9

Distribution of poles and zeros in the complex z^+−plane via the Argand plot for complex harmonics $z^+ = \exp(2i\pi\nu^+)$ in polar coordinates for the noiseless FID with input data from Table 3.1. Symbols $\circ$ and $\bullet$ show the poles and zeros, respectively, reconstructed by the $\mathrm{FPT}^{(+)}$ at the partial signal lengths $N_P = 180, 260$.

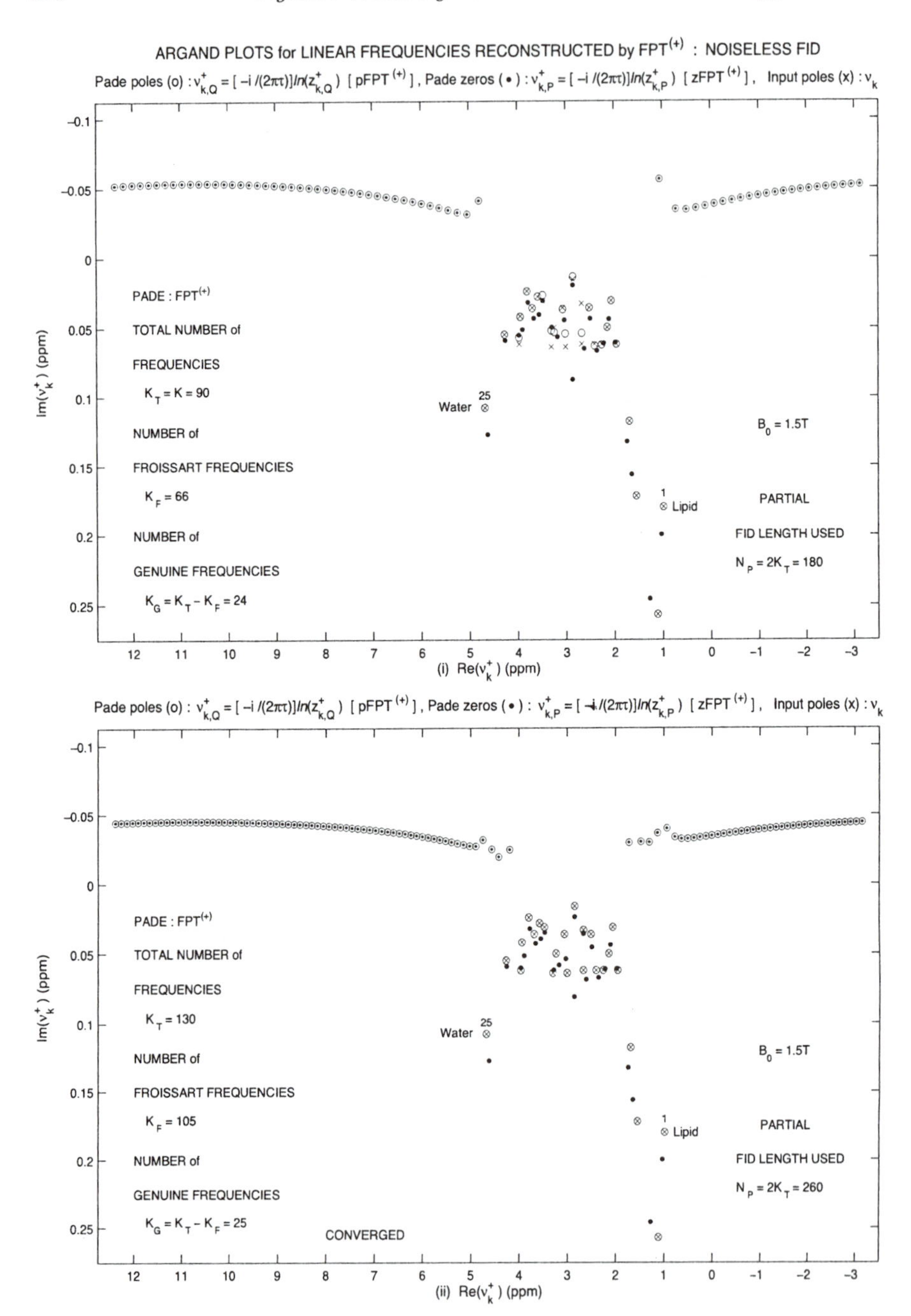

FIGURE 6.10
Distribution of poles and zeros in the complex ν^+−plane via the Argand plot for complex frequencies ν in rectangular coordinates for the noiseless FID with input data from Table 3.1. Exact input frequencies are denoted by $\times$. Symbols $\circ$ and $\bullet$ show the poles and zeros, respectively, reconstructed by the FPT$^{(+)}$ at the partial signal lengths $N_P = 180, 260$.

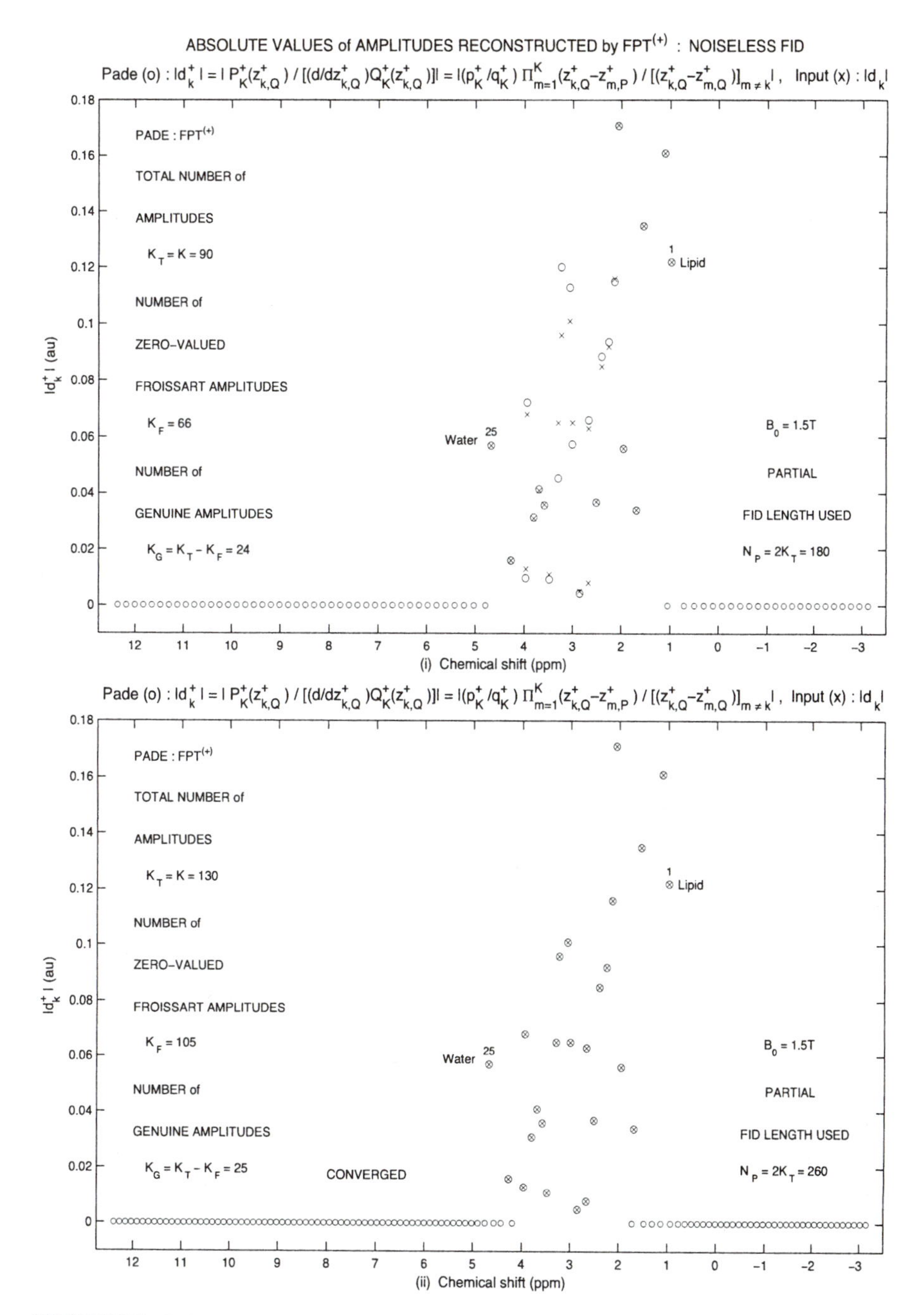

FIGURE 6.11

Absolute values of amplitudes as a function of chemical shifts, $\mathrm{Re}(\nu^+)$, for the noiseless FID with input data from Table 3.1. Exact input data $|d_k|$ are denoted by $\times$. Symbols $\circ$ represent the corresponding absolute values of amplitudes $\{|d_k^+|\}$ reconstructed by the $\mathrm{FPT}^{(+)}$ at the partial signal lengths $N_P = 180, 260$.

in the $\mathrm{FPT}^{(+)}$ at $N_P = 180$ and 260 are depicted in Fig. 6.10 via the Argand plots in Descartes rectangular coordinates. Likewise, Fig. 6.13 is devoted to $\{\nu_k^-\}$ in the $\mathrm{FPT}^{(-)}$ at $N_P = 180$ and 220. Finally, Fig. 6.11 displays the absolute values of amplitudes $\{d_k^+\}$ in the $\mathrm{FPT}^{(+)}$ at $N_P = 180$ and 260 as a function of chemical shifts in the whole Nyquist range. The same type of graph for the corresponding complete set $\{d_k^-\}$ retrieved by the $\mathrm{FPT}^{(-)}$ at $N_P = 180$ and 220 is illustrated in Fig. 6.14.

In both versions of the FPT, there is a striking difference between the Argand plots in Euler and Descartes coordinates when passing from the stage prior to convergence to full convergence of genuine harmonic variables and linear frequencies. Thus, in Fig. 6.9 for the $\mathrm{FPT}^{(+)}$, the only perceivable difference to the naked eye between panels (i) and (ii) of the Argand-Euler plots for $\{z_k^+\}$ at $N_P = 180$ and 260, respectively, is the absence and presence of the data for the 11th harmonic as indicated by a small star symbol in the first quadrant near the circumference of the unit circle $|z| = 1$ in the direction of about 45° from the center of the circle C. Other poles and zeros do not seem to differ much (only visually, of course) when comparing the top ($N_P = 180$) and bottom ($N_P = 260$) sub-plots in Fig. 6.9. A similar observation is made while looking at Fig. 6.12 for $\{z_k^-\}$ at $N_P = 180$ and 220 in the $\mathrm{FPT}^{(-)}$, when comparing panels (i) and (ii), respectively. Here, of course, all genuine harmonic variables $\{z_k^-\}$ are in the 4th quadrant, and the approximate location of the mentioned small star symbol near $k = 11\,\&\,12$ is now placed at $\sim 360° - 45° = 315°$ relative to the center of the unit circle C).

Except for the 11th resonance, the remaining possible differences in the Argand plots in Euler polar coordinates at and near convergence of genuine resonances are hardly visible because of the logarithmic scale which is inherent in the exponential harmonic variables $\{z_k^\pm\}$. Thus, it is expected that more perceivable differences between genuine resonances at and prior to convergence will be observed using a linear scale. Indeed, this is precisely the case, as can be seen by inspecting the Argand plots in Descartes rectangular coordinates for linear frequencies $\{\nu_k^+\}$ at $N_P = 180$ and 260 for the $\mathrm{FPT}^{(+)}$ on panels (i) and (ii) in Fig. 6.10. Significant differences are obvious among a number of genuine harmonic variables from panels (i) and (ii) of this figure. As to Froissart doublets, it is seen that they do not converge at all. As more spurious poles and zeros emerge in passing from $N_P = 180$ to $N_P = 260$, they redistribute themselves along two major lines or wings that, via a careful inspection, can be seen to differ from each other throughout the entire Nyquist range[3]. A like situation is encountered in the $\mathrm{FPT}^{(-)}$ when comparing the corresponding panels (i) and (ii) at $N_P = 180$ and 220 in Fig. 6.13. Similar to linear frequencies, more pronounced differences between the reconstructions at and prior to convergence are also anticipated on the level

[3]This is even more evident when the corresponding tabular data for all the retrieved resonances are analyzed, as done in Refs. [6, 35]).

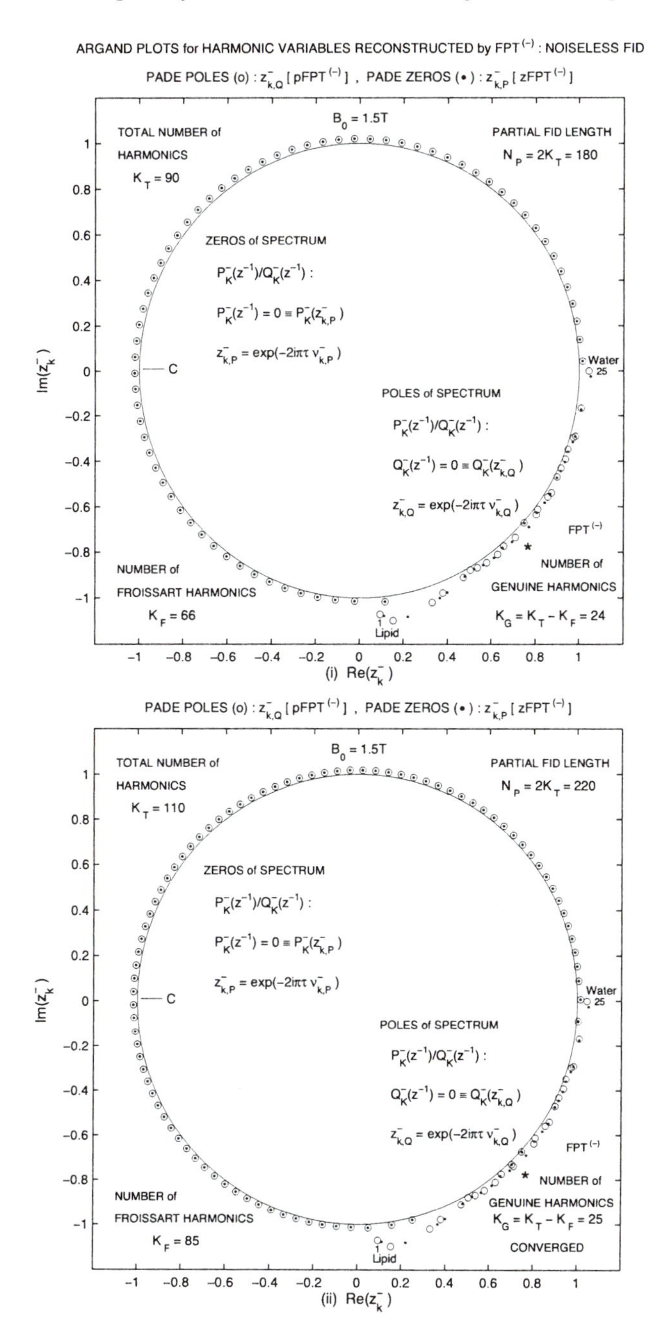

FIGURE 6.12

Distribution of poles and zeros in the complex z^--plane via the Argand plot for complex harmonics $z^{-1} = \exp(-2i\pi\nu^-)$ in polar coordinates for the noiseless FID with input data from Table 3.1. Symbols $\circ$ and $\bullet$ show the poles and zeros, respectively, reconstructed by the $\text{FPT}^{(-)}$ at the partial signal lengths $N_P = 180, 220$.

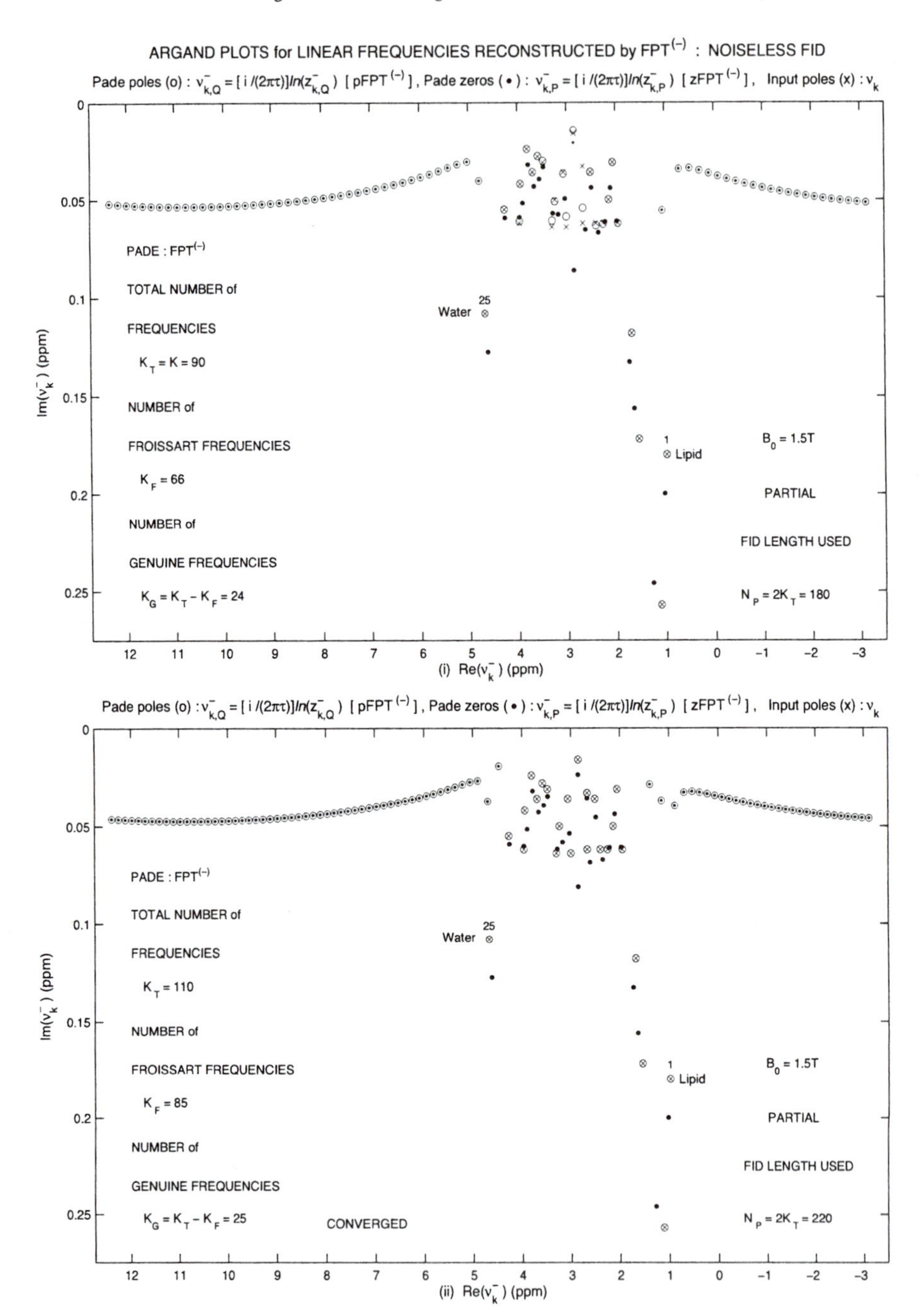

FIGURE 6.13
Distribution of poles and zeros in the complex ν^-−plane via the Argand plot for complex frequencies ν in rectangular coordinates for the noiseless FID with input data from Table 3.1. Exact input frequencies are denoted by $\times$. Symbols $\circ$ and $\bullet$ show the poles and zeros, respectively, reconstructed by the $\text{FPT}^{(-)}$ at the partial signal lengths $N_P = 180, 220$.

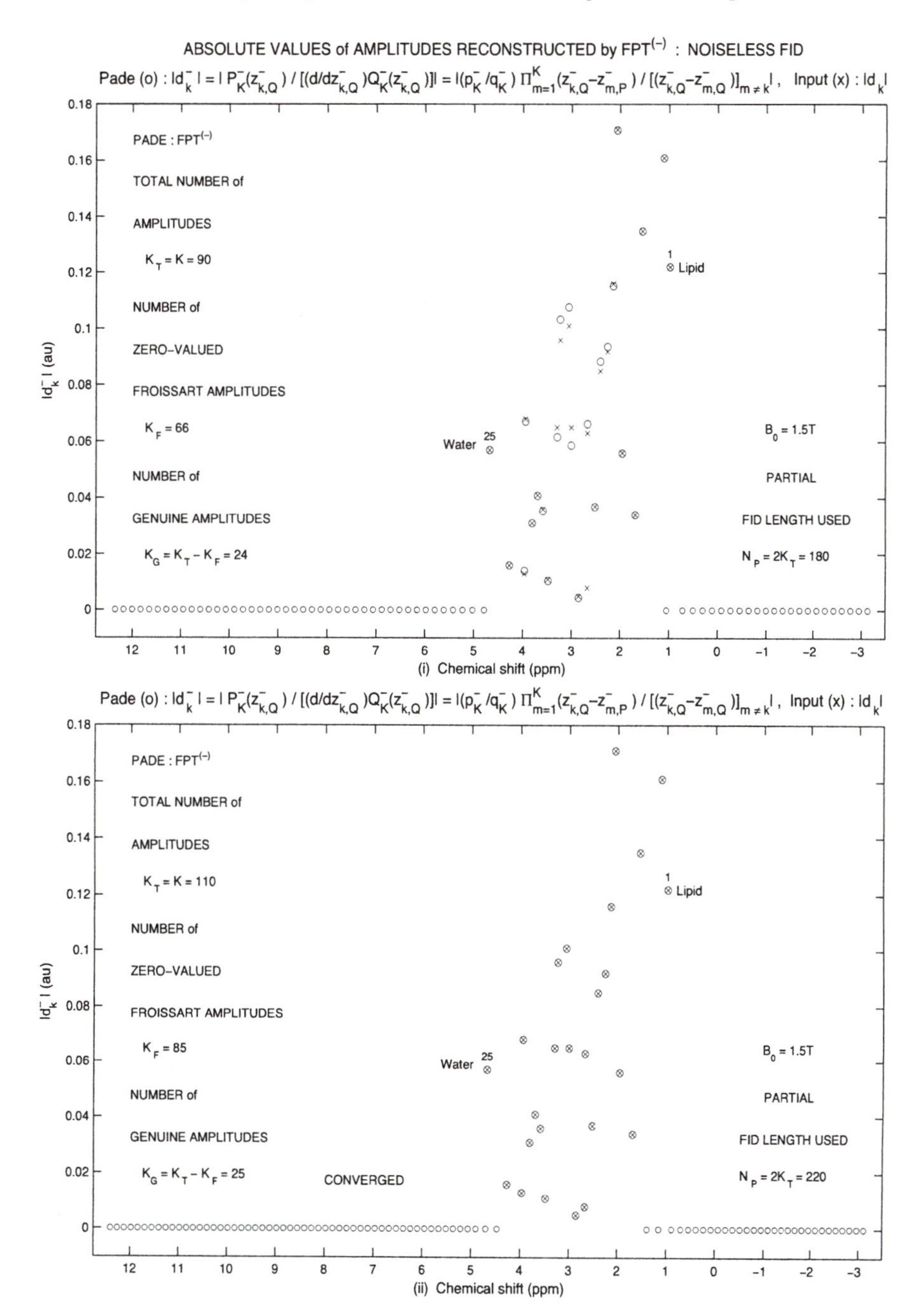

FIGURE 6.14

Absolute values of amplitudes as a function of chemical shifts, $\mathrm{Re}(\nu^-)$, for the noiseless FID with input data from Table 3.1. Exact input data $|d_k|$ are denoted by $\times$. Symbols $\circ$ represent the corresponding values $\{|d_k^-|\}$ reconstructed by the $\mathrm{FPT}^{(-)}$ at the partial signal lengths $N_P = 180, 220$.

of absolute values of amplitudes of genuine resonances than in the discussed case of harmonic variables. This is confirmed in Figs. 6.11 and 6.14 for the $\mathrm{FPT}^{(+)}$ and $\mathrm{FPT}^{(-)}$, respectively.

6.6 Practical significance of the Froissart filter for exact signal-noise separation

In chapter 4, we presented the theory of quantum-mechanical spectral analysis based upon the Padé approximant, the Lanczos algorithm as well as on their combination called the Padé-Lanczos approximant and Lanczos continued fractions. These are all equivalent. Lanczos continued fractions are of the type of contracted continued fractions that contain two times as many expansion terms as the ordinary continued fractions for the same order or rank. In signal processing, the Padé approximant is called the fast Padé transform. It has its two equivalent versions, the $\mathrm{FPT}^{(+)}$ and $\mathrm{FPT}^{(-)}$, that are initially defined to lie inside and outside the unit circle for complex harmonic variables. By the Cauchy analytical continuation, these two versions are also defined everywhere in the complex plane with the exception of the poles. This could be considered analogous to the typical outgoing and incoming boundary conditions ingrained in the standard Green function.

The $\mathrm{FPT}^{(+)}$ and $\mathrm{FPT}^{(-)}$ are equivalent to Lanczos continued fractions of the even and odd order, respectively. These two variants of the FPT can exactly reconstruct the input spectral parameters of every resonance from noiseless and noisy time signals. When convergence is achieved in both variants, the results of the $\mathrm{FPT}^{(+)}$ and $\mathrm{FPT}^{(-)}$ are identical. This provides an indispensable intrinsic check of the validity of the FPT. We have proven through illustrations that the FPT is an extremely reliable method for quantifying noise-corrupted FIDs reminiscent of those encoded by *in vivo* MRS.

A critical hurdle for spectral analysis is how to unambiguously separate genuine from spurious information in FIDs. We have shown that this exceedingly difficult problem can indeed be solved via the powerful concept of exact signal-noise-separation using Froissart doublets [44] or pole-zero cancellations [45]. This separation is unique to the FPT, due to its polynomial quotient form P_K/Q_K of the frequency-dependent response function, which is the total Green function of the investigated system. The true number K_G of genuine resonances, as the exact order or rank K_{ex} of the FPT with $K_G = K_{\mathrm{ex}}$, is reconstructed by reaching the constancy of P_K/Q_K when the polynomial degree K is systematically increased. By augmenting the 'running order' K above the plateau attained at $K = K_{\mathrm{ex}}$, the same values of P_K/Q_K are obtained via $P_{K_{\mathrm{ex}}+m}/Q_{K_{\mathrm{ex}}+m} = P_{K_{\mathrm{ex}}}/Q_{K_{\mathrm{ex}}}$ $(m = 1, 2, ...)$ as per (2.226). This can only be possible when, for $K > K_{\mathrm{ex}}$, all the new poles and zeros coincide with each other, so that they are canceled from the canonical representation of P_K/Q_K (see chapter 5). We have also shown that the same saturation $P_{K_{\rm ex}+m}/Q_{K_{\rm ex}+m} = P_{K_{\rm ex}}/Q_{K_{\rm ex}}$ $(m = 1, 2, ...)$ occurs in the equivalent Heaviside partial fraction representation of the Padé polynomial quotient through zero amplitudes for any $K > K_{\rm ex}$. Furthermore, pole-zero confluences can also appear at any $K \leq K_{\rm ex}$. However, the corresponding amplitudes are always found to be equal to zero. We have hereby demonstrated that all zero-valued amplitudes and the associated pole-zero coincidences represent the unambiguous signatures of non-physical information (noise and noise-like) that appears during spectral analysis. This is the essential feature of exact signal-noise separation by pole-zero cancellations in the FPT. It can certainly be anticipated that the presently-described new type of signal denoising through the Froissart filter will have important and broad applications throughout various fields within signal processing.

We have presented several illustrations concerning machine accurate reconstructions of all the fundamental frequencies and amplitudes, including the unambiguous retrieval of the exact number of resonances. Convergence under these stringently-imposed conditions (exact 12 digit output for exact 12 digit input data) has been accomplished. This further confirms the robustness of the FPT even with respect to round-off-errors. We achieve this by using two regular computational routines from MATLAB for solving a system of linear equations and rooting the characteristic equation. Thus, the robustness of the FPT is mainly due to the rational model for the response function, and not to some specially-designed algorithm. The Padé model of polynomial quotients is intrinsically robust. Above all, this is expected from the physics vantage point, as dictated by quantum mechanics through Green functions which invariably reduce to a ratio of two polynomials. The subsequent computations translate this solid theoretical basis into the exact numerical results.

Resonance is a special type of phenomenon in which phase transitions play a critical role. If we examine phase spectra, this is evidenced in the appearance of marked jumps by π at each resonant frequency, as the Levinson theorem prescribes. When there are numerous resonances in a spectrum, such as those studied herein, reconstruction of four machine accurate spectral parameters for each resonance (frequencies and amplitudes both complex-valued) becomes a daunting numerical challenge. Specifically, the 25 resonances need a numerically exact solution for 100 spectral parameters. We should thus consider this optimization problem as searching for the global minimum of an objective function in a hyperspace of 100 dimensions.

The density of states is one of the key determinants of resolution and convergence rate. The average number of fundamental frequencies in the window of interest defines the resolution of the FPT. This is also true for all other parametric estimators. In contrast, in the FFT, resolution is defined by the distance between the two adjacent frequencies. As is known from the Lanczos algorithm, convergence is first reached for the outermost frequencies in the range of investigation. The innermost and tightly congested frequencies converge last. The approach to the global minimum can indirectly be followed by looking for constancy of spectral parameters as a function of the truncated signal length at a fixed bandwidth. When all the spectral parameters stabilize such that the results remain unchanged with the addition of more signal points, the exact results for all the 100 parameters are obtained to within machine accuracy. However, this does not take place in a steady process, i.e., by obtaining the exact outer frequencies first and then awaiting for the remaining frequencies to attain their exact values one by one. Rather, the contrary actually occurs. While the physical outermost frequencies do appear first, this does not initially occur to machine accuracy. Machine accurate values for the outermost frequencies are obtained simultaneously with the machine accurate results for all the other genuine resonances. A very sharp transition occurs at that point. Even when extremely close to the global minimum of the objective function, all the spectral parameters still fluctuate around their true values. At convergence, all the parameters reach their exact values and for this to take place, it is sufficient that the running order K of the FPT changes by only one unit, i.e., two more signal points are needed. Subsequent to this stage, any further addition of signal points does not affect the results within the required 12 digit accuracy for noiseless FIDs. Thus, this process is seen to be a veritable phase transition in which all the complex-valued quantities are needed, with their phases being critical to the emergence of simultaneous resonance for all the fundamental harmonics. We can see this most dramatically if we fix the phases of the input amplitudes of each fundamental harmonics to zero, as done in the present analysis. The reconstructed phases fluctuate around zero, but no fully exact value of the remaining spectral parameters for any resonance is reached until all the machine accurate zero-valued phases are obtained. This is the signature of the algorithm's optimality attained in the simultaneous retrieval of 100 spectral parameters through undergoing a dynamic phase transition at which point all the sought quantities collectively synchronize their machine accurate numerical values.

The distributions of poles and zeros in complex frequency planes are essential to spectral analysis. Argand plots are particularly helpful for visualizing pole-zero cancellations of complex-valued frequencies in the polar as well as the rectangular planes. Zero-valued spurious amplitudes for non-physical resonances as a function of chemical shift are also displayed together with the positive values of the amplitudes for true resonances. Illustrations are given for both the $\mathrm{FPT}^{(+)}$ and $\mathrm{FPT}^{(-)}$. In the $\mathrm{FPT}^{(-)}$ physical and non-physical resonances are inter-mixed outside the unit circles. Nevertheless, the denoising Froissart filter completely disentangles one from the other according to the twofold signature: pole-zero coincidences and zero-valued amplitudes. In the $\mathrm{FPT}^{(+)}$, on the other hand, the true and spurious resonances are completely separated from each other in two disjoint portions of the complex frequency plane. This remarkable separation of genuine from spurious resonances represents a key feature for signal processing in biomedical applications and for other disciplines, as well.

7

Padé processing for MR spectra from in vivo time signals

We now will turn our attention to Padé processing of *in vivo* time signals. These FIDs were encoded using high-field MR scanners, and have excellent SNR, so that the converged shape spectra from the FFT with all N points (N =2048) can be used as reliable envelopes for comparisons.

7.1 Relative performance of the FPT and FFT for total shape spectra for encoded FIDs

In this chapter we shall use both the truncated $N/M(M > 1)$ and the full length N of the time signal $\{c_n\}$ to perform detailed comparisons of the $\mathrm{FPT}^{(\pm)}$ with the FFT. Our primary goal is to assess the resolving power, convergence rate and robustness of the Padé and the Fourier estimators relative to each other for encoded *in vivo* time signals.

To ensure complete fairness through strictly valid comparisons, all the FFT and $\mathrm{FPT}^{(\pm)}$ spectra are consistently computed by employing exactly the same number $N/M(1 \le M \le 32)$ of time signal points from either a truncated $(M > 1)$ or full $(M = 1)$ set of the originally *encoded* time signal points $\{c_n\}$. In the frequency domain, the Padé grid differs conceptually from the more customary Fourier grid.

Thus, for example, the absorption spectra in the $\mathrm{FPT}^{(\pm)}$ given by the real parts of the complex-valued polynomial quotients $P_K^{\pm}(z^{\pm 1})/Q_K^{\pm}(z^{\pm 1})$, where $z^{\pm 1} = \mathrm{e}^{\pm i\omega\tau}$, can be evaluated at any frequency ω. Therefore, the Padé grid is simply an arbitrary plotting mesh for frequency ω chosen independently of the employed number of time signal points $\{c_n\}$. The ability to possess such flexibility in the Padé grid is rooted deeply in the interpolation feature of the $\mathrm{FPT}^{(\pm)}$ as achieved via the rational functions $P_K^{\pm}(z^{\pm 1})/Q_K^{\pm}(z^{\pm 1})$.

Recall that these polynomial quotients are extrapolators, since they stem from the explicit summation of the Maclaurin series (i.e., an infinite sum) with the signal points $\{c_n\}$ as the expansion coefficients. Thus, the Padé spectra inherently contain inferences on the otherwise unmeasurable infinitely long time signals by relying solely upon the corresponding encoded data of finite lengths. Such extrapolation properties of the $\mathrm{FPT}^{(\pm)}$ lead directly to the resolution enhancement relative to the FFT.

As known, the FFT is not an extrapolator, since the Fourier resolving power is limited by the total acquisition time T. Moreover, the FFT is not an interpolator either, since the frequency mesh in the Fourier spectra is fixed in advance by the basic limitation of the Fourier grid $2\pi k/T(0 \le k \le N-1)$. Another related disadvantage of the FFT, especially when compared with the corresponding flexibility in the Padé grid of the $\mathrm{FPT}^{(\pm)}$, is that the number of frequency points in the Fourier spectrum is rigorously restricted to the number $N/M(M \ge 1)$ of employed time signal points. In computation for $M > 1$, the FFT employs $N - N/M$ zeros for completion to $N = 2048$, but the $\mathrm{FPT}^{(+)}$ and $\mathrm{FPT}^{(-)}$ do not.

We have noted that due to its non-linearity, as is apparent from the polynomial quotient for the spectrum, the FPT can effectively reduce noise. Noise reduction by the FPT is also due to the fact that short, i.e., truncated signals can be used with better SNR to extract the full spectral information. This is opposed to the FFT which, as a linear processor, brings the entire noise content from the time to the frequency domain. Furthermore, reduction of noise by the FFT is not possible by truncation of the encoded data, since the Fourier method must use the full signal length to achieve reasonable resolution, as will now be illustrated.

7.2 The FIDs, convergence regions and absorption spectra at full signal length encoded at high magnetic field strengths

In this chapter, the $\mathrm{FPT}^{(+)}$ and $\mathrm{FPT}^{(-)}$ are applied to two complex-valued time signals $\{c_n\}$ with the common bandwidth of 6001.5 Hz as encoded via MRS at 4T and 7T from the brain of a healthy volunteer [141]. These FIDs are long ($N = 2048$) with very good SNR, so that the shape spectra from the FFT using all N points are excellent and can be used as a gold standard.

The investigated FIDs and the corresponding absorption spectra in the FFT and $\mathrm{FPT}^{(\pm)}$ computed with the full signal length ($N = 2048$) are displayed in Figs. 7.1 and 7.2 for 4T and 7T, respectively. The real and the imaginary parts of the FIDs are displayed on the top and the middle panels of the left column, respectively. On the bottom left panels of Figs. 7.1 and 7.2, we show the "initial convergence regions" of the $\mathrm{FPT}^{(+)}$ and $\mathrm{FPT}^{(-)}$ that are located inside ($|z| < 1$) and outside ($|z| > 1$) the unit circle, respectively.

Recall as well that by the Cauchy concept of analytical continuation, the $\mathrm{FPT}^{(+)}$ and $\mathrm{FPT}^{(-)}$ are valid in the whole complex frequency plane with the exception of the poles. Both variants, the $\mathrm{FPT}^{(+)}$ and $\mathrm{FPT}^{(-)}$ of the fast

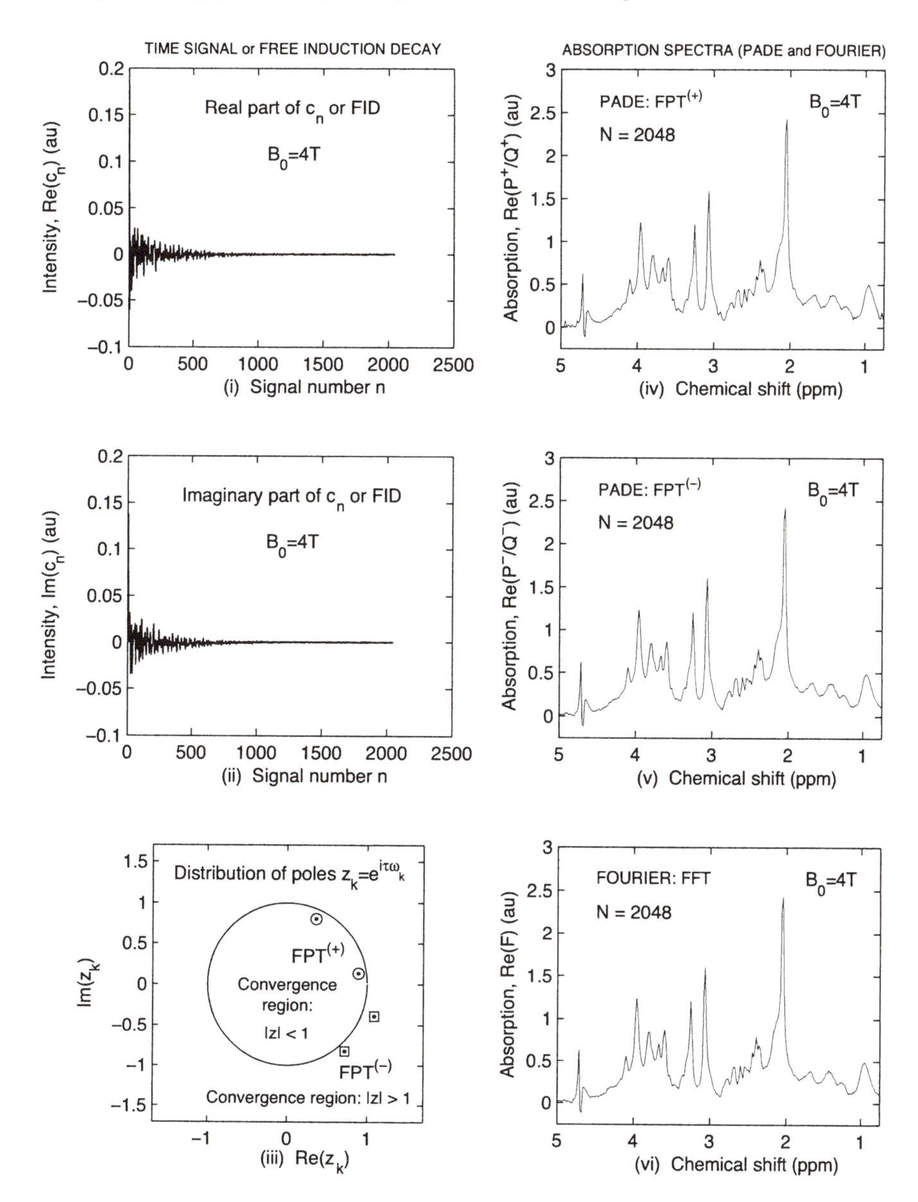

FIGURE 7.1
The real and imaginary parts of the complex-valued FID as encoded by Tkáč *et al.* [141] at 4T from occipital grey matter of a healthy volunteer: panels (i) and (ii). Each FID point is divided by 2048×10^2. Resonances in the $\text{FPT}^{(+)}$ and $\text{FPT}^{(-)}$ are seen to lie inside and outside the unit circle on panel (iii), as symbolized by small circles and squares with the dots in their centers. The absorption spectra at 4T computed using the $\text{FPT}^{(+)}$, $\text{FPT}^{(-)}$ and FFT with the full FID ($N = 2048$) are depicted on panels (iv), (v) and (vi), respectively.

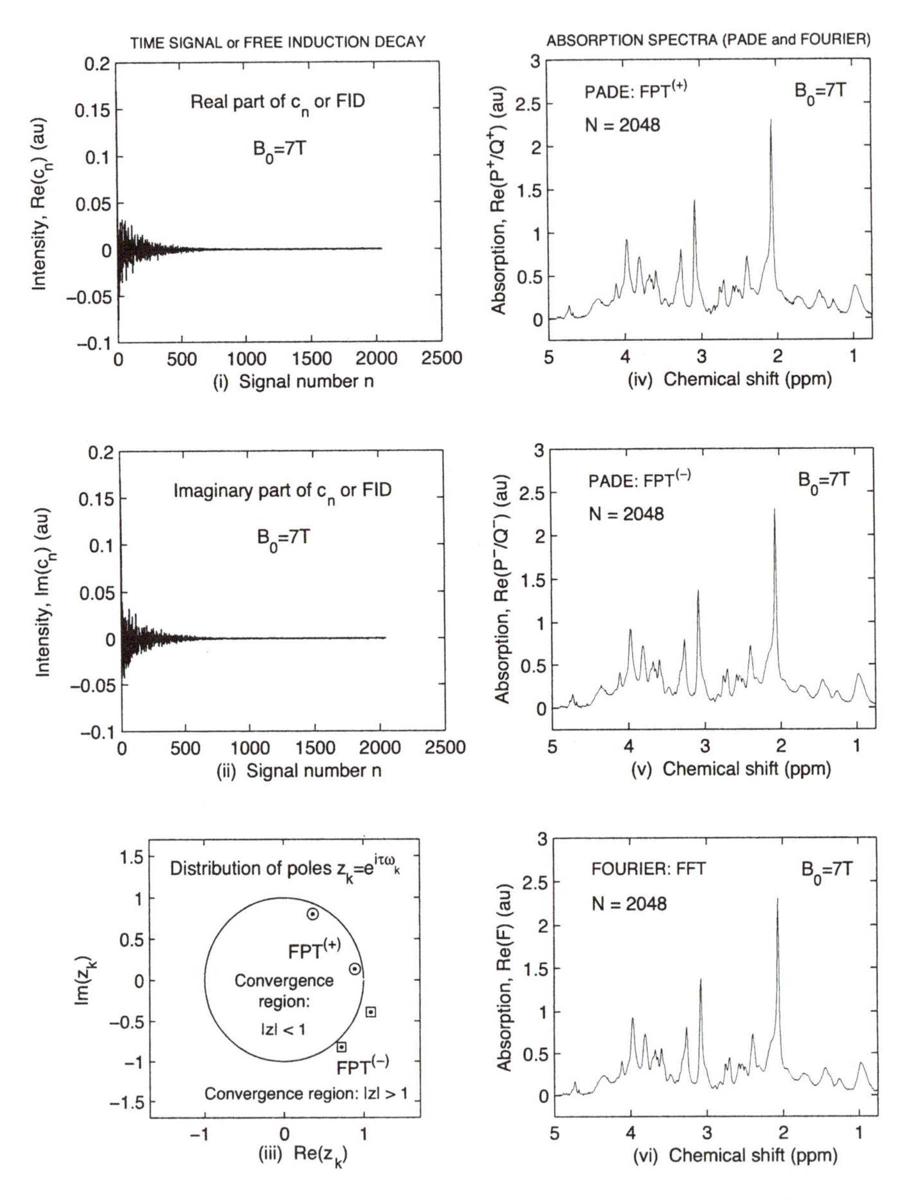

FIGURE 7.2
The real and imaginary parts of the complex-valued FID as encoded by Tkáč *et al.* [141] at 7T from occipital grey matter of a healthy volunteer: panels (i) and (ii). Each FID point is divided by 2048×10^3. Resonances in the $\text{FPT}^{(+)}$ and $\text{FPT}^{(-)}$ are seen to lie inside and outside the unit circle on panel (iii), as symbolized by small circles and squares with the dots in their centers. The absorption spectra at 7T computed using the $\text{FPT}^{(+)}$, $\text{FPT}^{(-)}$ and FFT with the full FID ($N = 2048$) are depicted on panels (iv), (v) and (vi), respectively.

Padé transform stem from the same input Maclaurin series which is divergent for $|z| < 1$ and convergent for $|z| > 1$. Hence, we have seen that the FPT$^{(+)}$ has a much more difficult task than FPT$^{(-)}$. This is because FPT$^{(+)}$ must induce convergence into the divergent series, while the FPT$^{(-)}$ ought to merely increase the convergence rate of the already convergent Maclaurin series.

On the right side of Figs. 7.1 and 7.2, the absorption spectra are depicted for the FPT$^{(+)}$ (top panels), FPT$^{(-)}$ (middle panels) and FFT (bottom panels) all using the full signal length, $N = 2048$. It can be observed that the FPT$^{(+)}$ and FPT$^{(-)}$ yield practically the same spectra. These two latter spectra are seen in Figs. 7.1 and 7.2 to coincide with the corresponding spectrum from the FFT to within the background noise. Such an excellent agreement among these three methods for total shape spectra is expected, since the FFT can serve as a gold standard for the investigated FID of high quality [141].

7.3 Convergence patterns of FPT$^{(-)}$ and FFT for absorption total shape spectra

Figures 7.3–7.6 at 4T show the convergence rates of the absorption spectra computed by the FFT and the FPT$^{(-)}$ at 5 signal fractions ($N/32 = 64, N/16 = 128, N/8 = 256, N/4 = 512, N/2 = 1024$) as well as at the full signal length $N = 2048$. Likewise, Figs. 7.7–7.10 display the corresponding spectra generated by the FFT and the FPT$^{(-)}$ at 7T. More specifically, Figs. 7.3, 7.4, 7.7 and 7.8 use separate graphs to show the convergence patterns pertinent solely to the FFT or the FPT$^{(-)}$. Additionally, with Figs. 7.5, 7.6, 7.9 and 7.10, that juxtapose the FFT and the FPT$^{(-)}$ on the same graphs at 4T and 7T, respectively, we provide a direct inspection of the relative convergence rates of these two signal processors.

As can be seen from Figs. 7.3 and 7.7, the FFT appears as a stable signal processor with no undesirable surprises when the partial signal length N/M is systematically augmented. This favorable feature of the FFT is amply shared by the FPT$^{(-)}$, which likewise produces no spikes or other spectral deformations, as is clear from Figs. 7.4 and 7.8. Moreover, it is observed in Figs. 7.4 and 7.8 that even with small fractions of the full time signal, the FPT$^{(-)}$ can reconstruct the main metabolites.

For example, at $N/32 = 64$ with the FPT$^{(-)}$, the NAA peak near 2.0 ppm, as well as the creatine (Cre) peak near 3.0 ppm are clearly visible (top left panels on Figs. 7.4 and 7.8). Over 60% of the NAA concentration (which is proportional to the area under the peak) is already predicted at this very short signal length. Furthermore, on the left middle panels in Figs. 7.4 and 7.8 ($N/16 = 128$), the FPT$^{(-)}$ is seen to yield nearly 90% of the NAA concentration, as given by this most prominent peak (after water suppression).

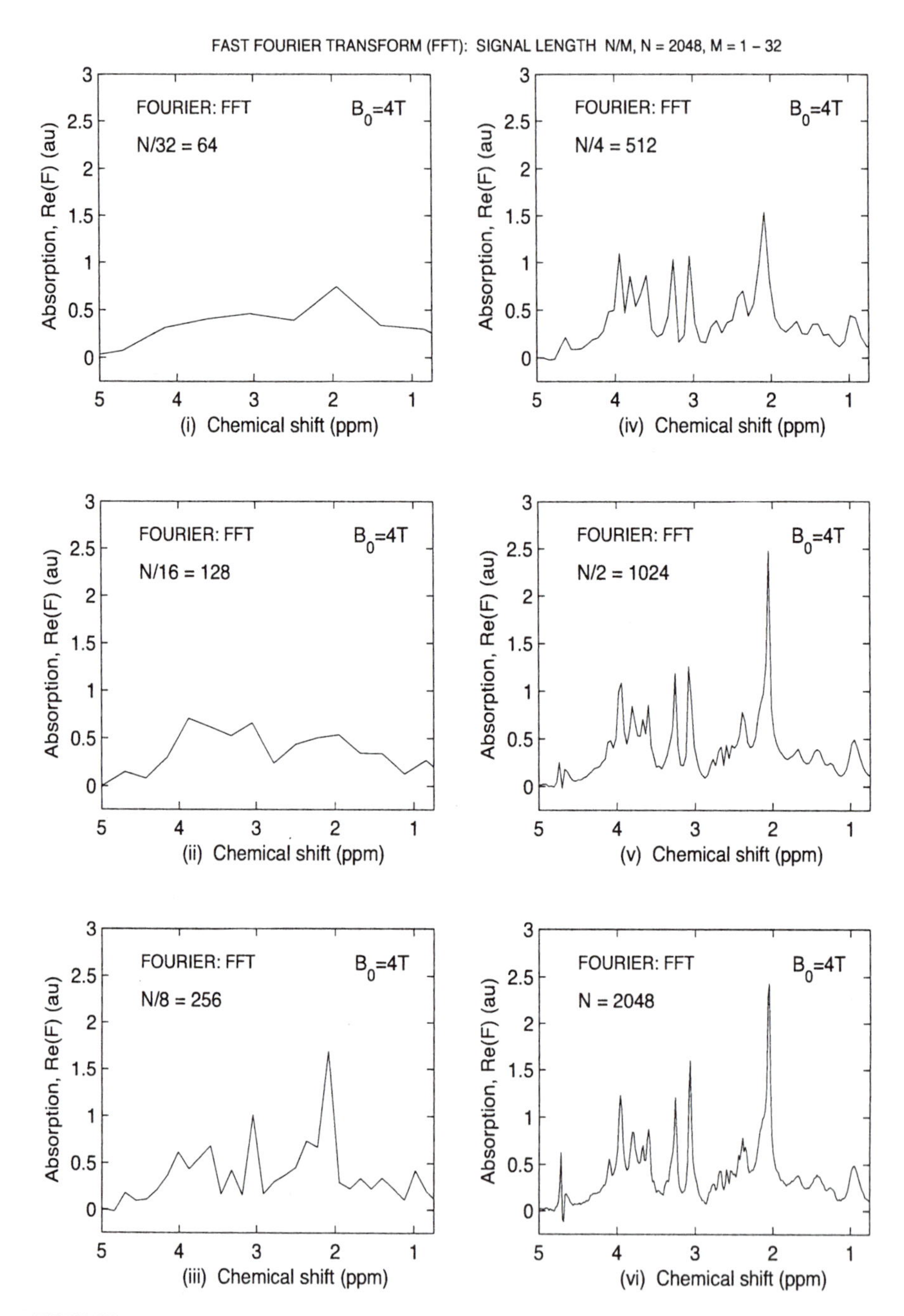

FIGURE 7.3

Fourier absorption spectra using the FID encoded at B_0=4T with 5 partial signal lengths ($N/32 = 64, N/16 = 128, N/8 = 256, N/4 = 512, N/2 = 1024$) and the full signal length $N = 2048$.

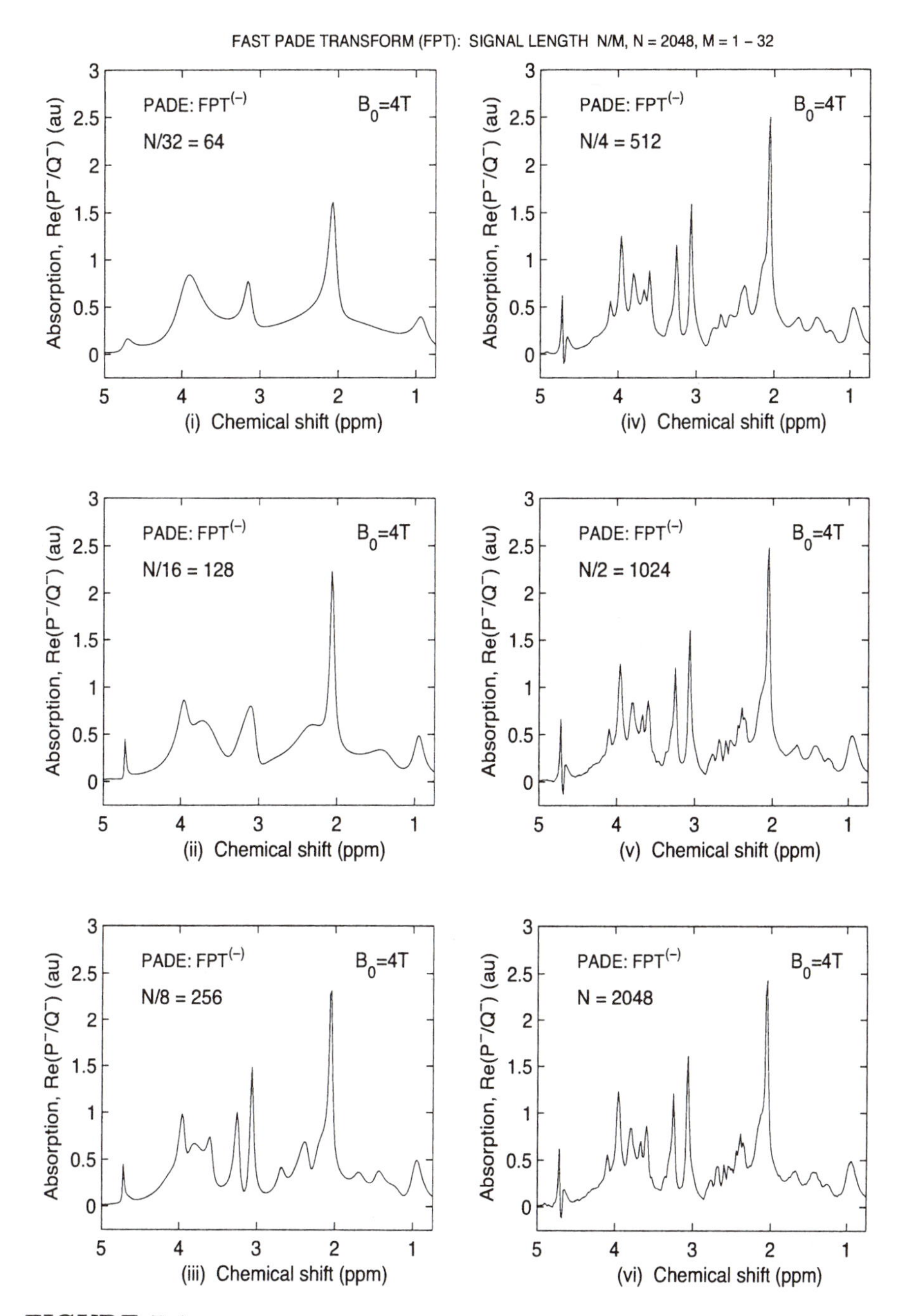

FIGURE 7.4

Padé absorption spectra using the $\mathrm{FPT}^{(-)}$ and the FID encoded at B_0=4T with 5 partial signal lengths ($N/32 = 64, N/16 = 128, N/8 = 256, N/4 = 512, N/2 = 1024$) and the full signal length $N = 2048$.

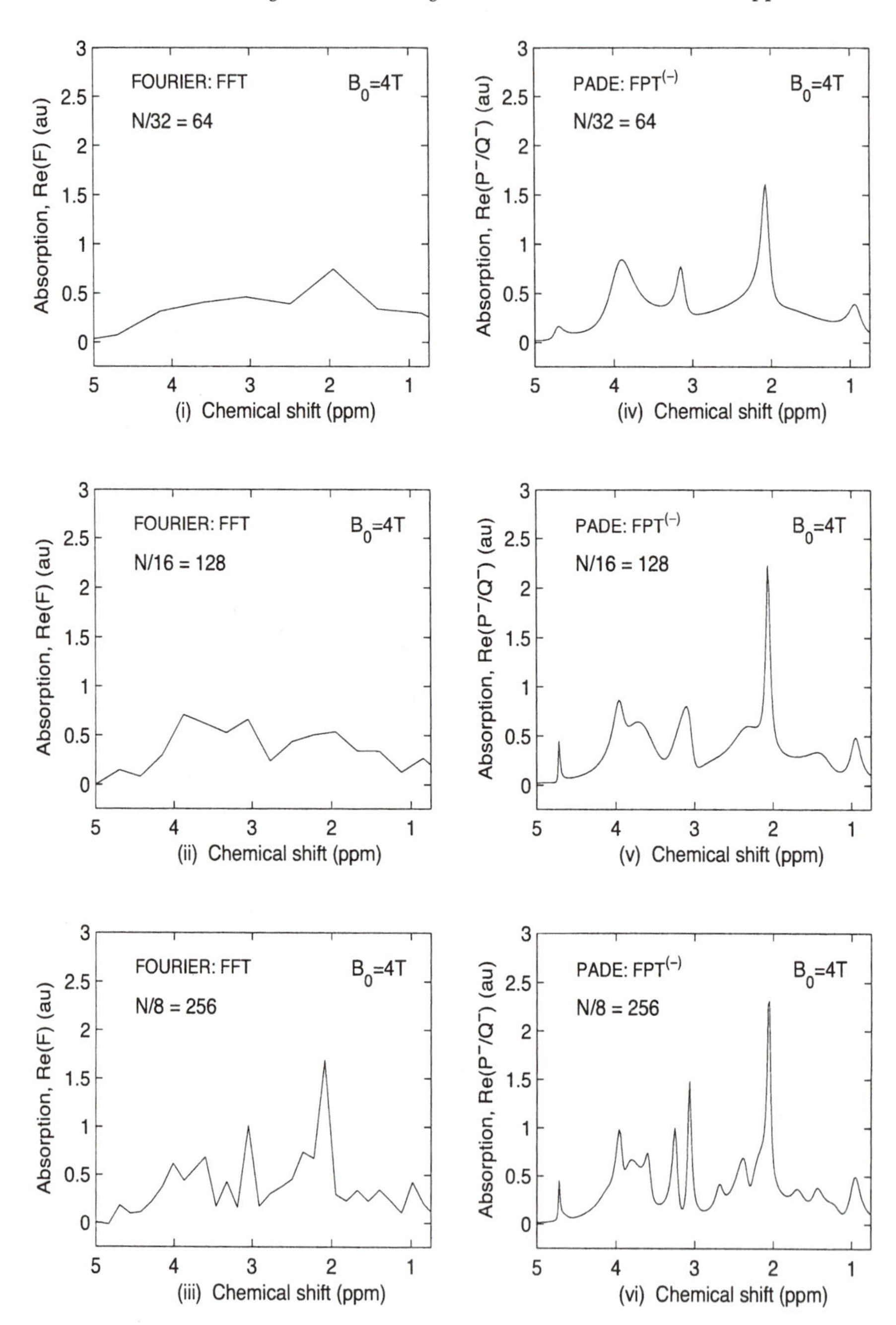

FIGURE 7.5
Fourier (left) and Padé (right) absorption spectra using the FFT and $\mathrm{FPT}^{(-)}$, respectively, from the FID encoded at B_0=4T [141], shown at 3 partial signal lengths ($N/32 = 64, N/16 = 128, N/8 = 256$).

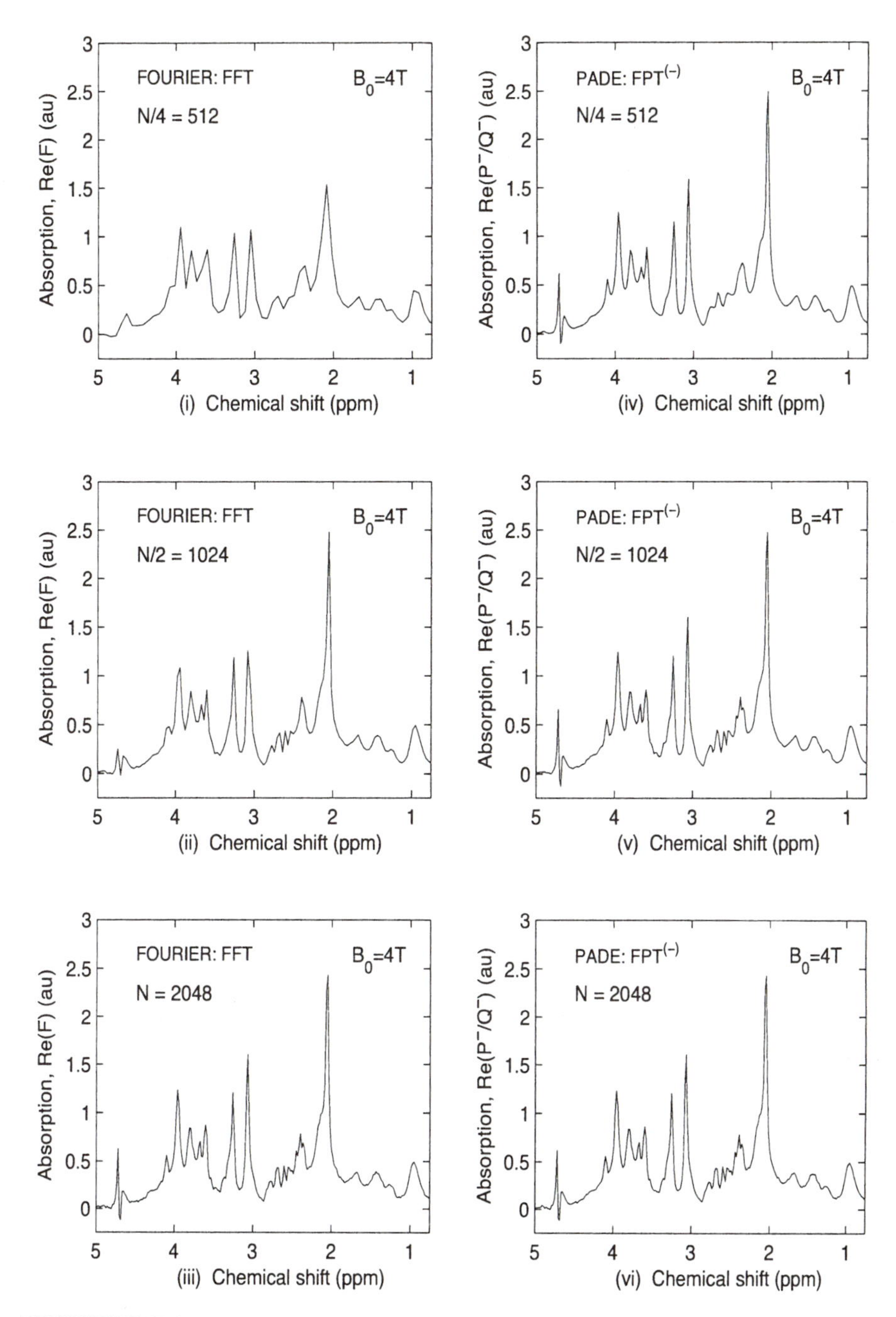

FIGURE 7.6

Fourier (left) and Padé (right) absorption spectra using the FFT and $\mathrm{FPT}^{(-)}$, respectively, from the FID encoded at B_0=4T [141], shown at 2 partial signal lengths ($N/4 = 512, N/2 = 1024$) as well as the full signal length $N = 2048$.

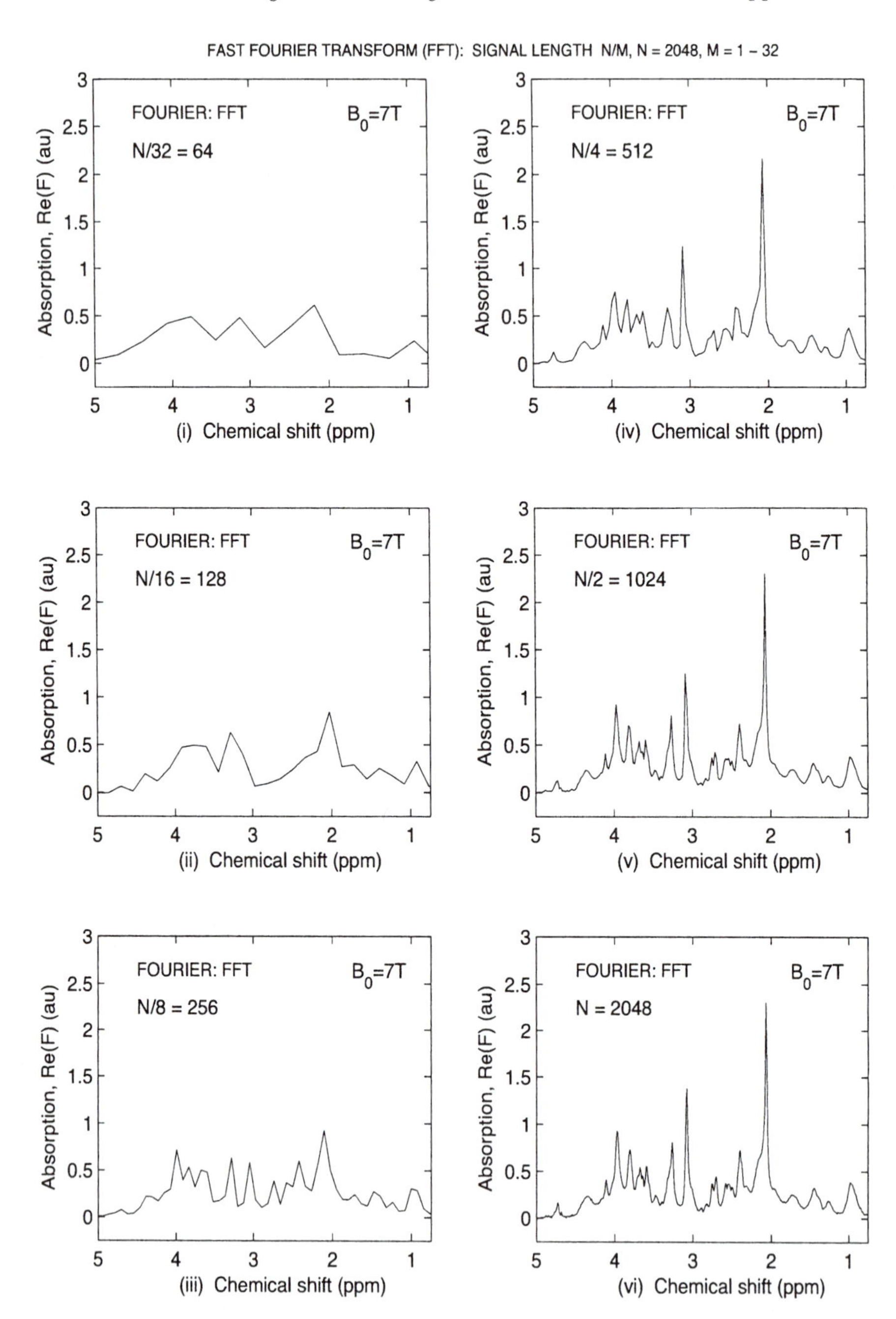

FIGURE 7.7
Fourier absorption spectra using the FFT and the FID encoded at B_0=7T [141] with 5 partial signal lengths ($N/32 = 64, N/16 = 128, N/8 = 256, N/4 = 512, N/2 = 1024$) and the full signal length $N = 2048$.

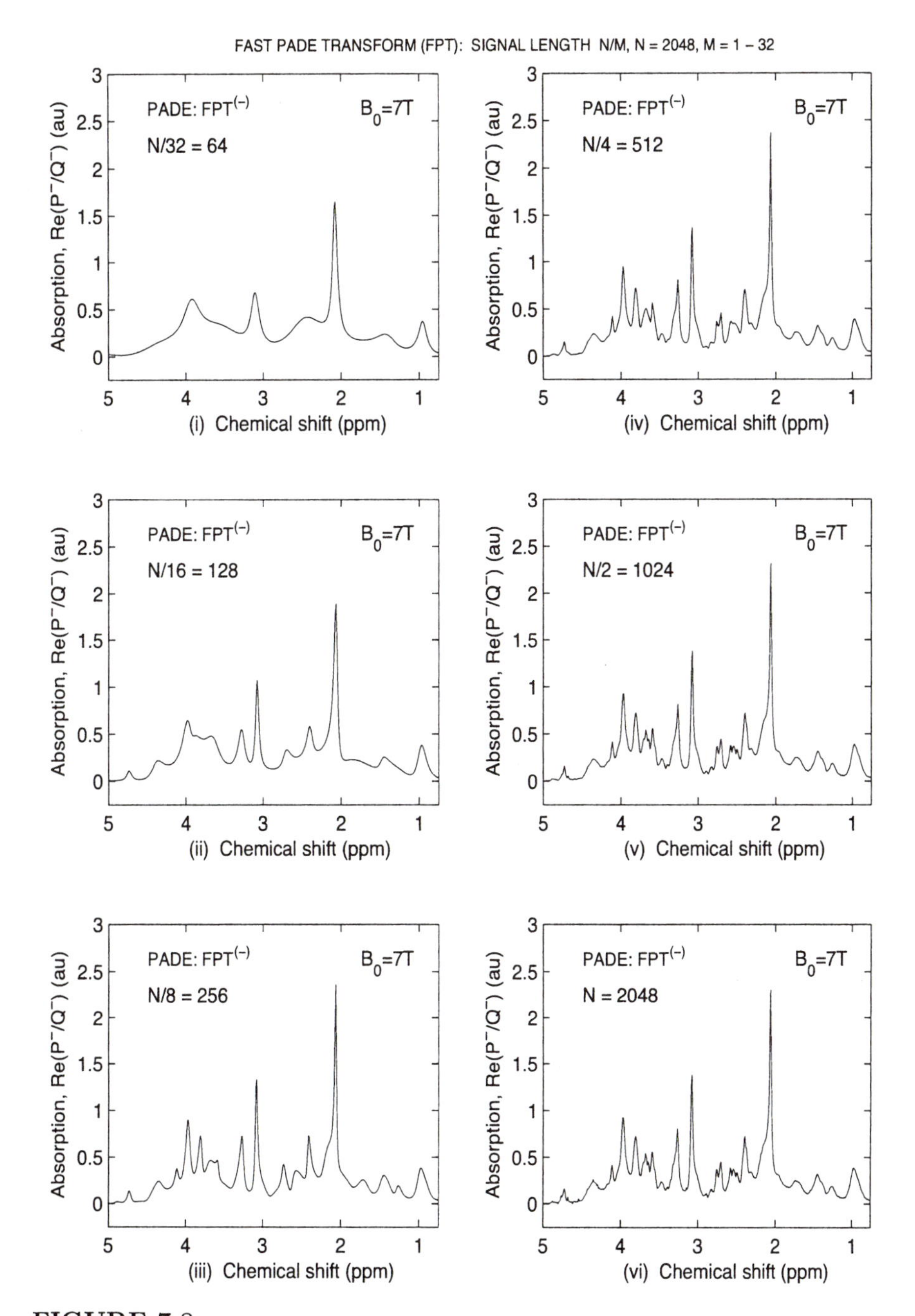

FIGURE 7.8

Padé absorption spectra using the $\text{FPT}^{(-)}$ and the FID encoded at B_0=7T [141] with 5 partial signal lengths ($N/32 = 64, N/16 = 128, N/8 = 256, N/4 = 512, N/2 = 1024$) and the full signal length $N = 2048$.

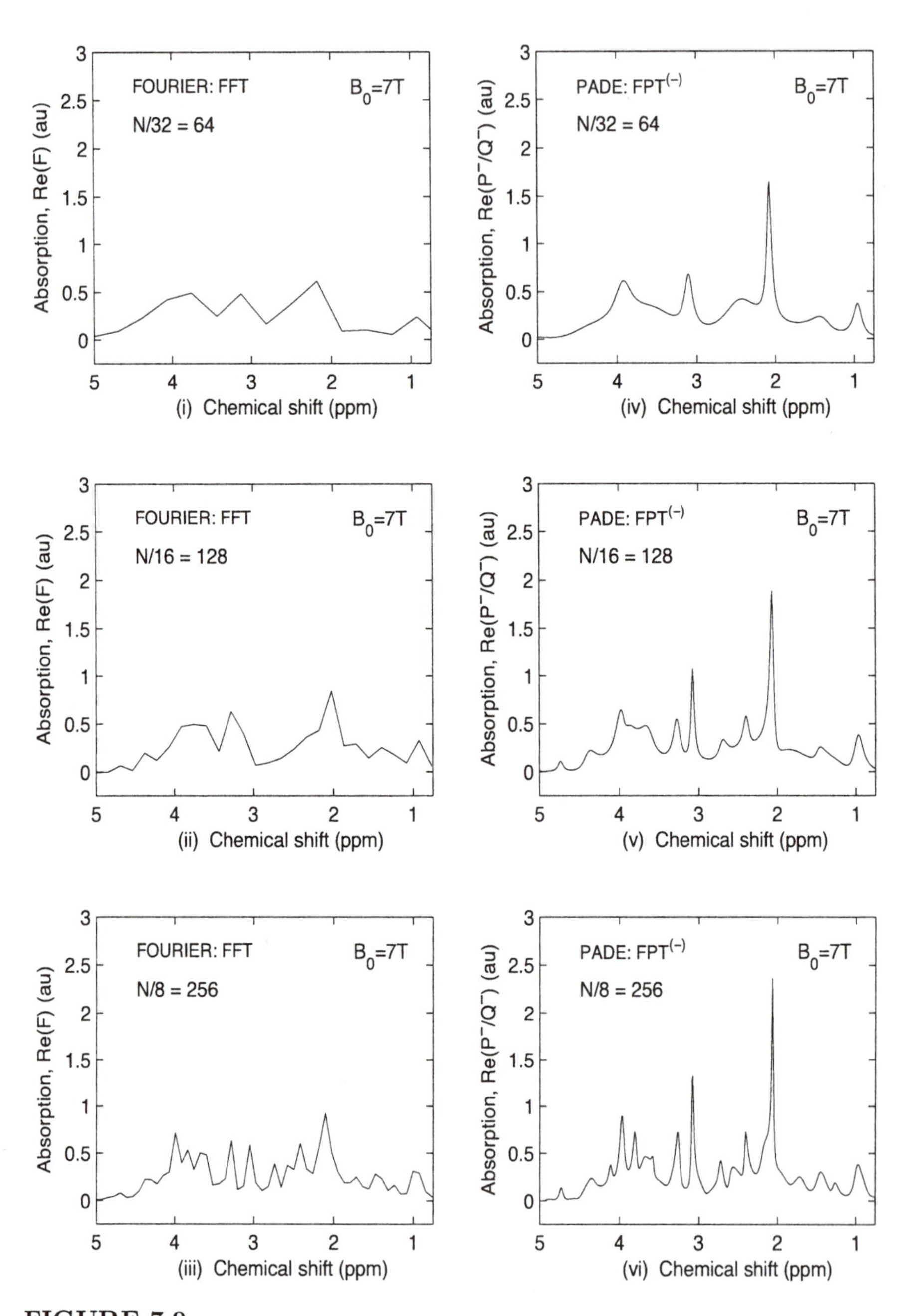

FIGURE 7.9
Fourier (left) and Padé (right) absorption spectra using the FFT and $\text{FPT}^{(-)}$, respectively, from the FID encoded at B_0=7T [141], shown at 3 partial signal lengths ($N/32 = 64, N/16 = 128, N/8 = 256$).

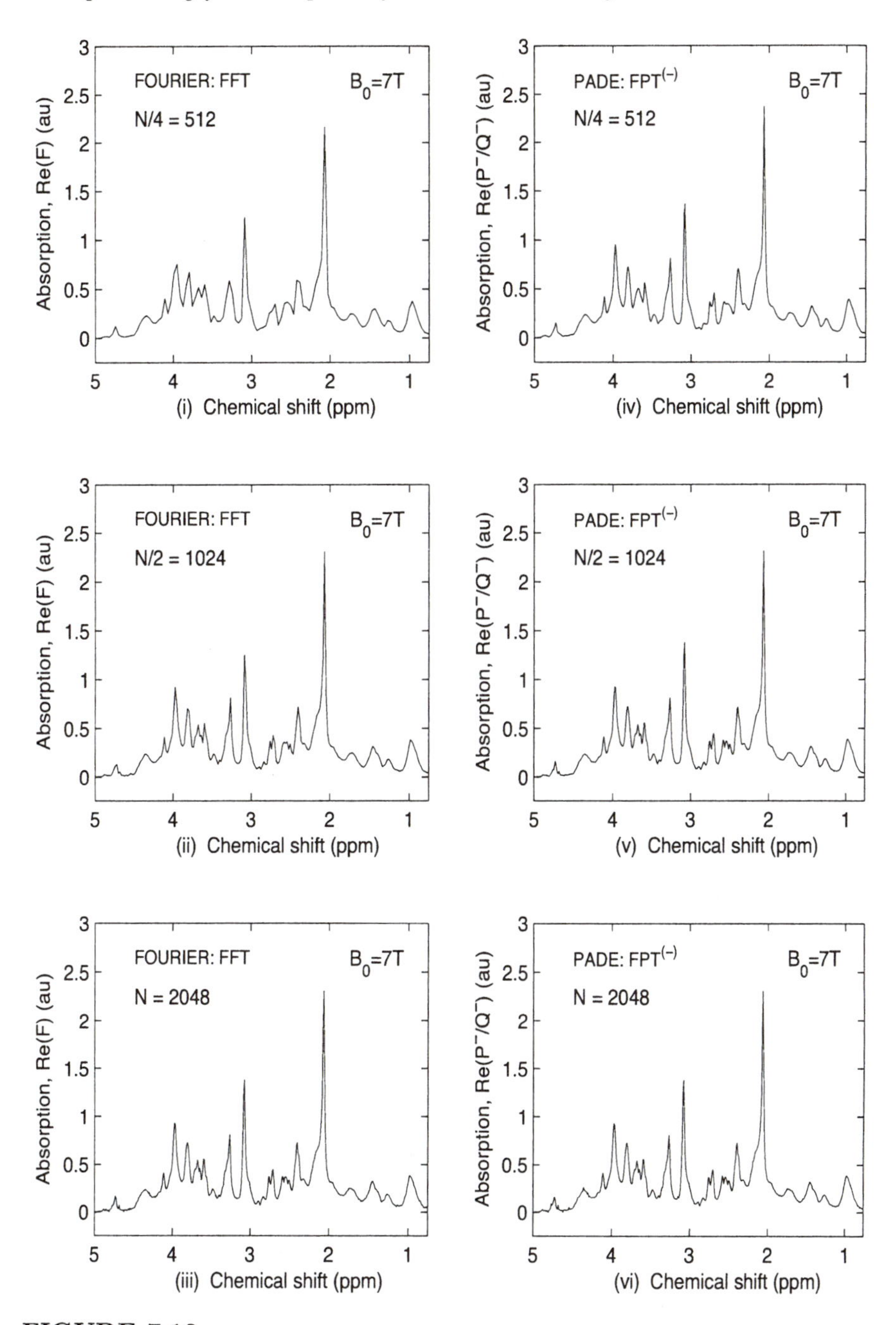

FIGURE 7.10
Fourier (left) and Padé (right) absorption spectra using the FFT and $\mathrm{FPT}^{(-)}$, respectively, from the FID encoded at B_0=7T [141], shown at 2 partial signal lengths ($N/4 = 512, N/2 = 1024$) as well as the full signal length $N = 2048$.

At $N/8 = 256$ shown on the left bottom panels of Figs. 7.4 and 7.8, the concentration of NAA predicted by the $\text{FPT}^{(-)}$ achieves nearly 100% of the correct NAA concentration. Moreover, at this one-eighth signal length, the $\text{FPT}^{(-)}$ yields nearly 90% of the true concentrations of the other two main metabolites that are creatine and choline, at approximately 3.0 ppm and 3.3 ppm, respectively.

A very important finding from Figs. 7.4 and 7.8 (left bottom panels) is that at 4T and 7T, respectively, already at $N/8 = 256$, the $\text{FPT}^{(-)}$ predicts the nearly correct peak height ratio of creatine and choline. With the FFT, the correct peak height ratio of creatine and choline at, e.g., 4T is not obtained even at $N/2 = 1024$ (middle right panel in Fig. 7.3).

The upper right panels of Figs. 7.4 and 7.8 show the Padé absorption spectrum at $N/4 = 512$. At a quarter of the FID at 4T, the $\text{FPT}^{(-)}$ is able to reconstruct practically all the major metabolites with the exception of the triplet glutamine-glutamate (near 2.4 ppm) which appears as a single, unresolved peak. On the middle right panel in Figs. 7.4 and 7.8, the spectrum computed by the $\text{FPT}^{(-)}$ is depicted at $N/2 = 1024$. Here, at one half of the FID, it is seen that the $\text{FPT}^{(-)}$ clearly resolves the remaining glutamine-glutamate triplet.

Furthermore, up to differences on the level of the background noise, the $\text{FPT}^{(-)}$ gives practically the same spectra at $N/2$ and N (middle and bottom right panels of 7.4 and 7.8). This means that the total shape spectra in the $\text{FPT}^{(-)}$ has fully converged at $N/2$, so that the second half of the examined FID might not be indispensable. By comparison with Figs. 7.3 and 7.7 (middle and bottom right panels), the FFT needs to exhaust the full FIDs ($N = 2048$) to completely resolve the glutamine-glutamate triplet near 2.4 ppm at both field strengths, 4T and 7T.

A direct comparison between the FFT and the $\text{FPT}^{(-)}$ carried out via the same graphs, as done on Figs. 7.5 and 7.6 at 4T as well as in Figs. 7.9 and 7.10 at 7T shows that the latter converges faster than the former. Moreover, it is clear from this juxtaposition that resolution is much better in the $\text{FPT}^{(-)}$ than in the FFT at any truncation $M > 1$.

Remarkably, at $N/32 = 64$, the overall shape of the spectrum in the $\text{FPT}^{(-)}$ has already emerged, whereas the corresponding FFT displays only broad bumps near some of the true resonances (the top left panels in Figs. 7.5 and 7.9). On the middle right panels in Figs. 7.5 and 7.9 ($N/16 = 128$), the $\text{FPT}^{(-)}$ is seen to be able to clearly resolve several among the major resonances such as NAA. In sharp contrast, it is seen in Fig. 7.5 that at 4T, the FFT does not even exhibit the presence of NAA at these two partial signal lengths. On the bottom right panels in Figs. 7.5 and 7.9 ($N/8 = 256$), the $\text{FPT}^{(-)}$ has further improved its prediction of the concentration of NAA at ≈ 2.02 ppm, Cre at ≈ 3.0 ppm and Cho at ≈ 3.3 ppm. On the other hand, at $N/8$ the FFT continues to greatly under-estimate the concentrations of all the metabolites, and the resulting Fourier spectrum is still seen to be very rudimentary.

In Figs. 7.6 and 7.10, we extend comparisons of the convergence patterns of the FFT and the $\mathrm{FPT}^{(-)}$ to $M = 1-4$. As stated earlier, full convergence of the $\mathrm{FPT}^{(-)}$ is obtained at half signal length ($N/2 = 1024$). This can also be observed by comparing the spectra at half and full signal length of the $\mathrm{FPT}^{(-)}$ (the middle and bottom right panels in Figs. 7.6 and 7.10).

These two spectra differ only by the background noise and, therefore, the total shape spectrum in the $\mathrm{FPT}^{(-)}$ at half signal length can be considered as fully converged. Notably, even the triplet of glutamine-glutamate near 2.5 ppm is completely resolved by the $\mathrm{FPT}^{(-)}$ at $N/2$. To achieve the same resolution for this triplet, the FFT requires the full signal length N.

Overall, due to rapid convergence, the $\mathrm{FPT}^{(-)}$ can extract vital information about short-lived metabolites that are conventionally detectable only at very short echo times. The obtained unprecedented, steady and fast convergence of the $\mathrm{FPT}^{(-)}$ sharply contrasts with most non-linear estimators that wildly oscillate before eventually stabilizing and only then perhaps converging. Moreover, the steadiness and convergence rate are markedly better in the $\mathrm{FPT}^{(-)}$ than in the FFT, as seen on Figs. 7.5, 7.6, 7.9 and 7.10.

7.4 Error Analysis for encoded *in vivo* time signals

No parametric method is of much practical value without error analysis. When the fully converged FFT does not qualify to be considered as a gold standard, as in most clinical MRS data at 1.5T, we resort to intrinsic testing within the two variants of the FPT and carry out the error analysis by computing the residual or error spectra. These are defined as the difference between the spectra in the $\mathrm{FPT}^{(+)}$ and $\mathrm{FPT}^{(-)}$ at a given signal length N. Such spectra are used to help answer the question: where does one stop in the convergence process within the fast Padé transform?

The first part of the answer is: one stops when the residual spectra as a function of the signal length become indistinguishable from the background noise, e.g., the RMS. The second part of the answer is: one stops when constancy is reached in the values of all four spectral parameters (position, height, width and phase) of each genuine, physical resonance for varying signal length.

Hence, these two components of the present strategy applied while varying the signal length, i.e., (i) indistinguishability of the residual spectra from the background noise, and (ii) stabilization of spectral parameters, constitute error analysis of proven validity [5, 20].

Here, we shall begin our error analysis by using the FFT as a gold standard for these spectra from the FIDs encoded at high magnetic field strengths.

7.4.1 Residual spectra as the difference between the fully converged Fourier and Padé spectra at various partial signal lengths

Figures 7.11 and 7.12 corresponding to 4T and 7T, respectively, show the residual absorption spectra $\mathrm{Re}(F) - \mathrm{Re}(P_K^-/Q_K^-)$ computed at the Fourier grid points with the $\mathrm{FPT}^{(-)}$ being taken at the fractions $N/32 = 64, N/16 = 128, N/8 = 256, N/4 = 512, N/2 = 1024$ as well as at the full signal length $N = 2048$, whereas the FFT is obtained at $N = 2048$ throughout.

The extremely stable and fast convergence pattern of the $\mathrm{FPT}^{(-)}$ seen in Figs. 7.4 and 7.8 is also reflected on the residual or error spectra depicted in Figs.7.11 and 7.12 at 4T and 7T, respectively. At $N/2 = 1024$ (the right middle panels in Figs. 7.11 and 7.12), the residual spectra are indistinguishable from the background noise. This demonstrates that the total shape spectra in the $\mathrm{FPT}^{(-)}$ has fully converged by exhausting at most one half of the full signal length, as discussed. In the residual spectra displayed in Figs. 7.11 and 7.12 as the difference between the two absorption spectra ("Fourier *minus* Padé"), the FFT is always taken at the full signal length ($N = 2048$). This is because the fully converged FFTs for the studied signals are of excellent SNR and, as such, can rightly be considered as a gold standard for the corresponding total shape spectra, as emphasized before. On the other hand, the Padé absorption spectra are taken at $N/M(M = 1 - 32)$ with no zero filling. Therefore, the mentioned difference "Fourier *minus* Padé" represents the error spectra that ideally should be indistinguishable from the background noise after the $\mathrm{FPT}^{(-)}$ has reached full convergence. This is indeed observed on Figs. 7.11 and 7.12 already at $N/2 = 1024$.

At this point, it is pertinent to return to Figs. 7.1 and 7.2 to recall that the second half of the time signal contains mainly noise. In such a case, it is natural to expect that only the first half of the signal should suffice to extract the entire physical information. Such an expectation is not fulfilled in the FFT as seen on the middle right panel in Figs. 7.3 and 7.7. Namely, the triplet of glutamine-glutamate around 2.4 ppm is not fully resolved by the FFT at $N/2 = 1024$, as opposed to the $\mathrm{FPT}^{(-)}$. The FFT can completely resolve this triplet only by exhausting the full signal length $N = 2048$ as seen on the bottom right panels in Figs. 7.3 and 7.7.

This illustrates the severity of the Rayleigh bound $2\pi/(N\tau)$ imposed on the resolution by the linearity feature of the Fourier transform. The FPT rescues the situation by means of its non-linearity to surpass the Rayleigh bound and, therefore, improves the Fourier resolution. Both non-linearity and extrapolation of the FPT are due to the very definition of the Padé spectrum as a ratio of two polynomials P/Q. The inverse Q^{-1} of the denominator polynomial Q is a series, i.e., an infinite expansion. Given a finite Riemann sum $\sum_{n=0}^{N-1} c_n z^{-n}$, the FPT obtains the unique polynomial quotient P/Q, which by means of an implicit series in Q^{-1}, effectively extrapolates the input time signal $\{c_n\}(0 \leq n \leq N-1)$ beyond the original acquisition time $T = N\tau$.

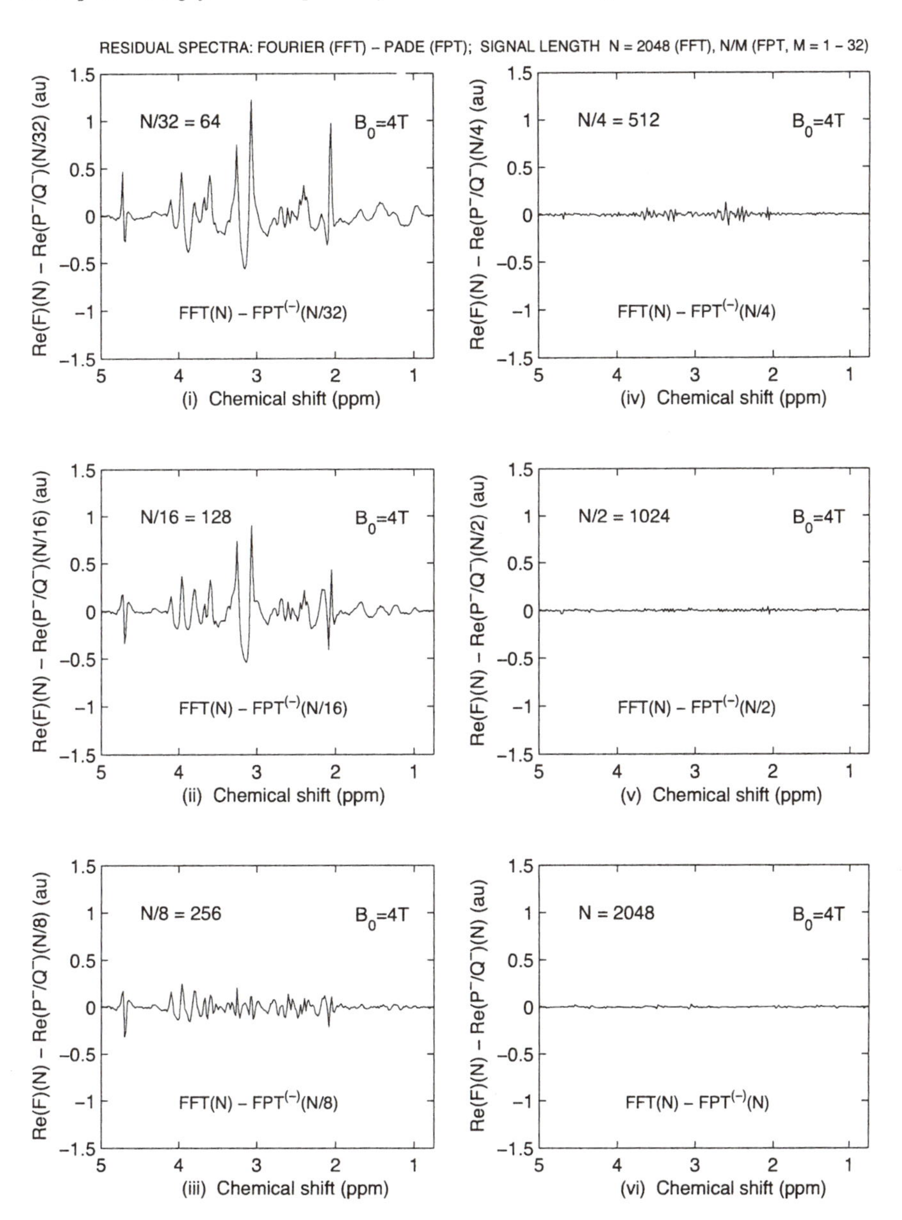

FIGURE 7.11
The residual absorption spectra $\mathrm{Re}(F) - \mathrm{Re}(P_K^-/Q_K^-)$ for the partial FID of human brain metabolites as computed by applying the FFT and $\mathrm{FPT}^{(-)}$ to the complex-valued time signal $\{c_n\}$ which has been encoded in Ref. [141] at 4T. The FFT is evaluated only with the full signal length $N = 2048$, whereas the $\mathrm{FPT}^{(-)}$ is computed at the partial signal lengths $N/M (M > 1)$ with no zero filling $(N/32 = 64, N/16 = 128, N/8 = 256, N/4 = 512, N/2 = 1024)$ and the full signal length $N = 2048$.

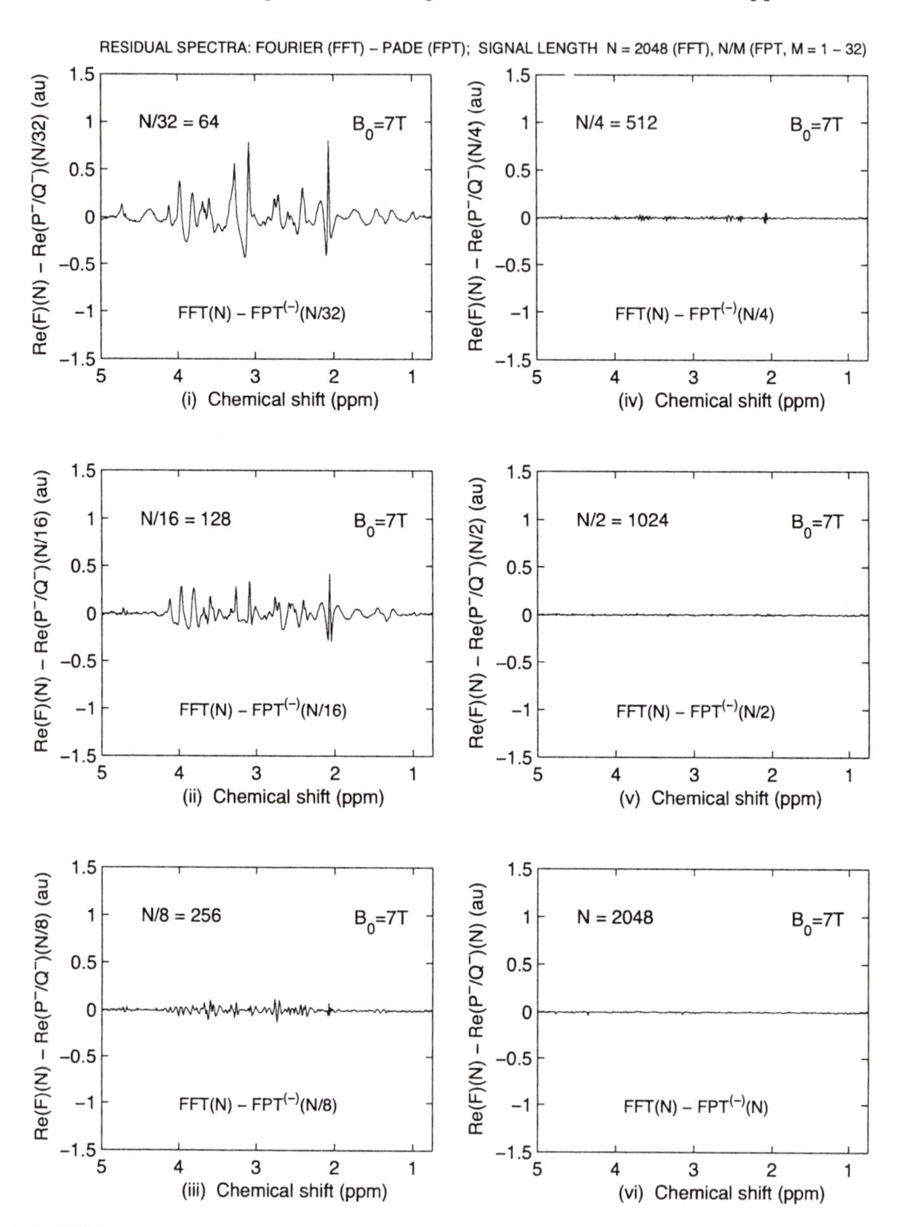

FIGURE 7.12
The residual absorption spectra $\mathrm{Re}(F) - \mathrm{Re}(P_K^-/Q_K^-)$ for the partial FID of human brain metabolites as computed by applying the FFT and $\mathrm{FPT}^{(-)}$ to the complex-valued time signal $\{c_n\}$ which has been encoded at 7T in Ref. [141]. The FFT is evaluated only with the full signal length $N = 2048$, whereas the $\mathrm{FPT}^{(-)}$ is computed at the partial signal lengths $N/M(M > 1)$ with no zero filling ($N/32 = 64, N/16 = 128, N/8 = 256, N/4 = 512, N/2 = 1024$) and the full signal length $N = 2048$.

7.4.2 Self-contained Padé error analysis: Consecutive difference spectra

Now we will present consecutive difference spectra within the $\mathrm{FPT}^{(-)}$ in order to illustrate that self-contained Padé error analysis can be achieved. In Figs. 7.13 and 7.14 at 4T and 7T, respectively, six consecutive difference spectra are displayed. The left upper panels begin with $\mathrm{Re}(P_K^-/Q_K^-)[N/32] - \mathrm{Re}(P_K^-/Q_K^-)[N/64]$. Next follows $\mathrm{Re}(P_K^-/Q_K^-)[N/16] - \mathrm{Re}(P_K^-/Q_K^-)[N/32]$ on the middle left panels and then $\mathrm{Re}(P_K^-/Q_K^-)[N/8] - \mathrm{Re}(P_K^-/Q_K^-)[N/16]$ on the bottom left panels. The right upper panels show the subsequent difference spectra: $\mathrm{Re}(P_K^-/Q_K^-)[N/4] - \mathrm{Re}(P_K^-/Q_K^-)[N/8]$, then on the right middle panels: $\mathrm{Re}(P_K^-/Q_K^-)[N/2] - \mathrm{Re}(P_K^-/Q_K^-)[N/4]$ and then $\mathrm{Re}(P_K^-/Q_K^-)[N] - \mathrm{Re}(P_K^-/Q_K^-)[N/2]$ on the right lower panels.

Finally, we present Figs. 7.15 and 7.16 where the left and the right columns convey two different types of information. Here, the left columns correspond to the right columns in Figs. 7.13 and 7.14, respectively, displaying three consecutive difference spectra $\mathrm{Re}(P_K^-/Q_K^-)[N/4] - \mathrm{Re}(P_K^-/Q_K^-)[N/8]$, $\mathrm{Re}(P_K^-/Q_K^-)[N/2] - \mathrm{Re}(P_K^-/Q_K^-)[N/4]$, $\mathrm{Re}(P_K^-/Q_K^-)[N] - \mathrm{Re}(P_K^-/Q_K^-)[N/2]$, where $N/4 = 512$, $N/2 = 1024$ and $N = 2048$. These three sub-plots of intrinsic error spectra demonstrate a steady decrease of the local error in the $\mathrm{FPT}^{(-)}$ when the number of signal points are systematically augmented. Of course, error spectra that concern the $\mathrm{FPT}^{(-)}$ can also be generated by reference to other estimators, e.g., the FFT as done in Figs. 7.11 and 7.12. Likewise, the error spectra can be constructed by using the two variants of the FPT, i.e., the $\mathrm{FPT}^{(+)}$ and $\mathrm{FPT}^{(-)}$, as done on the right columns of Figs. 7.15 and 7.16 showing the three residuals that are all related to the whole FIDs ($N = 2048$) : $\mathrm{Re}F[N] - \mathrm{Re}(P_K^+/Q_K^+)[N]$ (top panels), $\mathrm{Re}F[N] - \mathrm{Re}(P_K^-/Q_K^-)[N]$ (middle panels) and $\mathrm{Re}(P_K^+/Q_K^+)[N] - \mathrm{Re}(P_K^-/Q_K^-)[N]$ (bottom panels). All the three latter error spectra are seen to be indistinguishable from the background noise. Importantly, the entirely negligible values of the whole residual spectrum $\mathrm{Re}(P_K^+/Q_K^+)[N] - \mathrm{Re}(P_K^-/Q_K^-)[N]$ are seen at any of the considered frequencies on the right bottom panels in Figs. 7.15 and 7.16. It is thereby shown that the differences steadily diminish, and at $\mathrm{Re}(P_K^+/Q_K^+)[N] - \mathrm{Re}(P_K^-/Q_K^-)[N]$, become essentially indistinguishable from background noise levels.

This is a very important internal cross-validation within the FPT itself. It shows that the two variants, the $\mathrm{FPT}^{(+)}$ and $\mathrm{FPT}^{(-)}$, are sufficient for establishing consistency of this processor without necessitating external checking against, e.g., the FFT or other estimators. In non-parametric processing, the $\mathrm{FPT}^{(+)}$ and $\mathrm{FPT}^{(-)}$ yield, respectively, the upper and lower bounds to envelopes of the shape spectra separated from each other by residual values of the order of the background noise, as evidenced in Figs. 7.15 and 7.16. Likewise, in parametric processing via the fast Padé transform, the $\mathrm{FPT}^{(+)}$ and $\mathrm{FPT}^{(-)}$ provide respectively the upper and lower variational bounds of the estimated complex frequencies $\{\omega_k^\pm\}$ and complex amplitudes $\{d_k^\pm\}$. It should be recalled that the FPT is a variational method [5].

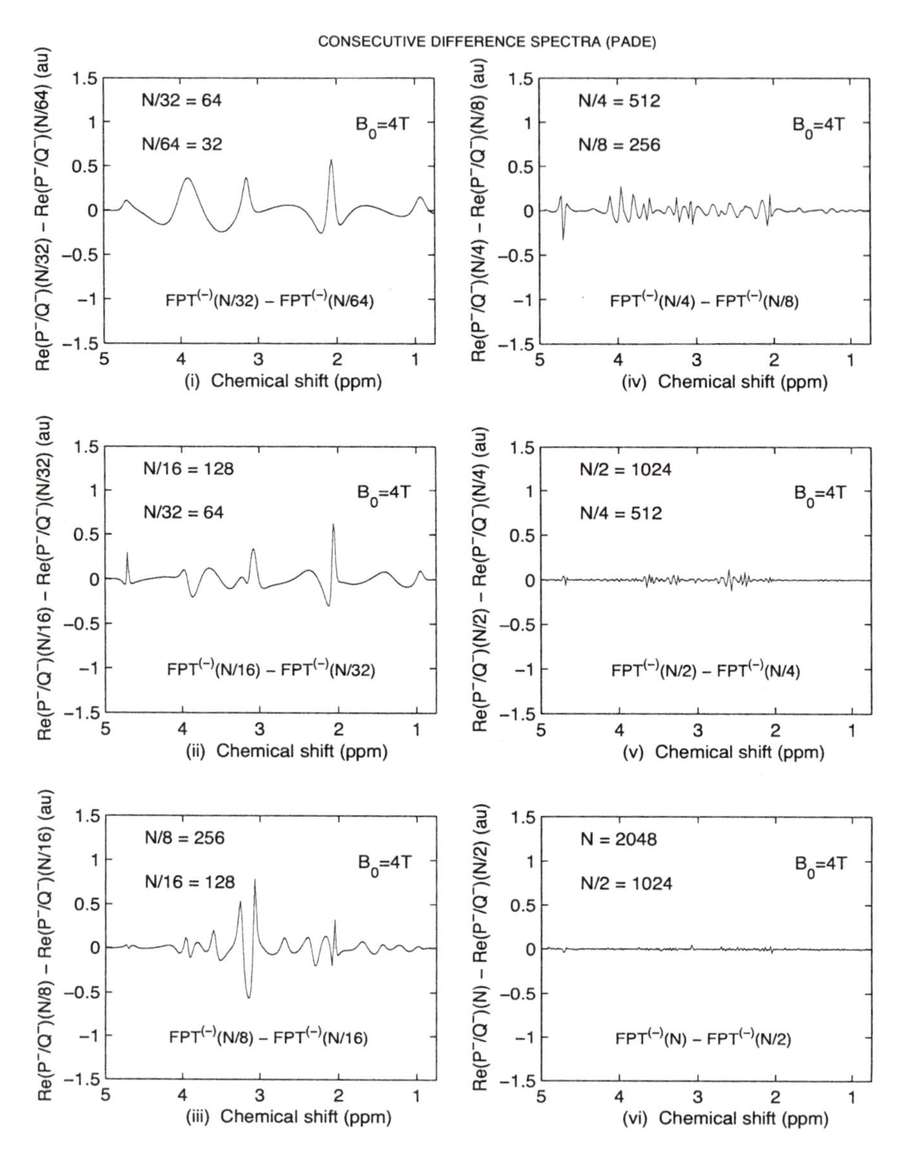

FIGURE 7.13

The six consecutive absorption difference spectra at 4T from the $\text{FPT}^{(-)}$ computed via $\text{Re}(P^-/Q^-)[N/32] - \text{Re}(P^-/Q^-)[N/64]$ (top left panel), $\text{Re}(P^-/Q^-)[N/16] - \text{Re}(P^-/Q^-)[N/32]$ (middle left panel), $\text{Re}(P^-/Q^-)[N/8] - \text{Re}(P^-/Q^-)[N/16]$ (bottom left panel), $\text{Re}(P^-/Q^-)[N/4] - \text{Re}(P^-/Q^-)[N/8]$ (top right panel), $\text{Re}(P^-/Q^-)[N/2] - \text{Re}(P^-/Q^-)[N/4]$ (middle right panel) and $\text{Re}(P^-/Q^-)[N] - \text{Re}(P^-/Q^-)[N/2]$ (bottom right panel), where N is the full length ($N = 2048$) of the FID which has been encoded in Ref. [141].

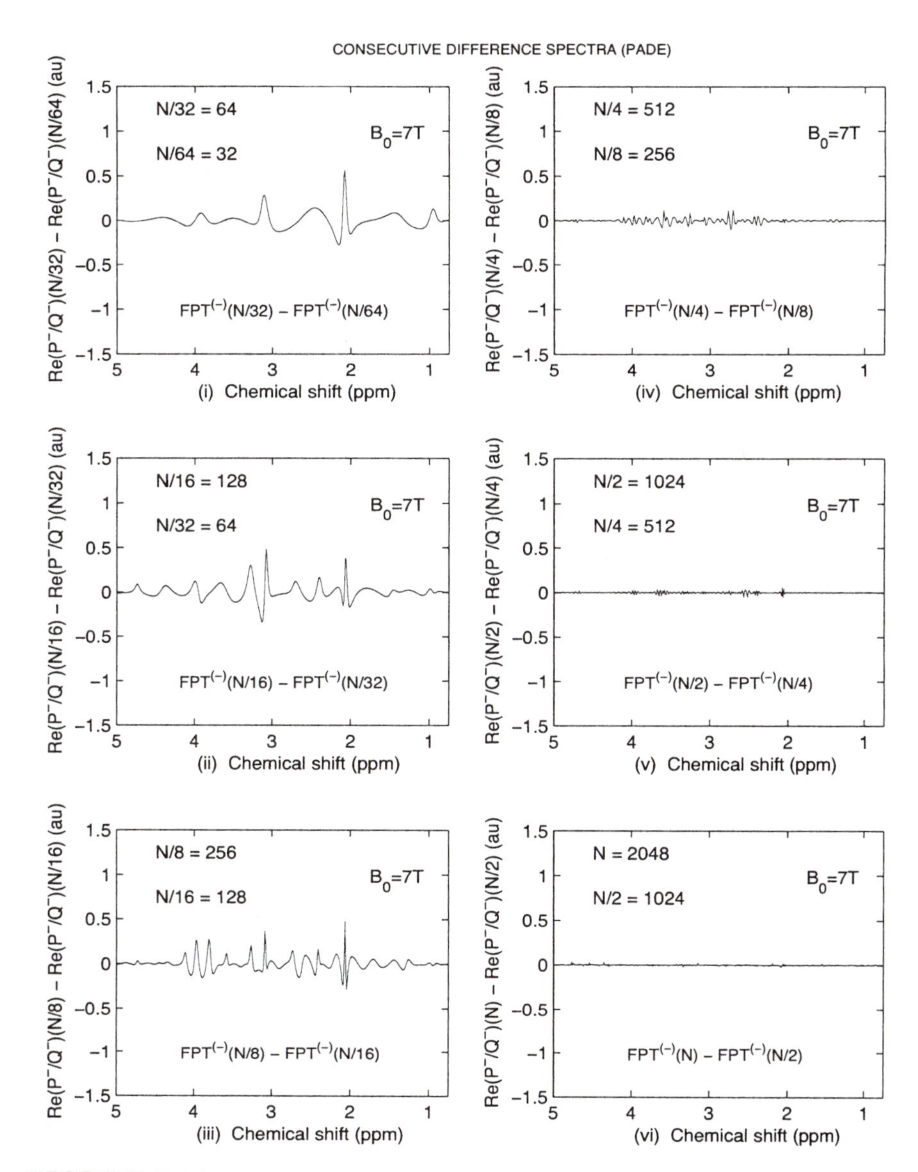

FIGURE 7.14

The six consecutive absorption difference spectra at 7T from the $\mathrm{FPT}^{(-)}$ computed via $\mathrm{Re}(P^-/Q^-)[N/32]$ – $\mathrm{Re}(P^-/Q^-)[N/64]$ (top left panel), $\mathrm{Re}(P^-/Q^-)[N/16]$ – $\mathrm{Re}(P^-/Q^-)[N/32]$ (middle left panel), $\mathrm{Re}(P^-/Q^-)[N/8]$ – $\mathrm{Re}(P^-/Q^-)[N/16]$ (bottom left panel), $\mathrm{Re}(P^-/Q^-)[N/4]$ – $\mathrm{Re}(P^-/Q^-)[N/8]$ (top right panel), $\mathrm{Re}(P^-/Q^-)[N/2]$ – $\mathrm{Re}(P^-/Q^-)[N/4]$ (middle right panel) and $\mathrm{Re}(P^-/Q^-)[N]$ – $\mathrm{Re}(P^-/Q^-)[N/2]$ (bottom right panel), where N is the full length ($N = 2048$) of the FID which has been encoded in Ref. [141].

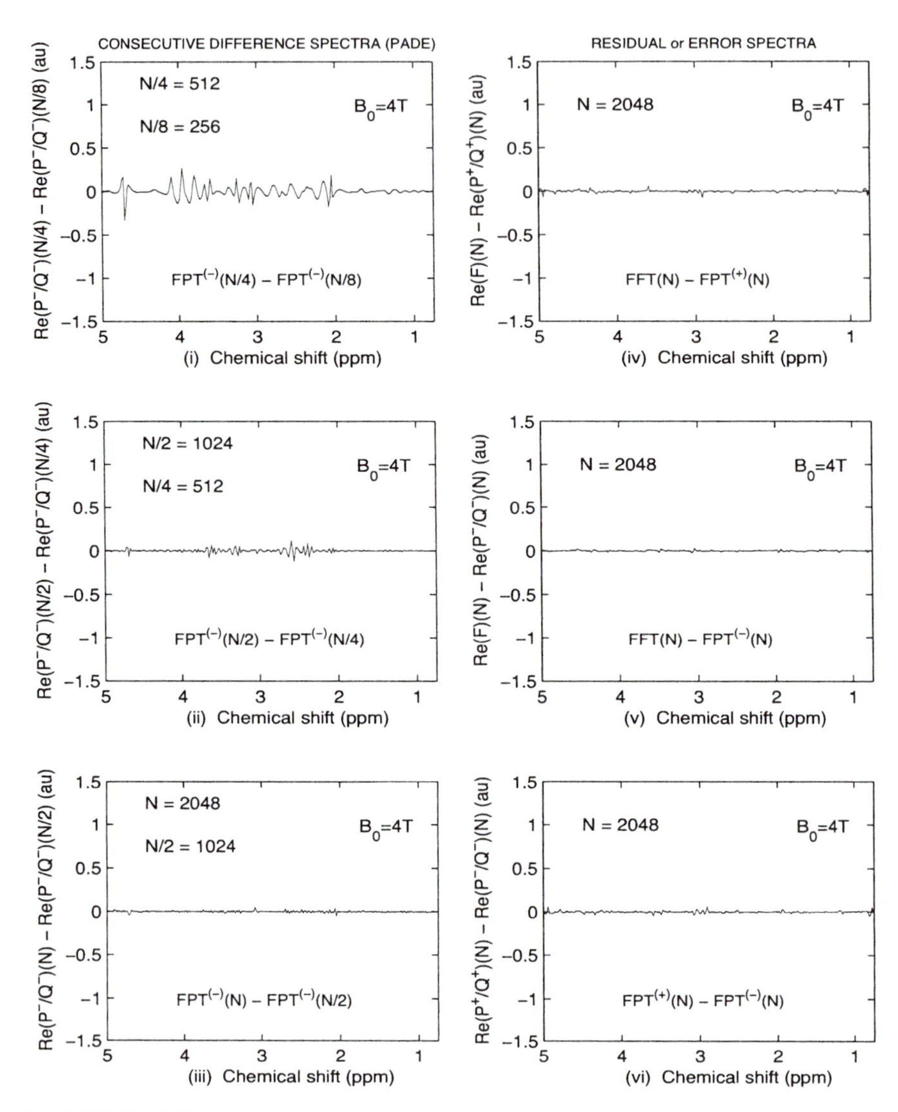

FIGURE 7.15

The three consecutive absorption difference spectra at 4T from the $\mathrm{FPT}^{(-)}$ computed via $\mathrm{Re}(P^-/Q^-)[N/4] - \mathrm{Re}(P^-/Q^-)[N/8]$ (top left panel), $\mathrm{Re}(P^-/Q^-)[N/2] - \mathrm{Re}(P^-/Q^-)[N/4]$ (middle left panel) and $\mathrm{Re}(P^-/Q^-)[N] - \mathrm{Re}(P^-/Q^-)[N/2]$ (bottom left panel), where N is the full length ($N = 2048$) of the time signal which has been encoded in Ref. [141]. Also shown here are the three residual absorption spectra computed at the full signal length $N = 2048$ for $\mathrm{Re}(F) - \mathrm{Re}(P_K^+/Q_K^+)$ (top right panel), $\mathrm{Re}(F) - \mathrm{Re}(P_K^-/Q_K^-)$ (middle right panel) and $\mathrm{Re}(P^+/Q^+) - \mathrm{Re}(P^-/Q^-)$ (bottom right panel).

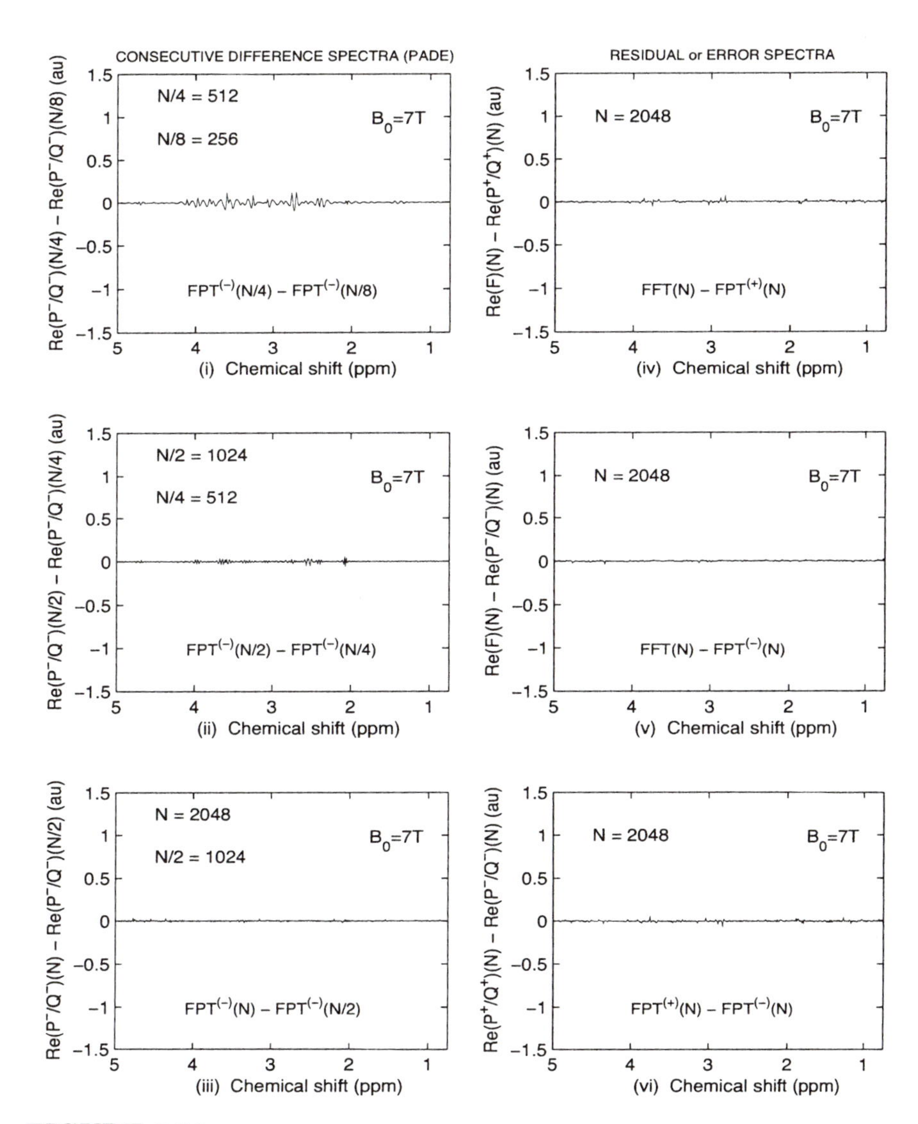

FIGURE 7.16

The three consecutive absorption difference spectra at 7T from the $\text{FPT}^{(-)}$ computed via $\text{Re}(P^-/Q^-)[N/4] - \text{Re}(P^-/Q^-)[N/8]$ (top left panel), $\text{Re}(P^-/Q^-)[N/2] - \text{Re}(P^-/Q^-)[N/4]$ (middle left panel) and $\text{Re}(P^-/Q^-)[N] - \text{Re}(P^-/Q^-)[N/2]$ (bottom left panel), where N is the full length ($N = 2048$) of the time signal which has been encoded in Ref. [141]. Also shown here are the three residual absorption spectra computed at the full signal length $N = 2048$ for $\text{Re}(F) - \text{Re}(P_K^+/Q_K^+)$ (top right panel), $\text{Re}(F) - \text{Re}(P_K^-/Q_K^-)$ (middle right panel) and $\text{Re}(P^+/Q^+) - \text{Re}(P^-/Q^-)$ (bottom right panel).

7.5 Prospects for comprehensive applications of the fast Padé transform to *in vivo* MR time signals encoded from the human brain

In this chapter, we assessed the utility of the FPT for estimations of time signals encoded by means of *in vivo* MRS. Convergence performance of the FPT is demonstrated using two raw time signals encoded via MRS at the magnetic field strengths of 4T and 7T from the brain of a healthy volunteer. All the computed spectra are presented and analyzed in the absorption mode. The employed signals of a sufficiently long length ($N = 2048$) with the associated bandwidth of 6001.5 Hz are of excellent SNR, so that the shape spectrum from the FFT, using all N points, can give a reference spectrum.

The convergence rate of the FPT was monitored by gradually including more and more signal points in the computations. It was observed that the results of the FPT for one half and the full signal length were indistinguishable from the background noise. Moreover, the convergence pattern of the spectra predicted by the FPT was strikingly stable without spikes or any other spectral deformations. Further, the FPT yielded excellent estimates of the main metabolites even for a quarter of the full signal length. This points to the fast and accurate convergence of the FPT as a function of the signal length. The residual or error spectra between the FPT (with either $N/2$ or N signal points) and the FFT which utilizes the full FID are totally embedded in the background noise.

The fidelity of the FPT was also cross-validated internally using the two equivalent variants, the $\text{FPT}^{(+)}$ and $\text{FPT}^{(-)}$. Due to the uniqueness of the FPT, all the physical resonances from the $\text{FPT}^{(+)}$ and $\text{FPT}^{(-)}$ must coincide after convergence has been reached in both variants. Moreover, by employing the full signal length $N = 2048$, the residual/error spectra for the difference between the results from the $\text{FPT}^{(+)}$ and $\text{FPT}^{(-)}$ were found to be entirely indistinguishable from background noise. Further, we computed the consecutive difference spectra as the difference between two consecutive values of the partial signal lengths and found a remarkably fast convergence of the $\text{FPT}^{(-)}$.

Overall, the FPT emerges as a powerful, stable processor with robust and self-contained error analysis for MR *in vivo* time signals. On top of providing the shape spectra in any desired mode (absorption, dispersion, magnitude, power), it should also be recalled that the FPT can simultaneously perform quantification without fitting the FFT spectra or any other spectra. This is clinically important for the needed accurate quantitative assessment of *in vivo* time signals encoded via MRS from brain. A number of reasons are highlighted for which the most recent advances in signal processing via the paradigm shift initiated by the fast Padé transform could be of critical value in early tumor diagnostics.

8

Magnetic resonance in neuro-oncology: Achievements and challenges

Recently, much attention has been paid in leading investigative clinical journals to *in vivo* MRS and MRSI as a potentially key non-invasive modality for cancer diagnostics [101]–[113]. Anatomic localization and metabolic information are frequently of decisive importance for identifying cancer. Metabolic changes often precede anatomic or morphologic alterations. Along with PET and SPECT, it is well-known that MRS and MRSI have played a critical role especially in difficult cases such as differentiating recurrent tumor from radiation necrosis or post-operative changes [7, 112, 206].

8.1 MRS and MRSI as a key non-invasive diagnostic modality for neuro-oncology

The advantages of MR-based diagnostics have become particularly clear for neuro-oncology. There has been a literal "explosion" of information in recent years on MRS and MRSI for brain tumor diagnostics [101, 108],[113]–[115],[207]. It is here that the most delicate of clinical decisions are made, and these require maximal information of the highest possible reliability. In no other area of oncology have MRS and MRSI become so widely incorporated into clinical practice. They now represent a key modality for nearly all aspects of brain tumor diagnostics.

8.1.1 MRI for brain tumor diagnostics

Magnetic resonance imaging offers a number of important advantages in neurodiagnostics. Its multi-planar capabilities improve identification of the origin of the tumor and extension to adjacent structures. With T_2 weighted MRI alone, brain tumors usually appear hyper-intense [208]. Rapidly growing tumors frequently are accompanied by edema, which also appears hyper-lucent with T_2 weighting [209]. However, T_2 hyper-intensity is not unique to brain tumors. A T_2 weighted signal increase is also seen in many other abnormalities of the central nervous system, including ischemic or hyper-acute as well as sub-acute hemorrhagic stroke, inflammation, infection, and a number degenerative brain disorders [210, 211]. The MRI patterns of bilateral thalamic astrocytomas seen in children are very similar to those of neuro-metabolic disorders and of encephalitis [212]. Not all brain tumors appear T_2 hyper-intense. Meningiomas, for example, are often iso-intense to the surrounding brain tissue, when viewed on MRI without contrast [208]. Small brain metastases may not appear at all on T_2 weighted imaging [210]. Moreover, the peri-tumoral region often appears unchanged on MRI, but represents an "uncertain" zone where tumor may be present in patients with cerebral glioma [213]. Gadolinium contrast is used in MRI work-ups of suspected brain tumors, with the majority of brain tumors showing contrast enhancement [208]. However, since disruption of the blood brain barrier is not unique to brain tumors, contrast enhancement on MRI is not pathognomonic for malignancy. Cerebral infarcts, brain abscesses, multiple sclerosis (acute plaques) also show contrast enhancement with gadolinium [210]. While the pattern of contrast enhancement is often useful in identifying tumors, this is not always the case. For example, ring-like enhancement, which is considered characteristic of brain abscesses, can also be seen with cystic brain tumors [214].

It should also be emphasized that malignant lesions from areas that are T_2 hyper-intense, but outside the contrast-enhancing region have been frequently reported [215]. Although the contrast enhancing volume is usually considered to define the region of tumor, Vigneron *et al.* [216], note that the contrast-enhancing region does not always reflect the actual extent of the tumor. The authors from Ref. [216] point out that this can be due to tumor which does not show contrast enhancement or to necrosis which does take up contrast. Thus, a lesion may appear larger on the T_2 weighted imaging because it encompasses not only the tumor, but also "non-specific effects such as edema and inflammation. These factors produce considerable uncertainty concerning the reliability of MRI techniques for defining the treatment target and evaluating therapeutic success" (p. 89).

Another way of attaining contrast is via diffusion-weighted imaging (DWI), based upon the molecular motion of water. This technique has been helpful in providing earlier detection of pathology, including brain tumors. Diffusion-weighted imaging can also help distinguish malignancy from secondary effects of treatment on the tumor. However, the findings on DWI are not pathognomonic for neoplasia [209, 217, 218]. The role of MRI versus stereotactic biopsy in brain tumor diagnostics has been described succinctly by Howe and Opstad [207]. They emphasize that accurate diagnosis is vital for optimum clinical management of patients with intra-cranial tumors. Accessible tumors are generally surgically resected, but there is "a balance between removing as much tumor tissue as possible, whilst maintaining vital brain functions" (p.123). Thus, radiation therapy (RT) is frequently also used to treat residual tumor. Magnetic resonance imaging is widely applied to determine tumor extent for surgical and RT planning, as well as for post-therapy monitoring of tumor recurrence or progression to higher grade. The initial diagnosis of an intra-cranial mass lesion is accurately made in 30-90% of cases, depending on tumor type. However, a biopsy is still considered the gold standard. Furthermore, Howe and Opstad [207] point out that brain biopsy has a 1.7% mortality rate, and note that in a study of 550 patients undergoing stereotactic biopsy: 8% had abscesses or inflammatory processes, 2.2% had other lesions, 3.4% of the biopsies were non-diagnostic, and 8% of the patients had complications. Thus, "a non-invasive and accurate prediction of lesion type would reduce unnecessary surgical biopsy procedures for non-cancerous lesions and for less accessible tumors that would be treated by radio- or chemotherapy rather than surgical resection" (p.123).

8.1.2 Primary diagnosis of brain tumors by MRS & MRSI

The combination of anatomic plus molecular imaging via MRI, MRS and MRSI is particularly promising for primary diagnosis of brain tumors. Brandão and Domingues [211] note that MRI yields excellent spatial and contrast resolution to assess structure. However, the need to assess more than the purely anatomic aspects, such as biochemistry and tissue physiology, required the development of functional techniques. "As a non-invasive method providing metabolic information about the brain, MRS enables tissue characterization at a biochemical level surpassing that of conventional magnetic resonance imaging. MRS is also able to detect abnormalities that are invisible to MRI, because metabolic abnormalities often precede structural changes" (p. 2).

8.1.2.1 Total choline

The hallmark for identifying neoplastic lesions using proton MRS has been increased levels of choline; these are associated with high cell membrane turnover and high cell density, as occurs with proliferation of brain tumor cells. Sijens and Oudkerk [219] report a significantly higher mean choline peak area in brain tumors (18 gliomas and 5 metastatic lesions) versus uninvolved brain tissue. Similarly, Utriainen *et al.* [220] found significantly elevated choline in twelve patients with suspected brain tumors compared to MRS recordings from eleven healthy volunteers.

However, choline can be low in very small brain tumors and with necrosis as in, e.g., glioblastoma multiforme. Pediatric brain tumors can also show low choline [211, 221]. Low choline has also been reported in mixed tissues that contained some brain tumor [222]. Nagar *et al.* [223] reported that choline levels in intracranial tumors normalized to uninvolved brain tissue in the same twenty patients were significantly higher than in fifteen patients with non-neoplastic brain processes. However, in two of the fifteen patients with non-neoplastic brain lesions the normalized choline levels were also elevated (false positive rate 13%). High choline in a non-cancerous lesion reportedly also led to a false positive for malignancy in one of five children undergoing MRS for workup of brainstem lesions [224].

8.1.2.2 Nitrogen acetyl aspartate

Since NAA at ~ 2.0 ppm is a marker of number and viability of neurons, a normal level of NAA is generally considered to be incompatible with tumor, because most brain tumors are of non-neuronal origin [113, 211]. Sijens and Oudkerk [219] reported significantly higher NAA in unaffected brain tissue compared to sites of brain tumors in the above-mentioned study. On the other hand, the NAA level may be normal in small or low grade tumors. It should also be noted that nitrogen acetyl aspartate can sometimes also be normal with infiltrative brain tumors [211].

8.1.2.3 Ratios of choline to NAA or choline to creatine

Ratios of metabolites have also been used for identifying brain tumors. In a comparison of MRI-localized proton MR spectra from the tumor bed in fifty-one patients with high-grade gliomas with those from thirty healthy volunteers, significantly higher choline to NAA as well as choline to creatine ratios in the former were found by Tarnawski *et al.* [225].

Similarly, Lin *et al.* [226] reported that the choline to creatine ratio was significantly higher among forty nine patients with biopsy-proven brain tumors compared with fourteen healthy persons. The mean choline to NAA ratios from brain regions without tumor were 1.7 ± 1.2 versus 5.3 ± 2.1 to 6.8 ± 3.7 in gliomas, depending upon the grade, in the study by McKnight *et al.* [215], in which all findings were confirmed histopathologically. Increased choline and decreased NAA were found by Utriainen *et al.* [220] to distinguish brain tumors in twelve patients from corresponding regions in the normal brain recorded in eleven healthy referents.

Hourani *et al.* [227] found that the NAA/Cho ≤ 0.61 together with relative cerebral blood flow ≥ 1.50 distinguished sixty-nine patients with primary brain tumors of various grades from thirty-six patients with diverse non-malignant brain lesions with a sensitivity of 72.2% and specificity of 91.7%. Choline to NAA ratios > 1.9 provided significant help for neuro-radiologists to distinguish tumorous from pseudo-tumorous lesions in a study of eighty-four patients with solid brain masses [228].

8.1.2.4 Guidance for stereotactic biopsy

Guidance via MRSI based upon areas of increased choline to creatine ratios has improved the yield of diagnostic tissue obtained with stereotactic biopsy [113, 229, 230]. It has been suggested that by using multi-voxel MRS the area of highest choline levels or choline to NAA ratio can be identified, and that would be the optimal site for biopsy [113, 211, 231]. Other metabolites, such as lipid might also be instructive for locating the best biopsy site [232].

8.1.2.5 Other metabolites often seen in brain tumors: Lactate, lipids, glutamine – glutamate, myoinositol

Some other metabolites or metabolite ratios have also been helpful in detecting brain tumors via MRS and MRSI. Due to the predominance of anaerobic glycolysis, lactate is seen often in malignancy. Lactate appears as a doublet due to J-coupling, and is centered at 1.33 ppm [113]. Lactate is said to be found in most pediatric brain tumors [211]. A lactate peak was detected in three of ten patients with high-grade gliomas, as reported in a study by Saindane *et al.* [233]. In another study, lactate was registered among eighteen patients with gliomas and five patients with metastatic brain tumors, but was totally absent from all the unaffected brain tissue investigated [219].

Depending upon the presence or absence of necrosis, lipids in the region around 0.9 ppm to 1.3 ppm may or may not be detected in brain tumors. Lipids are generally described as absent in normal brain tissue [207, 211]. In a study by Tarnawski *et al.* [225] there was no detected lipid whatsoever in brain MRS recordings from thirty healthy volunteers, and a significantly higher lipid to NAA ratio was found in the tumor bed of fifty-one patients compared to the contra-lateral, non-involved lobe. Smith *et al.* [234] reported that the lipid to creatine ratio was significantly lower among five healthy volunteers compared to twenty patients with brain-stem tumors. It has also been suggested [232] that targeting the regions with highest lipid content could improve the diagnostic yield of stereotactic biopsy [232].

In a study by Fan *et al.* [235] glutamine – glutamate at 2.3 ppm to 2.5 ppm was found significantly more often at the site of brain tumors (high-grade gliomas and metastases) among twenty-two patients compared to normal parts of the brain. These authors observe that glutamate may act as an excito-toxin associated with accelerated cell proliferation of brain tumors. Myoinositol at around 3.56 ppm is involved in the activation of protein C kinase, which leads to the generation of proteolytic enzymes seen in brain neoplasms [113]. As a precursor in lipid metabolism, myoinositol also may be elevated with cell proliferation [207]. A number of reports indicate that myoinositol is elevated in various brain tumors [220, 225, 236, 237]. A myoinositol to NAA ratio > 0.9 was also found to help distinguish malignant from pseudo-tumorous lesions in the earlier cited study of eighty four patients with solid brain masses [228]. However, Brandão and Domingues [211] state just the opposite, namely, that intra-cerebral neoplasms are one of the conditions associated with *decreased* myoinositol, whereas increased myoinositol is seen in multiple sclerosis, Alzheimer's and Pick's disease, *inter alia.* Further elucidation has been provided in a recent study by Hattingen *et al.* [238] using 2D MRS with a short TE of 30 ms among fifty-six patients with glial brain tumors. These investigators found that myoinositol concentrations were significantly higher in all tumor tissues compared to normal white matter, with the highest concentrations found in association with marked astrogliosis.

8.1.2.6 Identification of brain tumors in children

The choline to NAA ratio is considered to be especially helpful in identifying low-grade neoplasms in children, in whom the choline concentration might be low or normal within the tumor. Since active myelination is ongoing in the normal parts of the brain, the choline to NAA ratio would be increased at the tumor site [211]. High choline, low NAA together with a minimal lactate peak on proton MRS were helpful in differentiating bilateral thalamic astrocytomas from encephalitis and neuro-metabolic disorders in a study by Gudowius *et al.* [212]. As noted, these disorders are difficult to distinguish among children on the basis of MRI. Similar findings were reported by Möller-Hartmann *et al.* [239]: high choline with low creatine and NAA were characteristic of brain tumors, whereas non-cancerous lesions such as cerebral infarction and brain abscesses showed decreased choline, creatine and NAA. However, high choline in a non-cancerous lesion also led to a false positive MRS finding of malignancy in a child undergoing MRS for workup of brainstem lesions [224].

8.1.2.7 Metabolites that may help rule-out malignancy: acetate and succinate

Brain abscesses often show acetate (1.92 ppm) and succinate (2.4 ppm), whereas this is generally not the case for brain neoplasms [113, 211, 214, 240]. The appearance of these metabolites is thus considered to distinguish brain abscesses from cystic or necrotic tumors [211]. In an investigation of fifty patients with intra-cranial cystic mass lesions, Lai *et al.* [241] reported identifying acetate or succinate only in brain abscesses, but not in any of the twenty-three patients with brain neoplasms. Altogether, however, acetate or succinate were present respectively in only seven and four of the twenty-one brain abscesses, suggesting that these metabolites were rather infrequently detected in the pyogenic lesions.

8.1.3 Grading of primary brain tumors by MRS & MRSI

Complete surgical resection is the aim for successful treatment of brain tumors. Realization of this goal vitally depends upon the grade and histopathologic characteristics of the tumor and anatomical location. If accessible, low-grade astrocytomas are usually surgically resected with RT often also employed. Protocols vary substantially across centers, and strategies also differ greatly in relation to the clinical considerations. Since high-grade gliomas only rarely have clearly defined margins, total surgical resection is not possible in most cases. Partial resection to control mass effect is often performed, as well as RT, chemotherapy and glucocorticoids. Overall survival is poor, generally below 1 year. Total surgical resection represents a curative treatment for meningiomas. If the resection is sub-total, local RT is usually given and reduces recurrence rates to fewer than 10%. If the meningioma is not surgically accessible, targeted radio surgery with the gamma knife or heavy particle radiation may be considered. Accurate grading of intra-cerebral neoplasms can be very difficult, due especially to tumor heterogeneity. For this reason, brain biopsy may not provide the definitive answer, besides being associated with substantial morbidity, as noted. Thus, there has been interest in using *in vivo* MRSI with full volumetric coverage for grading brain tumors.

Indeed, MRSI has been helpful in this endeavor. For example, it is often found that higher grade gliomas show the greatest elevation in the ratios of choline to creatine and choline to NAA [211, 242]. In a study of fifteen patients with grade II gliomas, seventeen patients with grade III and seventeen patients with grade IV, the maximum choline to NAA ratio was found to be a much better indicator of glioma tumor grade than were either contrast enhancement or T_2 hyper-intensity on MRI [242]. In another study [243], among eighteen patients with Grade II or Grade III gliomas, the choline to creatine ratios were found to best distinguish these two grades. A choline to creatine ratio > 2.33 was reported to be the most accurate for detecting nine high grade malignancies from a total of twenty-two patients with oligodendroglial tumors [244]. A choline to creatine ratio > 1.55 identified high grade brain tumors with a sensitivity of 92% and specificity of 80% in another study which included twenty-four patients [245].

Lactate levels are often found to parallel tumor grade [211, 240], [246]–[248]. This is biologically plausible, since the appearance of lactate caused by anaerobic glycolysis is the main metabolic pathway for glucose utilization by anaplastic brain tumors [249]. Histopathologic evidence is also corroborative, in that a positive correlation has been reported between the height of the lactate peak and the degree of heterogeneity of the nuclear roundness factor in anaplastic gliomas [250]. Lactate to creatine ratios provided the best distinction between Grade III and Grade IV gliomas in the earlier-cited study [243], which included ten patients with these two grades of tumor. Necrosis is a frequent characteristic of higher-grade brain tumors. The spectra of glioblastomas are reportedly often dominated by large lipid – lactate resonances [251]. Several investigations show that the presence or greater concentration of lipids is associated with higher-grade intra-cerebral tumors [232, 239, 242, 252, 253].

8.1.3.1 Grading of pediatric brain tumors

Magnetic resonance spectroscopy and spectroscopic imaging have also been helpful for assessing brain tumor grade in the pediatric population. Low total choline was found to be associated with low proliferative tumors in an investigation of fourteen children with diffuse brainstem gliomas [254]. High choline levels have been reported to be associated with high-grade brain tumors among children [255]. In a study [256] of twenty-four children with brain tumors, those who responded to chemotherapy or RT also showed significantly lower choline and higher total creatine. Lipids and lactate were also lower among the responders than among those treated only by surgery or who did not respond to treatment. The only metabolite to independently predict active tumor growth was total creatine.

8.1.3.2 Tumor grading based upon a combination of metabolites via MRS & MRSI

Pattern recognition based upon a larger number of metabolites has also been used for tumor grading, since it is appreciated that relying upon single metabolites can be unreliable [113]. A study of forty eight patients using a 3T scanner by Di Costanzo *et al.* [257] corroborates that several metabolites and their ratios yielded better accuracy in classifying high versus low grade gliomas. Herminghaus *et al.* [249] employed a combination of metabolites to aid in grading histopathologically confirmed glial brain tumors among ninety patients. With single-voxel proton MRS, linear discriminant analysis based upon normalized NAA, total creatine, choline, lactate and lipid to total Cr ratio was used to classify tumor grades. Accurate tumor grading was achieved in 77% to 94% of cases. The most serious grading errors occurred in two patients with Grade I/II who were considered to be Grade III/IV and for three patients with Grade III/IV who were judged as I/II. These authors [249] consider that spectral pattern analysis yielded a high level of accuracy with the possibility of identifying cases that cannot be reliably graded, and "thereby avoids therapeutic decisions based on ambiguous data" (p. 79). An automated pattern recognition system based upon a data base of over 300 tumor spectra and using several metabolites has been developed within the European Union, to aid in brain tumor diagnostics with MRS and MRSI. However, tumor grading using that system cannot yet be considered reliable since there is considerable overlap due to tumor heterogeneity [113].

8.1.4 Characterization of brain tumors by MRS & MRSI

Brain tumor types can sometimes be identified by using MRS and MRSI [258].

8.1.4.1 Identification of meningiomas

Distinguishing meningiomas, which are usually slow growing from more aggressive brain tumors, is most often achieved by MRI and CT. When morphologic characteristics are indeterminate, MRS and MRSI have been helpful. The appearance of alanine at 1.48 ppm on MRS or MRSI is considered by some authors to be highly suggestive of meningioma [113, 211, 259]. Alanine was found to be present in fifteen of nineteen patients with meningiomas, but was not detected in eleven Schwannomas, or in any of the eight metastatic tumors examined in Ref. [144]. Higher alanine and glutamine – glutamate best discriminated between twenty-two patients with meningiomas and nine patients with primitive neuro-ectodermal tumors in another study [260]. In another paper by Majós *et al.* [261], alanine was found to be elevated in atypical meningiomas, compared to other tumors types. In a third paper, these authors [262] described alanine as the most characteristic resonance of meningiomas, with alanine being rarely seen in non-meningeal tumors. However, mean alanine concentrations were lower in eight patients with meningiomas compared to those of six patients with metastatic brain lesions in a study by Howe *et al.* [263]. An alanine peak was also observed in each of three patients with central neurocytomas who underwent proton MRS [264].

Besides alanine, the appearance of glutamine – glutamate in the spectral region between 2.1 ppm and 2.55 ppm is also described as "highly suggestive of meningioma" [211]. In Ref. [144], increased glutamine – glutamate was detected in twelve of the nineteen patients with meningiomas, but this was also found in four of eight patients with metastatic brain tumors, though in none of the patients with Schwannomas. Majós *et al.* [262] reported increased glutamine – glutamate in the patients with meningiomas compared to astrocytomas. These authors found that, in addition to alanine, glutamine – glutamate was the best metabolite for distinguishing patients with meningiomas from those with primitive neuro-ectodermal tumors. Opstad *et al.* [145] found that six patients with meningiomas had elevated glutamine – glutamate concentrations compared to six patients with astrocytomas as well as in comparison to normal white matter. The reduced glutathione resonance at 2.9 ppm was also significantly greater in the patients with meningioma compared to those with astrocytomas, as well as in the normal white matter in Ref. [145]. Some authors [144, 262, 265] have used various combinations of metabolites to identify meningiomas, with the presence of alanine and glutamine – glutamate usually playing an important role.

8.1.4.2 Distinguishing primary brain tumors from brain metastases

Another clinically important distinction that has been aided by MRSI is between primary brain tumors and metastatic lesions. Again, morphologic characteristics are often sufficient to distinguish these entities, but when this is not the case, MRSI has sometimes been helpful. The most consistent distinction is in the peri-tumoral region. Since the peri-tumoral region of metastatic lesions is typically comprised of vasogenic edema rather than cellular infiltration, choline is usually low in the area surrounding metastases. On the other hand, elevated choline in the peri-tumoral region is suggestive of a primary brain tumor [211]. Fan *et al.* [235] reported significantly lower choline-to-creatine ratios in the peri-tumoral region in eight patients with brain metastases compared to fourteen patients with high-grade gliomas. These authors consider infiltration of adjacent brain parenchyma a unique feature of high-grade glioma. Similar findings have been reported by Law *et al.* [266]. A different pattern was reported, however, by Ricci *et al.* [267] in their study of thirteen patients with intra-cerebral tumors. The peri-lesional edema showed normal choline to creatine ratios. Included in this series were seven patients with glioblastomas.

8.1.4.3 Characterization of pediatric brain tumors

Proton MRS has also provided some help in characterizing childhood brain tumors. Thirty five children with cerebellar tumors were examined in a single-voxel study using a 1.5T scanner and short TE [268]. Eighteen non-glial tumors (medulloblastomas) were identified on the basis of high taurine, phosphocholine and glutamate together with low glutamine. High myoinositol and glycerophosphocholine characterized five ependymomas. *In vitro* MRS was used for corroboration [268, 269]. In an earlier report of these children from that same group of investigators [270], the combination of a creatine to choline ratio < 0.75 and a myoinositol to NAA ratio < 2.1 distinguished the medulloblastomas from the ependymomas.

8.1.5 MRSI for target planning for brain tumors

It has been suggested that incorporation of MRS and MRSI into RT planning for high-grade gliomas could improve control while reducing complications [271]. Most often, target definition in planning radiation therapy has been based upon ratios of choline to creatine or choline to NAA. The clinical target volume for RT of gliomas was previously generated by adding uniform margins of 2 to 3 cm to the area of T_2 hyper-intensity [272]. This is because the peri-tumoral "uncertain" zone, while appearing normal on MRI is often infiltrated by tumor when examined histopathologically [235]. Using MRSI to determine areas of high choline to NAA ratio, the shape and size of the clinical target volume can be better identified, with more confident sparing of uninvolved brain tissue [273]. Choline to creatine ratios have been used to tailor radiation dose levels to the glioma grade within a map of clinical target volumes [274]. Not only have maps of choline to creatine ratios as well as choline to NAA ratios been helpful in refining RT dose contouring, they have also improved the effectiveness of surgical planning to treat brain tumors [207]. A phase I dose intensification trial for gliomas using intensity modulated radiation therapy (IMRT) is being planned using the additional information provided by choline to creatine ratios from MRSI as well as functional MRI [275].

8.1.6 Assessing response of brain tumors to therapy and prognosis via MRSI

8.1.6.1 Distinguishing recurrent tumor from response to therapy

Brain tumors frequently return after treatment, and often do so at a higher grade. In addition, the treatment itself, especially RT, provokes changes in brain tissue that are difficult to distinguish from tumor recurrence using MRI. Radiation necrosis, local response to immuno-therapy, as well as the tumor itself all can show contrast enhancement. MRSI improves specificity by helping to differentiate radiation necrosis or enhancement phenomena after local immuno-therapy, from tumor recurrence. Distinction between recurrent glioma and radiation injury with the appearance of new contrast-enhancing lesions has been facilitated by ratios of choline to creatine and choline to NAA, particularly with the addition of diffusion weighted imaging [276]. Weybright *et al.* [277] performed MRSI in twenty-nine consecutive patients who had a new contrast-enhancing lesion in the vicinity of a previously diagnosed and treated brain neoplasm. These authors [277] found that the ratios of choline to creatine and of choline to NAA were significantly higher in tumor than in radiation injury. In turn, the ratios of choline to creatine and of choline to NAA were significantly higher in radiation injury than in normal-appearing white matter. However, these metabolite ratios showed substantial overlap, so that the distinction among recurrent tumor, radiation injury and normal appearing white matter was not absolute. Smith *et al.* [278] have recently reported that among thirty-three patients who had a new contrast enhancing lesion after RT for primary brain tumors, an elevated choline to NAA ratio predicted tumor recurrence with a sensitivity of 85% and specificity of 69%. The importance of an increase in choline to NAA ratio in an area of contrast enhancement on post-operative follow-up MRS as an indicator of tumor recurrence as opposed to post-radiation lesions was also reported in a study of nine patients with high grade gliomas [279]. Sankar *et al.* [280] used MRSI to assess response to tamoxifen chemotherapy among sixteen patients treated for recurrent high-grade glioma. The metabolite intensities were found to be stable in all responders, but changed prior to disease progression. Choline, lipid, as well as ratios of choline to NAA and lactate to NAA were significantly increased, while creatine was significantly decreased, compared to stabilized levels. In one patient who continued to respond to tamoxifen at the conclusion of the trial, metabolite levels remained stable. These authors conclude: "characteristic global intra-tumoral metabolic changes, detectable on serial (proton) MRSI studies, occur in response to chemotherapy for malignant glioma and may predict imminent treatment failure before actual clinical and radiological disease progression". They emphasize that their study was motivated by the aim of "early prediction of imminent failure during chemotherapy for malignant glioma which has the potential to guide proactive alterations in treatment before frank tumor progression" (p. 63) [280]. A case report of a patient with low grade glioma treated with temozolomide chemotherapy and with several follow-up MRSI examinations revealed that choline levels within the tumor paralleled shrinkage of the tumor [281]. In contrast, however, monovoxel MRS assessment of NAA to choline ratios provided very poor sensitivity (28%) for assessing tumor progression under chemotherapy and/or RT in a small series of patients treated for low grade gliomas [282].

8.1.6.2 Predicting likelihood of proliferation and relapse of brain tumors in adults

In a prospective study among eighty-two patients with grade II gliomas, a comparison was made between spectroscopic and histological data including the proliferation index assessed via Ki-67 immuno-chemistry. The presence of lactate at 1.36 ppm and free lipids between 0.8 ppm and 1.2 ppm was found to have a significant multivariate association with proliferation index (both intermediate and high) [283]. Regions with choline to NAA ratios ≥ 2 prior to combination chemotherapy and RT were reported to be predictive of relapse in a study of nine patients with newly diagnosed glioblastoma followed bimonthly [284]. On the other hand, the metabolic characteristics of suprasellar tumors assessed in forty patients by single-voxel proton MRS were not found to predict recurrence following initial surgical resection [285].

8.1.6.3 Prognostic indicators via MRS & MRSI for pediatric brain tumors

Tzika and colleagues [286] suggest that *in vivo* MRSI can be used as a prognostic indicator for pediatric brain tumors. These authors examined twenty-seven children with neuroglial brain tumors, and found that the percent change in choline to NAA ratio on MRSI, as well as relative tumor blood volume were significant predictors of tumor progression. A more recent study from the same center revealed that among seventy-six children with tumors of the central nervous system, a combined index of choline and lactate provided better prediction of survival than standard histopathology [287]. Similarly, increased total choline was also found to precede clinical deterioration in the earlier cited study of fourteen children with diffuse intrinsic brainstem gliomas [254]. In a series of twenty-seven children who were afflicted with pilocytic astrocytomas, those tumors that progressed were found to have initially lower myoinositol levels and decreased levels over time [288]. This information is particularly important because it can impact upon management strategies when tumors occur in brain regions that are not amenable to surgery [270].

8.2 Major limitations and dilemmas in MRS & MRSI for neuro-oncology due to FFT envelopes and fittings

As summarized thus far in this chapter, MRS and MRSI have shown great promise in relation to various aspects of brain tumor diagnostics, and tremendous strides have been made in the most recent period with these molecular imaging modalities. It is also clear that further improvements are still needed. In this section we analyze many of the problems and dilemmas encountered using MRS and MRSI in neuro-oncology and their relation to reliance upon the conventional Fourier-based framework for data analysis.

8.2.1 Poor resolution and SNR

Poor resolution and low SNR represent a critical limitation which has hampered wider use of MRS for clinical oncology. As stated, the FFT is a low-resolution estimator, lacking both extrapolation and interpolation capacities. Moreover, with attempts to enhance resolution by increasing acquisition time, SNR deteriorates. This is due to the fact that mainly noise is recorded later in the time signal, particularly with 1.5T clinical scanners. Moreover, the FFT is a linear transform and therefore imports noise from the time to the frequency domain, further contributing to poor SNR.

8.2.1.1 Limited capability to detect small brain tumors

A number of the problems with current applications of MRS and MRSI in brain tumor diagnostics are related to resolution and SNR issues. Particularly troublesome in this regard is the limited possibility of MRS and MRSI to detect very small brain tumors [211], at the very time at which therapeutic interventions would have the best chance for success. As noted by Huang *et al.* [289] *in vivo* MRS has low SNR, which severely limits capabilities to determine critical characteristics such as the grade of brain tumors.

Attempts to improve the SNR have most often entailed either increasing the acquisition time, or increasing the volume of brain tissue from which data is acquired. The latter approach frequently results in a heterogeneous voxel with a mixture of tissue types. There are special problems in achieving adequate SNR using MRSI within scan times of reasonable length [4]. Because of the vital importance of achieving volumetric coverage of brain tumors, which are often heterogeneous, the SNR issues specifically related to MRSI are of particular concern for neuro-oncology, as well as for cancer diagnostics in general [290].

8.2.1.2 Attempts to improve resolution and SNR via stronger fields and short echo times

Resolution and SNR for brain tumor diagnostics have been improved by using MR scanners with higher magnetic field strength [291]. However, the detection of small residual foci of brain tumor has been reported to be problematic even when a 3T scanner was used [291]. Some recent data indicate that a multichanneled phased-array head coil may improve the possibility to predict the spatial extent of brain tumors [113, 292]. Another strategy has been to use short TE in an attempt to capture the larger signal intensities observed early in the recording, and thereby also capture clinically important metabolites for brain tumor diagnostics such as myoinositol, lipids and glutamine – glutamate that decay rapidly and can therefore only be seen at short TE [293, 294]. However, at short TE the problems of relying upon fitting can be exacerbated for certain metabolites. In addition, if problems with the baseline occur, longer echo times are recommended [293]. Moreover, since T_2 relaxation times of various metabolites differ, not only peak heights, but also peak ratios can be affected by changes in TE [252, 259]. Thereby, reliance upon metabolite ratios also becomes more problematic. Thus, for example, Kaminogo *et al.* [252] report that the NAA to choline ratios assessed from MR spectra obtained at short TE were of limited value in brain tumor grading. These authors [252] note that besides NAA, "the peak around 2.0 ppm may contain other fractions of metabolites, including lipids at 2.05 ppm, and glutamate and glutamine at 2.1 ppm and 2.5 ppm ... Especially at short TE spectroscopy, Glx and lipids may contribute to the spectral area at 2.0 ppm and possibly affect grading" (p. 361). The problem of assessing NAA levels at short TE is thus related to the presence of overlapping resonances that cannot be identified unequivocally by Fourier processing and post-processing fitting.

8.2.1.3 Advantages for brain tumor diagnostics by the high resolution of the fast Padé transform

As discussed, one of the key advantages of the fast Padé transform relative to the FFT is that convergence is not only stable, but also rapid. This means that even at short signal lengths, the FPT is still capable of determining the true concentrations of the main metabolites that remain undetected by the FFT. A spectrum in the FPT does not use the fixed Fourier mesh in the frequency domain, and can be computed at any frequency. Thus, resolution is not pre-determined by the total acquisition time T. The conundrum between increasing acquisition time for improved resolution and increasing noise is thereby obviated by the FPT. This is especially important for accurate detection of short-lived metabolites. In brain tumor diagnostics, we have seen that a number of informative metabolites, notably, lipids, glutamine – glutamate, and myoinositol decay rapidly and, therefore, require shorter echo times in order to be adequately detected.

We have also reviewed another very important advantage of the FPT, namely, its power of extrapolation. The FFT is limited by a sharp cut-off of the time signal at the end of the acquisition time T, replacing any extension of the signal by zeros or using the signal periodic extension, with no new information in either case. However, the FPT uses its polynomial quotient to extrapolate beyond the given T, and this is the main contributor to the markedly improved resolution [18]. As noted, the FFT has a poor SNR in part due to the need for long acquisition times. The poor SNR of the FFT is also related to the fact that it is a linear mapping where the transformation coefficients are *independent* of the time signal points. The FPT is a non-linear mapping, such that its coefficients are *dependent* upon the time signal points. Thus, while the linearity of the FFT preserves noise from the time signal, the non-linearity of the FPT permits noise suppression. Numerical computations from Ref. [35] show that the FPT is powerful for noise suppression especially in the vicinity of genuine signal poles. Furthermore, the FFT has a linear convergence $(1/N)$ with increased signal length N, whereas the convergence of the FPT is quadratic ($\approx 1/N^2$) or better [17].

By way of an extract from the detailed analysis presented in chapter 7, we now recapitulate the comparison of the resolution performance of the FPT and the FFT on two complex-valued time signals c_n $(0 \le n \le N-1, N = 2048)$ with bandwidth = 6001.5 Hz encoded at 4T via MRS by Tkáč *et al.* [141]. We show three partial signal lengths at a fixed bandwidth. These data of full signal length $N = 2048$ encoded by the group at the Center for Magnetic Resonance Research, University of Minnesota, Minneapolis, USA [141] have been kindly made available to us. In Fig. 8.1, we present the absorption total shape spectra computed by the FFT (left column) and FPT (right column) at three signal lengths. At the top of Fig. 8.1 the most dramatic difference between the FFT and FPT is seen at the shortest signal length ($N/16 = 128$). Here, the FFT essentially presents no meaningful spectroscopic information. In contrast, with the FPT, at $N/16 = 128$ nearly 90% of the NAA concentration is predicted by the peak at around 2.0 ppm.

On the middle panel at $N/4 = 512$ the FFT has still not predicted even 70% of the NAA concentration at 2.0 ppm. Moreover, the ratio of creatine at about 3.0 ppm to choline at about 3.2 ppm appears to be approximately equal to one, and thus is incorrect. On the other hand, with the FPT at $N/4 = 512$, these three major peaks are nearly identical to those at full signal length. At half signal length $N/2 = 1024$ on the bottom panel, the FFT has still not demonstrated the accurate ratio between creatine and choline at 3.0 ppm and 3.2 ppm, respectively. These two latter metabolites are still incorrectly appearing as being almost of equal intensity. Furthermore, the triplet of glutamine and glutamate near 2.4 ppm can be discerned at half signal length only by the FPT, and not by the FFT.

By contrast, it is seen that at half signal length ($N/2 = 1024$) the FPT resolves with fidelity more than twenty metabolites. Furthermore, while the FFT requires the total signal length ($N = 2048$) to fully resolve all the metabolites, the difference between the two FPT spectra at $N = 1024$ and $N = 2048$ is buried entirely in the background noise [20]. In other words, the FPT total shape spectra at half-signal length can be treated as fully converged. In chapters 3 and 7, detailed presentations are given of the intrinsic and robust error analysis of the FPT.

It is also important to point out that the FPT is shown in this and many other examples in chapters 3 and 7 to produce no Gibbs ringing in the process of converging in a steady fashion as a function of the increased signal length. This is in sharp contrast to other existing parametric estimators that are usually unstable as a function of N, typically undergoing wide oscillations with unacceptable results before eventually converging, if at all.

We have noted that besides the computational efficiency of automatic software, the main reason for which the FFT gained attractiveness among users is that it presents no surprises, while steadily converging with increasing signal length. The FPT shares such an advantageous property of the FFT. In addition, the FPT provides a much faster convergence rate than in the FFT

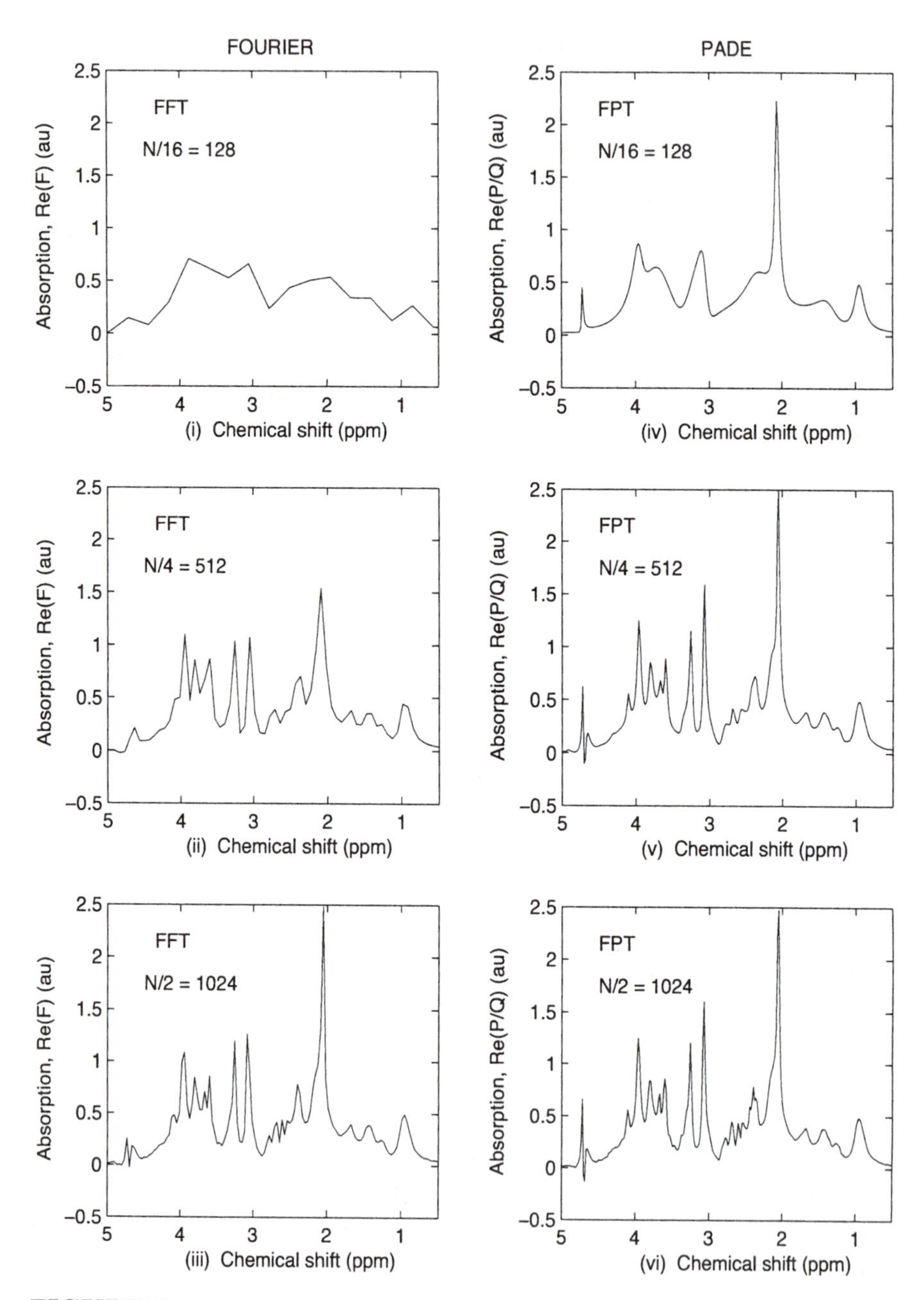

FIGURE 8.1

Fourier (FFT) and Padé ($\mathrm{FPT}^{(-)}$) absorption spectra computed using the time signal (divided by 10000) at three partial signal lengths ($N/16 = 128, N/4 = 512, N/2 = 1024$), where the full signal length is $N = 2048$, as encoded in Ref. [141] at 4T from occipital grey matter of a healthy volunteer.

with increasing signal length, as clearly demonstrated herein, and in other examinations of *in vivo* MRS signals encoded at high magnetic field strength as seen in chapter 7. The favorable convergence rate of the FPT is one of the main reasons for the obtained resolution enhancement relative to the FFT.

8.2.2 Unreliable quantifications by fitting FFT spectra

As elaborated in our systematic review [114] through the year 2004, very few of the metabolite concentrations or their ratios estimated via the FFT unequivocally distinguished brain tumors from normal tissue, nor were these specific for malignancy. In MRS and MRSI, with conventional Fourier processing, infection, demyelinating disorders and infarction often showed spectral changes that were identical to those of brain tumors [114]. Considering the more recent data included in the present chapter, this conclusion still holds.

Thus, for example, choline can be elevated not only in brain tumors, but also in numerous other cerebral pathologies including sub-acute and chronic ischemia, infiltrative processes, encephalitis, demyelinization (including the acute phase and with the chronic plaques of multiple sclerosis), organizing hematoma, Alzheimer's disease, Down's syndrome, as well as depression, the reactive astrocytosis of epilepsy, to name a few [211].

Low NAA can reflect loss of neurons with infiltration of brain tissue by tumor, but as a marker of neuronal viability and density, NAA can also be decreased in almost any brain insult [211]. On the other hand, as discussed earlier in this chapter, choline can be low in very small brain tumors in adults, in necrotic brain tumors or otherwise containing mixed tissues, as well as in brain tumors in children [211, 221]. Levels of NAA can also sometimes be normal in small brain tumors. Even alanine, which has been found to be helpful for identifying meningioma can be elevated in processes such as brain abscess and neuro-cystocercosis, as well [211].

The cut-points for metabolite ratios and choice of the denominator metabolite used to indicate the presence of brain tumors vary substantially among different investigators. For example, McKnight *et al.* [215] used a cut-off value of 2.5 for the choline to NAA ratio, reporting that biopsy samples containing tumor (Grade II to Grade IV gliomas) were distinguished from those with normal, edematous, gliotic or necrotic tissue with 90% sensitivity and 86% specificity, comparing pre-operative MRSI and histopathology. These authors also reported that up to half of the T_2 hyper-intense regions outside the contrast-enhancing lesion contained areas with choline to NAA ratios > 2.5. However, choline to creatine ratios ≥ 1.7 were reported by Murphy *et al.* [251] to be associated with unequivocal tumor presence, and that ratios ≤ 1.3 were seen in normal brain tissue. Among thirty-one patients with malignant brain tumors, all regions with confirmed cancer showed significant choline levels and a choline to NAA ratio > 1.3 in an investigation by Vigneron *et al.* [216].

In the study of Smith *et al.* [234], choline to creatine ratios in the brain of healthy persons were found to be as high as 2.14, while choline to creatine ratios as low as 0.36 could be seen at sites of brain tumors. As presented, more recent investigations have also used varying cut-points of metabolite ratios to define the presence of brain tumors. For example, choline to NAA ratios > 1.9 at long TE were used to define tumorous from pseudo-tumorous lesions in a study of eighty-four patients with solid brain masses by Majós *et al.* [228]. On the other hand, in a study by Hourani *et al.* [227] a cut-point of NAA to choline of 0.61 (whose reciprocal is 1.64) was used to distinguish various grades of brain tumor from non-malignant lesions.

Metabolite ratios can be affected by many confounding factors including cancer treatment itself. The use of ratios is problematic for accurate assignment of spectral changes to specific disease processes [295], since they often vary for reasons unrelated to the oncologic process of interest. Notably, the ratio of choline to NAA, upon which detection of brain tumor is heavily based in proton MRS, may also be due to loss of NAA, which is seen in a wide variety of neurological disorders including epilepsy, multiple sclerosis, as well as in cerebrovascular accidents. Metabolite ratios can be affected by contamination from adjacent tissues. High NAA can occur with contamination from adjacent tissues within brain tumors [216]. Metabolite ratios can also be affected by regional differences. There is a wide variation in metabolite ratios in normal brain, especially comparing white and grey matter [211, 259].

We have discussed earlier in this chapter that treatment-related changes can affect metabolite ratios. Alteration in the ratios of choline to NAA and to creatine may also occur as a result of sub-clinical neuro-toxicity related to chemotherapy, possibly due to neuronal loss and/or inhibition of cell metabolism, as well as to radiation effects [291, 296]. Moreover, metabolite ratios are affected by measurement parameters. Since different metabolites have different relaxation times, metabolite ratios are dependent upon echo time. Changes in TE were noted to impact upon the usefulness of metabolite ratios for, e.g., distinguishing various grades of brain tumors [252].

Earlier in this chapter, we reviewed the data showing that brain tumor grading has been substantially aided by *in vivo* MRS. Nevertheless, there are quite a few contradictory findings in the literature. Thus, for example, while higher mean choline levels and lower mean NAA levels are generally associated with high-grade tumors, there are large standard deviations, which are considered to preclude accurate tumor grading [216]. This is due, at least in part, to the fact that high-grade brain neoplasms frequently contain necrosis, such that choline levels and their ratios to NAA and creatine may be similar to or even less than those of low-grade brain tumors [211, 242, 252, 297].

In one of these studies [297], choline concentrations were reported to be significantly lower in 13 high-grade compared to 10 low-grade gliomas. We also noted that pattern recognition based upon a larger number of metabolites has also been used for tumor grading, because it was recognized that relying upon single metabolites can be unreliable [113]. However, tumor grading using that system was not considered sufficiently reliable either, since there was substantial overlap of findings due to tumor heterogeneity [113].

8.2.3 Fitting estimates for concentrations of a small number of metabolites

There is a limited number of metabolites identified with *in vivo* MRS processed via FFT. Danielsen and Ross [259] stated that *in vivo* MRS has introduced a "new neuro-chemistry", providing the clinician with the possibility of non-invasively accessing neuro-physiologic and neuro-chemical information. This, together with the exquisite detail of neuro-anatomical imaging provided by MRI, has been said to offer "an indispensable tool for state-of-the-art neuro-diagnosis" (p. 23). However, the current Fourier-based applications of MRS and MRSI in brain tumor diagnostics yield a relatively small number of metabolites (low molecular weight, high concentration) that are observable on clinical scanners. In other words, in fact, we are provided with an exceedingly limited view of normal and abnormal brain chemistry.

Thus, we have seen for example that definitions of the presence or absence of brain tumor are very frequently based solely upon the ratios of choline to NAA and choline to creatine. Besides the above-discussed problems of varying cut-points and numerous extraneous factors that could affect these ratios, the fundamental issue is that MRS and MRSI have made great strides in neuro-oncology by relying upon a mere handful of metabolites. This restricted metabolite window stems directly from the limitations of conventional data analysis based upon the FFT and accompanying post-processing fitting and other related *phenomenological* approaches.

8.2.4 Lack of component spectra of clinically important overlapping resonances for brain tumor diagnostics

As we have elaborated in detail earlier, the FFT is a non-parametric method which provides only the shape spectrum, but not the parameters that define the underlying components. Two-dimensional correlated and *in vitro* in MRS studies reveal that there is a much richer store of potentially informative metabolites for brain tumor diagnostics than is currently extracted using *in vivo* 1D MRS and MRSI. Many of the most diagnostically important components overlap [295]. For example, in the lipid – lactate region with localized 2D J-resolved MRS, Thomas *et al.* [89] were able to distinguish the lactate peak from the overlapping lipid in a patient with glioblastoma. They conclude: "*in vivo* J-resolved plots of human brain tumors indicate the exciting potential of this technique in extracting additional information from the conventional MR spectrum" (p. 459). The content of the resonances at 0.9 ppm and 1.3 ppm is further elucidated by *in vitro* MRS. The ratios between the 0.93 ppm and 1.3 ppm peaks have been highly variable with *in vivo* MRS [298]. These high-resolution studies reveal that the CH_3 signal at 0.9 ppm arises from protein residues at 0.92 ppm and from lipids at 0.88 ppm. This is due to a non-lipid contribution super-imposed on the 0.9 ppm resonance. The large lipid content of the normal human brain is not visible with *in vivo*

MRS. When lipid peaks are seen, this often reflects necrosis. Identification of specific lipids has heretofore required the high resolution of *in vitro* MRS applied to biopsy specimens from brain tumors [259].

In vitro NMR shows that triglycerides are typical spectral features of actively growing non-necrotic tumors. High-grade gliomas have prominent neovascularity and high-esterified cholesteryl. No cholesteryl esters were detected in healthy brain tissue or in low-grade and benign tumors [299]. Highly esterified cholesteryl could be a biochemical marker of malignancy. The presence of cholesteryl esters and triglycerides has been found [300] to be correlated with the degree of vascular proliferation in high-grade brain tumors.

The overlapping components of total choline within the spectral region of 3.2 ppm to 3.3 ppm can also be informative regarding brain tumor characteristics. The ratio of phosphocholine to choline is reported to be correlated with the percentage of highly cellular glioma [301]. Similarly, the ratio of phosphocholine to choline assessed via *in vitro* MRS was correlated with the percentage of malignant tissue in pediatric brain tumors in an investigation by Tzika *et al.* [302]. A more recent study [268] among children also corroborates that significantly higher levels of phosphocholine are found in various types of cerebellar tumors compared to non-involved cerebellum. On the other hand, glycerophosphocholine levels did not differ significantly.

Deeper insights into clinically important molecular dynamics in brain tumors are also often provided by *in vitro* MRS when assessed in direct relation to detailed histopathological analysis. For example, using high-resolution magic-angle spinning (HRMAS), Opstad *et al.* [303] have reported that in astrocytomas taurine at around 3.3 ppm to 3.4 ppm showed a highly significant correlation with apoptotic cell density. Moreover, unlike lipids at 0.9 ppm and 1.3 ppm, this association was independent of necrotic tissue. These authors [303] suggest that taurine may represent an MR-visible biomarker of apoptosis that is unrelated to the presence of tumor necrosis. Furthermore, they note that assessment of taurine levels *in vivo* has been difficult due to overlap with myoinositol, choline and glucose in that spectral region.

8.2.5 The number of metabolites & non-uniqueness of fitting

As explained earlier, the FFT relies upon post-processing fitting in attempts to provide quantification. We elaborated the reasons why this is non-unique. In short, the number of resonances can only be surmised in this way. Within cancer diagnostics using *in vivo* MRS, the consequences of this limitation of the FFT have been noteworthy. These become particularly troublesome for closely located and overlapping resonances. With respect to brain metabolites, most authors have attributed peaks between 3.8 ppm and 4.0 ppm to glutamine in the alpha region, and to the second creatine peak. However, Opstad *et al.* [145] included glutathione at 2.9 ppm into their LCModel, and obtained a better fit. They reported that reduced glutathione was significantly greater in patients with meningiomas compared to those with astrocytomas.

The authors of Ref. [144] found a resonance at 3.8 ppm at long TE, to which they did not give an assignment, but which was also characteristic of meningiomas, and was absent from other brain tumors. Further complicating the situation is that when patients are treated with mannitol, a resonance at 3.8 ppm will often appear, although in Ref. [144] only two of the nineteen patients with meningiomas were receiving mannitol at the time of the MRS examination. Contrary to claims of objectivity [76], the subjectivity of fitting procedures has been clearly demonstrated and this can undermine diagnostic accuracy in neuro-oncology.

Another important example of the problem of fitting when overlapping resonances are present is illustrated in the study of Kaminogo *et al.* [252]. As noted, these authors reported that the NAA to choline ratio at short TE was of limited value in grading gliomas. They suggest that in gliomas the peak around 2 ppm could contain other metabolites besides NAA, namely lipids at 2.05 ppm, and glutamine - glutamate at 2.1 ppm.

The data concerning myoinositol are contradictory for distinguishing tumors from normal brain tissue and from non-neoplastic processes, as well as for brain tumor grading and histopathological typing. Most authors have assigned the peak at 3.56 ppm to myoinositol alone, while some have viewed this as a combined myoinositol-glycine peak. Glycine is an inhibitory neurotransmitter, which also has a proton signal at this position. After release, glycine is taken up by astrocytes, the site where myoinositol is also located [211, 304]. It remains to be determined whether better distinction between these two overlapping resonances of glycine and myoinositol would help clarify some of the above-mentioned diagnostic dilemmas.

Fitting procedures such as the LCModel have been noted to be particularly problematic in the presence of large amounts of mobile lipids. Auer *et al.* [298] point out that prominent broad resonances at 0.9 ppm and 1.3 ppm are not fully modeled by the baseline spline functions of the LCModel and, as such, lead to incorrect estimates of lactate and alanine. Inappropriate fit of the entire spectrum with substantial phasing errors can in some cases be the final, *and wrong*, result. These authors [298] suggest some procedures to tackle this latter problem, with the aim of improving the accuracy of tumor grading using *in vivo* proton MRS. They note, however, that limited spectral resolution and SNR, as well as a lack of prior knowledge about the specific lipid constituents of various disease states, presently limit the accuracy of quantifying lipid and macromolecular constituents.

In their study using a long TE (272 ms), Kuznetsov *et al.* [305] omitted the peak at 0.9 ppm attributed to lipids, because they considered it could be biased by baseline fluctuations. Furthermore, they stated that limitations in resolution prevented them from directly ascertaining the contribution of lipid versus lactate to the 1.3 ppm peak. In Ref. [144] mixed forms of lipid plus lactate doublet at 1.3 ppm were reported, using an even longer TE (288 ms). This peak was prominent in the patients with metastatic lesions. As discussed in chapter 2 (section 2.6), the off-diagonal FPT can adequately describe this background contribution.

Dilemmas at both long and short echo times concerning assignment and quantification, especially of overlapping resonances, have also been reported [235] when clinical MRS findings from patients with high grade gliomas and solitary brain metastases were compared. The finding that patients with brain tumor recurrences had elevated lipids and glutamine – glutamate underscores the importance of proper assignment, as well as quantification.

8.3 Accurate extraction of clinically-relevant metabolite concentrations for neuro-diagnostics via MRS

8.3.1 Methodological strategy: The need for standards in quantification

As discussed, by relying upon the FFT with post-processing fitting, there are many discrepancies in the neuro-oncologic literature. These discrepancies were particularly flagrant with regard to tumor-defining cut-points for metabolite ratios that varied widely among various centers. One of the key obstacles to greater use of MRS and MRSI in neuro-oncology has been the lack of a uniform approach to data analysis. Data compatibility is needed for multi-center comparisons that are indispensable for wider routine application of MRS in neuro-oncology and elsewhere [306]. Differences in data processing methods, rather than real differences, are considered as one of the major contributors to deviations among various clinical results in MRS [259].

At odds with the need for data compatibility is the requirement within the FFT for fitting, which, as discussed, can lead both to spurious peaks (over-modeling or over-fitting) and true metabolites being undetected (under-modeling or under-fitting) and it also renders inter-study comparisons tenuous, at best. This is, of course, particularly unacceptable in the area of brain tumor diagnostics in which the most delicate of clinical decisions are made, since these decisions require maximal information of the highest possible reliability. It is precisely here that the advances in computational methods are uniquely contextualized – for the promise they hold in helping the fight against one of the most difficult cancers known to humankind [114].

8.3.2 High-resolution quantification of brain MR signals in a clinician-friendly format

As reviewed in detail in the previous chapters in this book, benchmark studies have now been performed showing that the fast Padé transform can yield highly accurate metabolite concentrations from MRS signals. We have emphasized the clinical urgency of this information, particularly for cancer diagnostics. We now demonstrate how this methodology could be implemented in practice, as a hands-on tool for clinicians.

In Fig. 8.2 we give an example of how this can be achieved, from the analysis performed in chapter 3. This was for an MRS time signal that closely matches FIDs encoded via proton MRS from the brain of a healthy volunteer [88]. We first show the parameters (position, width and absolute value of the amplitude) for each of the 25 reconstructed peaks on the upper left panel (i) in Fig. 8.2. On the upper right panel (iv) in Fig. 8.2, we give the metabolite assignments. The T_2 relaxation times and concentrations for each of these 25 metabolites are also shown, as computed from the Padé-reconstructed amplitudes. Here, the Larmor frequency is 63.864 MHz which corresponds to the static magnetic field strength B_0 =1.5T. The fraction of the concentration of a given metabolite is also displayed relative to that which is most abundant. In this example, peak number 6 corresponding to NAA at 2.065 ppm has the highest concentration. Thus, for instance, peak number 16 associated with choline at 3.239 ppm, has a concentration of 6.240 mM/g ww and, therefore, its fraction relative to NAA is 0.56.

The component spectrum is presented with the peak numbers on the left middle panel (ii) in Fig. 8.2 and the total shape spectrum on the right middle panel (v). It is important to compare these two middle panels, from which it can be seen that many of the peaks are closely overlapping, and could not be identified from the total shape spectrum. On the bottom left panel (iii), the peak assignments plus peak numbers are given for each of the 25 components, with the localization of the peaks. The absorption total shape spectrum (as the sum of all the absorption component shape spectra with all the peak assignments) is shown on the bottom right panel (vi).

The advantages of the Padé analysis are particularly apparent when comparing the left and right bottom panels (iii) and (vi) in Fig. 8.2. There, it can be seen how tenuous it is to make any attempt whatsoever to surmise which components are actually hidden underneath a spectral structure. Thus, rather than reconstructing the mobile lipids under the two broad structures in the range 1 - 2 ppm, as done unambiguously by the FPT, equally acceptable (in the least-square sense) results of fitting by the usual methods from MRS (e.g., VARPRO, AMARES, LCModel, etc.) could "reconstruct" two, three, four or more peaks that would all give the same absorption total shape spectrum from 1 ppm to 2 ppm, similarly to the Lanczos paradox [78, 148]. Even more serious problems with clinically unacceptable uncertainties stemming from fittings are found to occur in many other parts of the spectrum from panel (vi) in Fig. 8.2. Notably, any attempts to use fitting in order to ascertain that the peaks close to 2.7 ppm are, in fact, nearly degenerate (each comprised to two components with exceedingly close chemical shifts differing from each other by 0.001 ppm) would be virtually impossible.

We consider that Fig. 8.2 would be very helpful for clinicians, since it provides both a graphic and a quantitative overview of MRS. By enabling repeated cross-checking between these two presentations, the clinician can

EXACT RECONSTRUCTION of SPECTRAL PARAMETERS, CONCENTRATIONS and ABSORPTION COMPONENT SPECTRA : $FPT^{(-)}$

Peak #	Position	Width	|Amplitude|
1	0.985	0.180	0.122
2	1.112	0.257	0.161
3	1.548	0.172	0.135
4	1.689	0.118	0.034
5	1.959	0.062	0.056
6	2.065	0.031	0.171
7	2.145	0.050	0.116
8	2.261	0.062	0.092
9	2.411	0.062	0.085
10	2.519	0.036	0.037
11	2.675	0.033	0.008
12	2.676	0.062	0.063
13	2.855	0.016	0.005
14	3.009	0.064	0.065
15	3.067	0.036	0.101
16	3.239	0.050	0.096
17	3.301	0.064	0.065
18	3.481	0.031	0.011
19	3.584	0.028	0.036
20	3.694	0.036	0.041
21	3.803	0.024	0.031
22	3.944	0.042	0.068
23	3.965	0.062	0.013
24	4.271	0.055	0.016
25	4.680	0.108	0.057

(i) N_P=256: Position (ppm), Width (ppm), |Amplitude| (au)

Peak #	Chem. Sh.	Relax. t.	Concentr.	Fraction	Metabol.
1	0.985	0.087	7.930	0.71	Lip
2	1.112	0.061	10.46	0.94	Lip
3	1.548	0.091	8.775	0.79	Lip
4	1.689	0.133	2.210	0.20	Lip
5	1.959	0.253	3.640	0.33	Gaba
6	2.065	0.505	11.12	1.00	NAA
7	2.145	0.313	7.540	0.68	NAAG
8	2.261	0.253	5.980	0.54	Gaba
9	2.411	0.253	5.525	0.50	Glu
10	2.519	0.435	2.405	0.22	Gln
11	2.675	0.474	0.520	0.05	Asp
12	2.676	0.253	4.095	0.37	NAA
13	2.855	0.979	0.325	0.03	Asp
14	3.009	0.245	4.225	0.38	Cr
15	3.067	0.435	6.565	0.59	PCr
16	3.239	0.313	6.240	0.56	Cho
17	3.301	0.245	4.225	0.38	PCho
18	3.481	0.505	0.715	0.06	Tau
19	3.584	0.559	2.340	0.21	m–Ins
20	3.694	0.435	2.665	0.24	Glu
21	3.803	0.652	2.015	0.18	Gln
22	3.944	0.373	4.420	0.40	Cr
23	3.965	0.253	0.845	0.08	PCr
24	4.271	0.285	1.040	0.09	PCho
25	4.680	0.145	3.705	0.33	Water

(iv) N_P=256: Ch. Sh. (ppm), Relax. T_2 (s), Conc. (mM/g ww)

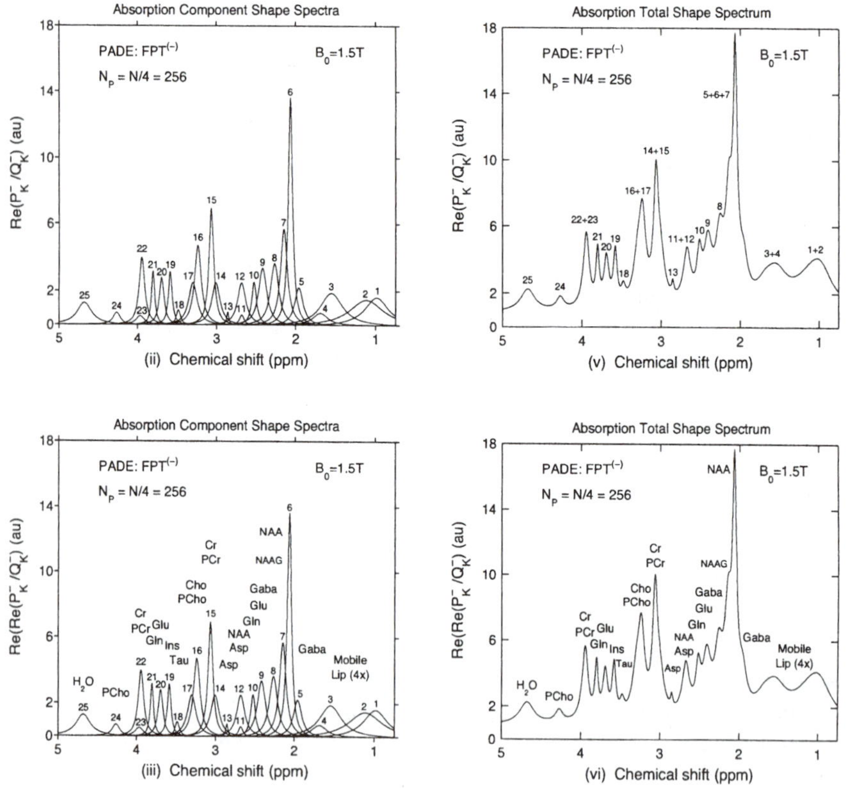

(ii) Chemical shift (ppm)

(v) Chemical shift (ppm)

(iii) Chemical shift (ppm)

(vi) Chemical shift (ppm)

FIGURE 8.2

Spectral peak parameters [panel (i)], metabolite concentrations [panel (iv)] and absorption component [panels (ii), (iii)] and total [panels (v), (vi)] shape spectra for a synthesized FID from chapter 3 reminiscent of time signals typically encoded clinically from healthy human brain via MRS at B_0 =1.5T, TE = 20 ms, full length $N = 1024$, bandwidth 1000 Hz and $\tau = 1$ ms [88].

readily acquire a deeper grasp of the method, together with acumen in interpretation of patterns typical of malignancy versus benign pathologies.

This approach is illustrated in the next part of this book, with the spectral features, metabolite concentrations and assignments from MRS, comparing cancerous and benign pathologies and normal tissue. The benchmark studies described earlier in this book, together with the practical implementation described in this chapter represent a valid, exact, and long awaited approach to quantification of MRS signals. With this type of implementation, we foresee that Padé-optimized MRS is set to very soon become a standard diagnostic tool for various branches oncology, including neuro-oncology.

8.3.3 Padé-reconstructed lipids in the MR brain spectrum

Let us now return in more detail to the example of lipids within the frequency range of 0.9 ppm and 1.7 ppm. Depending upon the presence of necrosis, lipids may or may not be detected in brain tumors. Lipids have generally been described as absent in normal brain tissue [207, 211]. In the study by Tarnawski *et al.* [225] there was no detected lipid whatsoever in brain MRS recordings from thirty healthy volunteers, and significantly higher lipid/NAA was found in the tumor bed of fifty-one patients (3.33 ± 3.79) compared to 0.73 ± 0.85 on the contra lateral, non-involved lobe.

However, in the study of Smith *et al.* [234], while the lipid to creatine ratio was lower among 5 healthy volunteer compared to twenty patients with brainstem neoplasms, two of the five healthy participants had some lipid detected at 1.3 ppm as opposed to eighteen of the twenty patients with tumors. Since the appearance of lipids often indicates necrosis or disruption of the myelin sheath, lipids are commonly detected in non-neoplastic intra-cerebral processes such as abscesses, hypoxia and infarction, demyelinization, toxoplasmosis, as well as for about 1 week post-epileptic seizures [211, 214]. Lipids at 1.3 ppm were found in thirteen of fourteen patients with non-neoplastic lesions in the study by Smith *et al.* [234]. Lipid concentration may also be an indicator of brain tumor grade. As stated by Murphy *et al.* [251]: "the spectra of glioblastomas are dominated by huge lipid - lactate resonances, and with few other features" (p. 330). Several investigations show that the presence or greater concentration of lipids is associated with higher-grade intra-cerebral tumors [239, 242, 251, 252].

In light of this particular example of lipids in the region between 0.9 ppm and 1.7 ppm, it becomes clear how vital it is to have trustworthy information about the number of resonances and all their spectral parameters. Continuing with this example, we can see that from the total shape spectrum (bottom right panel in Fig. 8.2), one can only guess about the number of underlying resonances within the region between 0.9 ppm and 1.7 ppm. However, with the FPT, it can be unequivocally determined that in the present example there are exactly four resonances, as per the input parameters defined earlier. All these parameters: position, width and amplitude are precisely and unequivocally reconstructed.

From the peak parameters presented on the top panel in Fig. 8.2, the computed metabolite concentrations of the four mobile lipid components are also presented therein. It can be clearly seen on the bottom left panel (iii) in Fig. 8.2 that there are four components, whereas only two can be appreciated from the total shape spectrum. This unambiguous quantitative information extracted by the FPT via parametric analysis could be used to develop normative data bases for mobile lipid concentrations in the normal brain versus the corresponding findings seen in brain tumors, in order to provide needed standards to aid in neuro-diagnostics, identifying gliomas of various grades versus non-malignant pathology with specific patterns of departures from normal concentrations of these lipid components.

8.3.4 Padé reconstruction of the components of total choline at 3.2 ppm to 3.3 ppm on the MR brain spectrum

As seen in Fig. 8.2 in the spectral region between 3.2 ppm and 3.3 ppm there are two closely lying peaks of choline and of phosphocholine, at 3.239 ppm and 3.301 ppm. On the absorption total shape spectra on the bottom right panel in Fig. 8.2, these appear as a single peak. In contrast, these two resonances are clearly distinguished on the absorption component spectra (bottom left panel in Fig. 8.2). The importance for oncologic diagnostics of identifying and exactly quantifying the components of choline has been repeatedly emphasized, not only within the realm of neuro-oncology [268, 301, 302], but also for diagnostics within breast, prostate and other cancers [32, 33], [307]–[310].

8.3.5 Padé reconstruction in the region between 3.6 ppm and 4.0 ppm on the MR brain spectrum

A similar approach can be also helpful for other regions of the spectrum in which there have been major, clinically-important dilemmas in neuro-diagnostics. These regions include 3.6 ppm to 4.0 ppm, where, as discussed, there has been uncertainty as to the number of components. Here, the FPT was shown to precisely identify the three resonances, assigned as glutamine 3.803 ppm, creatine 3.944 ppm and phosphocreatine (PCr) 3.965 ppm. This was achieved even though PCr was completely underlying creatine, as seen on the absorption component shape spectrum on the bottom left panel in Fig. 8.2. Once again, from the total shape spectrum these two resonances cannot be distinguished (bottom right panel of Fig. 8.2).

Overall, it is remarkable that thus far MRS and MRSI have made tremendous strides in neuro-diagnostics by relying merely upon a handful of metabolites or their ratios. We can anticipate that Padé-optimized MRS could indeed be of vital important to further strides.

9

Padé quantification of malignant and benign ovarian MRS data

Ovarian cancer is estimated to be the seventh leading cause of death from cancer among women worldwide [311]. It is the most common cause of death from gynecological cancer in many developed countries and has the highest mortality rate of all gynecological cancers [312, 313]. Despite progress in treatment, mortality rates from ovarian cancer have not substantially improved over the last 30 years [314]. This is due mainly to late detection after the disease has spread beyond the true pelvis, since when confined within the ovary (stage Ia) the 5-year survival rate is well over 90% [315]–[317]. Survival, therefore, depends upon early detection of ovarian cancer.

However, this is still beyond current reach with standard diagnostic methods [7, 318]. For example, although useful for following patients with established ovarian cancer, the tumor marker CA-125[1] does not yield improved early detection when used alone [318]–[320]. Some recent studies have suggested certain genomic and proteomic patterns as possible complementary biomarkers for early ovarian cancer [321]–[323]. However, more investigation is certainly needed to determine whether these provide enough improvement in sensitivity and specificity to support their routine use for ovarian cancer screening [313, 315].

Transvaginal ultrasound (TVUS) is sensitive but not sufficiently specific to distinguish benign from malignant adnexal lesions. The high rate of false positive findings with TVUS leads to many surgical procedures that do not detect any cancer [315, 318].

Large-scale trials are ongoing to determine whether the combination of TVUS plus CA-125 could provide acceptable levels of diagnostic accuracy for ovarian cancer screening. Unfortunately, however, this does not appear to be the case [314]. In the Prostate, Lung, Colorectal and Ovarian (PLCO) Trial in which approximately 40000 women participate, the positive predictive value of abnormal TVUS plus CA-125 was very low: 23.5%. Thus, to find 26 cases of ovarian cancer (plus 3 other cancers), 535 women underwent surgical procedures [324].

[1]CA-125 is a protein whose presence is often associated with ovarian cancer. However, it has poor sensitivity for early stage malignancy and is also non-specific, being present in other cancers, as well as several non-malignant conditions, including pregnancy.

In a randomized multi-center study from Japan, this strategy did not lead to a significant increase in stage I ovarian cancer detection among asymptomatic women who had passed menopause [325]. There were untoward consequences of such poor specificity.

In the PLCO Trial study, e.g., false-positive findings were significantly associated with lower adherence to the Trial and with emotional distress [326, 327]. In light of the high false positive rates of the existing screening methods, the U.S. Preventive Services Task Force [328] has recommended against routine screening for ovarian cancer.

Magnetic resonance based modalities have provided some improvement in the accuracy with which ovarian cancer is diagnosed. For example, the positive predictive value of TVUS can be enhanced by combination with MRI, as well as with Doppler flow imaging or CT. MRI is considered to be the most accurate of these other morphological imaging techniques for assessing adnexal masses prior to surgery, as well as for distinguishing benign from cancerous lesions and in some cases for making a specific diagnosis [329, 330]. In a meta-analysis comparing the three morphological imaging modalities, MRI was of greatest incremental value in identifying ovarian cancer when the findings on TVUS were indeterminate [331]. Nevertheless, with contrast-enhanced MRI, which provided the best results, there were still sixty of 241 (24.9%) false positive findings from initial TVUS that were not identified as being benign [331].

In contradistinction to CT, it should also be taken into account that MRI and other MR-based diagnostic techniques entail no exposure to ionizing radiation. This could be a particularly relevant consideration for groups at increased risk of ovarian cancer, for whom there is evidence that exposure to diagnostic medical radiation may be associated with further elevation in risk for radiation-induced cancer [332, 333].

Magnetic resonance imaging yields high spatial resolution to examine morphology. However, in order to assess more than the anatomic aspects, namely, to assess biochemistry and tissue physiology of the ovary, functional methods are needed. As a non-invasive technique providing metabolic information, MRS enables ovarian tissue characterization at a biochemical level, and therefore complements MRI for ovarian diagnostics. Via MRS, abnormalities can also be detected that are not seen MRI, because, as discussed, metabolic changes frequently precede morphological alterations.

9.1 Studies to date using *in vivo* proton MRS to evaluate benign and malignant ovarian lesions

There have been several published investigations using *in vivo* proton MRS to assess ovarian lesions [334]–[339]. As of early 2009, there have been in total

39 cancerous and 53 to 57 benign lesions[2] of various histopathology, as well as one borderline cancerous adnexal mass that have been examined by means of *in vivo* MRS.

The two most recent investigations [334, 339] of benign and malignant ovarian lesions were performed using a 3T MR scanner, whereas the other studies [335]–[338] utilized 1.5T scanners. A limited number of peaks were visible and these were assigned as follows: lipid at 1.3 ppm and an inverted lactate doublet also at around 1.3 ppm, creatine at 3.0 ppm, choline at 3.2 ppm and lipid at 5.2 ppm. A prominent but unassigned resonance at 2.07 ppm was reported by Stanwell *et al.* [339]. The metabolite concentrations were estimated qualitatively (present or absent) or semi-quantitatively (strongly present, present, possibly present or absent)[3]. Ratios were reported in some of the studies.

Choline at 3.2 ppm was detected in 27 of the 32 malignant ovarian lesions (84.4%) versus 14 to 18 of 30 to 34 benign ovarian lesions (41.2% - 60%). Stanwell *et al.* [339] reported that choline to creatine ratios above 3 were found for all seven malignant ovarian lesions, whereas this ratio was below 1.5 for six of the seven benign lesions (in the 7th benign lesion, a serous cystadenofibroma, the Cho/Cr ratio was 3.13).

In eleven of the eighteen cancerous ovarian lesions from Refs. [335, 337, 338], lipid at 1.3 ppm was reported (61.1%) and in twenty of thirty-three non-malignant lesions of the ovary (60.6%). One of the studies [335] also noted the presence of a lipid peak at 5.2 ppm only among patients with certain non-malignant processes affecting the ovary. The peak at 5.2 ppm was seen in 1 of 4 patients with endometriosis, 2 of 11 patients with teratomas, and in the single patient with an ectopic pregnancy, but in none of the 6 patients with benign epithelial tumors nor in the single patient with salpingitis.

Data concerning the inverted lactate doublet at 1.3 ppm were provided in three of the studies [337]–[339], altogether this doublet was observed in 11 of 18 (61.1%) and in 7 of 27 (25.9%) of the malignant and benign ovarian lesions, respectively. In Refs. [338, 339] the amplitude of the lactate doublet was observed to be larger in the malignant compared to the benign lesions. The unassigned resonance at 2.07 ppm was detected in two of seven (28.6%) ovarian cancers, and in four of the seven (57.1%) benign lesions (teratomas and serous cystadenomas) [339].

Based upon this summary of the reported findings to date from *in vivo* MRS, it can be concluded that none of the qualitatively or semi-quantitatively assessed metabolites provided sufficient distinction between malignant and benign ovarian lesions. Metabolite ratios (choline to creatine) were reported in a total of only fourteen lesions [339] and yet there was an overlap between

[2]In Ref. [334] the total number of benign gynecological lesions is given, and the various types described. From this description it can be deduced that at least two and at most six of these lesions were ovarian.

[3]Not all the studies reported on all of the mentioned metabolites. For example, choline was not described in any way with respect to the thirteen patients examined in Ref. [335].

the lowest value (3.09) for a high-grade ovarian cancer and the highest value (3.15) for a benign ovarian lesion.

It is still technically difficult to encode good quality time signals *in vivo* from the ovary, due to motion artefacts from respiratory and peristaltic movements [335]. Because of the small size and motion of this organ, *in vivo* MRS of the ovary is mired by problems of resolution and SNR. Nevertheless, initial results, particularly at higher magnetic field strength and with meticulous attention to technical considerations such as voxel placement and localized shimming suggest that *in vivo* MRS [339] could potentially yield clinically important information for ovarian cancer diagnostics.

9.2 Insights for ovarian cancer diagnostics from *in vitro* MRS

There are quite a bit more published data on ovarian cancer using *in vitro* MRS. The findings from these investigations generally indicate a better distinction between malignant from benign ovarian lesions. Moreover, insights can be gleaned about molecular mechanisms.

Wallace *et al.* [340] analyzed 19 normal ovarian samples, 3 that had borderline pathology and 37 ovarian carcinomas. Their analysis was based upon amplitude ratios of peaks at 0.9 ppm (lipid methyl), 1.3 ppm (lipid methylene), 1.7 ppm (lysine and polyamines) and 3.2 ppm (choline). The normal or benign samples were distinguished from borderline and neoplastic ovarian samples with a sensitivity of 95% and specificity of 86% [340].

Smith and Blandford [341] distinguished normal and benign from borderline and cancerous ovary with 95% sensitivity and 86% specificity. They used linear discriminant analysis training using leave-one-out (12 normal, 22 cancer) for analysis of 7 normal and 15 cancer specimens from the ovary. There were six discriminating peaks: 1.47 ppm (fatty acid), 1.68 ppm (lysine), 2.80 ppm (fatty acid), 2.97 ppm (creatine), 3.17 ppm (choline) and 3.34 ppm (taurine).

In their study of fluid samples from 9 malignant and 19 ovarian cysts, Massouger *et al.* [342] reported higher concentrations of lactate, isoleucine, valine, methionine and alanine in the cancerous specimens, but generally with wide, overlapping ranges. These authors also found higher 3-hydroxybutyrate and pyruvic acid in the malignant ovarian cyst fluid, noting that rapid cellular metabolism will lead to elevated 3-hydroxybutyric acid. The high concentrations of branched chain amino acids (isoleucine, leucine, and valine) were considered to be protein breakdown products related to proteolysis and necrosis. This investigation [342] also included an endometrioma located in the adnexal region, whose fluid showed much higher levels of isoleucine, valine, threonine, alanine, lysine, methionine, and glycine than did the cancerous ovarian cysts.

Boss *et al.* [343] carried out an extensive analysis of the spectroscopic features of cancerous in comparison benign ovarian cysts. The cystic fluid samples were from 12 patients with malignancy and 23 with benign ovarian fluid. Both one- and two-dimensional *in vitro* MRS analyses were performed. There were many differences in metabolite concentrations distinguishing the cancerous and benign cysts in that study [343]. For instance, the concentrations of isoleucine (1.02 ppm), valine (1.04 ppm), threonine (1.33 ppm), lactate (1.41 ppm), alanine (1.51 ppm), lysine (1.67 ppm – 1.78 ppm), methionine (2.13 ppm), glutamine (2.42 ppm – 2.52 ppm) as well as choline (3.19 ppm) were all significantly higher in the malignant samples. However, no metabolites were identified that yielded complete distinction between malignant and non-malignant cyst fluid.

Even though there have been noteworthy results for ovarian cancer diagnosis, major problems remain that hinder broader use of *in vitro* MRS in this clinical area. Due to these problems, *in vitro* findings still cannot be considered the "gold standard" with which MRS signals encoded *in vivo* from the ovary could be compared. Mountford et al. [344] suggested that via the statistical classification strategy (SCS) highly accurate distinction could be made between malignant and benign tissue, based upon identification of specific spectral regions of key diagnostic importance. For ovarian cancer, 2D MRS was particularly important because of overlapping resonances, and was reported to "provide unequivocal assignment of resonances from chemical species that contribute to the various pathological states defined during (ovarian) tumor development and progression" (p. 3692). However, in his comments on the SCS used in the above-described *in vitro* MRS analyses, Gluch [345] questions the suitability of this methodology for more complex pathological entities, stating: "A classifier can more readily be developed when the likelihood is high of belonging to a class of either 'yes' or 'no', but when a tissue undergoes numerous stages in evolution from normal to malignant, SCS shows no superiority over conventional pathology" (p. 467). Notably, the high diagnostic accuracy was achieved by excluding the fuzzy samples.

In corroboration with the statement of Gluch [345], it should be pointed out that even in the study by Boss *et al.* [343], which has, to date, yielded the most extensive *in vitro* analysis of MRS signals from benign and cancerous ovarian samples, the cited ranges for each of the metabolites were wide and overlapping. As a clinician, Gluch [345] enumerates a number of key limitations of MRS as a diagnostic tool in oncology. These include lack of reliable *in vitro* databases and lack of specific findings. For example, narrow lipid resonances can also be seen with necrosis, inflammation and other non-malignant processes. He emphasizes that difficulties arise not so much for large lesions with suspicious imaging characteristics, but instead for small or *in situ* lesions. It is precisely for these early stages of cancer that maximally sensitive and specific diagnostic methods are most urgently sought. Putting this into the real-world clinical perspective, Gluch [345] notes the "chance of missing a cancer of the order of only 1% would translate into a significant medicolegal concern, and for this reason clinicians have to err on the side of caution. Substituting one doubtful test with another leads to no greater certainty in clinical decision making" (p. 467).

We emphasize here that estimates of metabolite concentrations, including those of Ref. [343], were mainly obtained via integration of the areas under the peaks in the Fourier absorption spectra. This procedure is subjective since lower and upper integration limits are not unequivocally defined. Integration of peak areas is particularly troublesome when the peaks overlap. This has been termed "spectral crowding", which is known to create quantification problems [346]. It should be pointed out that concentrations of adjacent resonances such as threonine (1.33 ppm), lactate (1.41 ppm) and alanine (1.51 ppm) and the nearly overlapping resonances isoleucine and valine in the region of 1.02 ppm to 1.04 ppm appear to be of major importance for distinguishing benign and cancerous ovarian specimens. Thus, with respect to MRS applied to ovarian cancer diagnostics, exact, unequivocal quantification is needed not only for spectra whose peaks are completely distinct, but also especially for spectra with overlapping resonances. We will now examine how the fast Padé transform can advantageously be used to tackle this clinically urgent task.

9.3 Padé versus Fourier for *in vitro* MRS data derived from benign and malignant ovarian cyst fluid

In Refs. [27, 29] we examined the performance of the FPT applied to time signals as encoded *in vitro* from benign and malignant ovarian cyst fluid at a magnetic field strength of $B_0 \approx 14.1\text{T}$ in a 600 MHz NMR spectrometer from Ref. [343]. We synthesized two FIDs of the type $c_n = \sum_{k=1}^{K} d_k \mathrm{e}^{in\tau\omega_k}$ via a sum of $K = 12$ damped complex exponentials $\exp(in\tau\omega_k)$ $(1 \le k \le 12)$, as in (3.1), with the time-independent amplitudes d_k. These time signals were subsequently quantified using the FPT, as described in [10].

The input data for the spectral parameters were derived from the median concentrations $\{\mathrm{C}_k\}_{k=1}^{12}$ (expressed in μM/L) of twelve metabolites characteristic of benign ovarian cyst fluid from Ref. [343]. These concentrations reported in Ref. [343] were based upon twenty-three patients with benign ovarian cysts. On the other hand, the concentrations related to malignant ovarian cysts correspond to the median values from twelve patients, as given in Ref. [343]. These latter quantities are used as the input data for the malignant case. Table 9.1 displays all the mentioned input data.

In Ref. [343] the time signals were recorded at a bandwidth of 6667 Hz. The inverse of this bandwidth is used for the sampling time τ in the presently synthesized FIDs. The total signal length N from Ref. [343] was selected in accordance with the Fourier resolving power $\Delta\omega_{\min} = 2\pi/T$. This latter

TABLE 9.1
Input data from theoretically generated FIDs reminiscent of *in vitro* MRS data as encoded from benign and malignant ovarian cyst fluid [343]. Hereafter, $\mathrm{Re}(\nu_k)$ denotes the chemical shift as the relative frequency in dimensionless units of parts per million, or ppm, and au denotes arbitrary units.

INPUT DATA : SPECTRAL PARAMETERS, CONCENTRATIONS and METABOLITE ASSIGNMENTS

(i) Benign

n_k^o (Metabolite # k)	$\mathrm{Re}(\nu_k)$ (ppm)	$\mathrm{Im}(\nu_k)$ (ppm)	$\vert d_k \vert$ (au)	C_k (μM/L ww)	M_k (Assignment)
1	1.020219	0.000818	0.003060	10	Isoleucine (Iso)
2	1.040048	0.000821	0.034578	113	Valine (Val)
3	1.330124	0.000822	0.027540	90	Threonine (Thr)
4	1.410235	0.000828	0.758570	2479	Lactate (Lac)
5	1.510318	0.000824	0.089657	293	Alanine (Ala)
6	1.720125	0.000823	0.030906	101	Lysine (Lys)
7	2.130246	0.000819	0.002142	7	Methionine (Met)
8	2.470118	0.000825	0.084149	275	Glutamine (Gln)
9	3.050039	0.000822	0.019278	63	Creatine (Cr)
10	3.130227	0.000821	0.020808	68	Creatinine (Crn)
11	3.190136	0.000820	0.004590	15	Choline (Cho)
12	5.220345	0.000829	0.424419	1387	Glucose (Glc)

(ii) Malignant

n_k^o (Metabolite # k)	$\mathrm{Re}(\nu_k)$ (ppm)	$\mathrm{Im}(\nu_k)$ (ppm)	$\vert d_k \vert$ (au)	C_k (μM/L ww)	M_k (Assignment)
1	1.020219	0.000828	0.024174	79	Isoleucine (Iso)
2	1.040048	0.000831	0.120869	395	Valine (Val)
3	1.330124	0.000832	0.075887	248	Threonine (Thr)
4	1.410235	0.000838	2.000000	6536	Lactate (Lac)
5	1.510318	0.000834	0.179315	586	Alanine (Ala)
6	1.720125	0.000833	0.149939	490	Lysine (Lys)
7	2.130246	0.000829	0.018972	62	Methionine (Met)
8	2.470118	0.000835	0.253366	828	Glutamine (Gln)
9	3.050039	0.000832	0.020196	66	Creatine (Cr)
10	3.130227	0.000831	0.024174	79	Creatinine (Crn)
11	3.190136	0.000830	0.012852	42	Choline (Cho)
12	5.220345	0.000839	0.079559	260	Glucose (Glc)

relation was used for the chosen bandwidth 6667 Hz with the aim of achieving a spectral resolution $\Delta\omega = 0.02$ ppm. This would delineate isoleucine and valine, i.e., the two most closely lying metabolites. The closest integer in the form 2^m for the FID length required in the FFT is thus $N = 2^{15} = 32768$ $= 32$K (K $= 1024$). The FFT spectrum in Ref. [343] was computed by zero-filling the two encoded FIDs each to 64K. It has been verified [27, 29] that 32K signal points augmented by 32K zeros is the first FID length, which simultaneously yields the converged absorption spectra and resolves all the metabolites in the Fourier absorption spectra.

As discussed elsewhere, the FPT resolution is not predetermined by $2\pi/T$ such that a shorter FID length could suffice and in the present problem, we used a total signal length $N = 1024$. Using (3.15) via $2\mathrm{C}_k/\mathrm{C}_{\mathrm{ref}}$ we extracted the input peak amplitudes from the tabulated data from Ref. [343]. The reference concentration $\mathrm{C}_{\mathrm{ref}}$ was taken to be the largest concentration (6536 μM/L) from Ref. [343]. This was the median lactate concentration in the malignant ovarian samples. The phases ϕ_k $(1 \le k \le 12)$ from d_k were all set to zero, so that every amplitude d_k becomes real, $d_k = |d_k|$. The line-widths in Ref. [343] were estimated to be approximately 1 Hz. We allowed the line-widths to have small variations within the interval $\{8.21, 8.32\} \times 10^{-4}$ppm (labeled as $\mathrm{Im}(\nu_k)$ in the Tables).

The diagonal $\mathrm{FPT}^{(-)}$ was used to analyze the FIDs. The expansion coefficients $\{p_r^-, q_s^-\}$ of the polynomials P_K^- and Q_K^- were computed by solving the systems of linear equations from chapter 4 (section 4.10) by treating the product in $G_N(z^{-1}) * Q_K(z^{-1}) = P_K(z^{-1})$, as a convolution [10, 11]. In order to extract the peak parameters, we solve the characteristic equation $Q_K^-(z^{-1}) = 0$. This leads to K unique roots z_k^- $(1 \le k \le K)$, so that the sought ω_k^- is deduced via $\omega_k^- = (i/\tau)\ln(z_k^-)$.

Peak area is proportional to the concentration of the metabolite, relative to the reference concentration, which here is 6536 μM/L, the median lactate concentration in the ovarian cancer fluid samples. This was the largest concentration from Ref. [343], as mentioned. The metabolite concentrations $\{\mathrm{C}_k^-\}$ are thus computed via $\mathrm{C}_k^- = 3268|d_k^-|\mu$M/L ww.

9.3.1 Padé versus Fourier for MRS data derived from benign ovarian cyst fluid

Table 9.2 shows the spectral parameters reconstructed by the FPT for the data derived from benign ovarian cysts, at three signal lengths, $N/32 = 32, N/16 =$ 64 and $N/8 = 128$. At $N/32 = 32$ signal points (upper panel), the results are shown for spectral parameters of the reconstructed nine peaks out of twelve that should be present. Isoleucine and threonine were undetected at $N/32$, and only one peak was identified in the region between 3.07 ppm and 5.22 ppm. The spectral parameters and computed concentration were fully correct only for glucose at 5.220345 ppm. This shows that 32 FID points are insufficient to converge to all the physical resonances.

However, the situation improves dramatically with $N/16 = 64$ signal points, shown on the middle panel in Table 9.2. By comparing Tables 9.1 and 9.2, we demonstrate that the FPT succeeded in reconstructing *exactly* to all six decimal places all the spectral parameters of each of the twelve peaks with $N/16 = 64$ signal points for data derived from benign ovarian cysts. In other words, all the parameters are identical to the input data at the signal length $N/16 = 64$. From the reconstructed spectral parameters at $N/32 = 32$ and $N/16 = 64$, the metabolite concentrations were computed. It can be seen from the comparison of Tables 9.1 and 9.2 that the concentrations retrieved by the FPT for $N/16 = 64$ are exactly equal to the input concentrations.

The bottom panel in Table 9.2 reveals that the convergence is stable with an increased number of signal points. Namely, at $N/8 = 128$, all of the spectral parameters remain identical to those at $N/16 = 64$. At even longer fractions N/M $(M < 8)$ of the full FID including $N = 1024$ $(M = 1)$, we verified that all the peak parameters reconstructed by the FPT remained unchanged.

Figure 9.1 compares the convergence performance of the FFT and the FPT for the absorption total shape spectra at three different signal lengths for the FID corresponding to the benign ovarian cyst data. The three panels on the left present the absorption spectra of the FFT at $N/32 = 32$ (top, (i)), $N/16 = 64$ (middle, (ii)) and $N/8 = 128$ (bottom, (iii)). These spectra generated via the FFT are obviously rough and yield no interpretable information, whatsoever. The panels on the right show the Padé-generated spectra at these same three signal lengths. Concordant with Table 9.2, at $N/32 = 32$, nine of the twelve metabolites are detected and identified via the FPT. To be quantitatively identified, the other three resonances: isoleucine, threonine and choline require 64 signal points. At $N/16 = 64$, all the peak heights are correct, and, in agreement with Table 9.2, the total absorption shape spectrum is fully converged in the FPT at that signal length. At the longer signal length of $N/8 = 128$, the convergence is fully maintained, i.e., it is stable.

The convergence patterns of the FFT and FPT are further compared in Fig. 9.2. The top two panels recapitulate the rapid convergence of the FPT attained at $N/16 = 64$ signal points (right upper panel). At signal lengths $N/32 = 32$ and $N/16 = 64$ the FFT yielded rough and uninformative spectra (middle panels in Fig. 9.2). The bottom panels in Fig. 9.2 depict the convergence pattern of the absorption spectra in the FFT at two large signal lengths ($N = 8\text{K} = 8192, N = 32\text{K} = 32768$) where K denotes the kilobyte (K = 1024). The first FID length for which the positive-definite Fourier absorption spectra are obtained is very high, ($N = 8\text{K}$). All the twelve resonances are seen to be resolved in the FFT at $N = 8\text{K}$ on the lower left panel in Fig. 9.2, but several peak heights are incorrect. This implies that some of the metabolite concentrations estimated from the Fourier spectra by either fitting or peak integrations will be insufficiently accurate even at $N = 8\text{K}$. Moreover, at $N = 8\text{K}$ there are significant baseline distortions which would further compromise both fitting and numerical peak integrations. Eventually,

TABLE 9.2

Padé-reconstructions from a theoretically generated FID reminiscent of *in vitro* MRS data as encoded from benign ovarian cyst fluid [343]. The recovered data converged at $N/16 = 64\,(N = 1024)$ on the middle panel (ii).

CONVERGENCE of SPECTRAL PARAMETERS & CONCENTRATIONS in FPT$^{(-)}$; SIGNAL LENGTH : N/M, N = 1024, M = 8 – 32

(i) N/32 = 32 (Benign)

n_k^o (Metabolite # k)	Re(ν_k^-) (ppm)	Im(ν_k^-) (ppm)	$\vert d_k^- \vert$ (au)	C_k^-(μM/L ww)	M_k (Assignment)
2	1.040033	0.001124	0.038143	124	Valine (Val)
4	1.409673	0.001117	0.796643	2603	Lactate (Lac)
5	1.482370	0.005257	0.081645	266	Alanine (Ala)
6	1.706966	0.009141	0.035360	115	Lysine (Lys)
7	2.207406	0.216536	0.005435	17	Methionine (Met)
8	2.470151	0.000826	0.082607	270	Glutamine (Gln)
9	3.058390	0.001607	0.020043	65	Creatine (Cr)
10	3.141312	0.002566	0.024686	80	Creatinine (Crn)
12	5.220345	0.000829	0.424418	1387	Glucose (Glc)

(ii) N/16 = 64 (Benign)

n_k^o (Metabolite # k)	Re(ν_k^-) (ppm)	Im(ν_k^-) (ppm)	$\vert d_k^- \vert$ (au)	C_k^-(μM/L ww)	M_k (Assignment)
1	1.020219	0.000818	0.003060	10	Isoleucine (Iso)
2	1.040048	0.000821	0.034578	113	Valine (Val)
3	1.330124	0.000822	0.027540	90	Threonine (Thr)
4	1.410235	0.000828	0.758570	2479	Lactate (Lac)
5	1.510318	0.000824	0.089657	293	Alanine (Ala)
6	1.720125	0.000823	0.030906	101	Lysine (Lys)
7	2.130246	0.000819	0.002142	7	Methionine (Met)
8	2.470118	0.000825	0.084149	275	Glutamine (Gln)
9	3.050039	0.000822	0.019278	63	Creatine (Cr)
10	3.130227	0.000821	0.020808	68	Creatinine (Crn)
11	3.190136	0.000820	0.004590	15	Choline (Cho)
12	5.220345	0.000829	0.424419	1387	Glucose (Glc)

(iii) N/8 = 128 (Benign)

n_k^o (Metabolite # k)	Re(ν_k^-) (ppm)	Im(ν_k^-) (ppm)	$\vert d_k^- \vert$ (au)	C_k^-(μM/L ww)	M_k (Assignment)
1	1.020219	0.000818	0.003060	10	Isoleucine (Iso)
2	1.040048	0.000821	0.034578	113	Valine (Val)
3	1.330124	0.000822	0.027540	90	Threonine (Thr)
4	1.410235	0.000828	0.758570	2479	Lactate (Lac)
5	1.510318	0.000824	0.089657	293	Alanine (Ala)
6	1.720125	0.000823	0.030906	101	Lysine (Lys)
7	2.130246	0.000819	0.002142	7	Methionine (Met)
8	2.470118	0.000825	0.084149	275	Glutamine (Gln)
9	3.050039	0.000822	0.019278	63	Creatine (Cr)
10	3.130227	0.000821	0.020808	68	Creatinine (Crn)
11	3.190136	0.000820	0.004590	15	Choline (Cho)
12	5.220345	0.000829	0.424419	1387	Glucose (Glc)

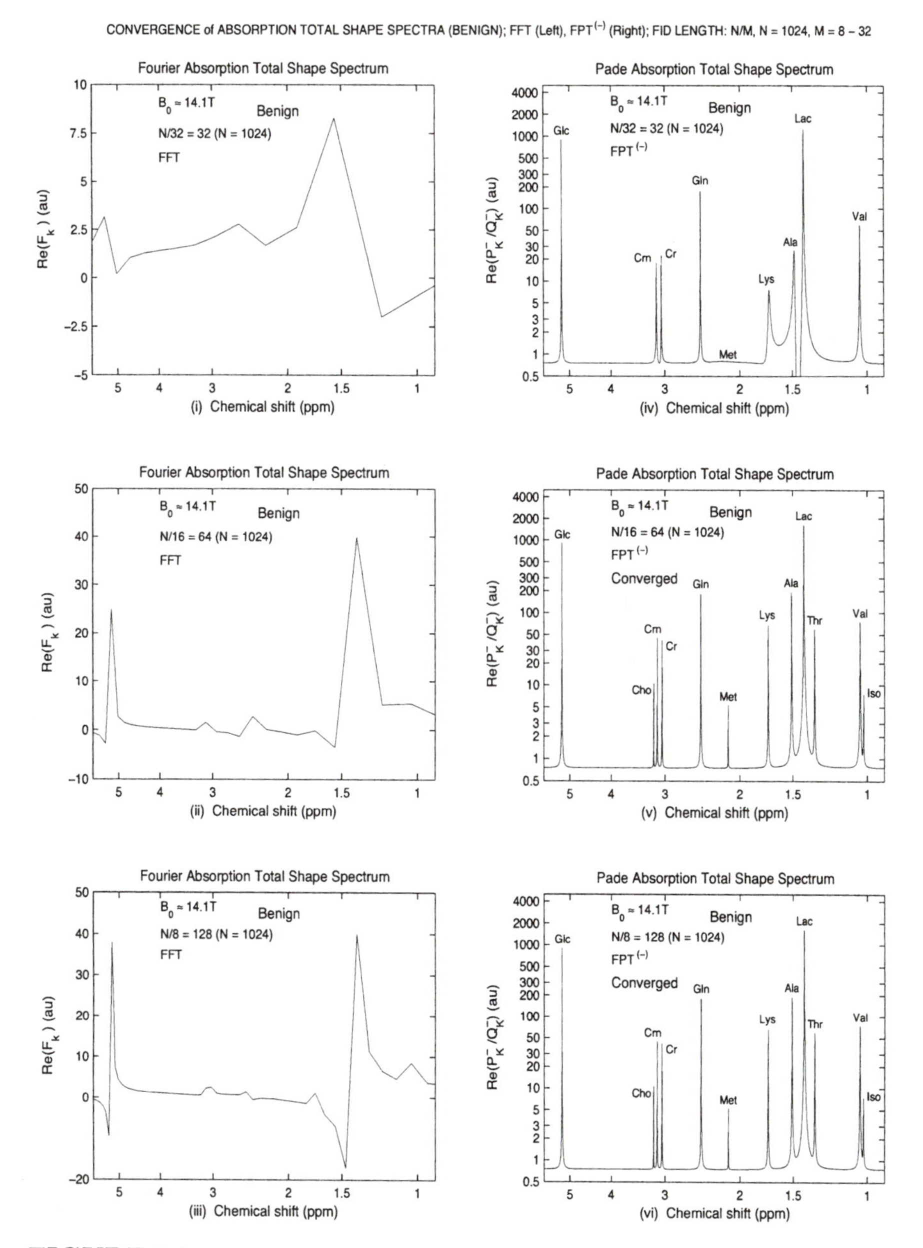

FIGURE 9.1

Convergence of absorption total shape spectra for the benign ovarian case using input data derived from Ref. [343]. The left panels correspond to the FFT and the right panels to the FPT. The signal lengths are $N/32 = 32$ (top panels (i), (iv)), $N/16 = 64$ (middle panels (ii), (v)), and $N/8 = 128$ (bottom panels (iii), (vi)), where N is the full signal length ($N = 1024$).

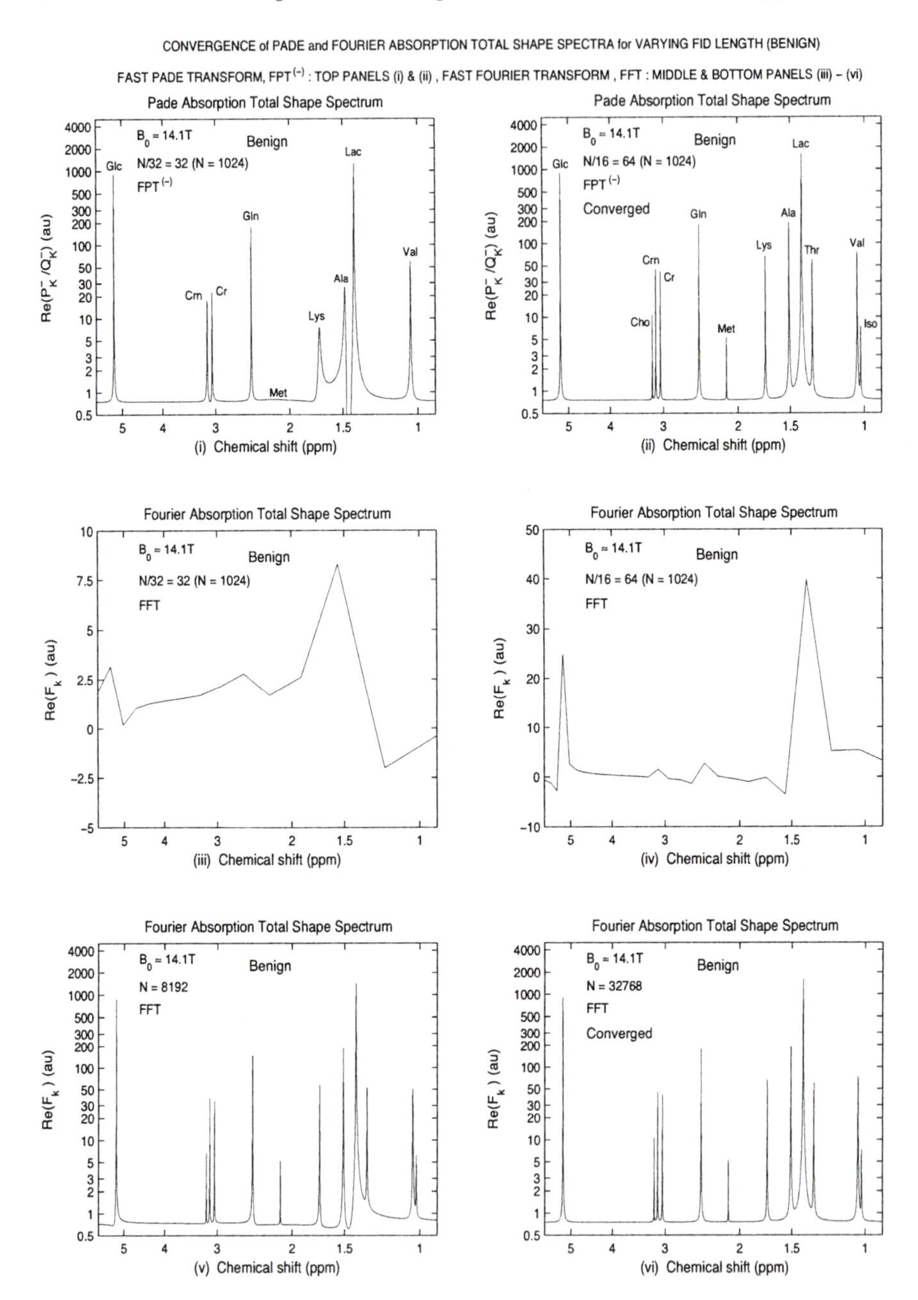

FIGURE 9.2

Absorption total shape spectra for the benign ovarian case using input data derived from Ref. [343]. Padé and Fourier convergence rates: the top panels (i) $[N/32 = 32]$ and (ii) $[N/16 = 64]$ correspond to the FPT, whereas the FFT is on the middle (iii) $[N/32 = 32]$ and (iv) $[N/16 = 64]$, as well as on the bottom panels (v) $[N = 8192]$ and (vi) $[N = 32768]$.

the FFT is found to converge at $N = 32\text{K} = 32768$. Nevertheless, it should be re-emphasized that this Fourier convergence is concerned only with the lineshapes, and not with quantification, which in this methodology necessitates post-processing via fittings or peak integrations that are non-unique and often ambiguous, especially when spectral crowding occurs.

9.3.2 Padé versus Fourier for MRS data derived from malignant ovarian cyst fluid

The spectral parameters reconstructed by the FPT for the data derived from malignant ovarian cysts are shown in Table 9.3, at three signal lengths, $N/32 = 32, N/16 = 64$ and $N/8 = 128$. Once again, at $N/32 = 32$ signal points as presented on the upper panel in Table 9.3, only nine of the twelve peaks were identified. Isoleucine and threonine were undetected at $N/32$, and only one peak was identified in the region between 3.07 ppm and 5.22 ppm. It was only for glucose at 5.220345 ppm that the spectral parameters and computed concentration for the malignant case were fully correct. Thus, 32 FID points were also insufficient to converge to all the physical resonances for the data derived from malignant ovarian cysts.

Shown on the middle panel in Table 9.3, the FPT reconstructed *exactly* to all six decimal places all the spectral parameters of each of the twelve peaks with $N/16 = 64$ signal points for data derived from malignant ovarian cysts. It can be seen from the comparison of Tables 9.1 and 9.3 that the spectral parameters and concentrations retrieved by the FPT for $N/16 = 64$ are exactly equal to the input data.

The bottom panel in Table 9.3 shows that the convergence is also stable at $N/8 = 128$, for which all of the spectral parameters remain identical to those at $N/16 = 64$. At even longer fractions N/M $(M < 8)$ of the full FID including $N = 1024\,(M = 1)$, we verified that all the peak parameters reconstructed by the FPT remained constant for the malignant ovarian data.

With respect to the absorption total shape spectra, a similar pattern is also seen for the malignant case, as illustrated in Fig. 9.3. Namely, the three panels on the left present the absorption spectra of the FFT at $N/32 = 32$ (top, (i)), $N/16 = 64$ (middle, (ii)) and $N/8 = 128$ (bottom, (iii)) which are all rough and uninterpretable. The panels on the right show the Padé-generated spectra at these same three signal lengths. Also concordant with Table 9.3, at $N/32 = 32$, nine of the twelve metabolites are detected and identified via the FPT. Isoleucine, threonine and choline need 64 signal points to be identified. At this latter signal length, all the peak heights are correct, and, in agreement with Table 9.3, at $N/16 = 64$ the total absorption shape spectrum is fully converged in the FPT. At the longer signal length of $N/8 = 128$, the convergence of the FPT is fully maintained, i.e., it is stable for the malignant case, as well.

The convergence patterns of the FFT and FPT for the malignant ovarian cyst data are further compared in Fig. 9.4. The top two panels once again

TABLE 9.3
Padé-reconstructions from a theoretically generated FID reminiscent of *in vitro* MRS data as encoded from malignant ovarian cyst fluid [343]. The retrieved data converged at $N/16 = 64\,(N = 1024)$ on the middle panel (ii).

CONVERGENCE of SPECTRAL PARAMETERS & CONCENTRATIONS in FPT$^{(-)}$; SIGNAL LENGTH: N/M, N = 1024, M = 8 – 32

(i) N/32 = 32 (Malignant)

n_k^o (Metabolite # k)	Re(ν_k^-) (ppm)	Im(ν_k^-) (ppm)	$\lvert d_k^- \rvert$ (au)	C_k^-(μM/L ww)	M_k (Assignment)
2	1.039257	0.001406	0.148343	484	Valine (Val)
4	1.409062	0.001372	2.046372	6688	Lactate (Lac)
5	1.446902	0.008300	0.196825	643	Alanine (Ala)
6	1.712040	0.004160	0.174575	570	Lysine (Lys)
7	2.101571	0.031100	0.028695	93	Methionine (Met)
8	2.470096	0.000838	0.251929	823	Glutamine (Gln)
9	3.069839	0.003200	0.028376	92	Creatine (Cr)
10	3.165277	0.003440	0.028617	93	Creatinine (Crn)
12	5.220345	0.000839	0.079557	260	Glucose (Glc)

(ii) N/16 = 64 (Malignant)

n_k^o (Metabolite # k)	Re(ν_k^-) (ppm)	Im(ν_k^-) (ppm)	$\lvert d_k^- \rvert$ (au)	C_k^-(μM/L ww)	M_k (Assignment)
1	1.020219	0.000828	0.024174	79	Isoleucine (Iso)
2	1.040048	0.000831	0.120869	395	Valine (Val)
3	1.330124	0.000832	0.075887	248	Threonine (Thr)
4	1.410235	0.000838	2.000000	6536	Lactate (Lac)
5	1.510318	0.000834	0.179315	586	Alanine (Ala)
6	1.720125	0.000833	0.149939	490	Lysine (Lys)
7	2.130246	0.000829	0.018972	62	Methionine (Met)
8	2.470118	0.000835	0.253366	828	Glutamine (Gln)
9	3.050039	0.000832	0.020196	66	Creatine (Cr)
10	3.130227	0.000831	0.024174	79	Creatinine (Crn)
11	3.190136	0.000830	0.012852	42	Choline (Cho)
12	5.220345	0.000839	0.079559	260	Glucose (Glc)

(iii) N/8 = 128 (Malignant)

n_k^o (Metabolite # k)	Re(ν_k^-) (ppm)	Im(ν_k^-) (ppm)	$\lvert d_k^- \rvert$ (au)	C_k^-(μM/L ww)	M_k (Assignment)
1	1.020219	0.000828	0.024174	79	Isoleucine (Iso)
2	1.040048	0.000831	0.120869	395	Valine (Val)
3	1.330124	0.000832	0.075887	248	Threonine (Thr)
4	1.410235	0.000838	2.000000	6536	Lactate (Lac)
5	1.510318	0.000834	0.179315	586	Alanine (Ala)
6	1.720125	0.000833	0.149939	490	Lysine (Lys)
7	2.130246	0.000829	0.018972	62	Methionine (Met)
8	2.470118	0.000835	0.253366	828	Glutamine (Gln)
9	3.050039	0.000832	0.020196	66	Creatine (Cr)
10	3.130227	0.000831	0.024174	79	Creatinine (Crn)
11	3.190136	0.000830	0.012852	42	Choline (Cho)
12	5.220345	0.000839	0.079559	260	Glucose (Glc)

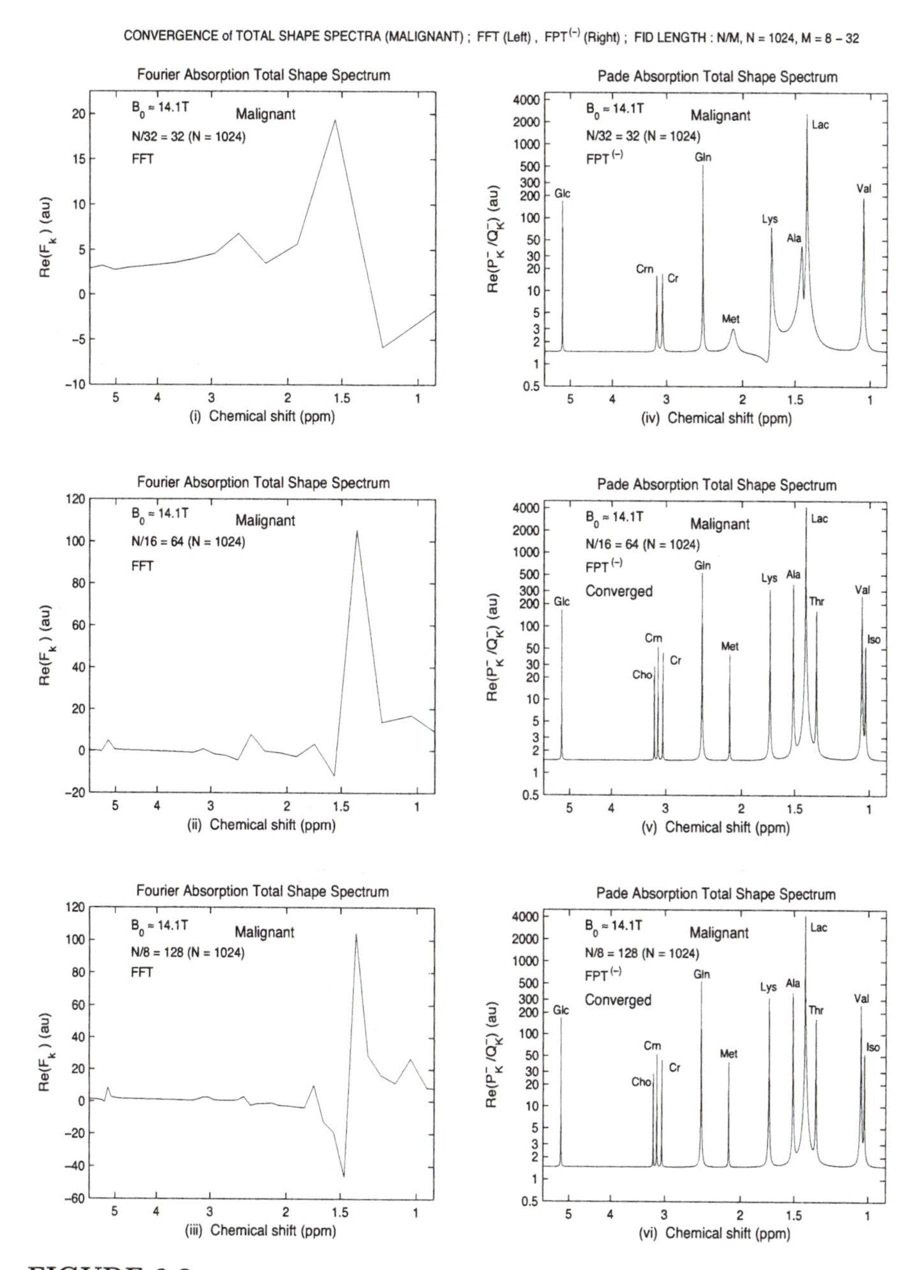

FIGURE 9.3

Convergence of absorption total shape spectra for the malignant ovarian case using input data derived from Ref. [343]. The left panels correspond to the FFT and the right panels to the FPT. The signal lengths are $N/32 = 32$ (top panels (i), (iv)), $N/16 = 64$ (middle panels (ii), (v)), and $N/8 = 128$ (bottom panels (iii), (vi)), where N is the full signal length ($N = 1024$).

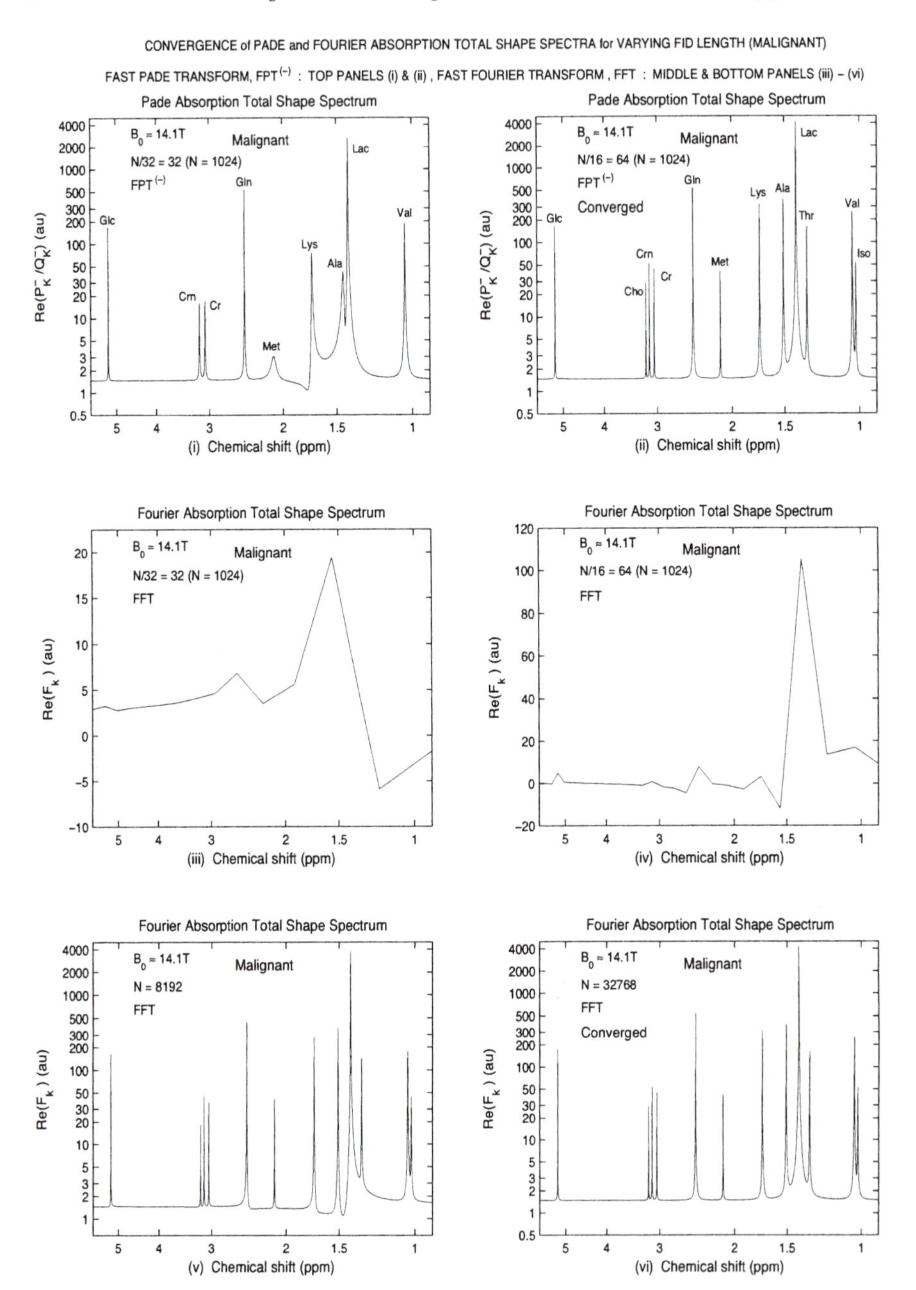

FIGURE 9.4

Absorption total shape spectra for the malignant ovarian case using input data derived from Ref. [343]. Padé and Fourier convergence rates: the top panels (i) $[N/32 = 32]$ and (ii) $[N/16 = 64]$ correspond to the FPT, whereas the FFT is on the middle (iii) $[N/32 = 32]$ and (iv) $[N/16 = 64]$, as well as bottom panels (v) $[N = 8192]$ and (vi) $[N = 32768]$.

illustrate the rapid convergence of the FPT achieved at $N/16 = 64$ signal points (right upper panel). At signal lengths $N/32 = 32$ and $N/16 = 64$ the FFT is also rough and uninformative for the malignant cyst data (middle panels in Fig. 9.4). The bottom panel in Fig. 9.4 shows the convergence pattern of the absorption spectra in the FFT at the two large signal lengths ($N = 8\text{K} = 8192, N = 32\text{K} = 32768$). The first FID length for which the positive-definite Fourier absorption spectra are obtained is at ($N = 8\text{K}$). All the twelve resonances are resolved in the FFT at $N = 8\text{K}$ on the lower left panel in Fig. 9.4. However, several peak heights are incorrect. At $N = 8\text{K}$ there are once again significant baseline distortions which, as noted, would invalidate both fitting and numerical peak integrations. The FFT converges at $N = 32\text{K} = 32768$, as shown on the lower right panel in Fig. 9.4 for the malignant case.

9.3.3 Summary comparisons of the performance of FPT and FFT for MRS data derived from benign and malignant ovarian cyst fluid

Table 9.4 provides a summary of the convergence performance of the FPT for the MRS data derived from benign (left panel) and malignant (right panel) ovarian cyst fluid at the partial signal lengths $N/32 = 32, N/16 = 64$ and $N/8 = 128$. The computed concentrations for both the benign and malignant cases are fully correct at $N/16 = 64$ (middle panels), since the spectral parameters for all twelve metabolites are exact to six decimal places. Stability of convergence is shown on the bottom panels at $N/8 = 128$.

Figure 9.5 shows the performance of the FFT for the absorption total shape spectra, at three much longer signal lengths: $N = 8\text{K} = 8192$ (top panels), $N = 16\text{K} = 16384$ (middle panels) and $N = 32\text{K} = 32768$ (bottom panels). The performance of the FFT for the benign ovarian cyst data is presented on the left column and for the malignant ovarian cyst data on the right column. As noted, at $N = 8\text{K} = 8192$ all the twelve resonances are resolved, but not all their peak heights are correct for the benign nor for the malignant ovarian case. At $N = 16\text{K} = 16384$ the peak heights are still not correct for either case. It is only at $N = 32\text{K} = 32768$ signal points that all the peaks have reached their correct heights for the benign and malignant ovarian cyst data, since convergence has been achieved by the FFT. This figure illustrates the importance of determining the point of convergence, since prior to that point, baseline distortions as well as incorrect peak heights would render all attempts at assessing metabolite concentrations completely tenuous.

The convergence patterns of the reconstructed absorption total shape spectra via the FPT are recapitulated for the benign (left column) and malignant (right column) cases in Fig. 9.6. The three signal lengths are $N/32 = 32$ (top panels), $N/16 = 64$ (middle panels) and $N/8 = 128$ (bottom panels), with convergence attained for both cases at $N/16 = 64$.

Figure 9.7 provides a summary comparison between the signal length at

TABLE 9.4
Comparison of the Padé-reconstructed spectral parameters from a theoretically generated FIDs reminiscent of *in vitro* MRS data as encoded from benign and malignant ovarian cyst fluids [343]. The reconstructed data converged at $N/16 = 64\,(N = 1024)$ on the middle panels (ii) and (v) for the benign and malignant case, respectively.

CONVERGENCE of SPECTRAL PARAMETERS and CONCENTRATIONS in FPT$^{(-)}$; SIGNAL LENGTH : N/M, N = 1024, M = 8 – 32

(i) N/32 = 32 (Benign)

n_k^o (# k)	Re(ν_k^-)(ppm)	Im(ν_k^-)(ppm)	$\vert d_k^-\vert$ (au)	C_k^-(μM/L ww)	M_k
2	1.040033	0.001124	0.038143	124	Val
4	1.409673	0.001117	0.796643	2603	Lac
5	1.482370	0.005257	0.081645	266	Ala
6	1.706966	0.009141	0.035360	115	Lys
7	2.207406	0.216536	0.005435	17	Met
8	2.470151	0.000826	0.082607	270	Gln
9	3.058390	0.001607	0.020043	65	Cr
10	3.141312	0.002566	0.024686	80	Crn
12	5.220345	0.000829	0.424418	1387	Glc

(ii) N/16 = 64 (Benign)

n_k^o (# k)	Re(ν_k^-)(ppm)	Im(ν_k^-)(ppm)	$\vert d_k^-\vert$ (au)	C_k^-(μM/L ww)	M_k
1	1.020219	0.000818	0.003060	10	Iso
2	1.040048	0.000821	0.034578	113	Val
3	1.330124	0.000822	0.027540	90	Thr
4	1.410235	0.000828	0.758570	2479	Lac
5	1.510318	0.000824	0.089657	293	Ala
6	1.720125	0.000823	0.030906	101	Lys
7	2.130246	0.000819	0.002142	7	Met
8	2.470118	0.000825	0.084149	275	Gln
9	3.050039	0.000822	0.019278	63	Cr
10	3.130227	0.000821	0.020808	68	Crn
11	3.190136	0.000820	0.004590	15	Cho
12	5.220345	0.000829	0.424419	1387	Glc

(iii) N/8 = 128 (Benign)

n_k^o (# k)	Re(ν_k^-)(ppm)	Im(ν_k^-)(ppm)	$\vert d_k^-\vert$ (au)	C_k^-(μM/L ww)	M_k
1	1.020219	0.000818	0.003060	10	Iso
2	1.040048	0.000821	0.034578	113	Val
3	1.330124	0.000822	0.027540	90	Thr
4	1.410235	0.000828	0.758570	2479	Lac
5	1.510318	0.000824	0.089657	293	Ala
6	1.720125	0.000823	0.030906	101	Lys
7	2.130246	0.000819	0.002142	7	Met
8	2.470118	0.000825	0.084149	275	Gln
9	3.050039	0.000822	0.019278	63	Cr
10	3.130227	0.000821	0.020808	68	Crn
11	3.190136	0.000820	0.004590	15	Cho
12	5.220345	0.000829	0.424419	1387	Glc

(iv) N/32 = 32 (Malignant)

n_k^o (# k)	Re(ν_k^-)(ppm)	Im(ν_k^-)(ppm)	$\vert d_k^-\vert$ (au)	C_k^-(μM/L ww)	M_k
2	1.039257	0.001406	0.148343	484	Val
4	1.409062	0.001372	2.046372	6688	Lac
5	1.446902	0.008300	0.196825	643	Ala
6	1.712040	0.004160	0.174575	570	Lys
7	2.101571	0.031100	0.028695	93	Met
8	2.470096	0.000838	0.251929	823	Gln
9	3.069839	0.003200	0.028376	92	Cr
10	3.165277	0.003440	0.028617	93	Crn
12	5.220345	0.000839	0.079557	260	Glc

(v) N/16 = 64 (Malignant)

n_k^o (# k)	Re(ν_k^-)(ppm)	Im(ν_k^-)(ppm)	$\vert d_k^-\vert$ (au)	C_k^-(μM/L ww)	M_k
1	1.020219	0.000828	0.024174	79	Iso
2	1.040048	0.000831	0.120869	395	Val
3	1.330124	0.000832	0.075887	248	Thr
4	1.410235	0.000838	2.000000	6536	Lac
5	1.510318	0.000834	0.179315	586	Ala
6	1.720125	0.000833	0.149939	490	Lys
7	2.130246	0.000829	0.018972	62	Met
8	2.470118	0.000835	0.253366	828	Gln
9	3.050039	0.000832	0.020196	66	Cr
10	3.130227	0.000831	0.024174	79	Crn
11	3.190136	0.000830	0.012852	42	Cho
12	5.220345	0.000839	0.079559	260	Glc

(vi) N/8 = 128 (Malignant)

n_k^o (# k)	Re(ν_k^-)(ppm)	Im(ν_k^-)(ppm)	$\vert d_k^-\vert$ (au)	C_k^-(μM/L ww)	M_k
1	1.020219	0.000828	0.024174	79	Iso
2	1.040048	0.000831	0.120869	395	Val
3	1.330124	0.000832	0.075887	248	Thr
4	1.410235	0.000838	2.000000	6536	Lac
5	1.510318	0.000834	0.179315	586	Ala
6	1.720125	0.000833	0.149939	490	Lys
7	2.130246	0.000829	0.018972	62	Met
8	2.470118	0.000835	0.253366	828	Gln
9	3.050039	0.000832	0.020196	66	Cr
10	3.130227	0.000831	0.024174	79	Crn
11	3.190136	0.000830	0.012852	42	Cho
12	5.220345	0.000839	0.079559	260	Glc

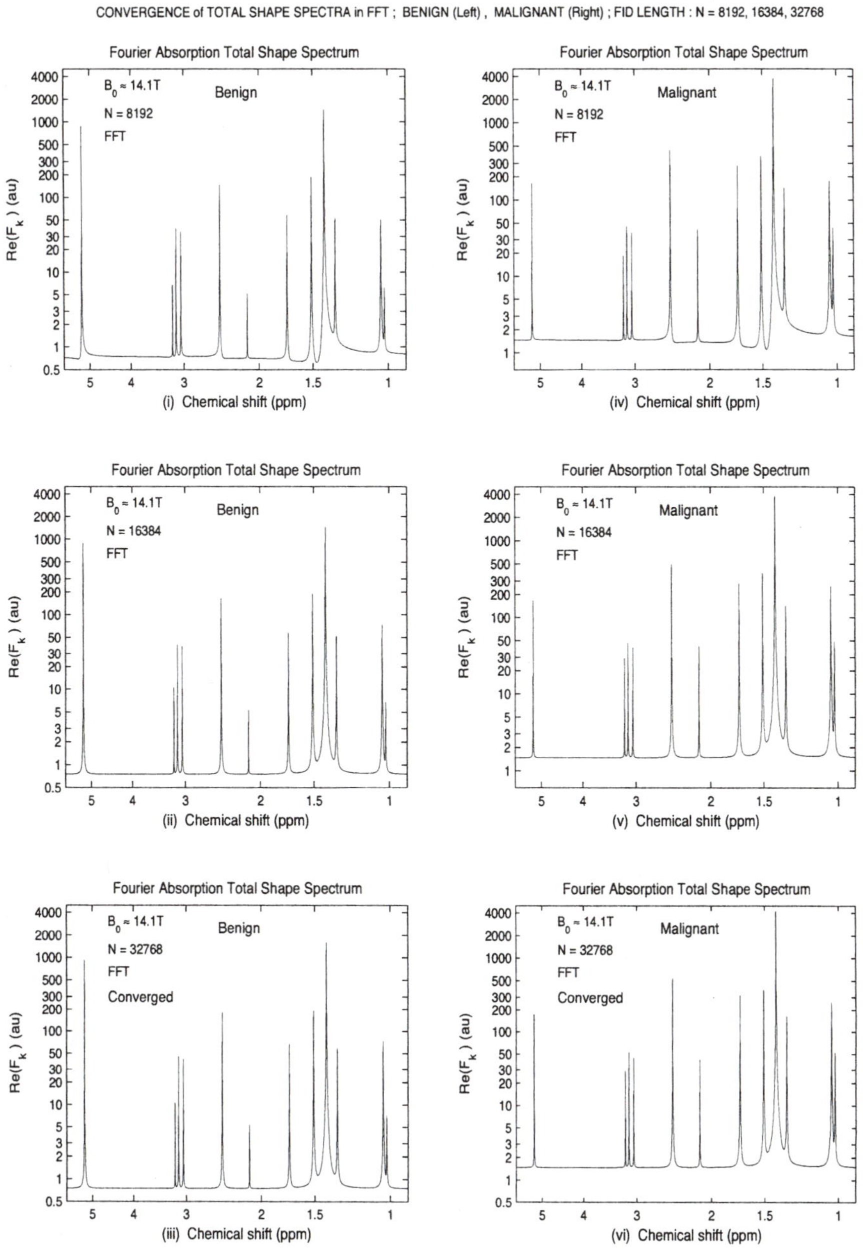

FIGURE 9.5

Convergence of absorption total shape spectra via the FFT for the benign and malignant ovarian cases using input data derived from Ref. [343]. The left and right panels correspond to the benign and malignant case, respectively. The signal lengths are $N = 8192$ (top panels), $N = 16384$ (middle panels) and $N = 32768$ (bottom panels).

which convergence is achieved via the FPT and the shape spectrum attained via the FFT at the same signal length, for both the benign and malignant ovarian data. Fully converged absorption total shape spectra were obtained using only $N/16 = 64$ signal points via the FPT, for the benign case (top right panel (iii)) and malignant case (bottom right panel (iv)). In sharp contradistinction, the spectra generated using the FFT at the latter signal length are completely uninterpretable (top left panel (i), benign case; bottom left panel (ii), malignant case).

The convergence pattern of the metabolite concentrations computed using the FPT for benign versus malignant ovarian cyst fluid is presented in Fig. 9.8. The chemical shifts are displayed along the abscissae of the six panels (i) - (vi), with concentrations as the ordinates. The input data are represented by the symbol x, whereas the Padé-reconstructed data are shown as open circles. The data corresponding to the benign and malignant cases are presented in the left ((i) - (iii)) and right panels ((iv) - (vi)), respectively. Prior to convergence, at $N/32 = 32$ (top panels (i) and (iv)), the only metabolite for which the fully correct concentrations were obtained in both the benign and malignant cases is glucose at 5.22 ppm (1387 μM/L (benign) and 260 μM/L (malignant), respectively). At $N/16 = 64$ (middle panels (ii) and (v)) and $N/8 = 128$ (bottom panels (iii) and (vi)), all of the reconstructed metabolite concentrations are completely correct, as seen both numerically and by the graphic representations. Thus, for $N/16$ and $N/8$, the x's are precisely centered within the open circles. This indicates that there was complete agreement between the input and reconstructed data.

Figure 9.9 is a recapitulation of the absorption spectra and the retrieved concentrations when full convergence is achieved by the FPT using $N/16 = 64$ signal points. This figure provides graphic illustration of the capabilities of the FPT, in providing both a shape estimation and quantification, without any post-processing and without reliance upon any other estimator. We consider Fig. 9.9 to be the most useful for clinicians, by giving both a graphic as well as a quantitative summary of the results obtained via the FPT, besides providing insight into how the FPT actually works. In other words, this figure illustrates the essential concepts of signal processing, by showing both lineshape estimation and quantification. Most importantly in the clinical setting, all the reconstructed concentrations for the benign and malignant cases, i.e., the diagnostically most relevant information can be readily deciphered.

Applying the fast Padé transform to MR data from benign and malignant ovarian cyst fluid clearly shows the powerful extrapolation features of the FPT. With only 64 data points, the Padé absorption spectra are fully converged, including a delineation of closely lying resonances such as alanine, lactate and threonine in the region between 1.3 ppm and 1.51 ppm, and even the nearly overlapping isoleucine and valine which are separated by only 0.02 ppm. In marked contrast, the FFT yielded entirely uninterpretable spectra at these short signal lengths. As discussed earlier in this book, the envelopes of MR time signals decay exponentially such that the signal intensity is the

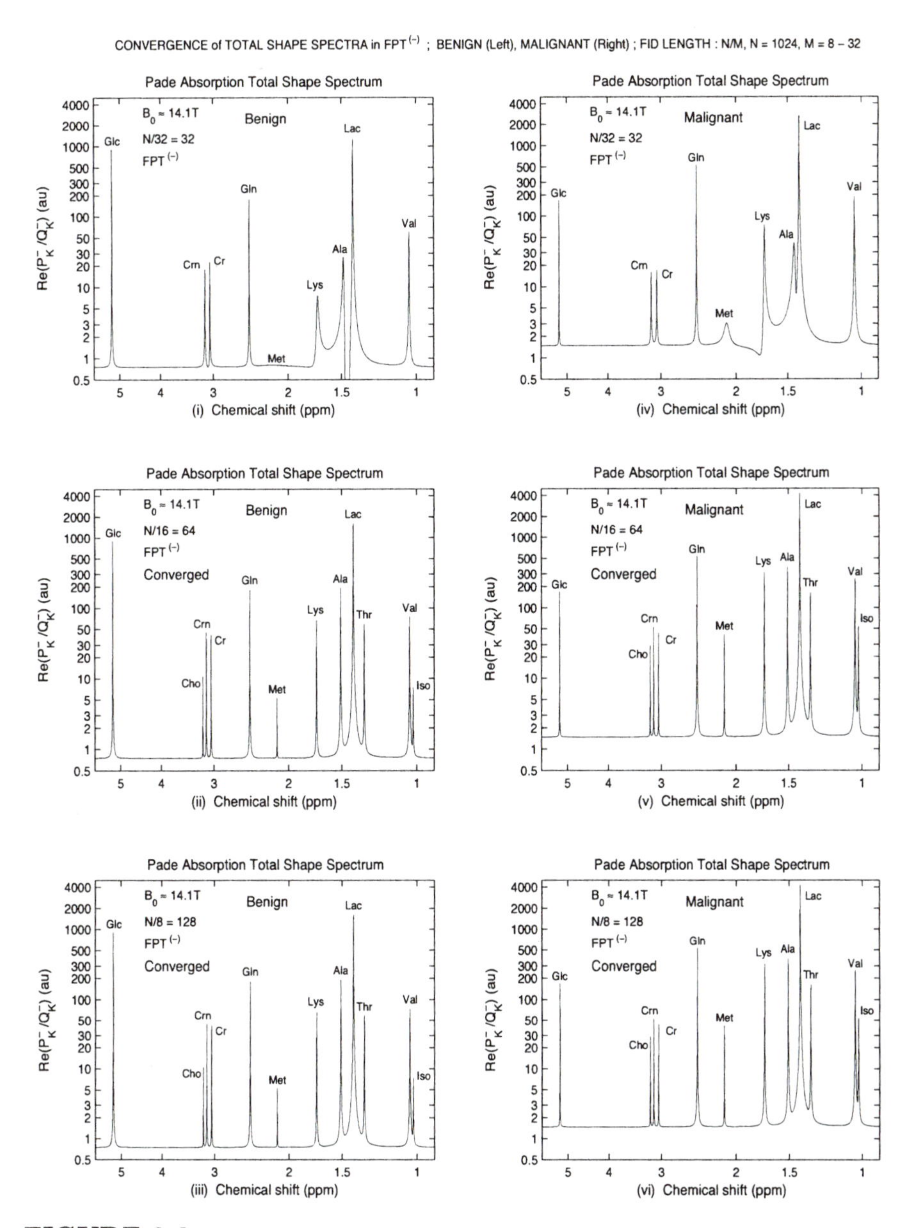

FIGURE 9.6
Convergence of absorption total shape spectra via the FPT for the benign and malignant ovarian cases using input data derived from Ref. [343]. The left panels correspond to the benign case and the right panels to the malignant case. The signal lengths are $N/32 = 32$ (top panels (i), (iv)), $N/16 = 64$ (middle panels (ii), (v)), and $N/8 = 128$ (bottom panels (iii), (vi)), where N is the full signal length ($N = 1024$).

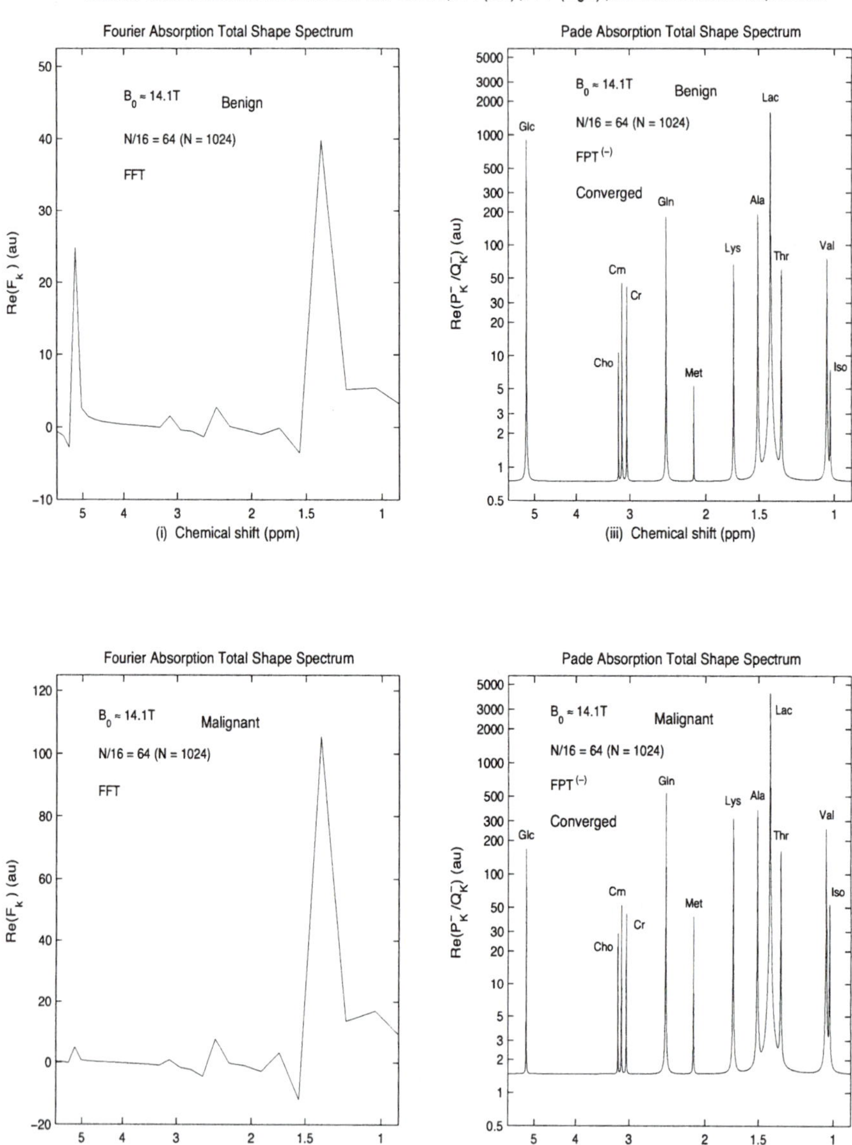

FIGURE 9.7

Absorption spectra for cases derived from benign and malignant ovarian cyst MRS *in vitro* encoded data from Ref. [343]. Top panels compare the performance of the FFT (left (i)) and FPT (right(iii)) at $N/16 = 64$ where $N = 1024$ for the benign case. The FPT is fully converged, while the corresponding FFT-generated spectra are uninterpretable. A similar pattern is seen in the malignant case (bottom panels, FFT (left (ii), FPT (right (iv)).

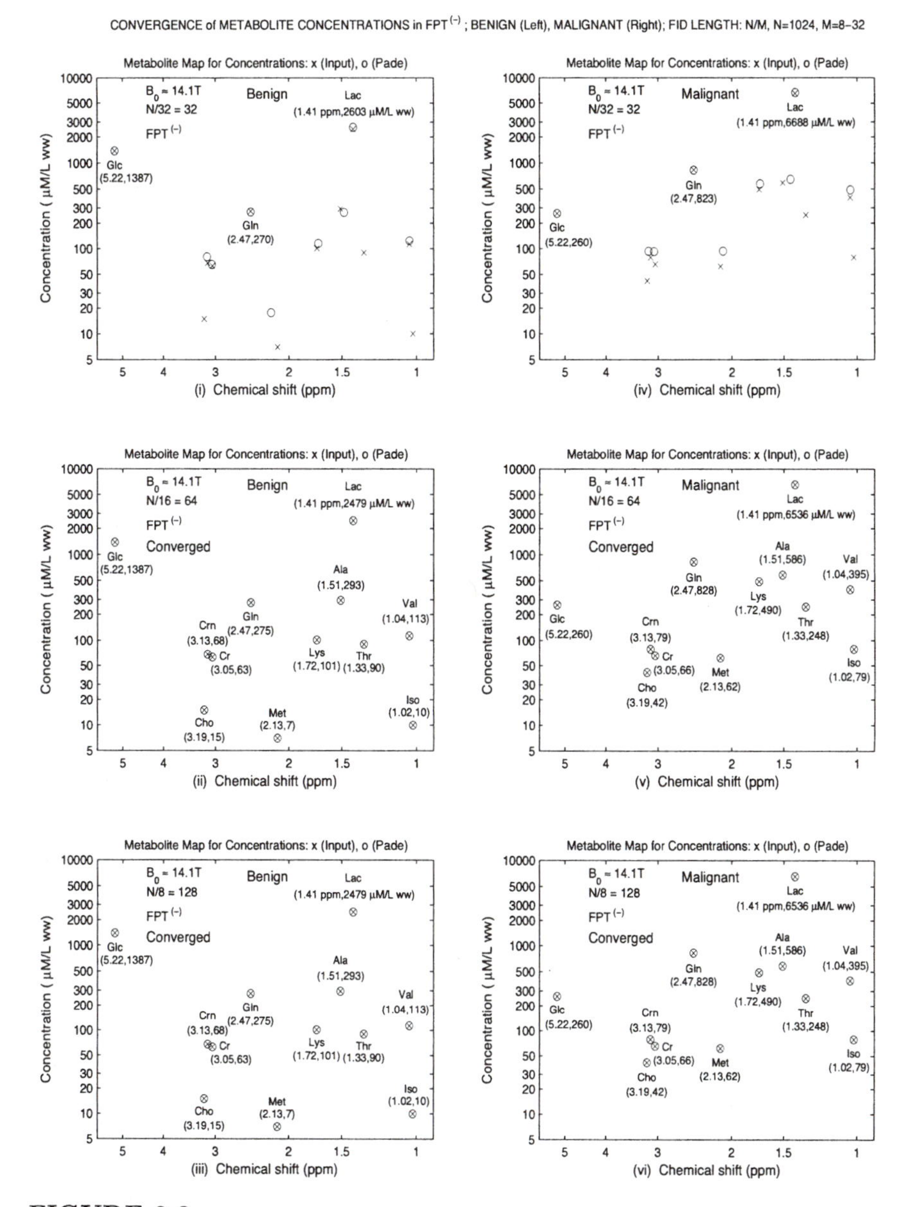

FIGURE 9.8
Convergence of the Padé-reconstructed concentrations of the metabolites for the non-malignant (left column) and cancerous (right column) ovarian cyst data from [343]. At $N/32 = 32$, convergence has not yet been achieved (top panels (i) and (iv)). At $N/16 = 64$, the reconstructed concentrations for all the 12 metabolites have converged (middle panels (ii) and (v)). This convergence is stable at longer signal lengths on bottom panels (iii) and (vi) at $N/8 = 128$, and beyond, as explicitly checked at $N/4 = 256, N/2 = 512$ and $N = 1024$.

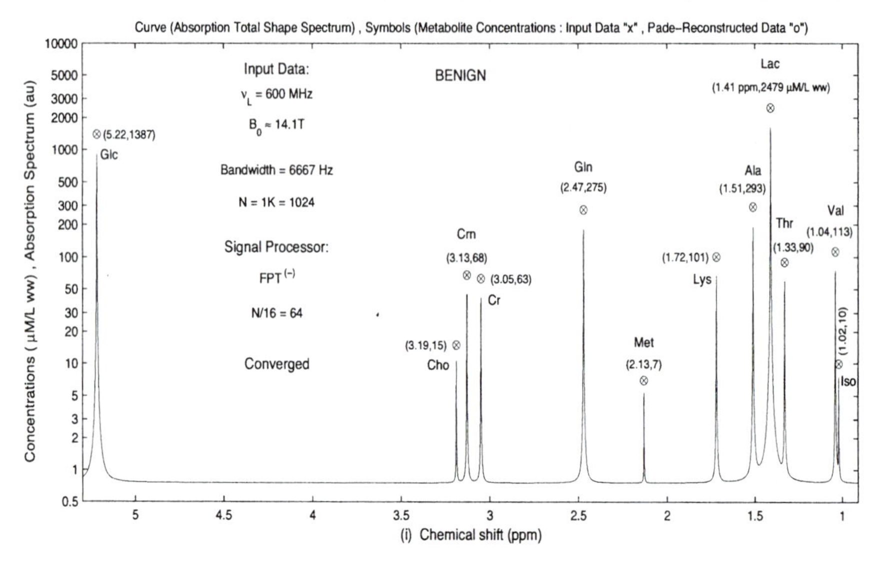

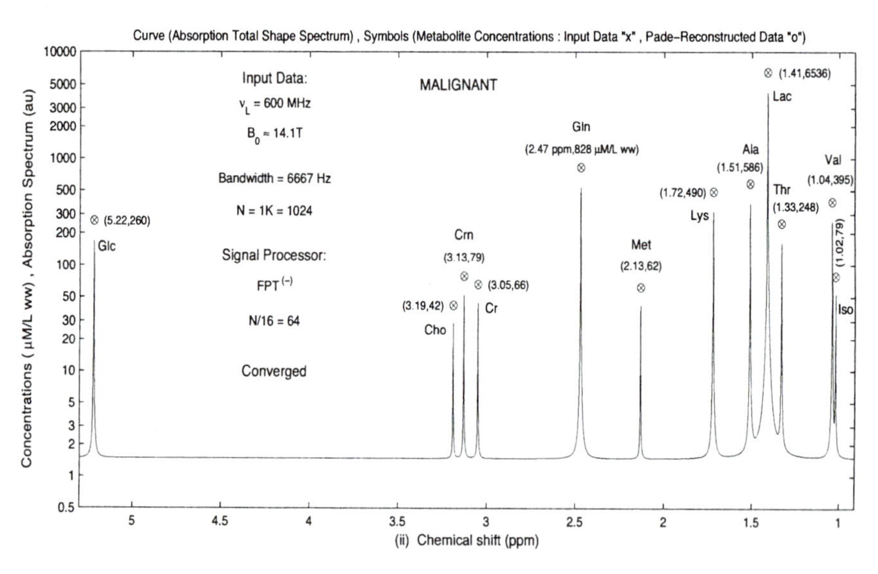

FIGURE 9.9
Absorption spectra at full convergence ($N/16 = 64$) achieved by the FPT for the benign (top panel (i)) and malignant (bottom panel (ii)) ovarian cyst fluid FIDs derived from the corresponding experimentally measured time signals [343], with the retrieved chemical shifts and concentrations of the 12 metabolites indicated above the corresponding peak.

highest early in the encoding. It is thus advantageous to encode the time signal as rapidly as possible, i.e., to avoid long T when mainly noise will be measured. This is particularly important for clinical signals encoded at lower magnetic field strengths.

As we have extensively reviewed, the FPT provides not only the shape spectra, but also the vitally important parametric analysis, i.e., the quantification from which the metabolite concentrations are obtained. Here, with only 64 data points out of 1024 sampled data, the FPT exactly reconstructed all the spectral parameters to an accuracy of six decimal places for all the twelve metabolite peaks. These parameters were then used to compute metabolite concentrations simply and unequivocally. With the standard Fourier approach, metabolite concentrations are estimated from the shape spectra by integrating the areas under the peaks or fitting the peaks to a subjectively chosen number of Lorentzians and/or Gaussians.

Even for peaks that are well separated, as noted, this procedure of numerical quadrature is subjective due to the uncertainty about the lower and upper integration limits. However, when the peaks overlap, this standard method for reconstructing metabolite concentrations is fraught with major difficulties and uncertainties; it is thus well recognized that "spectral crowding" creates quantification problems [346]. This "spectral crowding" is not a problem for the FPT, which via parametric analysis, yields reliable information about the concentrations not only of isolated resonances, but also of those that are overlapping [10, 11].

As discussed, several MR-observable compounds have been found to provide some distinction between benign and cancerous ovarian lesions when *in vitro* MRS is applied using high magnetic field strength and the conventional laboratory specimen processing techniques. Notably, concentrations of adjacent resonances such as threonine (1.33 ppm), lactate (1.41 ppm) and alanine (1.51 ppm) and the nearly overlapping resonances isoleucine and valine in the region of 1.02 ppm to 1.04 ppm differ significantly in these two types of lesions [343]. The high concentrations of these branched chain amino acids are seen as protein breakdown products due to necrosis and proteolysis.

However, none of these studies [341]–[343] reported any metabolite alone or in combination with other metabolites that unequivocally distinguished benign from cancerous ovaries. Even when the differences in concentrations were statistically highly significant such as in Ref. [343], the ranges were still overlapped.

One avenue for further investigation, and which is currently underway, is to apply the FPT to larger-scale *in vitro* experimental MRS data from benign and malignant ovarian lesions. In this way, we are exploring the possibilities of whether the FPT with its capacity to unequivocally yield exact quantifications could also specify metabolite concentrations that more clearly distinguish cancerous from non-neoplastic ovary. The FPT could thereby help establish the standards of MR-detectable metabolite concentrations for normal versus specific pathological entities of the ovary.

9.4 Prospects for Padé-optimized MRS for ovarian cancer diagnostics

The high resolution of the FPT also could be of benefit for *in vivo* MRS investigations, for which, as discussed, poor SNR has been an important obstacle hampering progress in ovarian cancer diagnostics via MRS. It has been suggested that *in vivo* MRS could become the method of choice for accurate detection of early stage ovarian cancer, insofar as the current obstacles hampering the acquisition of high quality time signals and the subsequent reliable analysis of spectra as well as their interpretation can be overcome [342]. The presented results clearly indicate that application of the FPT could be an important step towards achieving the stated goal, particularly insofar as *in vitro* and *in vivo* clinical correlations and histopathology were initially performed for verification. The metabolite concentration maps could be especially useful when Padé-optimized MRS becomes implemented in practice. These maps could then be a hands-on tool, facilitating not only a deeper understanding of this novel approach to signal processing, but also rapid and reliable interpretation of diagnostically relevant information in the clinical setting.

In the performed analyses, we used noise-free FIDs to set up the fully-controlled standard for the FPT in our initial application of this method to data within the realm of ovarian cancer diagnostics by MRS. This is methodologically justified [10]. The next steps will be to extend our analysis to both noise-corrupted synthesized ovarian data (still well-controlled) and to encoded FIDs such as those from Ref. [343]. These studies are currently underway.

The present results illustrate the capacity of the FPT to resolve and precisely quantify all the physical resonances as encountered in benign versus cancerous ovarian cystic fluids. Padé-based quantification as a parametric estimation is achieved with an exceedingly small number of signal points. This has practical implications, because the time signals' exponential tail with small amplitudes is avoided. Since the tail is embedded in the background noise, long signal lengths cause severe problems in quantification. Such a feature of the FPT is particularly advantageous relative to the FFT, which needs long signals (through which noise is inevitably invoked) in an attempt to improve resolution $2\pi/T\,(T = N\tau)$.

We have now shown that the fast Padé transform reliably and unequivocally yields the metabolite concentrations of major importance for distinguishing benign from malignant ovarian lesions. The FPT achieves this without any fitting or numerical integration of peak areas. These capabilities of the FPT are likely to be of benefit for ovarian cancer diagnostics via MRS. Further pursuit of this avenue of investigation should be a clinically urgent priority. The urgency is underscored by the fact that early ovarian cancer detection is a goal which is still elusive and achievement of which would confer a major survival benefit.

10

Breast cancer and non-malignant breast data: Quantification by FPT

10.1 Current challenges in breast cancer diagnostics

Among women worldwide, breast cancer is the most frequently diagnosed malignancy and the leading cause of death from cancer. Over one million annual new cases of breast cancer are reported in recent years, and over 400000 women die each year from breast cancer [347]–[351]. Relatively favorable overall survival rates for breast cancer have been attributed to early detection, which greatly increases the possibilities for effective treatment. Thus, currently, breast cancer is a highly prevalent malignancy, with an estimated 4.4 million survivors up to 5 years after the diagnosis [350].

Screening with mammography followed by appropriate diagnosis and management have been shown to significantly reduce deaths due to breast cancer [350], [352]–[354]. Mammography has been the mainstay of breast cancer screening. Calcifications indicative of cancer can be seen thereby, even at the earliest stage: ductal carcinoma *in situ*. However, mammography has relatively poor specificity. Moreover, for dense breasts seen particularly among young women, its sensitivity can also be low. Since invasive breast cancers that can spread to the lymph nodes and cause distant metastases are frequently non-calcified, they can be very difficult to detect mammographically, especially when the breast parenchyma is dense. It is precisely among younger women that breast cancer when it does occur, is often biologically more aggressive [353]. In addition, mammography entails exposure to ionizing radiation, which is of particular concern for the younger age group, and most especially for those already at high risk [351, 352, 355, 356].

It has been emphasized that a comprehensive breast cancer screening program should incorporate risk assessment and that screening strategies be tailored to the individual woman's risk [357]. However, reliance upon even the well-recognized risk factor of family history, despite the numerous algorithms, can sometimes be precarious. One obvious example is the case of adopted children. Denial, as well as fear and grief can also hinder disclosure about family risk [22, 358, 359]. Many countries have well-developed programs for detection of families at high risk for breast cancer. This is based strongly upon family history and use of population registries. However, for persons born outside the country and/or who have a "limited family structure" (LFS) [360, 361] this high risk may not be adequately detected. This LFS may possibly be related to trends toward smaller families, as well as lack of paternal aunts and premature mortality. The impact of missing family links has been shown to have a particularly dramatic impact when the family risk for disease is very high [362]. These considerations further complicate efforts to define high risk and underscore the need for maximal diagnostic accuracy and wide applicability of breast cancer screening methods.

There has been increasing interest in other modalities besides mammography for early breast cancer detection and screening, particularly for younger women at potentially high risk [363]. Ultrasound combined with mammography improves sensitivity for detecting breast cancer; however, this combination also substantially increases the false positive rate [353]. Thus, the usefulness of whole-breast ultrasonography has not been established for routine screening [355].

10.2 *In vivo* MR-based modalities for breast cancer diagnostics and clinical assessment

10.2.1 Magnetic resonance imaging applied to detection of breast cancer

Magnetic resonance imaging is gaining acceptance for screening women at increased risk for breast cancer. The American Cancer Society now recommends MRI for women with an estimated lifetime breast cancer risk at or above 20 to 25% [354]. One advantage of MR-based diagnostics is the lack of exposure to ionizing radiation to the breast, which is a radiosensitive tissue. This is important in view of the heightened radiosensitivity for women with genetic risk for breast cancer, i.e., with BRCA germline mutations, Li-Fraumeni syndrome (p53 tumor suppressor gene mutations), as well as those who are heterozygous for ataxia-telangiectasia [364, 365]. This concern about exposure to ionizing radiation is also based upon the need to begin screening women with genetic risk at a younger age and with increased frequency.

There is consistent evidence that MRI is more sensitive than mammography for detecting breast cancer among women with an increased risk [354, 366]. Breast MRI is valuable for identifying malignancy in dense breasts, in which cancers, unless calcified, are difficult to perceive using mammography [367, 368]. Dynamic contrast enhanced (DCE) MRI can detect a typical pattern of tumor microvasculature characterized by rapid wash-in and wash-out [369]. Breast MRI is also considered superior to mammography for identifying multifocal or multi-centric cancers, as well as tumor recurrence and response to chemotherapy. MRI is more effective than mammography in finding breast cancer at an earlier stage among women at high risk [354, 366]. Yet, false negative findings using MRI have been reported. This can occur, e.g., with small tumors, if they do not selectively take up the contrast agent. Moreover, MRI cannot reveal micro-calcifications [370] and occasionally misses invasive ductal and lobular carcinomas, although non-detection with MRI is more common with *in situ* ductal carcinoma [371]. Recent data suggest that MRI is more effective for women at high family risk of breast cancer compared to those with low family risk [372]. Based upon the data from an on-going surveillance trial centered in Bonn, Germany, recommendations have been made to discontinue mammographic screening in young women with BRCA1 mutations and instead to use MRI, in view of the vulnerability to ionizing radiation together with the fact that in the mentioned series, none of the BRCA1-associated breast cancers were found to be calcified [373].

The major problem with MRI, however, is that despite excellent spatial resolution and generally superior sensitivity, it has poor specificity, consistently lower than mammography [354]. Thus, while women at high risk undergoing intensive surveillance programs appear to be relieved by the greater sensitivity of MRS [374], a large number of false positive findings may unfavorably affect quality of life [22, 375]. Moreover, a missed cancer, albeit relatively less likely with MRI than mammography, can adversely affect prognosis [354]. In contradistinction to prophylactic mastectomy which while drastic, does effectively provide primary prevention for women with BRCA1 or BRCA2 or other high risk gene mutations, a significantly lowered mortality has not yet been demonstrated for surveillance with MRI among women at high breast cancer risk [376]. Thus, concerns remain about MRI as a breast cancer screening tool, with the need to improve specificity particularly underscored [363, 376].

10.2.2 Studies to date using *in vivo* MRS for distinguishing between benign and malignant breast lesions

In vivo MRS has been shown to increase the specificity of MRI with respect to the diagnosis of breast cancer [116, 117], [377]–[381]. Most of the studies using *in vivo* proton MRS have been based upon the composite choline signal, assessed either qualitatively or quantitatively. In a meta-analysis [116] of five clinical studies using *in vivo* single-voxel MRS to assess 100 cancerous and 53 non-malignant breast lesions, specificity was computed to be 85% (95% Confidence Intervals (CI) = 71 – 93%), and sensitivity was 83% (95% CI = 73% – 89%). In that meta-analysis [116] the diagnostic accuracy was higher for the twenty women who were age 40 or younger. Among those twenty younger women, all eleven breast cancers were accurately diagnosed. A subsequent publication describing single-voxel MRS among thirty women with a positive mammogram and rapid contrast enhancement on breast MRI similarly showed that the presence of choline (with SNR ≥ 2) improved the specificity for breast cancer to 87.5%, with three false positive results [382]. Another subsequent *in vivo* MRS study [383] employed a very long TE of 288 ms, as well as an inversion recovery sequence to diminish the fat signal, and water suppression by chemical selective saturation together with improved shimming of the static magnetic field. The reported results from this work [383] reveal that the presence or absence of choline accurately detected all nineteen malignancies and all sixteen non-cancerous breast lesions. The choline SNR varied from 2.4 to 12.7, with a mean value of 5.4 for the malignant lesions.

Case reports have been published describing the use of *in vivo* MRS to help rule out malignancies in unusual and suspicious appearing breast lesions. These have included a desmoid tumor, found to have absent choline [384] and a tubercular breast abscess also found to have absent choline, but with a prominent peak at 0.9 – 1.3 ppm considered to represent lipid-lactate [385]. Tse *et al.* [381] recently reviewed the various benign pathologies that have been evaluated using *in vivo* MRS. Of a total of 46 fibroadenomas, three were misdiagnosed as malignant. Of 20 fibrocystic lesions, two were false positives on *in vivo* MRS.

There have been some studies assessing the ratio between unsuppressed water to lipid in breast cancer and normal mammary tissue with *in vivo* MRS. The ratio of water at 4.7 ppm to the lipid peak at 1.3 ppm was significantly higher in breast voxels containing cancer compared to sites free of malignancy among seventeen patients. This water to lipid ratio was also higher at sites of malignancy compared to fourteen healthy referents [386]. However, there was a very large standard deviation among the malignant ratios (6.0 ± 6.9), which indicated overlap with the normal values (0.36 ± 0.25). When overlapping metabolites cannot be resolved in 1D, then 2D MRS has been applied in attempts to separate the resonances using cross-correlation plots. Using *in vivo* 2D MRS a significantly higher diagonal peak volume of unsuppressed water to fat (1.4 ppm) ratio was reported among twenty-one patients with invasive breast cancer compared to fourteen healthy women with fatty breasts, however no ranges were given [387]. The diagonal peak volume ratios of water to methyl fat at 0.9 ppm and to olefinic fat at 5.4 ppm were also significantly higher in the cancerous tissue using 2D MRS, as were the water to cross peak volumes of unsaturated fatty acid right (2.1 ppm, 5.4 ppm), unsaturated fatty acid left (2.9 ppm, 5.4 ppm) and triglyceride fat (4.3 ppm, 5.3 ppm) [387]. The elevated water to fat ratios observed in breast cancer may be related to edema, due to lymphatic blockage and involvement of sub-dermal lymphatics by the malignancy.

In the first reported investigation [388] of breast cancer via MRSI[1], using the choline SNR, a sensitivity of 87% and specificity of 85% were attained among eight patients with breast cancer and 7 with benign breast pathology. There were three other patients for whom MRSI was a technical failure. The

[1]This MRSI investigation was carried out at 1.5T, shimming was performed to optimize field homogeneity, 3 sequential Chemical Shift Selective (CHESS) pulses were used for water suppression, lipid signals were attenuated with inversion pulse and TE=272 ms was selected.

mean choline SNR among the patients with breast cancer was 6.2 ± 2.1 versus 2.4 ± 0.7 for those with benign breast lesions. The authors from Ref. [388] defined the distinguishing cut-point for choline SNR to be 4.1, with the highest benign value being 4, and lowest malignant value being 4.1. They remarked that unequivocal detection of the choline resonance was sometimes difficult, due to overlap with sidebands of residual water or lipid resonances. Improved data processing methods were cited as a necessary precondition for reliable quantification of choline in breast lesions [388].

Bolan *et al.* [389] employed a high static magnetic field (4T) and optimized surface coils with single voxel *in vivo* MRS, and attempted to quantify total choline by an algorithm which fitted the peaks one at a time over a narrow frequency band. The average of spectra at various TE values was employed to try to separate non-coupled metabolite resonances from the lipid-induced sidebands [390]. This method was applied to 500 spectra from patients with breast cancer, with benign breast pathology as well as with normal breast tissue. Total choline was significantly higher in the malignant compared to benign lesions, but there was also overlap in the ranges. Importantly, total choline was undetectable in several breast cancers, and high in some benign lesions, as well as in a few normal breasts. The total choline amplitude appeared to be influenced by the lipid content of the voxel, either through baseline artefacts not suppressed by TE averaging or by a true resonance at 3.25 ppm. In smaller voxels, the fitting error was also reportedly higher [389].

In the recent investigation of 124 patients by Sardanelli *et al.* [380] employing a 1.5T MR scanner, the cut-point for the integral under the total choline peak was 1.9 arbitrary units to classify a breast lesion as malignant. Using this threshold, sensitivity of 90% and specificity of 89% were obtained, with 3 patients excluded due to poor SNR. The authors [380] note that the diagnostic accuracy was better for the larger lesions (> 1 cm).

Other authors have also noted that, as currently applied, *in vivo* MRS is limited regarding the diagnosis of smaller tumors [101, 378, 381], [391]–[393]. We refer once again to the observation of surgical oncologist Gluch [345]: "difficulties arise not so much in regards to the large lesion with suspicious imaging characteristics, but rather the lesion < 1 cm in size, or ductal carcinoma *in situ*. *In vivo* studies have not satisfactorily addressed these entities. A chance of missing a breast cancer of the order of only 1% would translate into a significant medicolegal concern"(p. 467). Moreover, as pointed out by Bartella and Huang [378], it is usually necessary to evaluate several lesions on an MR image, so that in fact, this becomes a multiply important limitation.

10.2.3 *In vivo* MRS to assess response of breast cancer to therapy

In the earlier-described study by Jagannathan *et al.* [386] some (but not all) of the patients with breast cancer followed by *in vivo* MRS showed a drop in the unsuppressed water to lipid ratio after chemotherapy; this response appeared to correlate with response to the neoadjuvant therapy.

In a more recent study of 14 women with locally advanced breast cancer [118], the change in total choline concentrations assessed within 24 hours after the first dose of chemotherapy was correlated with change in the lesion size at day 64 ($p = 0.001$) and differed among those with and without a response to chemotherapy ($p = 0.007$). This study was done using a 4T static magnetic field. All the responders had a fall in total choline concentration of more than 0.4 mM/kg at 24 hours, whereas the non-responders all showed either an increase, no change or at most a fall of 0.2 mM/kg in total choline concentration at 24 hours.

Tozaki *et al.* [394] have compared the efficacy of proton MRS with fluorodeoxyglucose positron emission tomography (FDG-PET) for assessing response of seven patients during the first several days of neoadjuvant chemotherapy for breast cancer. The standardized uptake of FDG showed a significant correlation with the integral value of the choline peak; this correlation was particularly strong ($r = 0.99$, $p < 0.001$) after chemotherapy. The authors from Ref. [394] suggest that *in vivo* MRS may be considered as a potential alternative for sequential FDG-PET evaluation of response to therapy.

10.2.4 Special challenges of *in vivo* MRS for breast cancer diagnostics

Notwithstanding the important strides made by *in vivo* MRS for breast cancer diagnostics and for assessing response to therapy, there are a number of challenges that have precluded its more widespread application in this branch of clinical oncology and in breast radiology.

Proton MRS for breast cancer diagnostics has generally required suppression of lipid, since the MR spectra from the breast are usually dominated by fat. This hinders localized shimming and also produces sideband artefacts [392]. Lipid suppression has most often been achieved by increasing the echo time. However, this leads to diminished signal intensity. Hu *et al.* [391] have suggested the possibility of suppressing lipid via an echo-filter using short echo times. However, Stanwell and Mountford [392] emphasize that suppression of the lipid resonance eliminates the possibility of evaluating lipid that is part of the actual disease process.

Overall, use of various echo times impedes consistent interpretation of data from *in vivo* MRS of the breast [392], and this may be clinically important. Metabolites with short T_2 relaxation times will have decayed at longer TE, e.g., myoinositol whose estimated concentrations best distinguished breast cancer from fibroadenoma in our analysis [21, 22, 25] of *in vitro* MRS data [395] (see the next subsection).

Another major problem with the current applications of *in vivo* MRS for breast cancer diagnostics is the reliance upon the composite choline peak. This compromises diagnostic accuracy, since choline may also be observed in benign breast lesions, as noted in a number of the cited results. On the other hand, it has been found that choline is often undetected in small tumors that are then misclassified as benign [116]. Choline also appears in normal breast during lactation, although in the latter, a lactose resonance at 3.8 ppm is typically also seen [393]. It should, however, also be pointed out that breast cancer can coexist with lactation as well as with pregnancy, and these malignancies are often detected late [396]. Moreover, various authors have used different cut-points for total choline concentrations to define malignant versus benign breast tissue. This renders any attempt at standardization highly precarious.

10.3 Insights for breast cancer diagnostics from *in vitro* MRS

As opposed to *in vivo* MRS breast examinations based mainly upon one composite spectral entity (the total choline peak), the high resolution of *in vitro* MRS applied to extracted specimens provides a much greater insight into the metabolic activity of malignant breast tissue. Analysis of excised breast cancers shows that the composite choline peak contains phosphocholine (PCho), glycerophosphocholine (GPC), betaine, analogous compounds containing the ethanolamine head group and taurine, as well as free choline itself [397].

Using tracer kinetics and ^{13}C NMR and ^{31}P NMR to examine the biochemical mechanisms underlying the high levels of water-soluble choline metabolites seen in breast cancer, Katz-Brull *et al.* [397] identified two non-intersecting pathways: phosphorylation and oxidation of choline, that were augmented with malignant transformation of mammary cells, with increased synthesis of phosphocholine and betaine, and suppression of choline-derived ether lipids.

Gribbestad *et al.* [395] conducted an *in vitro* proton MRS study in which they compared fourteen extracts of cancerous breast tissue and one fibroadenoma to non-involved breast from the same group of patients. We subsequently performed logistic regression analysis of these data to determine the sensitivity and specificity of individual metabolite concentrations for identifying breast cancer [21, 22, 25]. We found that only lactate showed 100% diagnostic accuracy both with and without the fibroadenoma. The diagnostic accuracy of total choline[2] was marginally lower than several of the individual metabolites, including some of its own constituents. Paired t-test analyses revealed a significant difference in all metabolite concentrations when comparing non-infiltrated and malignant breast tissue (always higher in the latter).

In the study from Ref. [395] glycerophosphocholine, phosphocholine, phosphoethanolamine, total choline and lactate were elevated, as well, in the fi-

[2]The composite or total choline peak is comprised of choline (3.21 ppm), phosphocholine (3.22 ppm) and glycerophosphocholine (3.23 ppm) for *in vitro* proton MRS.

broadenoma compared to the non-infiltrated tissue of that patient. Moreover, most of the computed metabolite concentrations were at least one standard deviation greater than the mean values for normal breast tissue. In contrast, the computed concentration of myoinositol was nearly the same, i.e., 0.465 μM/g ww and 0.448 μM/g ww for the fibroadenoma and for the non-infiltrated tissue, respectively, of the same patient and showed the lowest difference from the mean for normal breast tissue (+ 0.52 μM/g ww of the SD) [22]. These analyses confirm that a rich "window" of information is provided by *in vitro* MRS study of metabolite concentrations in malignant versus non-cancerous breast tissue.

Corroboration of some of the above findings from Refs. [21, 22] is provided by Sharma *et al.* [398] who used *in vitro* 2D MRS to compare 11 involved and 12 uninvolved lymph nodes from patients with breast cancer. They reported that the concentrations of PCho and GPC were significantly higher in involved compared to non-involved nodes. This was attributed to increased membrane synthesis in cancer cells, suggesting that metastatic breast cancer cells were present in the lymph nodes. There was also a highly significant difference between the lactate concentrations in involved and non-involved nodes [398]. The high levels of lactate indicate the presence of cancer cells whose energy source is from the anaerobic glycolytic pathway.

There is also support from animal model studies of breast cancer concerning the importance of assessing the rate of glycolysis and lactate clearance with respect to the diagnosis and prognosis of breast cancer [399]. Thus far, however, neither one-dimensional nor two-dimensional clinical *in vivo* MRS has included lactate as a metabolic marker of breast cancer. In our analyses of the data of Ref. [395] a number of the metabolite concentrations in the malignant tissues showed significant correlation. However, alanine was not correlated with phosphoethanolamine or with glycerophosphocholine, nor was choline concentration correlated with those of several other metabolites in the malignant tissues. Furthermore, principle components analysis revealed that those metabolites with the strongest diagnostic accuracy did not consistently load with the others.

In addition, we also analyzed the phosphocholine to glycerophosphocholine ratio for the data from Ref. [395]. A significantly higher PCho/GPC ratio was found in the breast cancer samples compared to the normal tissue from the same patient. The breast cancer samples showed a significantly higher PCho/GPC ratio compared to the normal, non-infiltrated tissue [25]. Our analyses [25] corroborate human breast cell line research, indicating that malignant transformation is associated with a "glycerophosphocholine to phosphocholine switch" [307].

This is related to over-expression of the enzyme choline kinase, which is responsible for phosphocholine synthesis [308, 397]. It also reflects altered membrane choline phospholipid metabolism. The major steps in choline metabolism in mammalian cells are through the cytosine diphosphate (CDP)-choline pathway [308, 309, 400]. It should be noted that therein are several proton MR visible compounds, namely: choline (3.21 ppm), PCho (3.22 ppm) and GPC (3.23 ppm), thereby emphasizing the clinico-biological importance of analyzing the relationship among these closely overlapping resonances. On the other hand, by summing these three metabolites as "total choline", as is currently done with *in vivo* MRS, substantial information for breast cancer diagnostics is missed.

10.4 Performance of the FPT for MRS data from breast tissue

In the present analysis [32] the FPT is applied to time signals that are reminiscent of *in vitro* MRS data as encoded from extracted breast specimens [395]. We generated three FIDs of the type $c_n = \sum_{k=1}^{K} d_k e^{in\tau\omega_k}$ from (3.1), where $\mathrm{Im}(\omega_k) > 0$. This was via a sum of $K = 9$ damped complex exponentials $\exp(in\tau\omega_k)$ $(1 \le k \le 9)$ with stationary amplitudes $\{d_k\}$. We then quantified time signals using the FPT, as described in [10].

For the normal non-infiltrated breast, we derived the input data for the $|d_k|$'s using median concentrations $\{\mathrm{C}_k\}_{k=1}^{9}$ (expressed in μM/g ww) of metabolites from the data of Gribbestad *et al.* [395], based upon tissue samples from twelve patients. The input $|d_k|$'s for benign breast tissue were derived from the concentrations of the nine metabolites for a fibroadenoma from Ref. [395]. For breast cancer, we derived the input data for the $|d_k|$'s from median concentrations from the data of Ref. [395] for fourteen samples taken from twelve patients. Two samples each were taken from two of the patients. In Ref. [395] metabolite concentrations were computed in only six and nine malignant samples respectively for the metabolites $\beta-$glucose and myoinositol.

The time signals from Ref. [395] were recorded with a static magnetic field strength $B_0 \approx 14.1$T (Larmor frequency of 600 MHz). We used a bandwidth of 6 kHz (the inverse of this bandwidth is the sampling time τ) and set the total signal length $N = 2048$.

We grouped the resonances into two bands, the first from 1.3 ppm to 1.5 ppm and the second from 3.2 ppm to 3.3 ppm. There were seven resonances within the latter band, including two nearly degenerate resonances at 3.22 ppm. These were phosphocholine ($k = 4$) and phosphoethanolamine ($k = 5$) separated by only 2.03×10^{-4} ppm.

The input peak amplitudes $\{|d_k|\}$ were computed from the reported concentrations $\{\mathrm{C}_k\}$ using (3.15) via $|d_k| = 2\mathrm{C}_k/\mathrm{C}_{\mathrm{ref}}$ where $\mathrm{C}_{\mathrm{ref}} = 0.05$ μM/g ww. Also, TSP (3-(trimethylsilyl-) 3,3,2,2-tetradeutero-propionic acid) was used as the internal reference by Gribbestad *et al.* [395], such that $|d_k| = \mathrm{C}_k/(25\mu\mathrm{M/g}$ ww). The T_2 relaxation times were not reported in Ref. [395].

The line widths, or more precisely, the full widths at half maximum, were herein taken to be approximately 1 Hz, allowing slight variations, and also assuming that the peaks were Lorentzian. The line widths are proportional to $\mathrm{Im}(\nu_k)$. It is emphasized that the mentioned smallest chemical shift difference of $\approx 2\times 10^{-4}$ ppm is 4 times less than the typical line width of 8×10^{-4} ppm. The phases ϕ_k $(1 \le k \le 9)$ from complex-valued d_k were all set to zero, so that every d_k becomes real, $d_k = |d_k|$.

Table 10.1 displays the input data for the normal breast tissue (upper panel (i)), fibroadenoma (middle panel (ii)) and for breast cancer (bottom panel (iii)). The diagonal $\mathrm{FPT}^{(-)}$ was used to analyze the FIDs with the coefficients $\{p_r^-, q_s^-\}$ of the polynomials P_K^- and Q_K^- computed by solving a system of linear equations from chapter 4 (section 4.10) by treating the product in $G_N(z^{-1}) * Q_K^-(z^{-1}) = P_K^-(z^{-1})$ as a convolution. We solved the characteristic equation $Q_K^-(z^{-1}) = 0$ to extract the peak parameters. As discussed, this leads to K unique roots z_k^- $(1 \le k \le K)$, so that the sought ω_k^- is deduced via $\omega_k^- = (i/\tau)\ln(z_k^-)$. The kth metabolite concentration C_k^- was computed from the absolute value $|d_k^-|$ of the reconstructed amplitude d_k^- as $\mathrm{C}_k^- = 25|d_k^-|\mathrm{C}_{\mathrm{ref}}$ μM/g ww.

We systematically increased the signal length for the same bandwidth (i.e., 3 acquisition times) to demonstrate the constancy of the spectral parameters for all three signals. The spectral parameters were examined at total orders $K = 500$, 750 and 1000, where $2K = N_P$ and N_P denotes partial signal length. Convergence occurred at $K = 750$ for all three FIDs, and remained stable thereafter. We determined whether a given reconstructed resonance was true or spurious by using the concept of Froissart doublets throughout the whole Nyquist interval with a particular focus on the frequency range of interest from 1.3 ppm – 3.3 ppm. For all three examined FIDs, of the 750 total resonances, 741 were identified as spurious Froissart doublets by their zero amplitudes and the pole-zero coincidences; the nine remaining resonances were genuine.

10.4.1 Padé-reconstruction of MRS data for normal breast tissue

Table 10.2 shows the data reconstructed by the $\mathrm{FPT}^{(-)}$ for the normal breast tissue. Three illustrative partial signal lengths ($N_P = 1000, 1500$ and 2000) are shown therein. At $N_P = 1000$ (upper panel (i)), only eight of the nine resonances were identified. In the interval where there should be two peaks, PCho ($k = 4$) at 3.220012 ppm and PE ($k = 5$) at 3.220215 ppm, only a single resonance was identified at 3.220189 ppm. Since this peak was closer to PE, we gave it that assignment.

At $N_P = 1000$, the reconstructed amplitudes and, therefore, the concentrations for that single peak were approximately the sum of (PCho + PE). At $N_P = 1000$ all the reconstructed spectral parameters were fully exact for lactate (Lac, $k = 1$) at 1.330413 ppm, alanine (Ala, $k = 2$) at 1.470313 ppm,

TABLE 10.1
Input spectral parameters and metabolite concentrations for three cases: normal breast (top panel (i)), fibroadenoma (middle panel (ii)) and breast cancer (bottom panel (iii)) based upon *in vitro* data of Ref. [395]. As before, ppm denotes parts per million, au arbitrary units, ww wet weight, M_k denotes metabolite assignment, C_k denotes concentration of M_k, while Lac denotes lactate, Ala alanine, Cho choline, PCho phosphocholine, PE phosphoethanolamine, GPC glycerophosphocholine, $\beta-$Glc beta-glucose, Tau taurine, m-Ins myoinositol.

INPUT DATA: SPECTRAL PARAMETERS, CONCENTRATIONS and METABOLITE ASSIGNMENTS
Types of Tissue on Three Panels : (i) Normal, (ii) Fibroadenoma, (iii) Malignant

(i) NORMAL

n^o_k (Metabolite # k)	$Re(\nu_k)$ (ppm)	$Im(\nu_k)$ (ppm)	$\lvert d_k \rvert$ (au)	C_k(μM/g ww)	M_k (Assignment)
1	1.330413	0.000834	0.02016	0.5040	Lac
2	1.470313	0.000832	0.00350	0.0875	Ala
3	3.210124	0.000831	0.00068	0.0170	Cho
4	3.220012	0.000833	0.00076	0.0190	PCho
5	3.220215	0.000834	0.00516	0.1290	PE
6	3.230412	0.000832	0.00128	0.0320	GPC
7	3.250224	0.000833	0.01800	0.4500	β–Glc
8	3.270141	0.000831	0.00530	0.1325	Tau
9	3.280132	0.000832	0.01144	0.2860	m–Ins

(ii) FIBROADENOMA

n^o_k (Metabolite # k)	$Re(\nu_k)$ (ppm)	$Im(\nu_k)$ (ppm)	$\lvert d_k \rvert$ (au)	C_k(μM/g ww)	M_k (Assignment)
1	1.330413	0.000832	0.05928	1.4820	Lac
2	1.470313	0.000834	0.00440	0.1100	Ala
3	3.210124	0.000833	0.00088	0.0220	Cho
4	3.220012	0.000832	0.00432	0.1080	PCho
5	3.220215	0.000831	0.01476	0.3690	PE
6	3.230412	0.000833	0.00276	0.0690	GPC
7	3.250224	0.000832	0.03912	0.9780	β–Glc
8	3.270141	0.000834	0.01352	0.3380	Tau
9	3.280132	0.000831	0.01860	0.4650	m–Ins

(iii) MALIGNANT

n^o_k (Metabolite # k)	$Re(\nu_k)$ (ppm)	$Im(\nu_k)$ (ppm)	$\lvert d_k \rvert$ (au)	C_k(μM/g ww)	M_k (Assignment)
1	1.330413	0.000831	0.32474	8.1185	Lac
2	1.470313	0.000832	0.03156	0.7890	Ala
3	3.210124	0.000834	0.00446	0.1115	Cho
4	3.220012	0.000831	0.02448	0.6120	PCho
5	3.220215	0.000832	0.07776	1.9440	PE
6	3.230412	0.000833	0.00936	0.2340	GPC
7	3.250224	0.000832	0.02882	0.7205	β–Glc
8	3.270141	0.000831	0.11182	2.7955	Tau
9	3.280132	0.000833	0.03564	0.8910	m–Ins

β–glucose (β–Glc, $k = 7$) at 3.250224 ppm, taurine ($k = 8$) at 3.270141 ppm and myoinositol (m-Ins, $k = 9$) at 3.280132 ppm.

Although the reconstructed amplitude $|d_k^-|$ and metabolite concentration were completely correct for choline ($k = 3$), the chemical shift frequency, $\mathrm{Re}(\nu_k^-)$ and the line width, $\mathrm{Im}(\nu_k^-)$ for choline were exact for five of the six decimal places (3.210123 ppm instead of 3.210124 ppm, 0.000830 ppm instead of 0.000831 ppm respectively). At $N_P = 1000$ for glycerophosphocholine ($k = 6$) the line width, $\mathrm{Im}(\nu_k^-)$ was exact for five of the six digits (0.000831 ppm instead of 0.000832 ppm), whereas the reconstructed $\mathrm{Re}(\nu_k^-)$, $|d_k^-|$ and concentration were all completely correct.

Full convergence was achieved at $N_P = 1500$ in the reconstructed parameters for all nine resonances (middle panel (ii) of Table 10.2). We show that convergence was stable at a higher signal length $N_P = 2000$ on the bottom panel (iii) in Table 10.2, where it can be confirmed that all the reconstructed parameters remain fully correct. This stability continues at yet higher N_P and at the full signal length N, as theoretically predicted by the FPT.

In Fig. 10.1 the absorption component shape spectra and the total shape spectra as reconstructed by the $\mathrm{FPT}^{(-)}$ are displayed: $N_P = 1000, 1500$ and 2000 for the normal breast data. As can be seen on the right upper panel (iv), at $N_P = 1000$, the absorption total shape spectrum is converged. However, this was not so for the component shape spectrum (left upper panel (i)), where only one peak (phosphoethanolamine, $k = 5$) at 3.22 ppm is seen and is over-estimated, whereas phosphocholine, ($k = 4$) is unresolved.

At $N_P = 1500$ in the left middle panel (ii) of Fig. 10.1 the component shape spectrum is converged such that peaks $k = 4$ and 5 are resolved and have the true heights, as do all the other peaks. Therein, the small phosphocholine peak completely underlies phosphoethanolamine. This can be expected since, as noted, the difference between the chemical shifts of the peaks $k = 4$ and 5 is approximately four times less than the line widths. At $N_P = 2000$ convergence is confirmed to be stable, as seen in the lower panels for both the absorption component shape spectrum (iii) and the total shape spectrum (vi), and at still higher N_P including the full signal length N.

Figure 10.2 illustrates the convergence of metabolite concentrations for the normal breast data for the same three partial signal lengths: $N_P = 1000, 1500$ and 2000 as in Fig. 10.1 and Table 10.2. The input data are shown by the symbol "x" and the Padé-reconstructed data are represented by open circles. Before convergence, at $N_P = 1000$ neither the concentrations of PCho nor PE are correctly assessed in the reconstruction (top panel (i)), although the other metabolite concentrations are exact.

At $N_P = 1500$ (middle panel (ii)) and $N_P = 2000$ (bottom panel (iii)), all of the metabolite concentrations are fully correct, as seen both numerically and by the graphic representations. This is seen by the "x" being centered within the open circles, such that there is complete agreement between the input and reconstructed data.

TABLE 10.2
Reconstructed spectral parameters and metabolite concentrations via the $FPT^{(-)}$ for normal breast tissue with the input data based upon Ref. [395].

CONVERGENCE of SPECTRAL PARAMETERS and CONCENTRATIONS in $FPT^{(-)}$: PARTIAL SIGNAL LENGTHS N_P =1000, 1500, 2000

(i) Pade–Reconstructed Data (Normal): N_P = 1000 (# 4 PCho: Unresolved, # 5 PE: Overestimated)

n_k^o (Metabolite # k)	$Re(\nu_k^-)$ (ppm)	$Im(\nu_k^-)$ (ppm)	$\lvert d_k^- \rvert$ (au)	C_k^- (μM/g ww)	M_k (Assignment)
1	1.330413	0.000834	0.02016	0.5040	Lac
2	1.470313	0.000832	0.00350	0.0875	Ala
3	3.210123	0.000830	0.00068	0.0170	Cho
5	3.220189	0.000836	0.00592	0.1480	PE
6	3.230412	0.000831	0.00128	0.0320	GPC
7	3.250224	0.000833	0.01800	0.4500	β–Glc
8	3.270141	0.000831	0.00530	0.1325	Tau
9	3.280132	0.000832	0.01144	0.2860	m–Ins

(ii) Pade–Reconstructed Data (Normal): N_P = 1500 (Converged)

n_k^o (Metabolite # k)	$Re(\nu_k^-)$ (ppm)	$Im(\nu_k^-)$ (ppm)	$\lvert d_k^- \rvert$ (au)	C_k^- (μM/g ww)	M_k (Assignment)
1	1.330413	0.000834	0.02016	0.5040	Lac
2	1.470313	0.000832	0.00350	0.0875	Ala
3	3.210124	0.000831	0.00068	0.0170	Cho
4	3.220012	0.000833	0.00076	0.0190	PCho
5	3.220215	0.000834	0.00516	0.1290	PE
6	3.230412	0.000832	0.00128	0.0320	GPC
7	3.250224	0.000833	0.01800	0.4500	β–Glc
8	3.270141	0.000831	0.00530	0.1325	Tau
9	3.280132	0.000832	0.01144	0.2860	m–Ins

(iii) Pade–Reconstructed Data (Normal): N_P = 2000 (Converged)

n_k^o (Metabolite # k)	$Re(\nu_k^-)$ (ppm)	$Im(\nu_k^-)$ (ppm)	$\lvert d_k^- \rvert$ (au)	C_k^- (μM/g ww)	M_k (Assignment)
1	1.330413	0.000834	0.02016	0.5040	Lac
2	1.470313	0.000832	0.00350	0.0875	Ala
3	3.210124	0.000831	0.00068	0.0170	Cho
4	3.220012	0.000833	0.00076	0.0190	PCho
5	3.220215	0.000834	0.00516	0.1290	PE
6	3.230412	0.000832	0.00128	0.0320	GPC
7	3.250224	0.000833	0.01800	0.4500	β–Glc
8	3.270141	0.000831	0.00530	0.1325	Tau
9	3.280132	0.000832	0.01144	0.2860	m–Ins

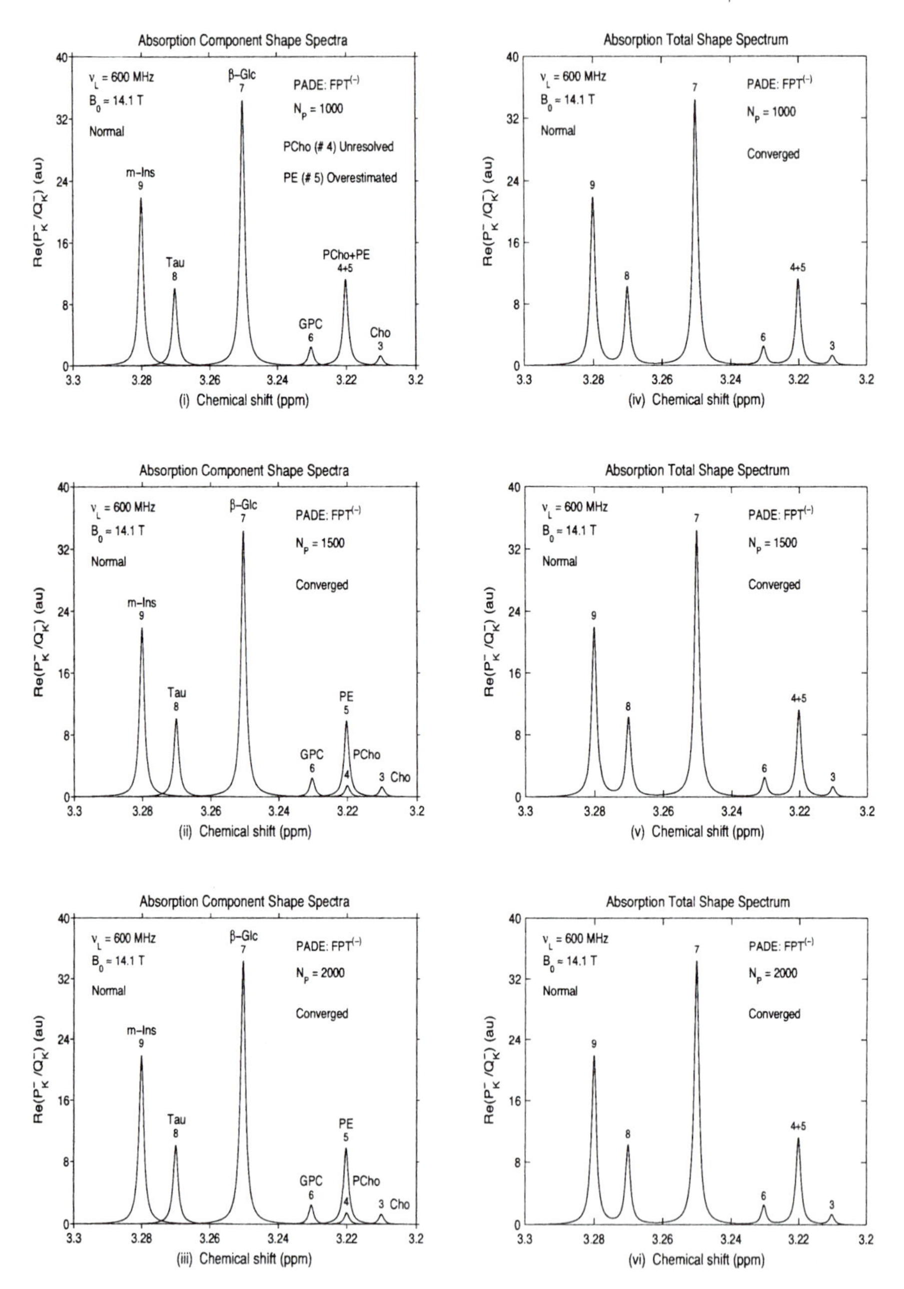

FIGURE 10.1
Padé-reconstructed absorption component spectra (left panels) and total shape spectra (right panels) for normal breast based upon *in vitro* MRS data from Ref. [395].

Convergence of Metabolite Concentrations Reconstructed by FPT$^{(-)}$: Partial Signal Lengths N_P = 1000, 1500, 2000
Two Sets of Independent Abscissae and Ordinates: [Bottom,Left] ## 3–9 (Cho,...,m–Ins) and [Top,Right] ## 1, 2 (Lac,Ala)

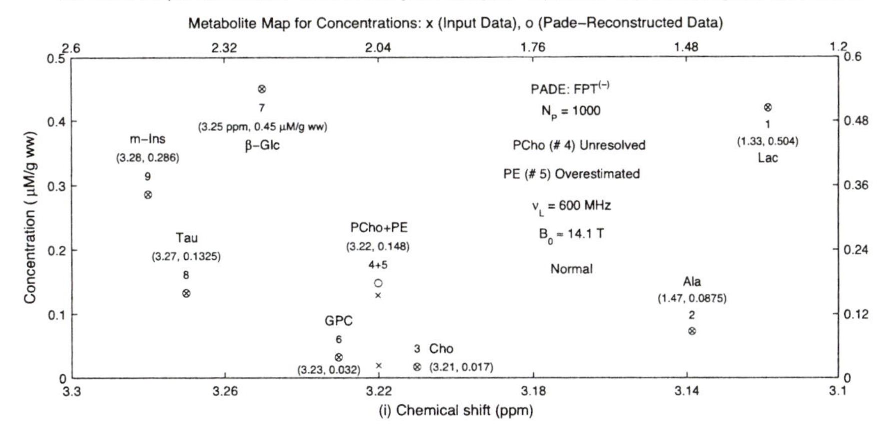

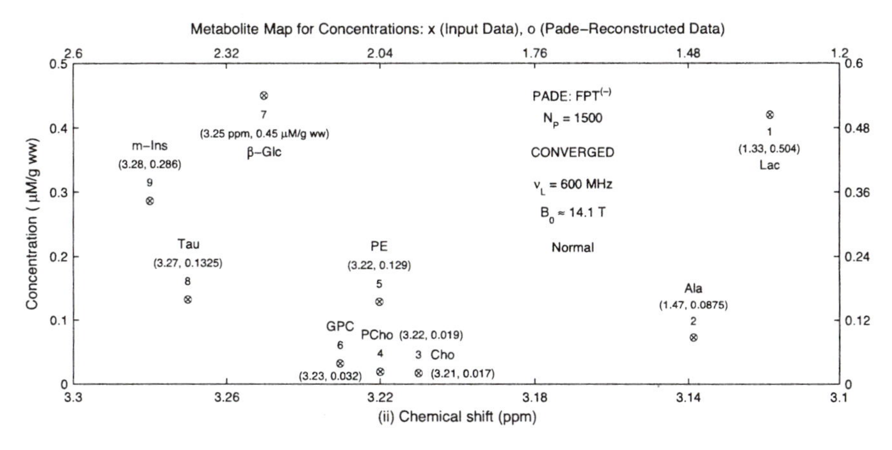

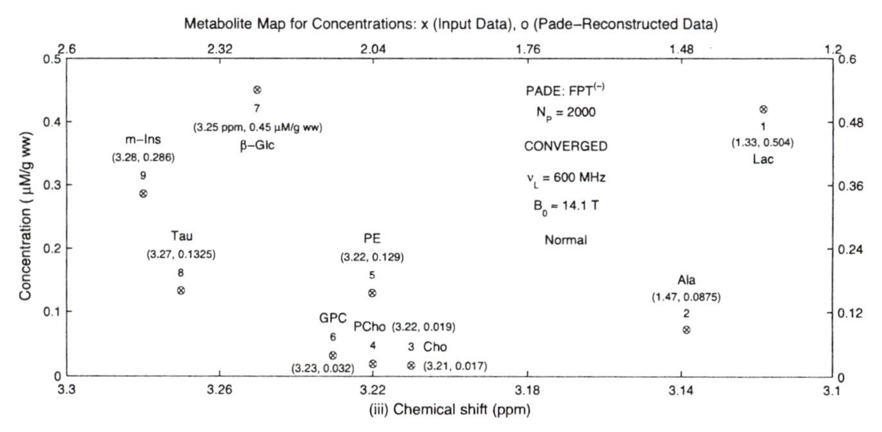

FIGURE 10.2
Padé-reconstructed metabolite concentrations in normal breast based upon *in vitro* MRS data from Ref. [395].

10.4.2 Padé-reconstruction of MRS data from fibroadenoma

For the fibroadenoma the data reconstructed by the $\mathrm{FPT}^{(-)}$ are shown in Table 10.3 for $N_P = 1000, 1500$ and 2000. For partial signal length $N_P = 1000$ (upper panel (i) of Table 10.3), only eight of the nine resonances were detected. In the interval where there should be two peaks, phosphocholine ($k = 4$) at 3.220012 ppm and phosphoethanolamine ($k = 5$) at 3.220215 ppm, there was only one resonance identified at 3.220169 ppm. Because this single peak was closer to PE, we gave it that assignment. The reconstructed amplitude and the concentration for that single resonance were approximately the sum of (PCho + PE) at $N_P = 1000$. At that partial signal length the reconstructed spectral parameters were fully exact for Lac ($k = 1$), Ala ($k = 2$), β−Glc ($k = 7$), Tau ($k = 8$) and m-Ins ($k = 9$). However, for Cho ($k = 3$) the chemical shift frequency, $\mathrm{Re}(\nu_k^-)$ and the line width, $\mathrm{Im}(\nu_k^-)$ were correct for five of the six digits (3.210122 ppm rather than 3.210124 ppm, 0.000830 ppm rather than 0.000833 ppm, respectively). The reconstructed amplitude $|d_k^-|$ for choline was fully correct, but the metabolite concentration was computed to be 0.0219 μM/g ww, rather than the correct value of 0.0220 μM/g ww. For glycerophosphocholine ($k = 6$), the chemical shift frequency, $\mathrm{Re}(\nu_k^-)$ and the line width, $\mathrm{Im}(\nu_k^-)$ were correct for five of the six digits (3.230413 ppm rather than 3.230412 ppm, 0.000831 ppm instead of 0.000833 ppm, respectively). The reconstructed amplitude $|d_k^-|$ was completely correct for GPC, although the computed metabolite concentration was 0.0689 μM/g ww, instead of the correct value of 0.0690 μM/g ww.

For the case of the data from the fibroadenoma, full convergence was achieved at $N_P = 1500$ for all the reconstructed parameters for all nine resonances (middle panel (ii) of Table 10.3). The stability of convergence at higher signal length $N_P = 2000$ is shown on the bottom panel (iii) of Table 10.3, where it is seen that all the reconstructed parameters are also fully exact. This stable convergence was seen at longer signal lengths, as well as the full signal length N.

The absorption component shape spectra and the total shape spectra reconstructed by the $\mathrm{FPT}^{(-)}$ at $N_P = 1000, 1500$ and 2000 for the fibroadenoma are shown in Fig. 10.3. At $N_P = 1000$ (right upper panel (iv)), the absorption total shape spectrum is converged. However, once again, this was not the case for the component shape spectrum (left upper panel (i)), which shows only one peak (PE, $k = 5$) at 3.22 ppm, which is over-estimated, whereas PCho ($k = 4$) is not resolved. At $N_P = 1500$ in the left middle panel (ii) of Fig. 10.3 the component shape spectrum is converged whereby peaks $k = 4$ and 5 are resolved and have the correct heights, as is the case for the other peaks. The small phosphocholine peak can be seen directly underneath PE. Stability of convergence is confirmed here, as well, at $N_P = 2000$ in the lower panels for both the absorption component shape spectrum (iii) and the total shape spectrum (vi) and this also holds true for longer signals lengths, including the full signal length N.

TABLE 10.3
Reconstructed spectral parameters and metabolite concentrations via the FPT for fibroadenoma with the input data based upon Ref. [395].

CONVERGENCE of SPECTRAL PARAMETERS and CONCENTRATIONS in $FPT^{(-)}$: PARTIAL SIGNAL LENGTHS N_P =1000, 1500, 2000

(i) Pade–Reconstructed Data (Fibroadenoma): N_P = 1000 (# 4 PCho: Unresolved, # 5 PE: Overestimated)

n_k^o (Metabolite # k)	$Re(\nu_k^-)$ (ppm)	$Im(\nu_k^-)$ (ppm)	$\|d_k^-\|$ (au)	C_k^-(μM/g ww)	M_k (Assignment)
1	1.330413	0.000832	0.05928	1.4820	Lac
2	1.470313	0.000834	0.00440	0.1100	Ala
3	3.210122	0.000830	0.00088	0.0219	Cho
5	3.220169	0.000835	0.01909	0.4773	PE
6	3.230413	0.000831	0.00276	0.0689	GPC
7	3.250224	0.000832	0.03912	0.9780	β–Glc
8	3.270141	0.000834	0.01352	0.3380	Tau
9	3.280132	0.000831	0.01860	0.4650	m–Ins

(ii) Pade–Reconstructed Data (Fibroadenoma): N_P = 1500 (Converged)

n_k^o (Metabolite # k)	$Re(\nu_k^-)$ (ppm)	$Im(\nu_k^-)$ (ppm)	$\|d_k^-\|$ (au)	C_k^-(μM/g ww)	M_k (Assignment)
1	1.330413	0.000832	0.05928	1.4820	Lac
2	1.470313	0.000834	0.00440	0.1100	Ala
3	3.210124	0.000833	0.00088	0.0220	Cho
4	3.220012	0.000832	0.00432	0.1080	PCho
5	3.220215	0.000831	0.01476	0.3690	PE
6	3.230412	0.000833	0.00276	0.0690	GPC
7	3.250224	0.000832	0.03912	0.9780	β–Glc
8	3.270141	0.000834	0.01352	0.3380	Tau
9	3.280132	0.000831	0.01860	0.4650	m–Ins

(iii) Pade–Reconstructed Data (Fibroadenoma): N_P = 2000 (Converged)

n_k^o (Metabolite # k)	$Re(\nu_k^-)$ (ppm)	$Im(\nu_k^-)$ (ppm)	$\|d_k^-\|$ (au)	C_k^-(μM/g ww)	M_k (Assignment)
1	1.330413	0.000832	0.05928	1.4820	Lac
2	1.470313	0.000834	0.00440	0.1100	Ala
3	3.210124	0.000833	0.00088	0.0220	Cho
4	3.220012	0.000832	0.00432	0.1080	PCho
5	3.220215	0.000831	0.01476	0.3690	PE
6	3.230412	0.000833	0.00276	0.0690	GPC
7	3.250224	0.000832	0.03912	0.9780	β–Glc
8	3.270141	0.000834	0.01352	0.3380	Tau
9	3.280132	0.000831	0.01860	0.4650	m–Ins

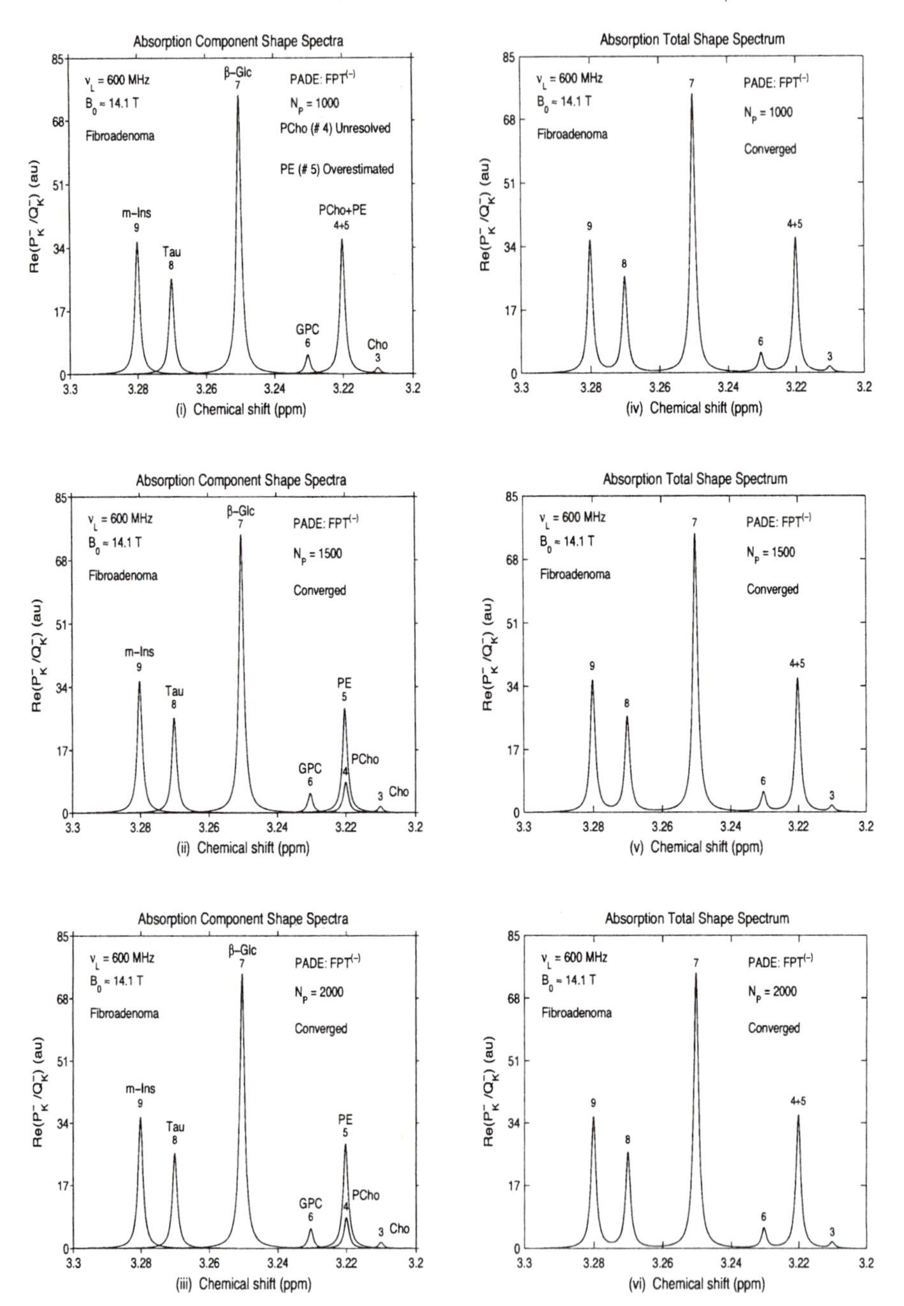

FIGURE 10.3
Padé-reconstructed absorption component spectra (left panels) and total shape spectra (right panels) for benign breast pathology (fibroadenoma) based upon *in vitro* MRS data from Ref. [395].

Convergence of Metabolite Concentrations Reconstructed by $\text{FPT}^{(-)}$: Partial Signal Lengths N_P = 1000, 1500, 2000
Two Sets of Independent Abscissae and Ordinates: [Bottom,Left] ## 3–9 (Cho,...,m–Ins) and [Top,Right] ## 1, 2 (Lac,Ala)

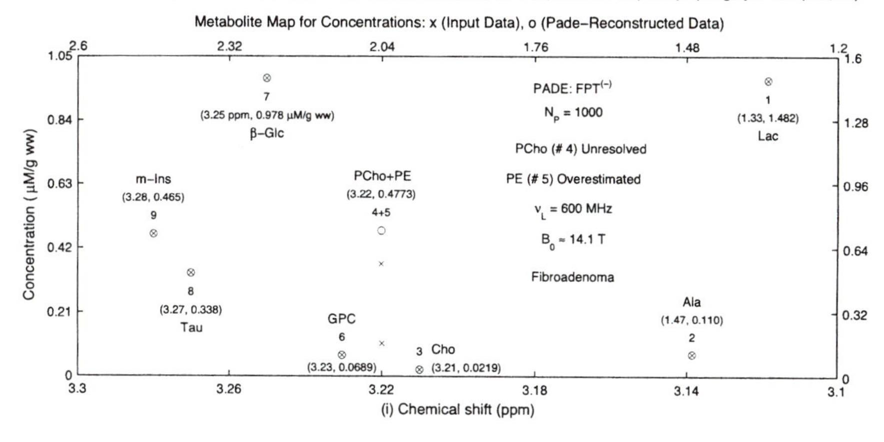

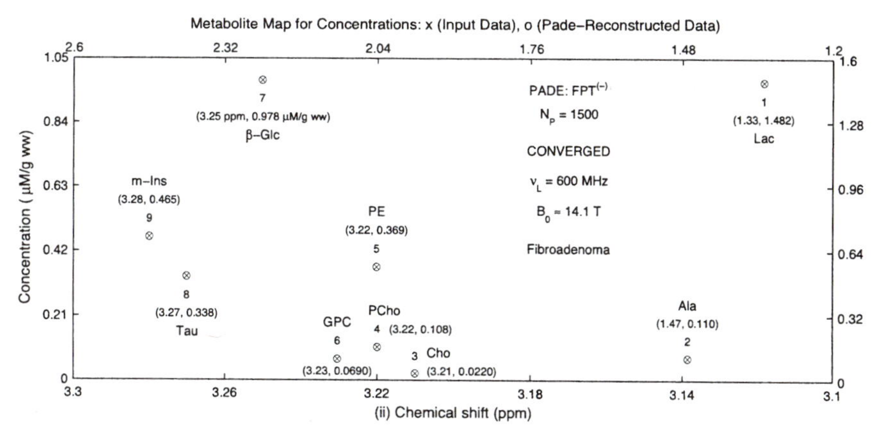

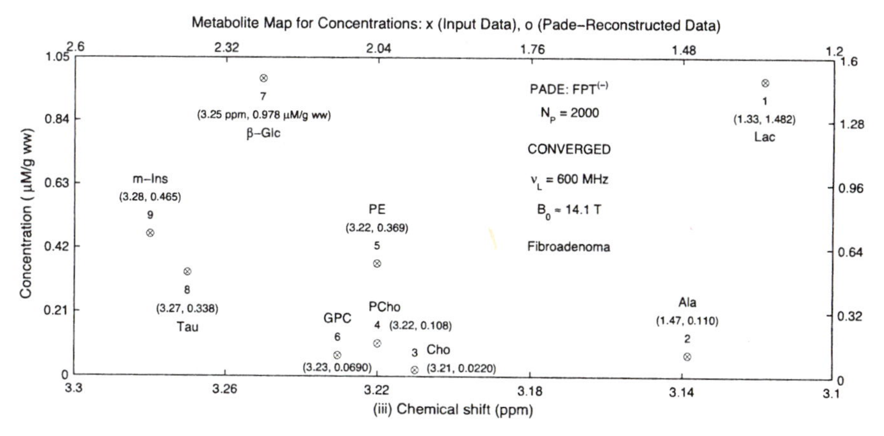

FIGURE 10.4
Padé-reconstructed metabolite concentrations in fibroadenoma based upon *in vitro* MRS data from Ref. [395].

For the fibroadenoma data the convergence of metabolite concentrations at $N_P = 1000, 1500$ and 2000 is displayed in Fig. 10.4. At $N_P = 1000$ neither the concentrations of phosphocholine nor phosphoethanolamine are accurately assessed in the reconstruction (top panel (i)), and there is a slight discrepancy in the concentrations of Cho ($k = 3$) and GPC ($k = 6$). At $N_P = 1500$ (middle panel (ii)) and $N_P = 2000$ (bottom panel (iii)), all of the metabolite concentrations are correct, as seen both numerically and by the graphic representations. The metabolite concentrations were verified to be correct at even higher partial signal lengths and at the full signal length.

10.4.3 Padé-reconstruction of MRS data from breast cancer

Table 10.4 shows the reconstructed data by the FPT$^{(-)}$ for breast cancer at the three partial signal lengths: $N_P = 1000, 1500$ and 2000. As was the case for the data corresponding to normal breast tissue and to fibroadenoma, only eight of the nine resonances were identified at $N_P = 1000$ (upper panel (i) of Table 10.4). In the interval where there should have been two resonances, phosphocholine at 3.220012 ppm and phosphoethanolamine at 3.220215 ppm, there was only one peak identified at 3.220166 ppm. The reconstructed amplitudes and concentrations for that single peak were approximately the sum of (PCho + PE). Once again, since that single peak was closer to phosphoethanolamine, it was assigned as PE. At $N_P = 1000$ all the reconstructed spectral parameters were exact for Lac ($k = 1$), Ala ($k = 2$) and m-Ins ($k = 9$). For peak $k = 3$ (Cho) the reconstructed $\text{Re}(\nu_k^-)$ and $\text{Im}(\nu_k^-)$ were exact to 5 of six decimal places (3.210122 ppm and 0.000830 ppm instead of the correct values 3.210124 ppm and 0.000834 ppm, respectively). The reconstructed $|d_k^-|$ for choline was 0.00444 au while the correct value is 0.00446 au. The concentration for choline was computed to be 0.1111 μM/g ww at $N_P = 1000$, although it should be 0.1115 μM/g ww. For peak $k = 6$ (GPC) at $N_P = 1000$ the reconstructed $\text{Re}(\nu_k^-)$ was correct to 5 of 6 decimal places, the $\text{Im}(\nu_k^-)$ to 4 of 6 decimal places, the $|d_k^-|$ was 0.00933 au rather than 0.00936 au, and the concentration was 0.2332 μM/g ww while the correct value is 0.2340 μM/g ww. For peak $k = 7$ (β−Glc) and peak $k = 8$ (Tau), the reconstructed $\text{Re}(\nu_k^-), \text{Im}(\nu_k^-)$ and $|d_k^-|$ were correct to all decimal places, but the computed concentrations were exact to 3 of the 4 decimal places.

Full convergence was attained at partial signal length $N_P = 1500$ for all the reconstructed parameters of all nine resonances corresponding to the data from breast cancer tissue (middle panel (ii) of Table 10.4). The stability of convergence at $N_P = 2000$ is also shown. On the bottom panel (iii) of Table 10.4, all the retrieved parameters are shown to be exact. This remains true even for higher partial signal lengths, as well as for the full signal length N.

The absorption component shape spectra and the total shape spectra reconstructed via the FPT are shown in Fig. 10.5 at $N_P = 1000, 1500$ and 2000 for the breast cancer data. At $N_P = 1000$ (right upper panel (iv)), the absorption total shape spectrum is converged. This was not the case for the component

TABLE 10.4
Reconstructed spectral parameters and metabolite concentrations via the FPT for malignant breast tissue with the input data based upon Ref. [395].

CONVERGENCE of SPECTRAL PARAMETERS and CONCENTRATIONS in $FPT^{(-)}$: PARTIAL SIGNAL LENGTHS N_P =1000, 1500, 2000

(i) Pade–Reconstructed Data (Malignant): N_P = 1000 (# 4 PCho: Unresolved, # 5 PE: Overestimated)

n_k^o (Metabolite # k)	$Re(\nu_k^-)$ (ppm)	$Im(\nu_k^-)$ (ppm)	$\lvert d_k^- \rvert$ (au)	C_k^-(μM/g ww)	M_k (Assignment)
1	1.330413	0.000831	0.32474	8.1185	Lac
2	1.470313	0.000832	0.03156	0.7890	Ala
3	3.210122	0.000830	0.00444	0.1111	Cho
5	3.220166	0.000836	0.10230	2.5575	PE
6	3.230414	0.000829	0.00933	0.2332	GPC
7	3.250224	0.000832	0.02882	0.7204	β–Glc
8	3.270141	0.000831	0.11182	2.7954	Tau
9	3.280132	0.000833	0.03564	0.8910	m–Ins

(ii) Pade–Reconstructed Data (Malignant): N_P = 1500 (Converged)

n_k^o (Metabolite # k)	$Re(\nu_k^-)$ (ppm)	$Im(\nu_k^-)$ (ppm)	$\lvert d_k^- \rvert$ (au)	C_k^-(μM/g ww)	M_k (Assignment)
1	1.330413	0.000831	0.32474	8.1185	Lac
2	1.470313	0.000832	0.03156	0.7890	Ala
3	3.210124	0.000834	0.00446	0.1115	Cho
4	3.220012	0.000831	0.02448	0.6120	PCho
5	3.220215	0.000832	0.07776	1.9440	PE
6	3.230412	0.000833	0.00936	0.2340	GPC
7	3.250224	0.000832	0.02882	0.7205	β–Glc
8	3.270141	0.000831	0.11182	2.7955	Tau
9	3.280132	0.000833	0.03564	0.8910	m–Ins

(iii) Pade–Reconstructed Data (Malignant): N_P = 2000 (Converged)

n_k^o (Metabolite # k)	$Re(\nu_k^-)$ (ppm)	$Im(\nu_k^-)$ (ppm)	$\lvert d_k^- \rvert$ (au)	C_k^-(μM/g ww)	M_k (Assignment)
1	1.330413	0.000831	0.32474	8.1185	Lac
2	1.470313	0.000832	0.03156	0.7890	Ala
3	3.210124	0.000834	0.00446	0.1115	Cho
4	3.220012	0.000831	0.02448	0.6120	PCho
5	3.220215	0.000832	0.07776	1.9440	PE
6	3.230412	0.000833	0.00936	0.2340	GPC
7	3.250224	0.000832	0.02882	0.7205	β–Glc
8	3.270141	0.000831	0.11182	2.7955	Tau
9	3.280132	0.000833	0.03564	0.8910	m–Ins

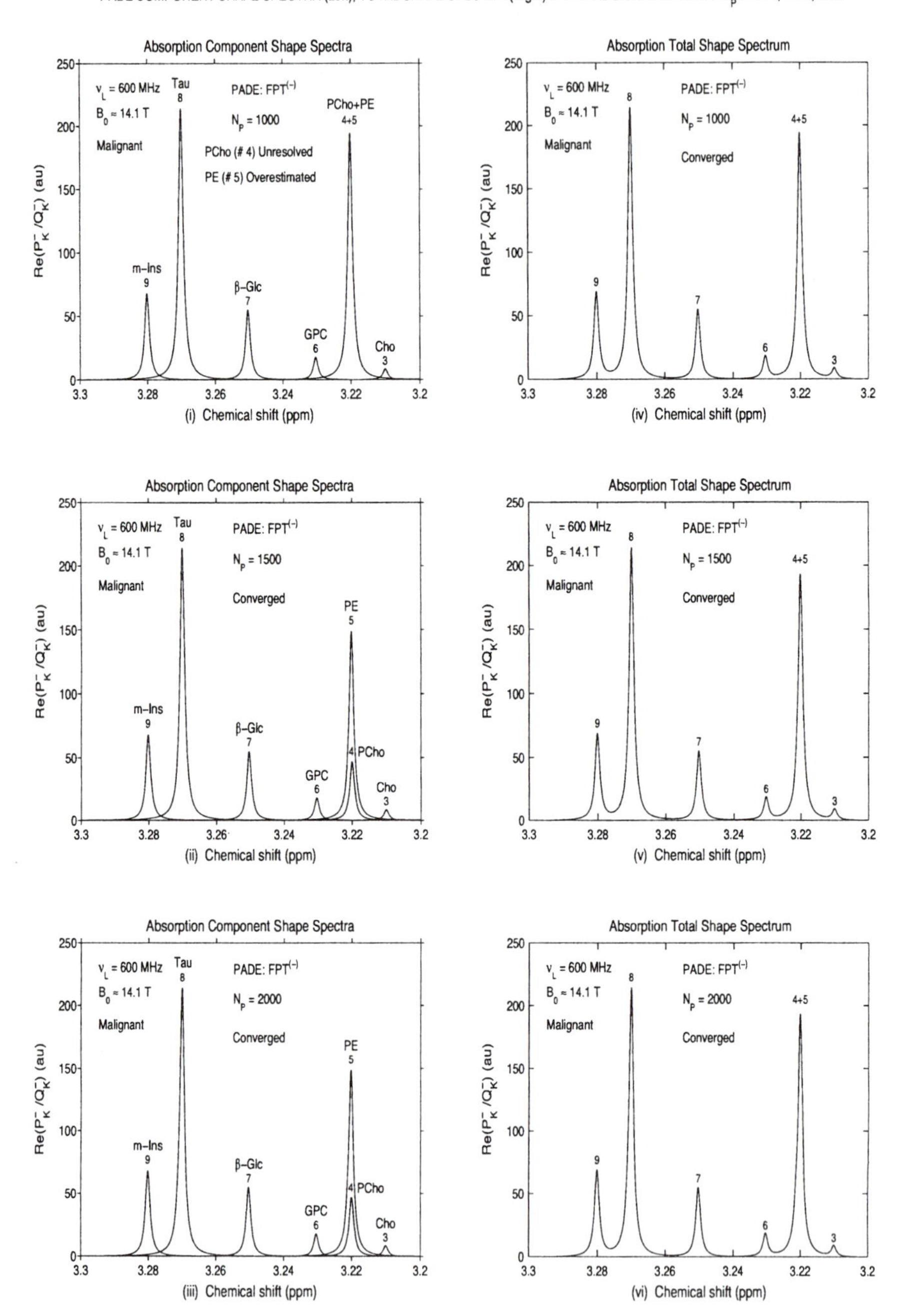

FIGURE 10.5
Padé-reconstructed absorption component spectra (left panels) and total shape spectra (right panels) for breast cancer based upon *in vitro* MRS data from Ref. [395].

Convergence of Metabolite Concentrations Reconstructed by FPT$^{(-)}$: Partial Signal Lengths N_P = 1000, 1500, 2000
Two Sets of Independent Abscissae and Ordinates: [Bottom,Left] ## 3–9 (Cho,...,m–Ins) and [Top,Right] ## 1, 2 (Lac,Ala)

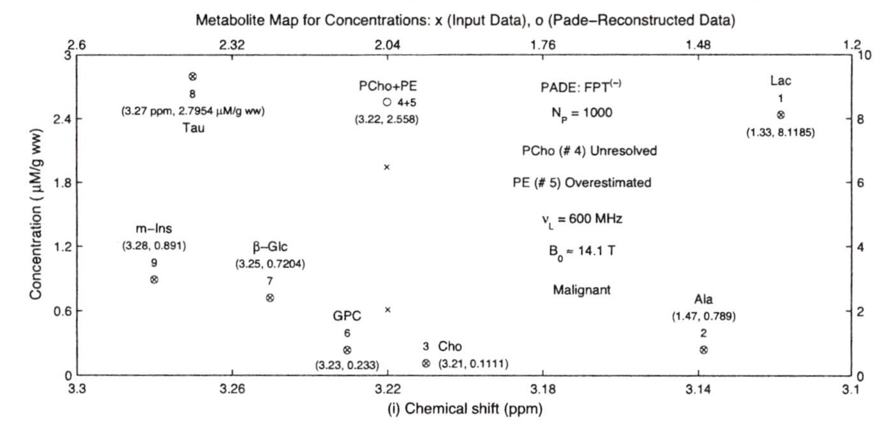

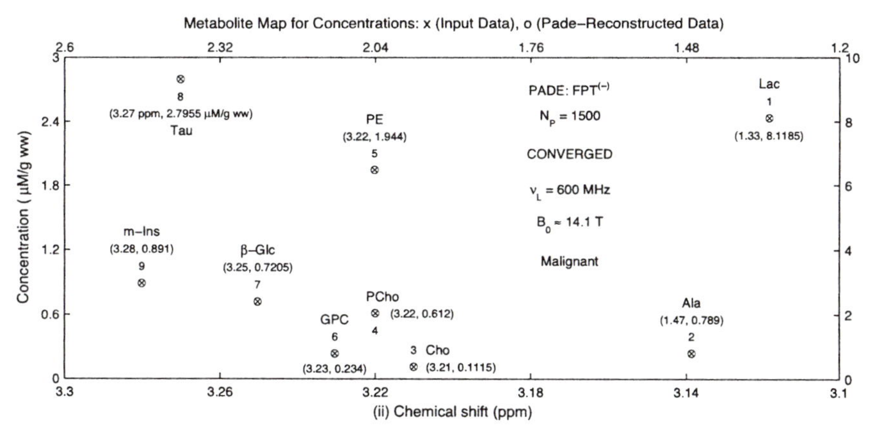

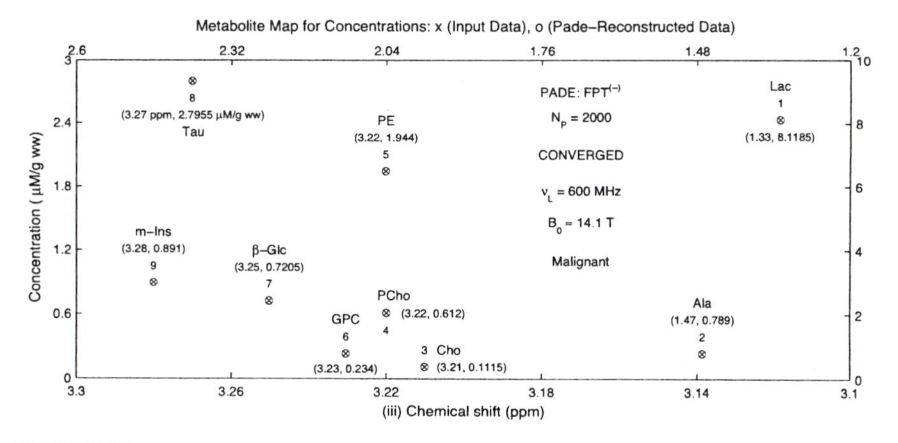

FIGURE 10.6
Padé-reconstructed metabolite concentrations in malignant breast based upon *in vitro* MRS data from Ref. [395].

shape spectrum (left upper panel (i)), which shows only a single resonance at 3.22 ppm. This was phosphoethanolamine ($k = 5$) which was over-estimated, whereas phosphocholine ($k = 4$) was unresolved. At $N_P = 1500$ in the left middle panel (ii) of Fig. 10.5 the component shape spectrum is converged with peaks $k = 4$ and 5 being resolved with correct heights, as was the case all the other peaks. Phosphocholine is seen to lie completely underneath PE. Stability of convergence is confirmed at $N_P = 2000$ in the lower panels for both the absorption component shape spectrum (iii) and the total shape spectrum (vi). For longer partial signal lengths and for the full signal length, this was also the case.

For the data corresponding to malignant breast tissue, the convergence of metabolite concentrations is illustrated on Fig. 10.6 for $N_P = 1000, 1500$ and 2000. Before convergence, at $N_P = 1000$ neither the concentrations of peaks $k = 4$ (PCho) nor $k = 5$ (PE) are correctly assessed in the reconstruction (top panel (i)) and there is a slight discrepancy in the concentrations of peaks $k = 3$ (Cho), $k = 6$ (GPC), $k = 7$ (β−Glc) and $k = 8$ (Tau). At $N_P = 1500$ (middle panel (ii)) and $N_P = 2000$ (bottom panel (iii)), all of the metabolite concentrations are observed to be correct, both numerically and by the graphic representations. The metabolite concentrations were verified as correct for higher N_P, as well as for the full signal length N.

10.4.4 Comparison of the Padé findings for normal breast, fibroadenoma and breast cancer

We compare the converged absorption component shape spectra and total shape spectra for normal breast, fibroadenoma and breast cancer. The converged Padé-reconstructed absorption component and total shape spectra within the range of 3.2 ppm to 3.3 ppm are shown for the normal breast data (Fig. 10.7, top panels (i) and (iv)), fibroadenoma (middle panels (ii) and (v)) and malignant breast data (bottom panels (iii) and (vi)). The amplitudes of all the metabolites within this frequency range are low for the normal data, β−glucose at 3.25 ppm ($k = 7$) is predominant, myoinositol ($k = 9$) at 3.28 ppm is the next most prominent peak. The phosphocholine peak is minimal.

The amplitudes of all the peaks are larger within the range of 3.2 ppm and 3.3 ppm for the fibroadenoma, in comparison to the spectra for the normal breast. Beta-glucose at 3.25 ppm ($k = 7$) is also the largest peak in the spectrum of the fibroadenoma within this frequency range. The difference between the total and component absorption shape spectrum is well-seen for the fibroadenoma, since the peak ($k = 4+5$) at 3.22 ppm is approximately the same height as myoinositol ($k = 9$ at 3.28 ppm) for the total shape spectrum, whereas myoinositol is obviously larger than the resolved peaks $k = 4$ (PCho) and $k = 5$ (PE) 3.22 ppm in the converged component shape spectrum.

For the malignant breast data the spectra have a notably different pattern within the range of 3.2 ppm and 3.3 ppm. The taurine peak ($k = 8$) at 3.27 ppm is the largest, with peak $k = 7$ (β−Glc) at 3.25 ppm among the smaller resonances. For the malignant data the difference is the most marked between the total shape spectrum in which the PCho+PE peak at 3.22 ppm is nearly as large as most abundant resonance, taurine, and the component spectrum in which phosphoethanolamine and phosphocholine are well delineated and each much smaller than taurine.

Figure 10.8 displays the converged metabolite maps reconstructed by the $\mathrm{FPT}^{(-)}$ for the normal breast tissue (top panel (i)), fibroadenoma (middle panel (ii)) and for the breast cancer (bottom panel (iii)). For the normal breast tissue it is observed that lactate ($k = 1$), has the largest concentration (0.5040 μM/g ww), slightly higher than $\beta-$glucose ($k = 7$) (0.4500 μM/g ww). The median lactate concentration in the normal breast is about 0.34 that in the fibroadenoma. For the malignant breast, the lactate concentration is over five times higher than in the fibroadenoma, and is clearly the largest resonance, nearly three time higher than taurine ($k = 8$).

In the present analysis the capacity of the FPT to resolve and precisely quantify very closely overlapping resonances with certainty is clearly demonstrated for the spectrally dense region between 3.21 ppm and 3.23 ppm. This spectral region encompasses the constituents of total choline: choline at 3.21 ppm, phosphocholine at 3.22 ppm and glycerophosphocholine at 3.23 ppm. Phosphocholine and phosphoethanolamine are nearly completely overlapping at 3.22 ppm, separated by only 0.000203 ppm which, as mentioned is about four times less than the line widths. Nevertheless, at convergence the $\mathrm{FPT}^{(-)}$ exactly reconstructs the input parameters for these two resonances with full fidelity. As presented earlier with MRS time signals that closely match FIDs encoded via proton MRS from the brain of a healthy volunteer, the FPT at convergence also exactly reconstructed all the resonances (25 in that case) including two that were nearly degenerate. It is also seen herein once again that convergence of the total shape spectrum does not necessarily imply that the component spectrum has done likewise.

As discussed, identification and quantification of these constituents of total choline within the tight spectral region of 3.21 ppm to 3.23 ppm have clinical relevance for breast cancer diagnostics. In our previous analysis, the ratio of PCho/GPC was significantly higher in the malignant versus the normal samples [25] of our input data based on Ref. [395]. This is concordant with human breast cell line research, indicating that malignant transformation is associated with the GPC to PCho switch [307]. Thus, for breast cancer diagnostics, the proton MR-visible compounds choline (3.21 ppm), phosphocholine (3.22 ppm) and glycerophosphocholine (3.23 ppm) can and should be quantified. We achieve this by the FPT, rather than summing these three metabolites via total choline, as is currently done with *in vivo* MRS.

Precise quantification of the metabolites within this tight spectral region may also help distinguish fibroadenoma from breast cancer. From the present input data that rely upon the work of Gribbestad *et al.* [395], the phosphocholine concentration was approximately 5.7 times lower in the fibroadenoma than the median phosphocholine concentration for breast cancer. This was

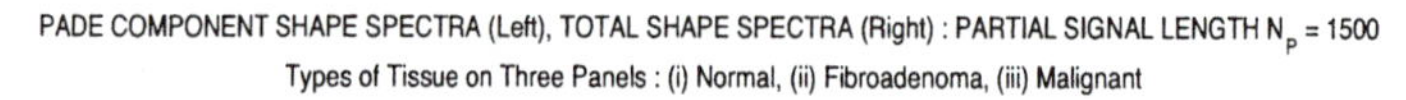

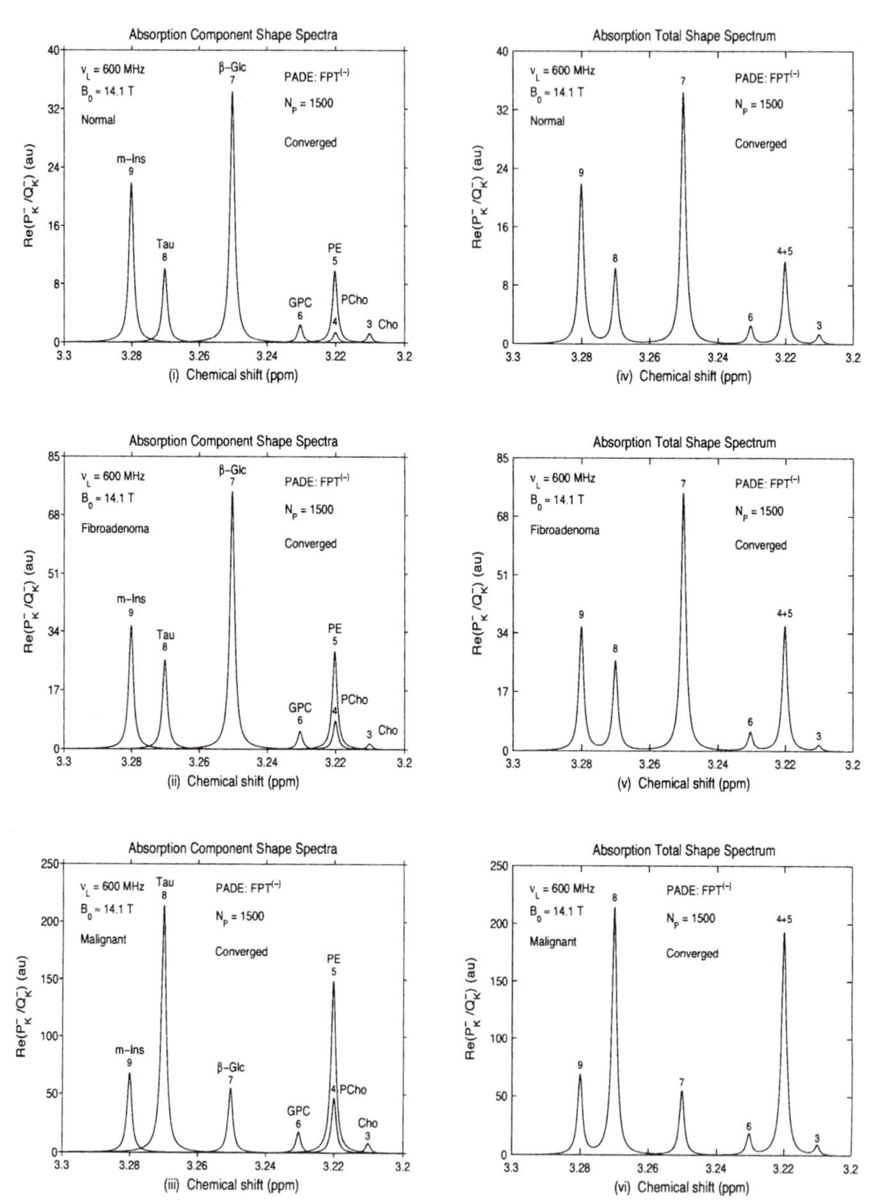

FIGURE 10.7
Padé-reconstructed converged absorption component spectra (left panels) and total shape spectra (right panels) at $N_P = 1500$ for normal breast tissue (top panels (i and iv)), benign breast pathology (fibroadenoma) (middle panels (ii and v) and breast cancer (bottom panels (iii) and (vi)) based upon data from *in vitro* data of Ref. [395].

Exact Reconstructions of Metabolite Concentrations by $FPT^{(-)}$: Partial Signal Length $N_P = 1500$
Types of Tissue on Three Panels : (i) Normal, (ii) Fibroadenoma, (iii) Malignant
Two Sets of Independent Abscissae and Ordinates: [Bottom,Left] ## 3-9 (Cho,...,m-Ins) and [Top,Right] ## 1, 2 (Lac,Ala)

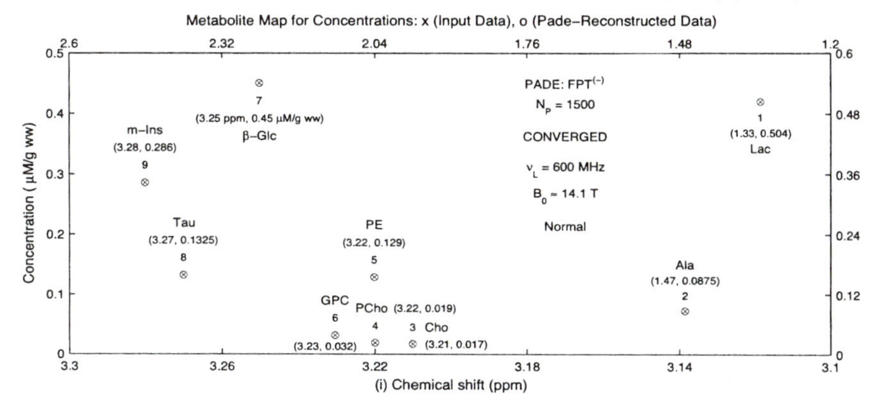

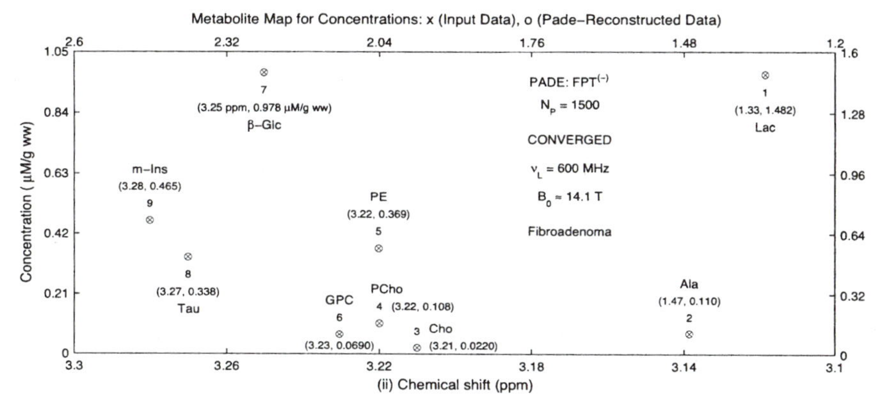

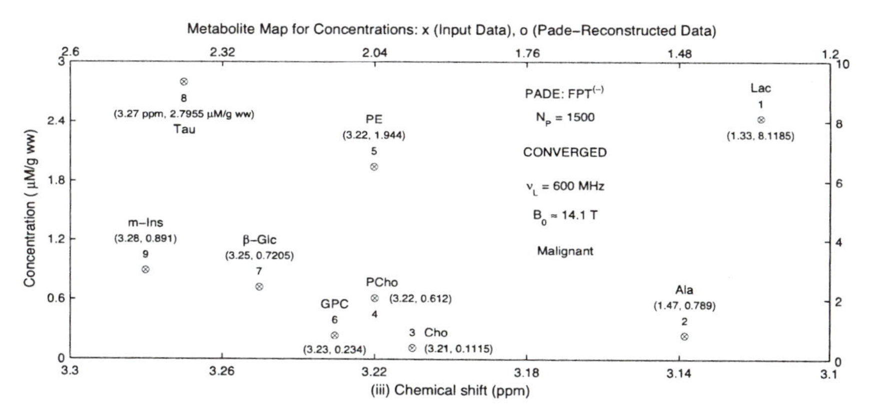

FIGURE 10.8
Converged Padé-reconstructed metabolite concentrations $N_P = 1500$ for normal breast tissue breast tissue (top panel (i)), benign breast pathology (fibroadenoma) (middle panel (ii)) and breast cancer (bottom panel (iii)) based upon *in vitro* data of Ref. [395].

percentage-wise a greater difference than for phosphoethanolamine or glycerophosphocholine. Clearly, from these input data based upon median values from a fairly small number of breast cancer samples and only one fibroadenoma, no definitive conclusions can be drawn about which metabolites best detect breast cancer and distinguish it most clearly from normal mammary tissue or benign lesions. Nevertheless, our multiple logistic regression analysis [21, 22, 25] of these data from Ref. [395] indicates that only lactate showed 100% diagnostic accuracy both with and without inclusion of the fibroadenoma. Thus, the lactate concentration from the present input data [395] was lower in the fibroadenoma than in all the individual breast cancer samples.

10.5 Prospects for Padé-optimized MRS for breast cancer diagnostics

Our results for Padé-reconstruction of MRS data employed noise-free FIDs, since we wanted to set up the fully-controlled standard for the FPT in the case of the initial application of this method to data within the realm of breast cancer diagnostics by MRS. As elaborated, this is methodologically justified. We are now taking the next steps to extend our analysis to both noise-corrupted synthesized data (still well-controlled) and to encoded FIDs similar to those from Ref. [395] as well as *in vivo* MRS data from the breast.

We therefore conclude that the demonstrated advantages of the FPT could definitely be of benefit for breast cancer diagnostics via MRS. This line of investigation will now continue with encoded data from benign and malignant breast tissue, *in vitro* and *in vivo*. Clinical correlations among the *in vitro* and *in vivo* findings and histopathology will be vital for initial verification of the studies in breast cancer diagnostics. We foresee that Padé-optimized MRS will reduce the false positive rates of MR-based modalities and further improve their sensitivity. Once this is achieved, and since MR entails no ionizing radiation, new horizons open up for screening and early detection. This could be especially important for risk groups. Thus, for example, we envision the possibility that Padé-optimized MRS could be used with greater surveillance frequency among younger women with deleterious BRCA mutations, Li-Fraumeni syndrome and for other women at increased breast cancer risk. It should also be recalled that there are promising data concerning the application of MRS for early assessment of response of breast cancer to chemotherapy [118, 386, 394]. Some of these were based upon attempts to estimate total choline concentrations from Fourier total shape spectra. It can be anticipated that Padé-based quantification of the components of choline and other diagnostically important metabolites might further contribute to this clinically vital step of optimizing therapeutic strategies for breast cancer.

11

Multiplet resonances in MRS data from normal and cancerous prostate

11.1 Dilemmas and difficulties in prostate cancer diagnostics and screening

Prostate cancer is an exceedingly common malignancy. In the U.S. carcinoma of the prostate is the most frequently diagnosed visceral cancer. Both in the U.S. and in the European Union it is cited as the most common cancer among men [401]–[403]. In the U.S. in 2006 the number of malignancy-related deaths due to prostate cancer exceeded those from colorectal cancer [403].

The number of detected cases has increased dramatically in the 1990s based on the widespread testing with prostate specific antigen (PSA) [404]. The highest rates are reported in the U.S. and Western Europe and lowest in the Asian countries. However, these findings should be interpreted in the light of diagnostic intensity and screening behavior. Namely, "incidence rates in some countries, the United States being a prime example, reflect the sum of clinical disease and latent disease, but in other countries only clinical disease" (p. 401) [405]. Nevertheless, even before PSA became available, there were marked geographic differences in incidence.

The death rate from prostate cancer has been declining in many countries of Western Europe and North America since the mid 1990s. This is thought to be related, to a substantial extent, to early detection [402, 406, 407], since the 5-year survival rate of local disease is cited to be "effectively 100%", whereas with distant metastases this falls to 34% [402].

Screening with PSA is probably responsible, at least in part, for these favorable developments. However, there are no specific cut-points that simultaneously yield optimal sensitivity and specificity [408]. Despite the lack of prospective, randomized controlled evidence demonstrating the benefit, screening with PSA plus digital rectal examination is recommended by, e.g. the American Cancer Society and the American Urological Association, although not by the American College of Preventive Medicine [403]. This more widespread screening has led to an increased detection of localized tumors, decreased frequency of nodal spread and of late stage detection, all of which improve survival dramatically. It should nevertheless also be pointed out that over-detection and over-treatment can entail morbidity and even mortality [404]. Recently, the U.S. Preventive Services Task Force has recommended against screening for prostate cancer among men over age 75 years, and considers the current evidence insufficient for making a decision regarding men younger than age seventy-five [409]. The rationale for this recommendation is that there is a lack of randomized, controlled studies, and that the cross-sectional and longitudinal evidence showed adverse psychological effects of a false-positive PSA screening result up to 1 year, while the benefits of screening are considered to be uncertain [410].

Currently, once prostate cancer is suspected on the basis of, e.g., abnormal PSA plus digital rectal exam, prostate biopsy guided by transurethral ultrasound (TRUS) is recommended. Since TRUS has limited sensitivity for visualizing prostate cancers, at least eight, and according to some authorities twelve cores should be obtained [401, 402]. Even with such measures, the sensitivity of TRUS-guided biopsy is approximately 80%; in other words, some 20% of prostate cancers are missed [402, 404]. On the other hand, given the limited specificity of PSA, up to 75% of patients undergoing biopsy for elevated PSA do not have cancer of the prostate [411].

Multi-planar T_2 weighted MRI has played an important role as a non-invasive tool for the diagnosis, staging as well as management of prostate cancer, and is particularly valuable because of its high sensitivity [102]. However, MRI poorly distinguishes benign from malignant lesions of the prostate, such that its specificity is low [412]. This high negative predictive value of MRI could help avoid unnecessary biopsies. However, because of its low positive predictive value, many unnecessary biopsies would still be performed based upon suspicious findings from MRI.

11.1.1 Initial detection of prostate cancer with *in vivo* MRS and MRSI

Spectroscopic imaging through *in vivo* magnetic resonance has made a major impact in a wide range of areas related to prostate cancer diagnostics and clinical decision making [102]–[104]. Compared to MRI alone, MRSI has substantially improved the accuracy with which prostatic tumor and extracapsular extension are detected, as well as helping to distinguish cancerous prostate from benign prostatic hypertrophy [413]. As stated recently by Casciani and Gualdi [414], "3D MRSI is emerging as the most specific tool for noninvasive evaluation of prostate cancer".

The major breakthroughs in prostate cancer diagnostics have relied heavily upon assessment of the metabolic characteristics of larger areas of suspected tumor, peritumoral regions and normal tissue. This is provided by MRSI. Complete volumetric coverage of a suspected lesion is critical for accurate diagnosis, and metabolically abnormal regions may not precisely correspond to much larger T_2 abnormalities [415]. Vigneron *et al.* [216] point out the weakness of single voxel techniques that "provide one spectrum as in MRS which is presumed to be representative of the entire tumor". They note that "this technique provides no information on the extent of the metabolic abnormality and suffers both from the inaccuracy of conventional MRI to define the exact extent of solid neoplasms and the considerable tissue heterogeneity within the tumor mass" (pp. 98, 99). Thus, these authors underscore the need for 3D spectroscopic imaging for "identification of viable tumor on the basis of both morphologic and metabolic parameters . . . for targeting and following local therapy" (p. 100) [216].

The ratio between two magnetic resonance observable metabolites, choline at 3.2 ppm and citrate (2.62 – 2.68 ppm), has been the cornerstone of applications of *in vivo* MRS and MRSI for prostate cancer detection. The normal prostate accumulates and secretes very high amounts of citrate produced by non-malignant prostate epithelial cells. Citrate is lowered with malignancy, since prostate cancer cells oxidize citrate [416]. Citrate helps distinguish benign prostatic hypertrophy (BPH) from cancer, since in the former there is high citrate production, whereas with malignant transformation, citrate is oxidized rather than synthesized [417]. Based primarily upon the ratio of choline to citrate, MRSI has substantially improved the accuracy with which prostatic malignancy and extracapsular extension are detected, as well as helping to distinguish cancerous prostate from benign prostatic hypertrophy [412, 413, 418]. Guidance as to the optimal site for biopsy has been substantially improved by MRSI, as well.

Wu *et al.* [419] note that regions of absent or low citrate concentration in the prostate can be visualized at a resolution of a few mm. They consider MRSI of the prostate as a potential tool for early diagnosis and screening.

Among 42 patients with PSA $>$ 4 ng/ml and without any prior treatment, a total of 201 benign sextants and 51 malignant sextants, all confirmed by biopsies, were analyzed via 3D MRSI by Chen *et al.* [420]. The ratio of (Choline +Creatine)/Citrate was assessed in each sextant. The mean normal value ($\pm$ standard deviation (SD)) of 0.42 $\pm$ 0.19 was used for scoring on a scale from 1 to 5, with the cutoff score being 5 denoting more than three SD above the mean. Using that criterion of MRSI for diagnosing prostate malignancy, the sensitivity and the specificity were 84.3% and 98.0%, respectively. The accuracy of the detection of prostate cancer was increased through a combination of T_2 weighted MRI and diffusion weighted imaging. Nevertheless, MRSI demonstrated higher accuracy compared to these other two methods [420].

In another series, the combination of MRI plus MRSI correctly indicated the peripheral zone sites containing prostate cancer in all 17 of 54 patients with elevated PSA and negative TRUS in whom biopsies subsequently confirmed the presence of malignancy [421]. The criteria for judging a site as malignant were (Choline + Creatine)/Citrate $>$ 0.86 or low intensity signal on MRI. However, the specificity was only 51.4% [421]. Other studies of MRI plus MRS among patients with these characteristics indicate a sensitivity ranging from 73% to 100% and specificity from 51% to 96% [422].

Of thirty six men with elevated PSA who had no malignant voxels on MRSI, none showed any prostate cancer on biopsy. The cut-point was chosen by the criterion: Citrate/(Choline + Creatine) < 1.4. Twenty six of these men were followed for at least 1.5 years. One patient with a persistently elevated PSA was diagnosed with prostate cancer over 2 years after the initial MRSI. Taken together, these findings suggest that prostate biopsy could perhaps be deferred in patients with an increased serum PSA if MRSI does not show any malignant voxels [423].

11.1.2 Distinguishing high from low risk prostate cancer

It is pertinent here to quote Cabrera *et al.* [424] for their statement: "the lifetime risk of receiving a diagnosis of prostate cancer is 16%, but the lifetime risk of dying of prostate cancer is ... 3%. The primary dilemma in the management of this malignancy is distinguishing patients with progressive disease that will become life-threatening from patients with indolent disease that may not require treatment" (p. 445) [424].

It has been noted that clinical nomograms that incorporate MRSI in addition to clinical and histopathological data may improve identification of patients with low risk prostate cancer who could be candidates for watchful waiting. This was defined as "no suspicious volume" apparently based upon (Choline + Creatine)/Citrate and decreased polyamines. Nevertheless, the level of certainty is currently considered insufficient to segregate low-risk patients from intermediate and high-risk patients, with full confidence [425].

On the other hand, MRSI may help identify aggressive prostate cancer. A strong linear correlation has been reported between with cancer aggressiveness, as gauged by the Gleason grade, a decrease in citrate and increased choline [426, 427].

11.1.3 Surveillance for residual disease or local recurrence after therapy

The time course and success of hormone-deprivation therapy can be assessed via MRSI. Metabolic atrophy, namely loss of all metabolic activity on the MR spectra signifies that this therapy has achieved its intended goal. The appearance of choline on MRSI in the setting of metabolic atrophy has been used as an indicator of recurrent or residual prostate cancer [428].

It has also been suggested that MRSI could improve assessment of recurrence after radiation therapy, which causes fibrotic and atrophic changes that distort the glandular anatomy. This is manifested in low T_2 weighted signal intensity on MRI, which is difficult to differentiate from prostate cancer. Here again, metabolic activity, notably the appearance of choline can be suggestive of local recurrence and help guide salvage therapy. Preliminary data indicate that MRSI detected recurrence of prostate cancer accurately after radiation therapy in over 80% of cases [429].

After therapy with transrectal high-intensity focused ultrasound in ten patients, MRI but not MRS, was found to provide some help in follow-up of patients in conjunction with elevated PSA levels [430]. Imaging via MRI used for surveillance after radical prostatectomy has also helped guide patient management strategies [431].

11.1.4 Treatment planning and other aspects of clinical management

Other areas of clinical decision-making with respect to prostate cancer have been impacted as well by MRSI. These include selection of therapeutic modality, timing, and other aspects of treatment planning.

Comparisons have been made using MRSI in order to assess the effectiveness of three-dimensional conformal external beam radiation therapy versus permanent prostate implantation among fifty patients with low risk prostate cancer. These comparisons showed the permanent implants to be more effective in generating metabolic atrophy [432].

Detection of dominant intra-prostatic lesions that should receive an escalated dose of radiation therapy with better sparing of surrounding normal tissue has also been aided by MRSI [433]. More recently, it was found among 67 patients with biopsy-proven prostate cancer who were followed up after external beam radiation, that MRI and MRSI findings prior to therapy were stronger independent predictors of outcome compared to clinical variables [434]. The criteria for considering a volume of tissue "having unequivocal malignant metabolism" on MRSI was that the choline peak was "clearly greater than the citrate peak" (p. 666) [434].

11.1.5 Limitations of current applications of *in vivo* MRSI directly relevant to prostate cancer

Notwithstanding the results heretofore achieved with MRSI in various aspects of prostate cancer diagnostics and clinical-decision making, shortcomings of current applications hamper wider implementation of the potential for molecular imaging through magnetic resonance in this area of oncology.

We will now briefly summarize some of the major limitations of MRSI based upon conventional, i.e., Fourier-processing that are directly relevant to prostate cancer. First of all, it should be noted that low citrate levels are not pathognomonic of prostate cancer. Low citrate is also characteristic of normal tissue in the stromal region of the prostate. In the face of metabolic atrophy, citrate levels are also low without cancer being present. On the other hand, hypertrophic prostate typically has high citrate levels, and this can still be the case despite coexistent malignancy [7]. Thus, it is clear that reliance upon ratios can be tenuous, since these can be affected by a number of processes unrelated to the presence or absence of malignant prostate tissue.

Moreover, the results are not entirely consistent that MRSI provides added value compared to MRI alone, and yet this is what was anticipated, given that metabolic information should provide greater insight than the morphological characteristics of the tissue. This, however, was not the case in a multi-center study whose aim was to localize peripheral zone sextants which contained prostate cancer among 110 patients prior to undergoing prostatectomy [435]. In that study [435], MRSI reportedly yielded no further diagnostically meaningful information compared to MRI alone. As mentioned, after therapy with transrectal high-intensity focused ultrasound, it was only MRI but not MRS, that provided diagnostically important insights during the follow-up of patients in conjunction with elevated PSA levels [430]. In yet another study [436], using the ratio of choline plus creatine to citrate, in 4 healthy volunteers and 13 patients with prostate cancer, a substantial overlap was found for various Gleason score levels and there were two false negative cases.

Despite advanced coil design and other technological solutions [419], resolution and SNR remain important challenges to the wider application of *in vivo* MRS and MRSI for prostate cancer diagnostics and management. Kundra and colleagues [402] have cited limitations in the resolution and interpretation of data from MRSI as important challenges with respect to prostate cancer diagnosis, staging and surveillance. Attempts to improve resolution and SNR by increasing the static magnetic field strength are noted to affect the spectral shape of citrate and its ratio to choline [437]. Kim *et al.* [437] state that with the limited existing data using 3T scanners, "it remains debatable as to whether the potential to improve the diagnostic performance for prostate cancer staging at 3T will translate to clear benefits" (p. 170).

Another very important drawback of MRSI for diagnostics of prostate cancer is its reportedly rather low sensitivity in the analysis of smaller lesions [417]. This could clearly limit the role of MRSI for early prostate cancer detection, since this is the way that this type of malignancy typically first presents. Concordant with these observations, Westphalen *et al.* [429] conclude that MR-based modalities are still not considered to have any role in screening for the general population.

11.2 Insights for prostate cancer diagnostics by means of 2D *in vivo* MRS and *in vitro* MRS

Besides total choline at 3.2 ppm and citrate at 2.6 ppm to 2.7 ppm, there are several other metabolites that may be indicators of prostate cancer. These have been identified by 2D *in vivo* MRS and by *in vitro* MRS.

Using *in vivo* 2D MRS, Thomas *et al.* [438] were able to detect multiplets from spermine which overlap with choline and creatinine on 1D *in vivo* MRS.

Their study [438] was carried out among seven healthy men with normal prostate and one man with benign prostatic hypertrophy. However, it should be recalled that while 2D MRS is frequently helpful in identifying overlapping peaks, attempts to extend 1D fitting algorithms to 2D MRS are virtually unfeasible [439]. Thus, these data from *in vivo* 2D MRS have provided essentially only qualitative information, due to reliance upon the Fourier-based signal processing.

A larger number of studies have been performed using *in vitro* MRS in the area of prostate cancer diagnostics. Smith *et al.* [341] compared biopsies from 66 patients with BPH and 21 with prostate cancer. They used multivariate analysis with a training set and then a test set, achieving 100% sensitivity and 95.5% specificity. The most discriminatory regions were centered on 3.49 ppm (taurine), 3.43 ppm, 2.53 ppm (citrate), 2.17 ppm, 1.87 ppm (glutamate), and 1.15 ppm. In addition, these authors were able to distinguish the two types of BPH (glandular and stromal), which require different therapy, on the basis of much higher citrate in the former.

In marked contrast to the usual basis for diagnosing prostate cancer by *in vivo* MRS, Swindle *et al.* [440] found in their study of 77 prostate specimens from the prostate, "depleted citrate and elevated choline levels alone were not accurate markers of malignancy, since citrate levels remain high when a small amount of malignant disease is present" (p. 144). On the other hand, they found that lipid, creatine and lysine were helpful. Cancerous prostate tissue showed a significantly higher lipid to lysine ratio (1.3:1.7) compared to stromal BPH. Compared to glandular BPH, the lipid to citrate ratio (1.3:2.5) was significantly higher in malignant prostate. They also found that spermine and spermidine (3.1 ppm) decreased with increased adenocarcinoma involvement. Although the choline-creatine ratio and lipid-lysine ratio were significantly higher in malignant prostate specimens compared to benign prostate tissue, there was "considerable overlap of ratios, resulting in poor separation of BPH from cancer" (p. 147). On the other hand, prostate cancers and prostate intra-epithelial neoplasia that were missed with routine histological exam, but found with histological step-slice analysis were correctly identified using *in vitro* MRS.

As noted, the normal prostate accumulates and secretes very high amounts of citrate. As well as in malignancy, citrate is also low with post-biopsy hemorrhage, prostatitis, and with therapy. Evaluation of citrate via MRSI is found to be helpful for following the progression or regression of prostate cancer after therapy. Costello *et al.* [416] propose that malignant transformation of prostate cells is associated with citrate oxidation, as opposed to citrate accumulation. Zinc also accumulates at very high levels in normal prostate cells; these authors note a relationship among citrate oxidation, loss of zinc and prostate cancer.

In a review of *in vitro* studies of prostate cancer through 2000, Koutcher and colleagues [441] consider that membrane structure may be important *vis-à-vis* spectroscopic findings for choline and related compounds that are precursors and degradation products of membrane phospholipids. They note that tumors with higher metastatic potential have higher bulk membrane rigidity with higher cholesterol and lower phospholipid. They also note that the peripheral zone is the site of 68% of prostate tumors and that region has an elevated citrate concentration. Citrate concentrations in the transition zone can increase with BPH if there is a high glandular fraction, and can decrease if the region of BPH has a high stromal content. With prostate cancer, the decreased glandular fraction can lead to decreased citrate concentration.

Further insights have subsequently been gleaned into the components of total choline with respect to prostate cancer. Significantly higher phosphocholine and glycerophosphocholine levels have been found in human prostate cells derived from metastases compared to normal prostate epithelial and stromal cells [310]. It is concluded: "the elevation of the choline peak observed clinically in prostate cancer is attributable to an alteration of phospholipid metabolism and not simply to increased cell density, doubling time or other non-specific effects" (p. 3599) [310].

Through HRMAS, Swanson *et al.* [442] assessed 54 post-surgical prostate samples obtained using MRI plus three-dimensional MRSI. Pre-surgical MRI plus MRSI identified healthy and cancerous prostate tissues with 81% accuracy. Healthy glandular tissue was distinguished from prostate cancer by significantly higher levels of citrate and polyamines, and lower choline, phosphocholine and glycerophosphocholine. Predominantly stromal tissue lacked citrate and polyamines, but had significantly lower levels of choline compounds compared to malignant tissue. Taurine, myoinositol and scylloinositol as well as choline compounds were higher in prostate cancer. More aggressive cancers showed higher choline and lower citrate and polyamines. The authors conclude: "the elucidation of spectral patterns associated with mixtures of different prostate tissue types and cancer grades, and the inclusion of new metabolic markers for prostate cancer may significantly improve the clinical interpretation of *in vivo* prostate MRSI data" (p. 944).

More recently, Swanson *et al.* [443] reported metabolite concentrations from sixty post-surgical samples of healthy glandular and stromal prostate tissue as well as prostate cancer using HRMAS. The malignant prostate samples were significantly distinguished from both these normal tissues by higher concentrations of lactate (1.34 ppm and 4.14 ppm), phosphocholine (3.23 ppm) plus glycerophosphocholine (3.24 ppm), as well as by total choline. However, it was only in comparison to healthy glandular tissue, that prostate cancer differed significantly with respect to lower citrate (2.54 ppm and 2.72 ppm) and polyamines (3.10 ppm and 3.14 ppm). On the other hand, the prostate cancer specimens had significantly higher alanine (1.49 ppm, 2.54 ppm, 2.72 ppm) and choline (3.21 ppm) concentrations than the healthy stromal tissue, but these concentrations differed only at a borderline level of significance for healthy glandular tissue versus prostate cancers.

The cited concentrations were computed using Lorentzian-Gaussian peak fitting within the Levenberg-Marquardt algorithm. These authors [443] reported that spectral overlap was a major problem in one-dimensional HRMAS studies of prostate cancer and other tissues. They noted, for example, that phosphocholine at 3.23 ppm and glycerophosphocholine at 3.24 ppm could not be completely resolved from each other; thus the sum of these two metabolites concentrations was given.

Most recently this group of authors [444] used two-dimensional HRMAS and total correlation spectroscopy (TOCSY). Phosphocholine was reportedly present in eleven of fifteen (73%) prostate cancer specimens whereas PCho was only seen in 6 of 32 (19%) of the non-malignant prostate samples. They also found that PCho/GPC ratios were significantly higher in the cancerous samples [444]. Another study performed by means of HRMAS [445] revealed that lactate and alanine were significantly lower in benign glandular prostate compared with malignant tissue.

It should be emphasized that a particular challenge within the spectra from the prostate are the numerous multiplets. Within the regions of greatest spectral density, the difficulties become even greater. In the study by Swanson *et al.* [443] of the polyamine resonances, only the contribution from the predominant spermine was included in fitting, whereas putrescine and spermidine could not be resolved. These difficulties underscore the need for exact, unequivocal quantification not only of spectra with completely distinct resonances, but also for multiplets with overlapping resonances. As elaborated earlier in this book, the FPT would appear to be ideally suited to this task.

11.3 Performance of the fast Padé transform for MRS data from prostate tissue

In the present spectral analysis, three FIDs were synthesized in the form $c_n = \sum_{k=1}^{K} d_k e^{in\tau\omega_k}$ from (3.1) with $\mathrm{Im}(\omega_k) > 0$, by using a sum of $K = 27$ damped complex exponentials $\exp(in\tau\omega_k)$ $(1 \le k \le 27)$ [33]. The complex amplitudes d_k were time-independent. As has been the case throughout this book, ω_k and d_k are the fundamental angular frequencies and amplitudes ($\omega_k = 2\pi\nu_k$, where ν_k is the linear frequency). We then quantified these time signals using the $\mathrm{FPT}^{(-)}$, as per Ref. [10]. The bandwidth was 6000 Hz, where the inverse of this bandwidth is the sampling time τ. The total signal length was set as $N = 1024$.

The input data for the absolute values of amplitudes were generated according to the given mean metabolite concentrations, as well as the description of multiplets and the total shape spectra for normal glandular prostate, normal stromal prostate and prostate cancer as per Ref. [443]. The total metabolite concentrations were split into multiplets to correspond to the spectra of Ref. [443] and also to fill in what was missing.

TABLE 11.1
Input spectral parameters and metabolite concentrations for normal glandular prostate derived from *in vitro* data of Ref. [443]. As usual, ppm denotes parts per million, au arbitrary units, M_k denotes metabolite assignment, C_k denotes concentration of M_k, while Lac denotes lactate, Ala alanine, Cit citrate, Cr creatine, PA polyamines, Cho choline, PCho phosphocholine, GPC glycerophosphocholine, s-Ins scylloinositol, Tau taurine, m-Ins myoinositol.

INPUT DATA (NORMAL GLANDULAR): SPECTRAL PARAMETERS, CONCENTRATIONS and METABOLITE ASSIGNMENTS

Phases of All Harmonics are set Equal to Zero: $\phi_k = 0$ (k = 1,..., K; K = 27)

n^o_k (Metabolite # k)	Re(ν_k) (ppm)	Im(ν_k) (ppm)	$\lvert d_k \rvert$ (au)	C_k (μM/g ww)	M_k (Assignment)
1	1.330148	0.013213	7.374199	40.3	Lac
2	1.490417	0.030361	1.579140	8.63	Ala
3	2.515197	0.008242	1.500457	8.20	Cit
4	2.540235	0.009624	2.616651	14.3	Cit
5	2.720146	0.010205	2.525160	13.8	Cit
6	2.750328	0.007014	1.244282	6.80	Cit
7	3.040309	0.006951	1.661482	9.08	Cr
8	3.100416	0.021432	2.104300	11.5	PA
9	3.140263	0.025760	1.280878	7.00	PA
10	3.210152	0.002367	0.644099	3.52	Cho
11	3.230242	0.010313	0.181153	0.99	PCho
12	3.240318	0.003014	0.468435	2.56	GPC
13	3.250298	0.004362	0.384263	2.10	Tau
14	3.260429	0.005921	0.823422	4.50	Tau
15	3.275351	0.005671	0.219579	1.20	Tau
16	3.350137	0.007324	0.331199	1.81	s–Ins
17	3.420319	0.004290	0.201281	1.10	Tau
18	3.430241	0.005014	0.329369	1.80	Tau
19	3.440448	0.004721	0.182983	1.00	Tau
20	3.533243	0.006432	0.534309	2.92	m–Ins
21	3.550326	0.009324	1.557182	8.51	m–Ins
22	3.563639	0.007291	0.795974	4.35	m–Ins
23	3.626378	0.006952	0.589204	3.22	m–Ins
24	3.640411	0.009850	1.374199	7.51	m–Ins
25	3.655389	0.005812	0.444648	2.43	m–Ins
26	4.070235	0.012823	0.962489	5.26	m–Ins
27	4.120142	0.019891	1.134492	6.20	Lac

TABLE 11.2
Input spectral parameters and metabolite concentrations for normal stromal prostate derived from *in vitro* data of Ref. [443].

INPUT DATA (NORMAL STROMAL): SPECTRAL PARAMETERS, CONCENTRATIONS and METABOLITE ASSIGNMENTS

Phases of All Harmonics are set Equal to Zero: $\phi_k = 0$ (k = 1,..., K; K = 27)

n^o_k (Metabolite # k)	$Re(\nu_k)$ (ppm)	$Im(\nu_k)$ (ppm)	$\lvert d_k \rvert$ (au)	C_k (μM/g ww)	M_k (Assignment)
1	1.330148	0.013214	7.154620	39.1	Lac
2	1.490417	0.030360	1.244282	6.80	Ala
3	2.515197	0.010242	0.559927	3.06	Cit
4	2.540235	0.012151	0.977127	5.34	Cit
5	2.720146	0.012203	0.944190	5.16	Cit
6	2.750328	0.009014	0.464776	2.54	Cit
7	3.040309	0.006132	1.396157	7.63	Cr
8	3.100416	0.018432	0.345837	1.89	PA
9	3.140263	0.022761	0.230558	1.26	PA
10	3.210152	0.002613	0.495883	2.71	Cho
11	3.230242	0.010314	0.221409	1.21	PCho
12	3.240318	0.005121	0.572736	3.13	GPC
13	3.250298	0.004132	0.329369	1.80	Tau
14	3.260429	0.003821	0.708143	3.87	Tau
15	3.275351	0.005670	0.188472	1.03	Tau
16	3.350137	0.005022	0.364135	1.99	s–Ins
17	3.420319	0.003690	0.175663	0.96	Tau
18	3.430241	0.003012	0.287283	1.57	Tau
19	3.440448	0.004123	0.159195	0.87	Tau
20	3.533243	0.006432	0.420860	2.30	m–Ins
21	3.550326	0.009324	1.225984	6.70	m–Ins
22	3.563639	0.007290	0.625801	3.42	m–Ins
23	3.626378	0.005951	0.462946	2.53	m–Ins
24	3.640411	0.007432	1.081427	5.91	m–Ins
25	3.655389	0.004814	0.351327	1.92	m–Ins
26	4.070235	0.010123	0.753888	4.12	m–Ins
27	4.120142	0.019891	1.097896	6.00	Lac

TABLE 11.3
Input spectral parameters and metabolite concentrations for prostate cancer derived from *in vitro* data of Ref. [443].

INPUT DATA (MALIGNANT): SPECTRAL PARAMETERS, CONCENTRATIONS and METABOLITE ASSIGNMENTS

Phases of All Harmonics are set Equal to Zero: $\phi_k = 0$ (k = 1,..., K; K = 27)

n_k^o (Metabolite # k)	Re(ν_k) (ppm)	Im(ν_k) (ppm)	$|d_k|$ (au)	C_k(μM/g ww)	M_k (Assignment)
1	1.330148	0.013213	11.07045	60.5	Lac
2	1.490417	0.030362	2.305581	12.6	Ala
3	2.515197	0.016240	0.680695	3.72	Cit
4	2.540235	0.020952	1.185727	6.48	Cit
5	2.720146	0.020204	1.152790	6.30	Cit
6	2.750328	0.018213	0.567246	3.10	Cit
7	3.040309	0.006951	1.780421	9.73	Cr
8	3.100416	0.021434	0.580055	3.17	PA
9	3.140263	0.032762	0.386093	2.11	PA
10	3.210152	0.002814	0.823422	4.50	Cho
11	3.230242	0.003952	1.235133	6.75	PCho
12	3.240318	0.005890	0.473925	2.59	GPC
13	3.250298	0.004123	0.364135	1.99	Tau
14	3.260429	0.003141	0.781336	4.27	Tau
15	3.275351	0.005672	0.208600	1.14	Tau
16	3.350137	0.004324	0.459286	2.51	s–Ins
17	3.420319	0.003626	0.190302	1.04	Tau
18	3.430241	0.001426	0.312900	1.71	Tau
19	3.440448	0.003821	0.173833	0.95	Tau
20	3.533243	0.006434	0.548948	3.00	m–Ins
21	3.550326	0.010323	1.601098	8.75	m–Ins
22	3.563639	0.007291	0.817932	4.47	m–Ins
23	3.626378	0.005952	0.611162	3.34	m–Ins
24	3.640411	0.009434	1.425435	7.79	m–Ins
25	3.655389	0.004812	0.461116	2.52	m–Ins
26	4.070235	0.012421	0.993596	5.43	m–Ins
27	4.120142	0.012890	1.701738	9.30	Lac

The absolute values $\{|d_k|\}$ of the input amplitudes $\{d_k\}$ were deduced from the reported metabolite concentrations $\{\mathrm{C}_k\}$ using (3.15) via $|d_k| = 2\mathrm{C}_k/\mathrm{C}_{\mathrm{ref}}$. Here, the reference material was taken to be the TSP molecule (3-(trimethylsilyl-) 3,3,2,2-tetradeutero-propionic acid). The reference concentration of TSP, $\mathrm{C}_{\mathrm{ref}}$ was computed from the data reported in Ref. [443]. Specifically, in [443] there was a total of (TSP + D_2O) = 3.79 mg, where TSP comprised 0.75% of that weight. There were altogether 0.02845 mg of TSP. Because the molecular weight of TSP = 172.23 g, there were 0.1652 μM of TSP and the mass of wet tissue = 15.11 mg, such that the concentration of TSP = 0.1652 μM/15.11 mg ww = 10.93 μM/g ww.

We set all the phases ϕ_k $(1 \le k \le 27)$ from complex-valued d_k to zero, and thereby every d_k becomes real $d_k = |d_k|$. Tables 11.1, 11.2 and 11.3 display the input data for the normal glandular, stromal prostate and for prostate cancer, respectively.

The diagonal FPT$^{(-)}$ was employed to analyze the FIDs. The coefficients $\{p_r^-, q_s^-\}$ of the polynomials P_K^- and Q_K^- were computed by solving the systems of linear equations from chapter 4 (section 4.10) by treating the product in $G_N(z^{-1}) * Q_K^-(z^{-1}) = P_K^-(z^{-1})$ as a convolution. To extract the peak parameters, we solved the characteristic equation $Q_K^-(z^{-1}) = 0$. This leads to K unique roots z_k^- $(1 \le k \le K)$, so that the sought ω_k^- is deduced via $\omega_k^- = (i/\tau)\ln(z_k^-)$.

The FPT$^{(-)}$ extracts the spectral parameters $\{\omega_k^-, d_k^-\}$ $(1 \le k \le K)$ of every physical resonance using only the input raw FID without any editing or modification. The kth metabolite concentration C_k^- of the tissue wet weight is computed from the absolute value $|d_k^-|$ of the reconstructed amplitude d_k^- as $\mathrm{C}_k^- = |d_k^-|\mathrm{C}_{\mathrm{ref}}/2 = 5.465|d_k^-|$ μM/g ww.

In order to confirm the constancy of the spectral parameters for all three signals, we progressively increased the signal length for the same bandwidth. We examined the spectral parameters at total orders $K = 250$, 300 and 350 where $2K = N_P$ and N_P denotes partial signal length. Convergence occurred at $K = 350$, 300 and 300 for all the FIDs in the respective cases of normal glandular and normal stromal prostate and for prostate cancer. All continued to be stable thereafter. In the parametric signal processing, we determined whether a given reconstructed resonance was true or spurious by computing signal poles and zeros, as prescribed by the concept of Froissart doublets within the fast Padé transform. Spurious resonances have zero amplitudes due to the coincident poles and zeros of the Padé complex-valued spectrum.

The constancy in non-parametric signal processing was also checked by computing a sequence of the Padé shape spectra $\{P_m^-/Q_m^-\}$ $(m = 1, 2, 3, ...)$ in the whole Nyquist range, which includes the frequency interval of interest from 1.3 ppm to 4.2 ppm.

For normal glandular prostate, of the 350 total resonances, 323 were found to be spurious due to zero amplitudes and the pole-zero coincidences, thus yielding the 27 genuine resonances. For normal stromal tissue and prostate cancer, respectively of the 300 total resonances, 273 were identified as spurious by their zero amplitudes and the pole-zero coincidences, yielding the 27 genuine resonances.

11.3.1 Normal glandular prostate tissue: MR spectral data reconstructed by FPT

For the normal glandular prostate the reconstructed data by the $\text{FPT}^{(-)}$ are presented in Table 11.4. The partial signal lengths shown are $N_P = 600$ and $N_P = 700$.

We emphasize that all the input concentrations are given with at most two decimals to preserve the accuracy of the corresponding data from Ref. [443]. The concentrations reported here are set to match this latter input. Nevertheless, all the reconstructed parameters are listed with 6 decimals.

At $N_P = 600$ (upper panel (i)), 25 of the 27 resonances were identified. Peak $k = 11$ (phosphocholine) at 3.230242 ppm and the component of the taurine triplet at 3.250298 ppm ($k = 13$) were missing. Few of the spectral parameters or concentrations were fully exact for any of the resonances. The closest to being totally correct at $N_P = 600$ were the spectral parameters and computed concentrations for the peaks at the outermost regions of the spectrum ($k = 1-8$ and $k = 25-27$). The reconstructed chemical shift, $\text{Im}(\nu_k^-)$ and computed concentration for lactate at 1.330148 ppm were exact at $N_P = 600$, but the $|d_k^-|$ was correct just to 4 of the 6 decimal places. The last digit of the chemical shift and two digits for $\text{Im}(\nu_k^-)$ were not exact for alanine, only two of the six decimal places were correct for the $|d_k^-|$ and its computed concentration was correct only to 1 decimal place at $N_P = 600$. For each of the doublets of doublets of citrate and for creatine the chemical shift, $\text{Im}(\nu_k^-)$ and concentrations were exact, although the $|d_k^-|$ were only correct to 3 or 4 of the six decimal places at $N_P = 600$. The computed concentration for the polyamine resonance peak $k = 8$ was fully correct. Here, however, the chemical shift, $\text{Im}(\nu_k^-)$ and $|d_k^-|$ were correct to 5, 4 and 2 decimal places, respectively at $N_P = 600$.

For the region from 3.14 ppm to 3.65 ppm which has a greater spectral density, none of the reconstructed spectral parameters of the identified resonances were fully exact at $N_P = 600$. However, the computed concentrations of choline, scylloinositol and the component within the myoinositol triplet centered at 3.6555394 ppm were completely correct. At $N_P = 600$ the chemical shift, $\text{Im}(\nu_k^-)$ and computed concentration of the myoinositol singlet at 4.070235 ppm were fully correct, but the $|d_k^-|$ was not. Also, $\text{Re}(\nu_k^-)$ of the lactate peak ($k = 27$) was correctly reconstructed to be 4.120142 ppm and the computed concentration was correct, but for the $\text{Im}(\nu_k^-)$ and $|d_k^-|$ the last one and three decimal places, respectively were not exact at $N_P = 600$. Although most of the computed concentrations had the correct integer values, this was not the case for GPC. The concentration of GPC was computed to be 5.83 μM/g ww at $N_P = 600$, rather than the correct concentration of 2.56 μM/g

TABLE 11.4
Padé-reconstructed spectral parameters and metabolite concentrations for normal glandular prostate with the input data as derived from Ref. [443].

CONVERGENCE of SPECTRAL PARAMETERS and CONCENTRATIONS in FPT$^{(-)}$: PARTIAL SIGNAL LENGTHS N_P = 600, 700

(i) Reconstructed Data (Normal Glandular) : N_P = 600 ; Missing Resonances : # 11 (PCho) and # 13 (Tau)

n_k^o (Metabolite # k)	$Re(\nu_k^-)$ (ppm)	$Im(\nu_k^-)$ (ppm)	$\lvert d_k^- \rvert$ (au)	C_k^-(μM/g ww)	M_k (Assignment)
1	1.330148	0.013213	7.374172	40.3	Lac
2	1.490419	0.030353	1.577285	8.62	Ala
3	2.515197	0.008242	1.500502	8.20	Cit
4	2.540235	0.009624	2.616647	14.3	Cit
5	2.720146	0.010205	2.525202	13.8	Cit
6	2.750328	0.007014	1.244305	6.80	Cit
7	3.040309	0.006951	1.661438	9.08	Cr
8	3.100415	0.021427	2.103129	11.5	PA
9	3.140287	0.025772	1.282285	7.01	PA
10	3.210129	0.002369	0.643592	3.52	Cho
12	3.242549	0.011163	1.066128	5.83	GPC
14	3.262437	0.006985	0.865424	4.73	Tau
15	3.274335	0.005416	0.254324	1.39	Tau
16	3.350141	0.007318	0.330869	1.81	s–Ins
17	3.420373	0.004365	0.208105	1.14	Tau
18	3.430250	0.004800	0.313167	1.71	Tau
19	3.440350	0.004816	0.191041	1.04	Tau
20	3.533241	0.006444	0.535570	2.93	m–Ins
21	3.550328	0.009301	1.551185	8.48	m–Ins
22	3.563630	0.007305	0.799499	4.37	m–Ins
23	3.626376	0.006939	0.587325	3.21	m–Ins
24	3.640406	0.009865	1.377351	7.53	m–Ins
25	3.655393	0.005808	0.443946	2.43	m–Ins
26	4.070235	0.012823	0.962450	5.26	m–Ins
27	4.120142	0.019890	1.134339	6.20	Lac

(ii) Reconstructed Data (Normal Glandular) : N_P = 700 ; Converged (All Resonances Resolved)

n_k^o (Metabolite # k)	$Re(\nu_k^-)$ (ppm)	$Im(\nu_k^-)$ (ppm)	$\lvert d_k^- \rvert$ (au)	C_k^-(μM/g ww)	M_k (Assignment)
1	1.330148	0.013213	7.374199	40.3	Lac
2	1.490417	0.030361	1.579140	8.63	Ala
3	2.515197	0.008242	1.500457	8.20	Cit
4	2.540235	0.009624	2.616651	14.3	Cit
5	2.720146	0.010205	2.525160	13.8	Cit
6	2.750328	0.007014	1.244282	6.80	Cit
7	3.040309	0.006951	1.661482	9.08	Cr
8	3.100416	0.021432	2.104300	11.5	PA
9	3.140263	0.025760	1.280878	7.00	PA
10	3.210152	0.002367	0.644099	3.52	Cho
11	3.230242	0.010313	0.181153	0.99	PCho
12	3.240318	0.003014	0.468435	2.56	GPC
13	3.250298	0.004362	0.384263	2.10	Tau
14	3.260429	0.005921	0.823422	4.50	Tau
15	3.275351	0.005671	0.219579	1.20	Tau
16	3.350137	0.007324	0.331199	1.81	s–Ins
17	3.420319	0.004290	0.201281	1.10	Tau
18	3.430241	0.005014	0.329369	1.80	Tau
19	3.440448	0.004721	0.182983	1.00	Tau
20	3.533243	0.006432	0.534309	2.92	m–Ins
21	3.550326	0.009324	1.557182	8.51	m–Ins
22	3.563639	0.007291	0.795974	4.35	m–Ins
23	3.626378	0.006952	0.589204	3.22	m–Ins
24	3.640411	0.009850	1.374199	7.51	m–Ins
25	3.655389	0.005812	0.444648	2.43	m–Ins
26	4.070235	0.012823	0.962489	5.26	m–Ins
27	4.120142	0.019891	1.134492	6.20	Lac

ww. While the correct value was 0.468435 au, the reconstructed $|d_k^-|$ was 1.066128 au for GPC at $N_P = 600$.

Full convergence was achieved at $N_P = 700$ for all the reconstructed parameters and computed concentrations for each of the 27 resonances (bottom panel (ii) of Table 11.4). The stability of convergence was verified at higher partial signal lengths as well as at the total signal length $N = 1024$.

In Fig. 11.1 the absorption component shape spectra as reconstructed by the $\mathrm{FPT}^{(-)}$ are shown at the two partial signal lengths: $N_P = 54$ and $N_P = 800$ for the normal glandular prostate data. At $N_P = 54$, fifteen of the twenty-seven resonances were missing (top panel). It was only at the two extremes of the spectrum that all the resonances were resolved and that heights were close to being correct (lactate and alanine at 1.33 ppm and 1.49 ppm and myoinositol and lactate at about 4.07 ppm and 4.12 ppm, respectively).

For the denser spectral region from $\approx$ 2.5 ppm to 3.70 ppm, only eight of the twenty-three peaks were resolved and an admixture of the absorption and dispersive modes are seen, the latter includes structures with negative intensities on the ordinate axis. The relative heights of the doublet citrate peak near 2.5 ppm are seen to be reversed, namely the peak at about 2.52 ppm is larger than the one at about 2.54 ppm. The second citrate peak near 2.75 ppm is seen to be a singlet, although it should be a doublet. The peaks near 3.55 ppm and 3.65 ppm corresponding to myoinositol also appear as singlets, when they should be triplets. Three broad peaks at $\approx$ 3.05 ppm, 3.22 ppm and 3.3 ppm are seen, but there should be ten peaks within that spectral region. These three broad peaks all have dispersive features with negative intensities. The lower panel of Fig. 11.1 at $N_P = 800$ shows that all 27 resonances were resolved with correct peak heights including all the multiplet resonances and the overlapping resonances of phosphocholine at 3.23 ppm and glycerophosphocholine at 3.24 ppm. The dispersive modes have disappeared at $N_P = 800$.

Herein, the case with $N_P = 54$ is shown since this is the least number of signal points which is theoretically needed to resolve 27 resonances (27 unknown frequencies and 27 unknown amplitudes considered in concert require 54 linear equations). However, it is seen from panel (i) of Fig. 11.1 that this algebraic condition of completeness ($N_P = 2K = 2 \times 27$) is totally insufficient, primarily due to the high density of states. Thus, not $2K$ but more than $10K$ signal points are needed for densely packed spectra such as these.

Figure 11.2 shows the total absorption shape spectra reconstructed by the $\mathrm{FPT}^{(-)}$ at the same two partial signal lengths $N_P = 54$ (top panel (i)) and $N_P = 800$ (bottom panel (ii)) for the normal glandular prostate data. As was the case for the component shape spectra at $N_P = 54$, the relative heights of the doublet citrate peak near 2.5 ppm are reversed, i.e., the peak at around 2.52 ppm is larger than the one at 2.54 ppm, and the second citrate peak near 2.75 ppm appears as a singlet, when it should be a doublet. The peaks near 3.55 ppm and 3.65 ppm corresponding to myoinositol, appear as singlets, although they should be triply serrated. It was just for those resonances in the

MULTIPLETS in MRS : PADE COMPONENT SHAPE SPECTRA with PARTIAL SIGNAL LENGTHS N_P = 54 (TOP), 800 (BOTTOM)
INPUT TIME SIGNAL CONTAINS 27 METABOLITES with SINGLETS, DOUBLETS and TRIPLETS

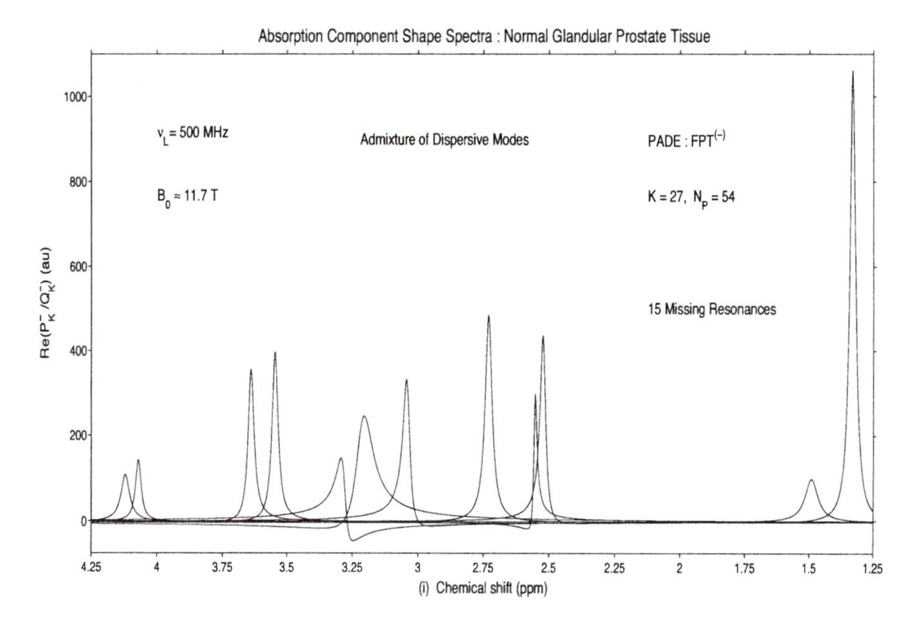

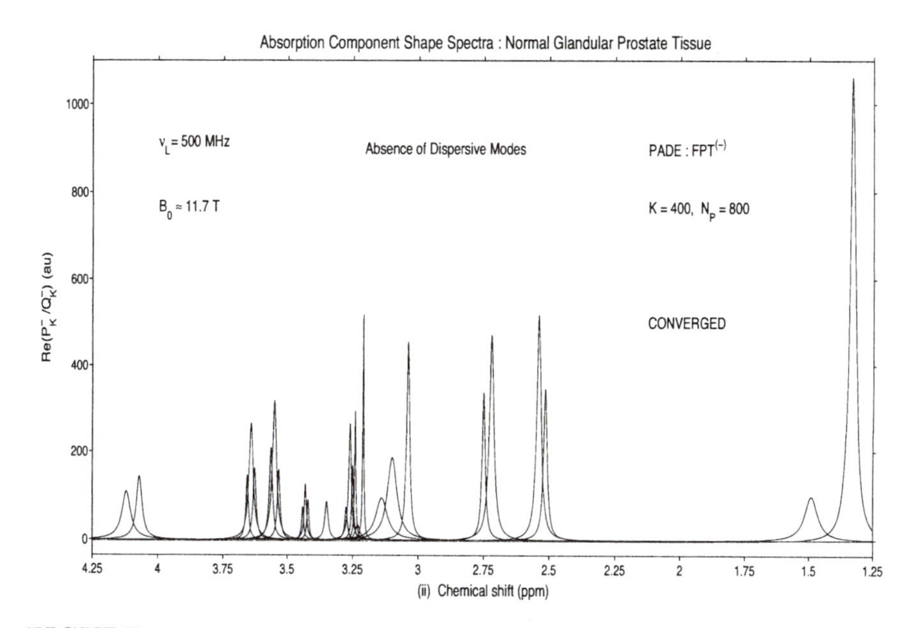

FIGURE 11.1
Convergence of the Padé absorption component spectra for normal glandular prostate using *in vitro* MRS data derived from Ref. [443].

MULTIPLETS in MRS : PADE TOTAL SHAPE SPECTRA with PARTIAL SIGNAL LENGTHS N_P = 54 (TOP), 800 (BOTTOM)
INPUT TIME SIGNAL CONTAINS 27 METABOLITES with SINGLETS, DOUBLETS and TRIPLETS

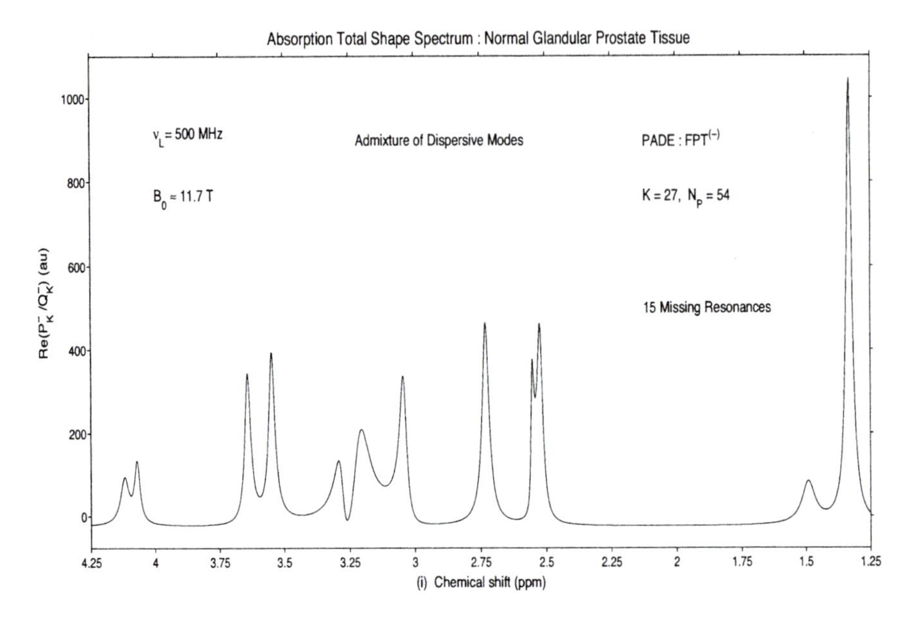

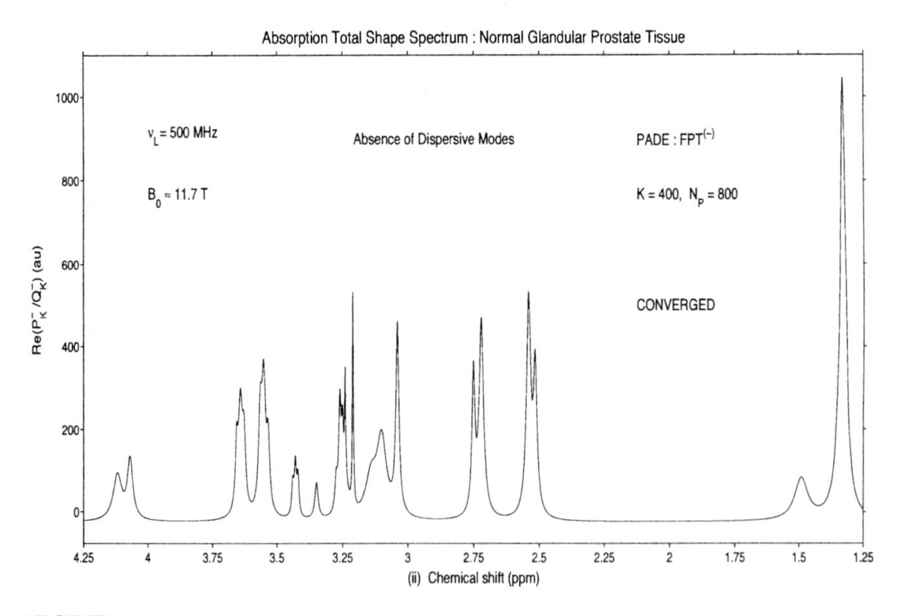

FIGURE 11.2

Padé absorption total shape spectra for normal glandular prostate using *in vitro* MRS data derived from Ref. [443]. Panel (i) is at $N_P = 54$ and the converged spectrum at $N_P = 800$ is shown on panel (ii).

MULTIPLETS in MRS : PADE SHAPE SPECTRA with PARTIAL SIGNAL LENGTH N_P = 800 (COMPONENTS: TOP, TOTAL: BOTTOM)
SPECTRAL CROWDING and OVERLAPS : FREQUENCY BANDS with FOCUS ON DOUBLETS and TRIPLETS

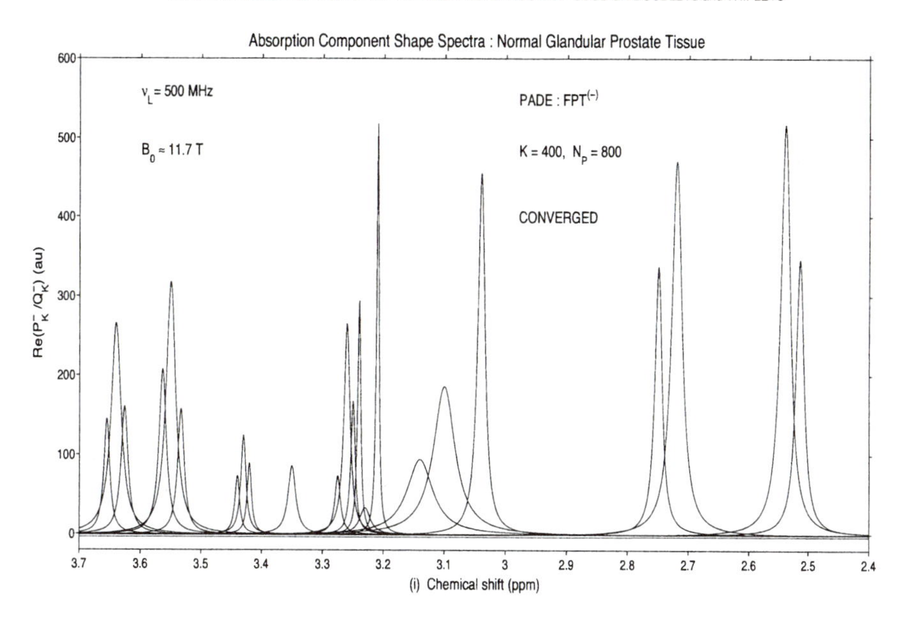

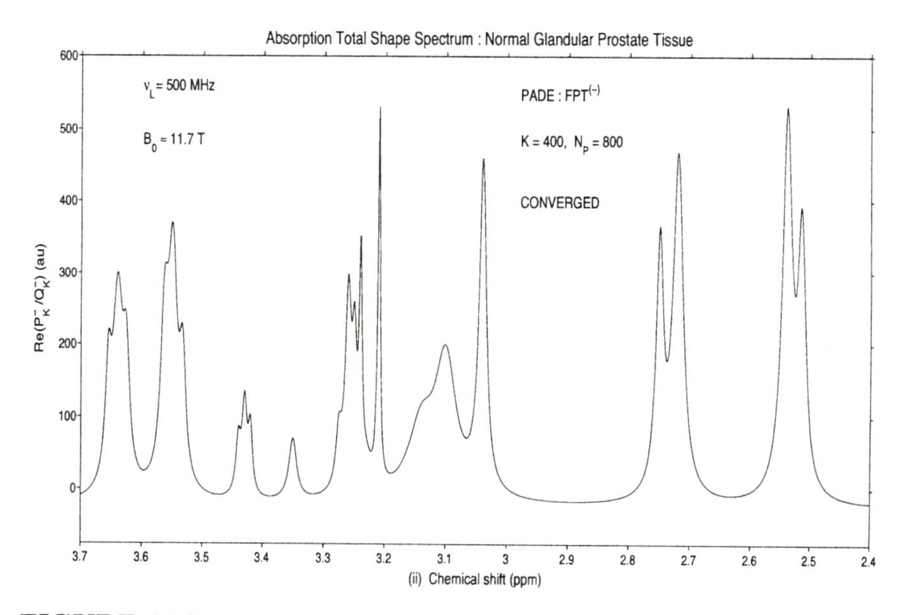

FIGURE 11.3

Converged Padé absorption component shape spectra (panel (i)) and converged absorption total shape spectra (panel (ii)) for normal glandular prostate using *in vitro* MRS data derived from Ref. [443].

two extremes of the total shape spectrum that approximately correct heights and widths (lactate and alanine at 1.33 ppm and 1.49 ppm respectively and myoinositol and lactate at $\approx$ 4.07 ppm and 4.12 ppm, respectively) were seen. The admixture of the absorption and dispersive mode is apparent in the region between $\approx$ 3.05 ppm and 3.3 ppm, reflected in the appearance of three peaks that are broadened with lowered heights and upward distortion. In the bottom panel of Fig. 11.2, the converged total absorption shape spectrum is shown at $N_P = 800$, and this was without any dispersive modes whatsoever.

The converged absorption component shape spectrum (top panel (i)) and total absorption shape spectrum (bottom panel (i)) at $N_P = 800$ are compared from 2.40 ppm to 3.70 ppm for normal glandular prostate (Fig. 11.3). Therein, phosphocholine at 3.23 ppm and glycerophosphocholine at 3.24 ppm are fully resolved in the component shape spectrum, while only a single peak appears on the total shape spectrum. From the total shape spectrum one can merely guess that the peaks are triplets assigned to myoinositol with three serrations centered respectively at $\approx$ 3.55 ppm and 3.66 ppm. In the Padé-reconstructed component shape spectrum, it is clearly seen there are three overlapping components at both of these chemical shift frequencies. Similarly, the two triply serrated peaks centered at $\approx$ 3.43 ppm and 3.26 ppm corresponding to taurine on the total shape spectrum, are shown in panel (i) to be comprised of three components each. For the peaks centered at $\approx$ 3.26 ppm, this would be practically impossible to ascertain from the total shape spectrum, especially towards the right-hand side which abuts against the GPC and PCho peaks. It would be just guesswork to even suggest from the total shape spectrum that the broad structure with a prominent peak centered at about 3.1 ppm has two components that correspond to polyamine.

11.3.2 Normal stromal prostate tissue: MR spectral data reconstructed by FPT

Table 11.5 shows the reconstructed data by the $\text{FPT}^{(-)}$ for normal stromal prostate at partial signal lengths $N_P = 500$ and $N_P = 600$. At the shorter signal length $N_P = 500$ (upper panel (i)) for the normal stromal tissue the pattern was quite similar to that described for the normal glandular prostate data. Two resonances were missing: peak $k = 11$ phosphocholine at 3.230242 ppm and the component of the taurine triplet at 3.250298 ppm ($k = 13$). At $N_P = 500$ all reconstructed spectral parameters and concentrations were fully exact for peak $k = 1$ (lactate) and peak $k = 5$ of the citrate doublet. The spectral parameters and computed concentrations were nearly correct for other resonances at the outermost regions of the spectrum ($k = 2, 3, 5-7$ and $k = 25-27$). For alanine ($k = 2$) other than the last two digits of the $|d_k^-|$ all the spectral parameters and computed concentration were exact at $N_P = 500$. For the other components of the doublets of doublets of citrate and for creatine ($k = 7$) the chemical shift, $\text{Im}(\nu_k^-)$ and concentrations were exact and the $|d_k^-|$ were correct to 4 or 5 of the six decimal places at $N_P = 500$. The

TABLE 11.5
Padé-reconstructed spectral parameters and metabolite concentrations for normal stromal prostate with the input data as derived from Ref. [443].

CONVERGENCE of SPECTRAL PARAMETERS and CONCENTRATIONS in FPT$^{(-)}$: PARTIAL SIGNAL LENGTHS N_P = 500, 600

(i) Reconstructed Data (Normal Stromal) : N_P = 500 ; Missing Resonances : # 11 (PCho) and # 13 (Tau)

n^o_k (Metabolite # k)	$Re(\nu_k^-)$ (ppm)	$Im(\nu_k^-)$ (ppm)	$\|d_k^-\|$ (au)	C_k^-(μM/g ww)	M_k (Assignment)
1	1.330148	0.013214	7.154620	39.1	Lac
2	1.490417	0.030360	1.244279	6.80	Ala
3	2.515197	0.010242	0.559928	3.06	Cit
4	2.540235	0.012151	0.977128	5.34	Cit
5	2.720146	0.012203	0.944190	5.16	Cit
6	2.750328	0.009014	0.464780	2.54	Cit
7	3.040309	0.006132	1.396156	7.63	Cr
8	3.100411	0.018433	0.345788	1.89	PA
9	3.140292	0.022726	0.229955	1.26	PA
10	3.210122	0.002639	0.498355	2.72	Cho
12	3.240382	0.010395	1.112238	6.08	GPC
14	3.261970	0.004502	0.765134	4.18	Tau
15	3.273710	0.005855	0.266296	1.46	Tau
16	3.350145	0.005025	0.364284	1.99	s–Ins
17	3.420144	0.003587	0.165519	0.91	Tau
18	3.430355	0.003221	0.304644	1.66	Tau
19	3.440427	0.003932	0.153322	0.84	Tau
20	3.533254	0.006419	0.420171	2.30	m–Ins
21	3.550303	0.009344	1.228695	6.71	m–Ins
22	3.563658	0.007292	0.624220	3.41	m–Ins
23	3.626374	0.005949	0.462618	2.53	m–Ins
24	3.640413	0.007434	1.081656	5.91	m–Ins
25	3.655388	0.004813	0.351285	1.92	m–Ins
26	4.070235	0.010123	0.753887	4.12	m–Ins
27	4.120142	0.019891	1.097886	6.00	Lac

(ii) Reconstructed Data (Normal Stromal) : N_P = 600 ; Converged (All Resonances Resolved)

n^o_k (Metabolite # k)	$Re(\nu_k^-)$ (ppm)	$Im(\nu_k^-)$ (ppm)	$\|d_k^-\|$ (au)	C_k^-(μM/g ww)	M_k (Assignment)
1	1.330148	0.013214	7.154620	39.1	Lac
2	1.490417	0.030360	1.244282	6.80	Ala
3	2.515197	0.010242	0.559927	3.06	Cit
4	2.540235	0.012151	0.977127	5.34	Cit
5	2.720146	0.012203	0.944190	5.16	Cit
6	2.750328	0.009014	0.464776	2.54	Cit
7	3.040309	0.006132	1.396157	7.63	Cr
8	3.100416	0.018432	0.345837	1.89	PA
9	3.140263	0.022761	0.230558	1.26	PA
10	3.210152	0.002613	0.495883	2.71	Cho
11	3.230242	0.010314	0.221409	1.21	PCho
12	3.240318	0.005121	0.572736	3.13	GPC
13	3.250298	0.004132	0.329369	1.80	Tau
14	3.260429	0.003821	0.708143	3.87	Tau
15	3.275351	0.005670	0.188472	1.03	Tau
16	3.350137	0.005022	0.364135	1.99	s–Ins
17	3.420319	0.003690	0.175663	0.96	Tau
18	3.430241	0.003012	0.287283	1.57	Tau
19	3.440448	0.004123	0.159195	0.87	Tau
20	3.533243	0.006432	0.420860	2.30	m–Ins
21	3.550326	0.009324	1.225984	6.70	m–Ins
22	3.563639	0.007290	0.625801	3.42	m–Ins
23	3.626378	0.005951	0.462946	2.53	m–Ins
24	3.640411	0.007432	1.081427	5.91	m–Ins
25	3.655389	0.004814	0.351327	1.92	m–Ins
26	4.070235	0.010123	0.753888	4.12	m–Ins
27	4.120142	0.019891	1.097896	6.00	Lac

computed concentration was completely correct for the polyamine resonance at 3.100416 ppm, although $\mathrm{Re}(\nu_k^-)$, $\mathrm{Im}(\nu_k^-)$ and $|d_k^-|$ were correct to 5, 5 and 4 decimal places, respectively at $N_P = 500$.

For the region between 3.14 ppm to 3.65 ppm, which is spectrally more dense, none of the reconstructed spectral parameters of the identified resonances were fully exact at $N_P = 500$, although the computed concentrations of scylloinositol ($k = 16$) and peaks $k = 23 - 25$ that were the components of one of the myoinositol triplets were correct. At $N_P = 500$ the reconstructed chemical shift, $\mathrm{Im}(\nu_k^-)$ and computed concentration of peak $k = 26$, the myoinositol singlet at 4.070235 ppm were completely correct, but not the last digit of the $|d_k^-|$. The $\mathrm{Re}(\nu_k^-)$ and $\mathrm{Im}(\nu_k^-)$ of the lactate peak $k = 27$ were reconstructed correctly at $N_P = 500$ and the computed concentration was also correct. However, the $|d_k^-|$ were exact only to 4 decimal places.

Most of the computed concentrations had the correct integer values, except for the two peaks immediately adjacent to the missed resonances, glycerophosphocholine ($k = 12$) and taurine ($k = 14$); these were both over-estimated. Rather than the correct concentration of 3.13 μM/g ww, the concentration of GPC was computed to be 6.08 μM/g ww, at $N_P = 500$ and the reconstructed $|d_k^-|$ was 1.112238 au, while the correct value was 0.572736 au. The concentration of the taurine peak $k = 14$ was computed as 4.18 μM/g ww at $N_P = 500$, whereas 3.87 μM/g ww is the correct value.

Full convergence was attained at $N_P = 600$ for all the reconstructed parameters and computed concentrations for all 27 resonances (bottom panel (ii) of Table 11.5). At higher partial signal length and the full signal length N all the reconstructed spectral parameters and computed concentrations were stable.

Figure 11.4 shows the Padé-reconstructed absorption component shape spectra at $N_P = 54$ and $N_P = 800$ for the normal stromal prostate data. A total of 16 the 27 peaks were missing at $N_P = 54$ (upper panel (i)). As was seen for normal glandular prostate, at $N_P = 54$ for the normal stromal tissue, only at the two extremes of the spectrum were the resonances resolved with nearly correct heights (lactate and alanine at 1.33 ppm and 1.49 ppm and myoinositol and lactate at 4.07 ppm and 4.12 ppm, respectively).

In the denser spectral region from $\approx$ 2.5 ppm to 3.65 ppm, only 7 of the 23 peaks were resolved and the absorption and dispersive modes were mixed. A singlet near 2.5 ppm was seen rather than the expected citrate doublet. The citrate doublet near 2.75 ppm appears mainly as a singlet but there is also a broad component with very small amplitude centered around 2.8 ppm. The peaks near 3.55 ppm and 3.65 ppm corresponding to myoinositol were seen as singlets, rather than triplets. There should have been ten peaks between 3.0 ppm and 3.35 ppm, but only two were seen. In the bottom panel (ii) of Fig. 11.4 at $N_P = 800$ all 27 resonances are resolved with the correct peak heights, including all the multiplets and the overlapping resonances of phosphocholine at $\approx$ 3.23 ppm and glycerophosphocholine at $\approx$ 3.24 ppm. The dispersive modes are absent at this converged partial signal length.

MULTIPLETS in MRS : PADE COMPONENT SHAPE SPECTRA with PARTIAL SIGNAL LENGTHS N_P = 54 (TOP), 800 (BOTTOM)
INPUT TIME SIGNAL CONTAINS 27 METABOLITES with SINGLETS, DOUBLETS and TRIPLETS

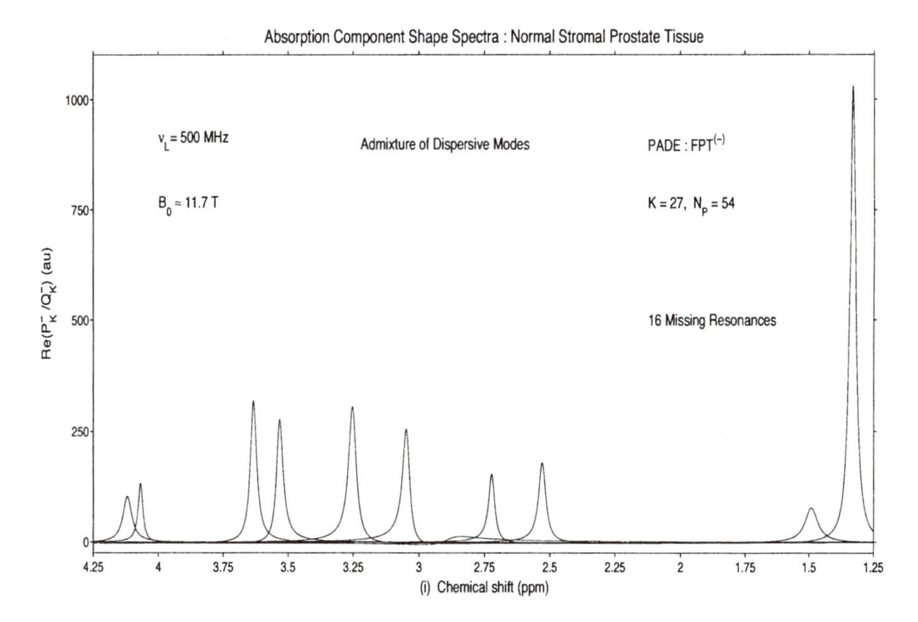

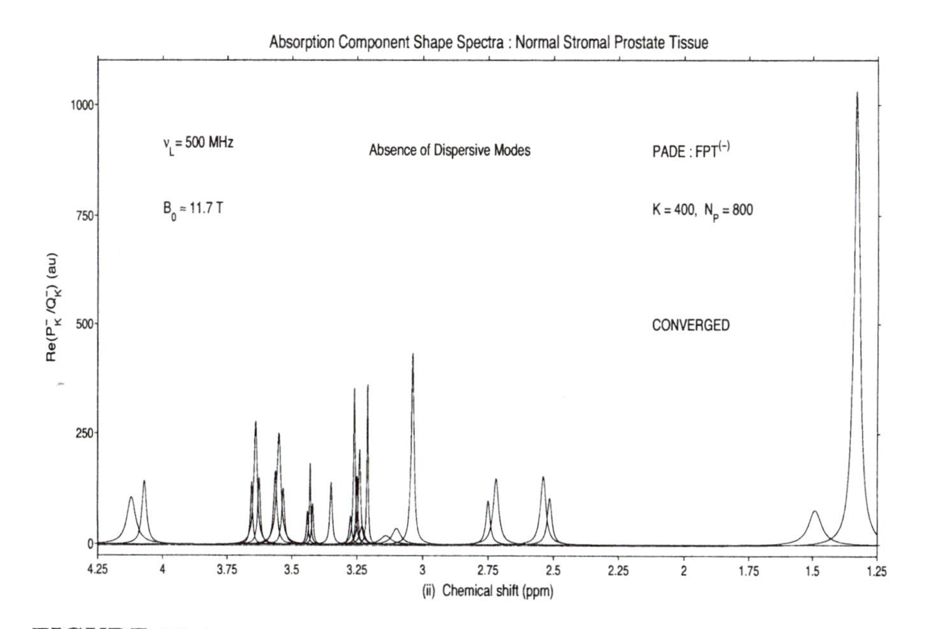

FIGURE 11.4
Convergence of the Padé absorption component spectra for normal stromal prostate using *in vitro* MRS data derived from Ref. [443].

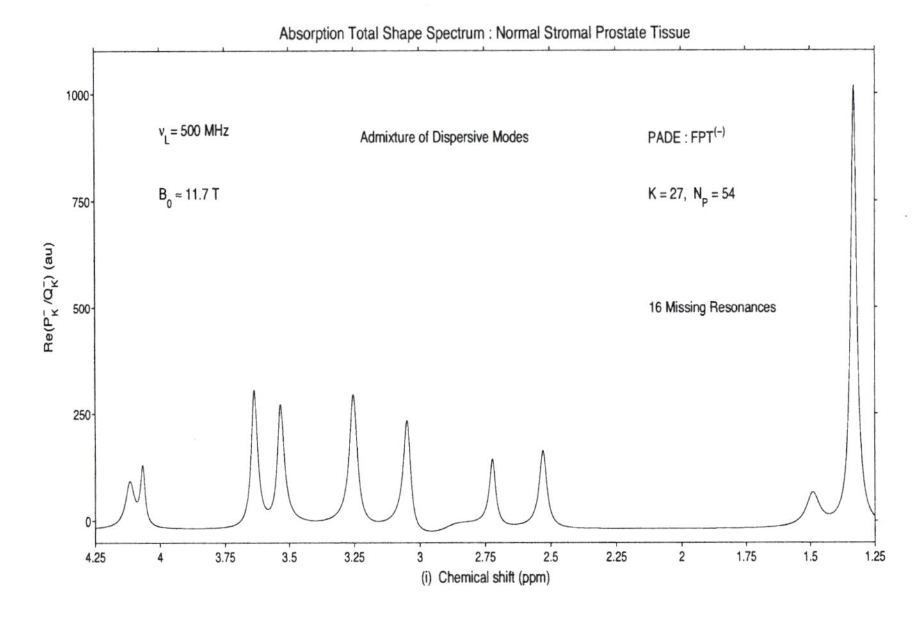

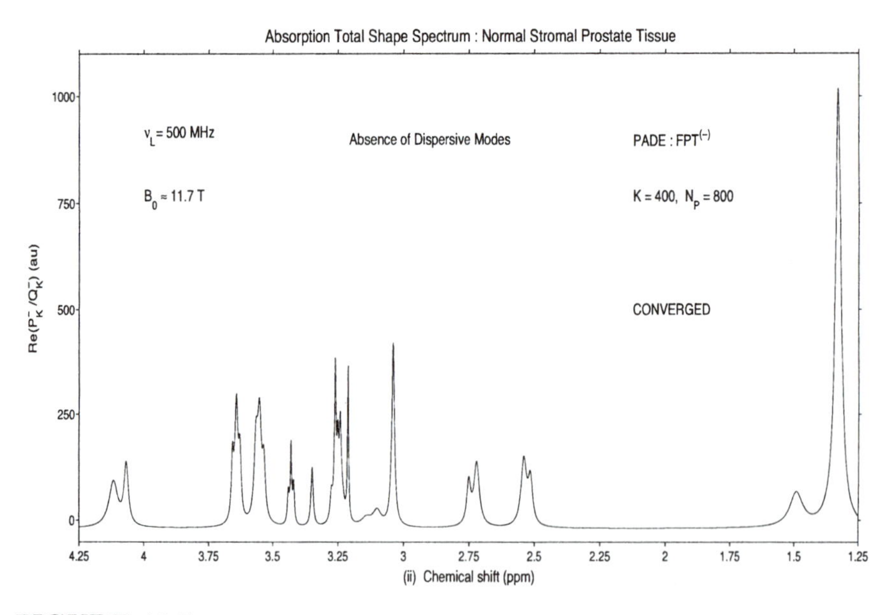

FIGURE 11.5

Padé absorption total shape spectra for normal stromal prostate using *in vitro* MRS data derived Ref. [443]. Panel (i) is at $N_P = 54$ and the converged spectrum at $N_P = 800$ is shown on panel (ii).

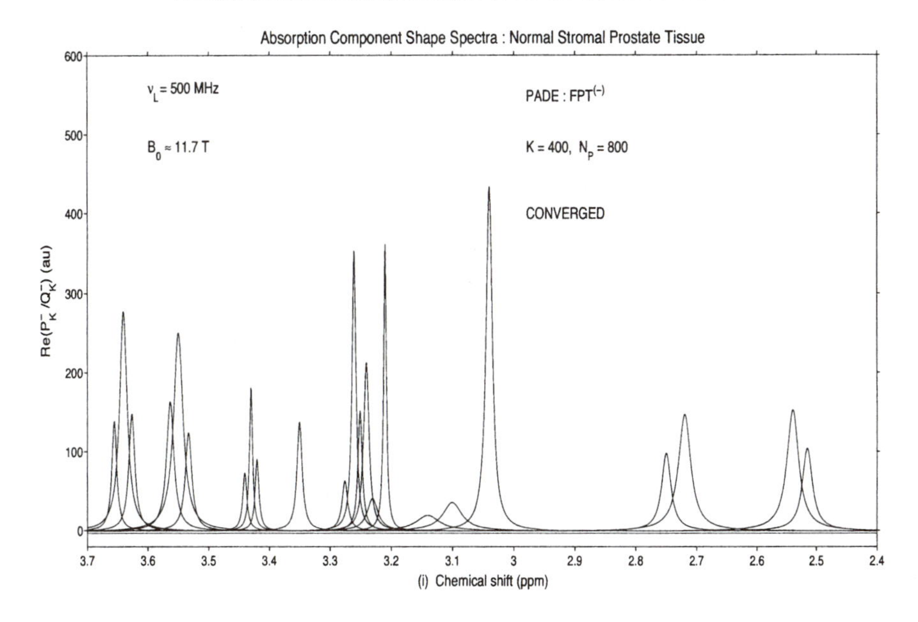

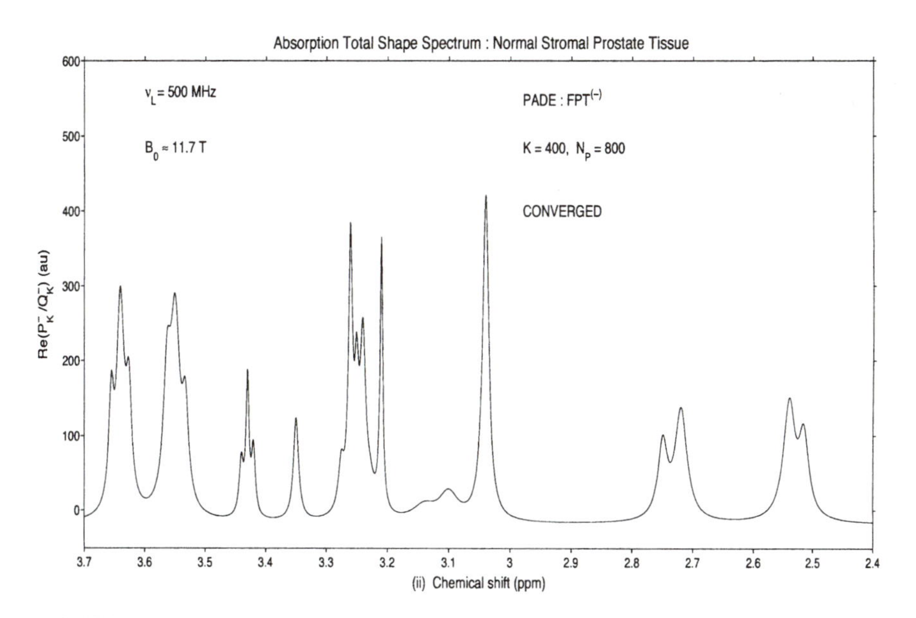

FIGURE 11.6
Converged Padé absorption component shape spectra (panel (i)) and converged absorption total shape spectra (panel (ii)) for normal stromal prostate using *in vitro* MRS data derived from Ref. [443].

The total absorption shape spectra reconstructed by the $\text{FPT}^{(-)}$ at the same two partial signal lengths: $N_P = 54$ (top panel (i)) and $N_P = 800$ (bottom panel (ii)) for the normal stromal prostate are presented in Fig. 11.5. Just as in the component spectrum, at $N_P = 54$ the peaks at the two extremes of the spectrum were resolved and the heights were close to being correct (lactate and alanine at 1.33 ppm and 1.49 ppm respectively and myoinositol and lactate at 4.07 ppm and 4.12 ppm, respectively). The peak near 2.5 ppm, however, appears as a singlet at $N_P = 54$, when it should have a double serration. A very low amplitude bump appears around 2.85 ppm, this is asymmetric with a higher right side. At around 3.05 ppm there is a creatine peak which is wider with a lower amplitude with respect to the creatine peak on the converged total shape spectrum. A single, wide peak is seen at about 3.25 ppm at $N_P = 54$, but there should be four peaks with a serrated structure to the left. On the total shape spectrum at $N_P = 54$ the peaks at 3.35 ppm (scylloinositol) and the triply serrated structure corresponding to taurine centered at 3.43 ppm are completely absent. There are two single peaks near 3.56 ppm and at 3.64 ppm corresponding to myoinositol, but they should be triply serrated. The polyamine structures expected at around 3.10 ppm and 3.14 ppm are absent. On the bottom panel of Fig. 11.5 the converged total absorption shape spectrum at $N_P = 800$ for normal stromal prostate shows that all the 27 resonances are resolved.

The converged absorption component shape spectrum (top panel (i)) and total absorption shape spectrum (bottom panel (i)) at $N_P = 800$ are compared within the expanded region from 2.40 ppm to 3.70 ppm for the normal stromal prostate (Fig. 11.6). As was true for normal glandular prostate, phosphocholine at 3.23 ppm and glycerophosphocholine at 3.24 ppm are fully resolved in the component shape spectrum. However, only one resonance is seen on the total shape spectrum. Only the converged component spectrum can be used to determine the actual number of resonances with confidence. The polyamine peaks are very small in the normal stromal prostate. From the total shape spectrum it would therefore be even more tenuous than for the glandular case to ascertain that there are precisely two polyamine components.

11.3.3 Malignant prostate tissue: MR spectral information reconstructed by FPT

Table 11.6 displays the Padé-reconstructed data for prostate cancer at the two partial signal lengths $N_P = 500$ and $N_P = 600$. At $N_P = 500$ (upper panel (i)), before convergence, 26 of the 27 resonances were identified. Peak $k = 13$, the component of the taurine triplet at 3.250298 ppm was missing. At $N_P = 500$ the computed concentrations for resonances $k = 1 - 8$ and $k = 23 - 27$ were fully correct and all the spectral parameters were exact for lactate at 1.330148 ppm ($k = 1$) and at 4.12042 ($k = 27$), and for the myoinositol singlet ($k = 26$) at 4.079235 ppm.

TABLE 11.6
Padé-reconstructed spectral parameters and metabolite concentrations for malignant prostate with the input data as derived from Ref. [443].

CONVERGENCE of SPECTRAL PARAMETERS and CONCENTRATIONS in $FPT^{(-)}$: PARTIAL SIGNAL LENGTHS N_P = 500, 600

(i) Reconstructed Data (Malignant) : N_P = 500 ; Missing Resonance : # 13 (Tau)

n_k^o (Metabolite # k)	$Re(\nu_k^-)$ (ppm)	$Im(\nu_k^-)$ (ppm)	$\lvert d_k^- \rvert$ (au)	C_k^-(μM/g ww)	M_k (Assignment)
1	1.330148	0.013213	11.07045	60.5	Lac
2	1.490417	0.030362	2.305586	12.6	Ala
3	2.515197	0.016240	0.680689	3.72	Cit
4	2.540235	0.020952	1.185733	6.48	Cit
5	2.720146	0.020204	1.152776	6.30	Cit
6	2.750328	0.018213	0.567253	3.10	Cit
7	3.040309	0.006951	1.780424	9.73	Cr
8	3.100418	0.021437	0.580318	3.17	PA
9	3.140233	0.032708	0.384660	2.10	PA
10	3.210153	0.002834	0.829223	4.53	Cho
11	3.229878	0.004204	1.252506	6.84	PCho
12	3.246399	0.012406	0.950890	5.20	GPC
14	3.261533	0.002729	0.625188	3.42	Tau
15	3.274760	0.006965	0.288772	1.58	Tau
16	3.350140	0.004321	0.458939	2.51	s–Ins
17	3.420142	0.003723	0.187071	1.02	Tau
18	3.430390	0.001404	0.309790	1.69	Tau
19	3.440276	0.003813	0.180274	0.98	Tau
20	3.533235	0.006428	0.547676	2.99	m–Ins
21	3.550337	0.010342	1.605352	8.77	m–Ins
22	3.563644	0.007281	0.815504	4.46	m–Ins
23	3.626378	0.005954	0.611484	3.34	m–Ins
24	3.640413	0.009432	1.424954	7.79	m–Ins
25	3.655388	0.004812	0.461176	2.52	m–Ins
26	4.070235	0.012421	0.993596	5.43	m–Ins
27	4.120142	0.012890	1.701738	9.30	Lac

(ii) Reconstructed Data (Malignant) : N_P = 600 ; Converged (All Resonances Resolved)

n_k^o (Metabolite # k)	$Re(\nu_k^-)$ (ppm)	$Im(\nu_k^-)$ (ppm)	$\lvert d_k^- \rvert$ (au)	C_k^-(μM/g ww)	M_k (Assignment)
1	1.330148	0.013213	11.07045	60.5	Lac
2	1.490417	0.030362	2.305581	12.6	Ala
3	2.515197	0.016240	0.680695	3.72	Cit
4	2.540235	0.020952	1.185727	6.48	Cit
5	2.720146	0.020204	1.152790	6.30	Cit
6	2.750328	0.018213	0.567246	3.10	Cit
7	3.040309	0.006951	1.780421	9.73	Cr
8	3.100416	0.021434	0.580055	3.17	PA
9	3.140263	0.032762	0.386093	2.11	PA
10	3.210152	0.002814	0.823422	4.50	Cho
11	3.230242	0.003952	1.235133	6.75	PCho
12	3.240318	0.005890	0.473925	2.59	GPC
13	3.250298	0.004123	0.364135	1.99	Tau
14	3.260429	0.003141	0.781336	4.27	Tau
15	3.275351	0.005672	0.208600	1.14	Tau
16	3.350137	0.004324	0.459286	2.51	s–Ins
17	3.420319	0.003626	0.190302	1.04	Tau
18	3.430241	0.001426	0.312900	1.71	Tau
19	3.440448	0.003821	0.173833	0.95	Tau
20	3.533243	0.006434	0.548948	3.00	m–Ins
21	3.550326	0.010323	1.601098	8.75	m–Ins
22	3.563639	0.007291	0.817932	4.47	m–Ins
23	3.626378	0.005952	0.611162	3.34	m–Ins
24	3.640411	0.009434	1.425435	7.79	m–Ins
25	3.655389	0.004812	0.461116	2.52	m–Ins
26	4.070235	0.012421	0.993596	5.43	m–Ins
27	4.120142	0.012890	1.701738	9.30	Lac

For the components of the myoinositol triplet ($k = 23 - 25$) the reconstructed chemical shifts, $\mathrm{Im}(\nu_k^-)$ and $|d_k^-|$ were either exact or were correct to 3-5 of the six decimal places (accounting for round-off). For alanine, the doublets of citrate doublets and creatine the chemical shifts and $\mathrm{Im}(\nu_k^-)$ were exactly reconstructed. Here, the $|d_k^-|$ for these resonances were correct to 4 to 5 of the six decimal places. For polyamine ($k = 9$) and choline ($k = 10$) the concentrations were correct to one decimal place.

The integer was correct for the computed concentration of phosphocholine ($k = 11$) at $N_P = 500$ and the spectral parameters were correct to $1-3$ decimal places. The concentration of glycerophosphocholine ($k = 12$) was twice higher than it should be. The computed concentration of peak $k = 14$ (taurine component) was under-estimated by about 20% lower than the correct value at $N_P = 500$. The concentration of the scylloinositol ($k = 16$) was correct. Peaks $k = 15$ and $k = 17-19$, i.e., the components of the two taurine triplets, had the correct integer values for the concentrations at $N_P = 500$. Taking into account round-off, the concentrations of the components of the myoinositol triplet ($k = 20 - 22$) were correct to at least one of two decimal places.

Full convergence was attained at $N_P = 600$ for all the reconstructed parameters for all twenty-seven resonances (lower panel (ii) of Table 11.6). As was true for the data from normal prostate, stability of convergence was maintained at higher partial signal length and at the full signal length N for the data corresponding to prostate cancer.

Figure 11.7 displays the Padé-reconstructed absorption component shape spectra at $N_P = 54$ and $N_P = 800$ for the malignant prostate data. At $N_P = 54$ (upper panel (i)) eleven of the twenty-seven peaks were resolved. Again, it was at the two extremes of the spectrum that all the resonances were resolved and with nearly correct heights for lactate and alanine at 1.33 ppm and 1.49 ppm, respectively, and for myoinositol and lactate at 4.07 ppm and 4.12 ppm, respectively. In the denser spectral region from ≈ 2.5 ppm to 3.65 ppm, seven of the 23 peaks were resolved.

Unlike what was seen for the normal prostate spectra at this signal length, there were no dispersive modes. Singlets near 2.5 ppm and 2.75 ppm were seen at $N_P = 54$, although doublets of citrate should appear at each of those regions.

The peaks near 3.55 ppm and 3.65 ppm corresponding to myoinositol are seen as singlets, but should be triplets. There should have been ten peaks between 3.0 ppm and 3.35 ppm, but only three peaks were seen. At $N_P = 800$ as shown on the bottom panel (ii) of Fig. 11.7, all twenty-seven resonances were resolved with the correct heights, widths, positions as well as the phases (zero).

Figure 11.8 shows the Padé-reconstructed total absorption shape spectra at the two partial signal lengths $N_P = 54$ (top panel (i)) and $N_P = 800$ (bottom panel (ii)) for the prostate cancer data. As was the case for the component absorption spectrum for malignant prostate data, lactate and alanine at 1.33 ppm and 1.49 ppm and myoinositol and lactate at 4.07 ppm and 4.12 ppm

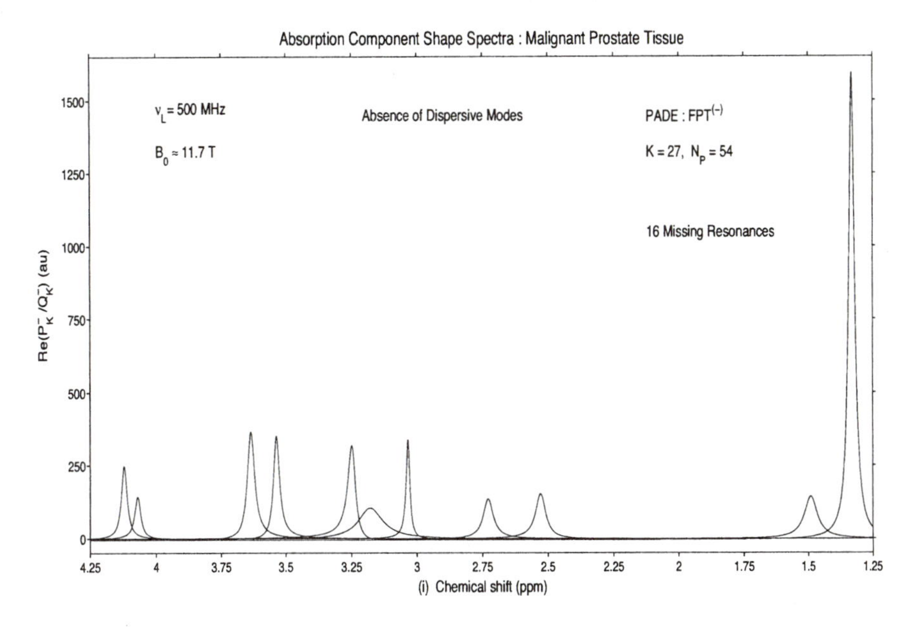

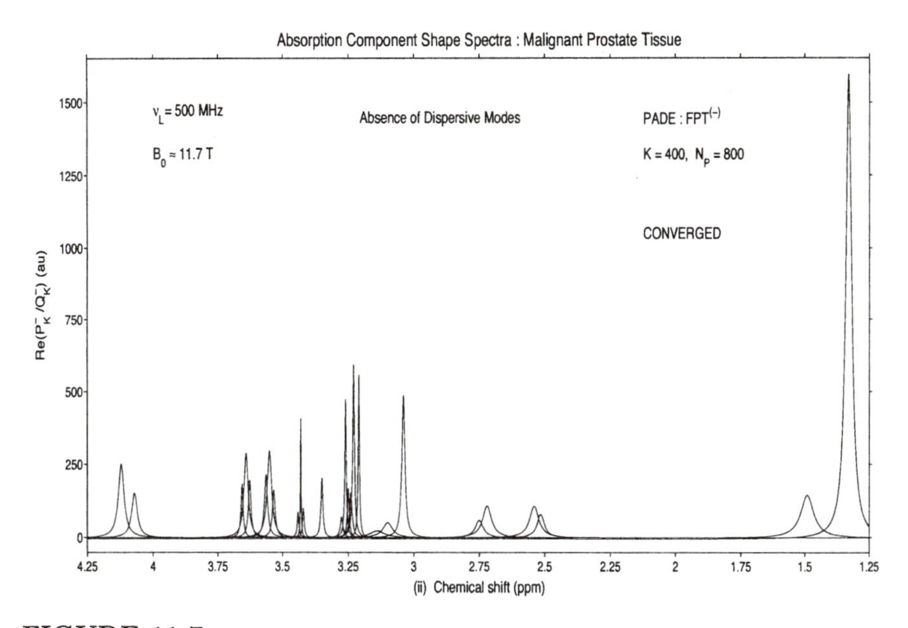

FIGURE 11.7
Convergence of the Padé absorption component spectra for malignant prostate using *in vitro* MRS data derived from Ref. [443].

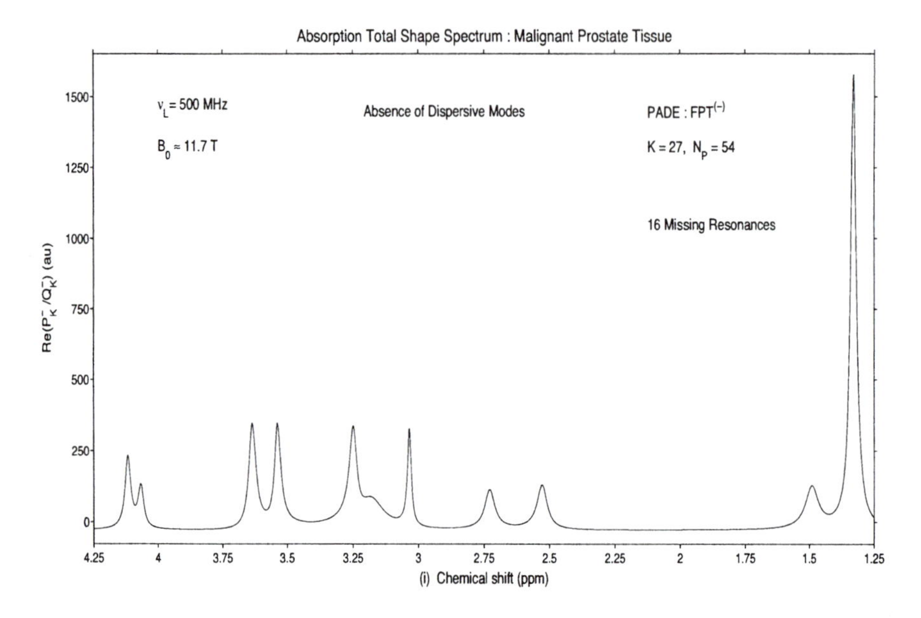

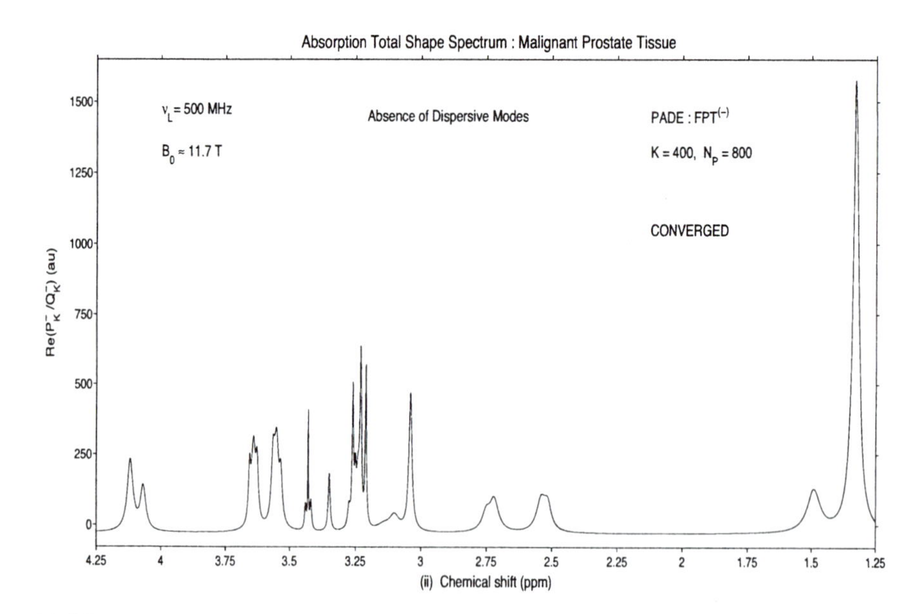

FIGURE 11.8
Padé absorption total shape spectra for malignant prostate using *in vitro* MRS data derived from Ref. [443]. Panel (i) is at $N_P = 54$ and the converged spectrum at $N_P = 800$ is shown on panel (ii).

MULTIPLETS in MRS : PADE SHAPE SPECTRA with PARTIAL SIGNAL LENGTH N_P = 800 (COMPONENTS: TOP, TOTAL: BOTTOM)
SPECTRAL CROWDING and OVERLAPS : FREQUENCY BANDS with FOCUS ON DOUBLETS and TRIPLETS

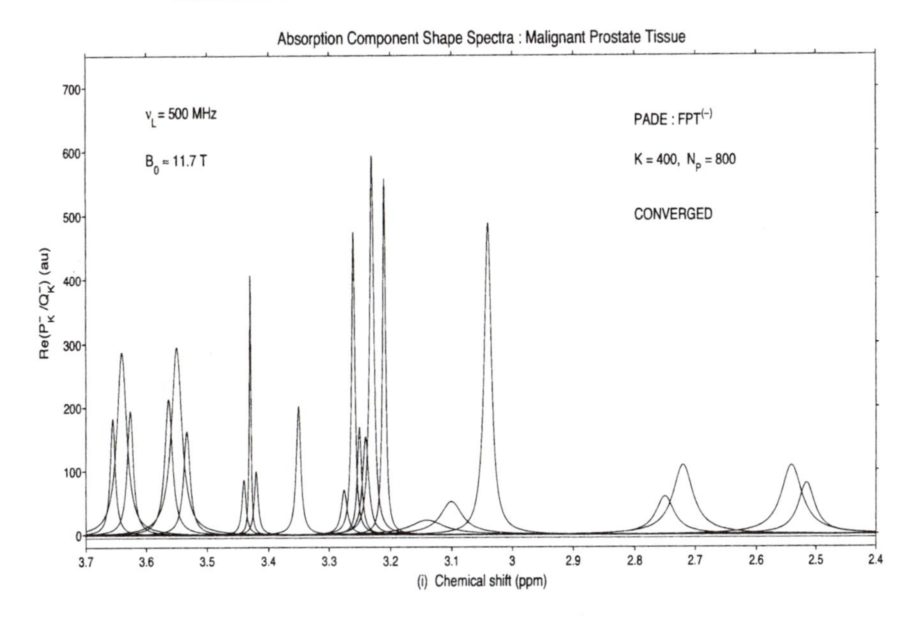

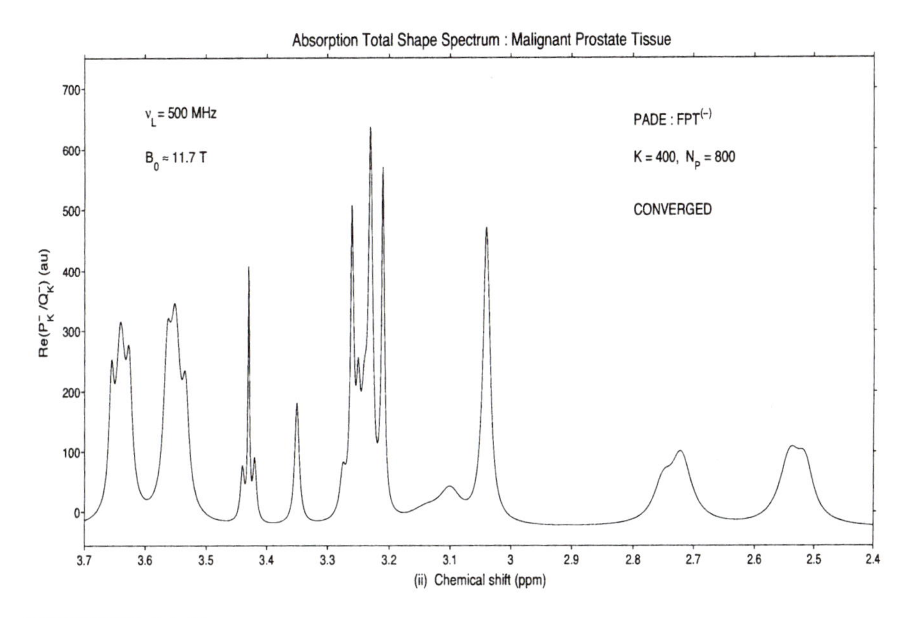

FIGURE 11.9
Converged Padé absorption component shape spectra (panel (i)) and converged absorption total shape spectra (panel (ii)) for malignant prostate using *in vitro* MRS data derived from Ref. [443].

were resolved and with heights that were nearly correct at $N_P = 54$.

The citrate peaks near 2.5 ppm and 2.75 ppm are single and narrower at $N_P = 54$, than on the converged total shape spectrum. The creatine peak centered at around 3.05 ppm is widened and has a lower amplitude compared to the peak corresponding to creatine on the converged total shape spectrum.

There is a single peak centered at 3.25 ppm at $N_P = 54$ instead of the two very narrow peaks and a triply serrated structure to the left. The peaks at 3.35 ppm corresponding to scylloinositol and the triply serrated structure corresponding to taurine centered at 3.43 ppm are not seen on the total shape spectrum at $N_P = 54$.

The two structures centered near 3.56 ppm and at 3.64 ppm corresponding to myoinositol are single peaks, although they should by triply serrated. The converged total absorption shape spectrum at $N_P = 800$ for the Padé-reconstructed data for malignant prostate is seen on the lower panel of Fig. 11.8.

Figure 11.9 shows the converged absorption component shape spectrum (top panel (i)) and total absorption shape spectrum (lower panel (ii)) for the Padé-reconstructed data for malignant prostate. The spectrum is displayed within the region between 2.40 ppm and 3.70 ppm at $N_P = 800$. Comparing the two panels, it can be seen that the triply and doubly serrated peaks on the total shape spectrum merely suggest the number of components, such that it is only from the converged component spectrum that the actual number of underlying resonances is revealed.

As was true for normal stromal prostate, the polyamine peaks are small in the case of prostate cancer. It would be much more difficult than for the glandular prostate case to ascertain from the total shape spectrum that there are precisely two PA components. On the total shape spectrum, the citrate doublets centered near 2.53 ppm and 2.73 ppm appear as broad peaks and there is only a hint of the doublet structures shown on the component spectrum. While phosphocholine at 3.23 ppm and glycerophosphocholine at 3.24 ppm are completely resolved in the component shape spectrum, that is not the case on the total shape spectrum.

11.3.4 Comparison of MRS retrievals from prostate tissue: Normal glandular, normal stromal and cancerous

Figure 11.10 presents the converged Padé-reconstructed absorption component shape spectra, for the normal glandular prostate (top panel (i)), normal stromal prostate (middle panel (ii)) and malignant prostate data (bottom panel (iii)). The converged total shape spectra for these three cases are shown in Fig. 11.11.

For the normal glandular prostate, the most prominent spectral structures besides the lactate resonance at 1.33 ppm, are the doublets of citrate doublets at 2.5 ppm and 2.75 ppm, creatine peak at 3.04 ppm and tall, narrow choline peak at 3.21 ppm.

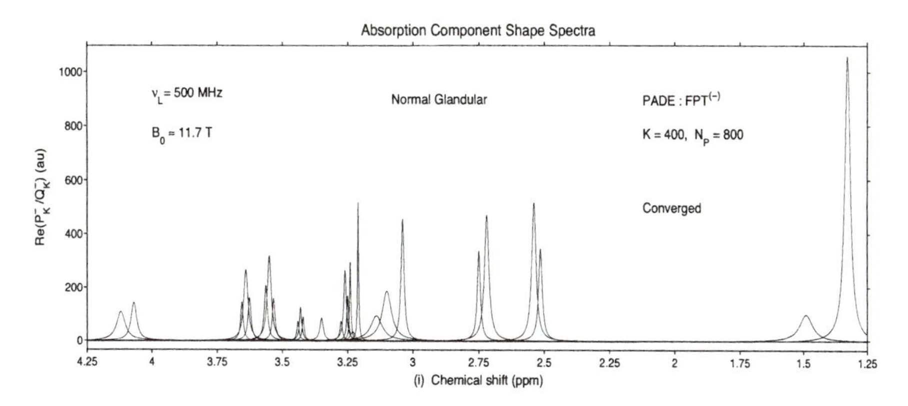

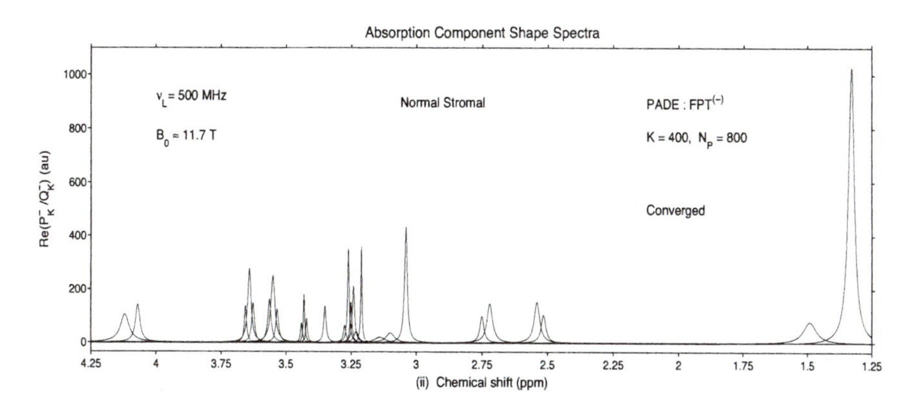

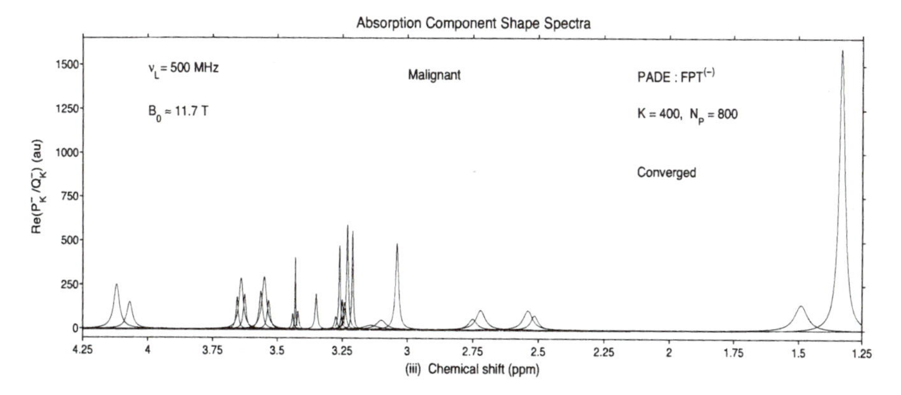

FIGURE 11.10

Converged Padé absorption component spectra at $N_P = 800$ for normal glandular prostate (panel (i)), normal stromal prostate (panel (ii)) and malignant prostate (panel (iii)) using *in vitro* MRS data derived from Ref. [443].

PADE TOTAL SHAPE SPECTRA with SINGLETS, DOUBLETS and TRIPLETS : PARTIAL SIGNAL LENGTH N_P = 800
THREE TYPES of PROSTATE TISSUE : NORMAL GLANDULAR (TOP), NORMAL STROMAL (MIDDLE) and MALIGNANT (BOTTOM)

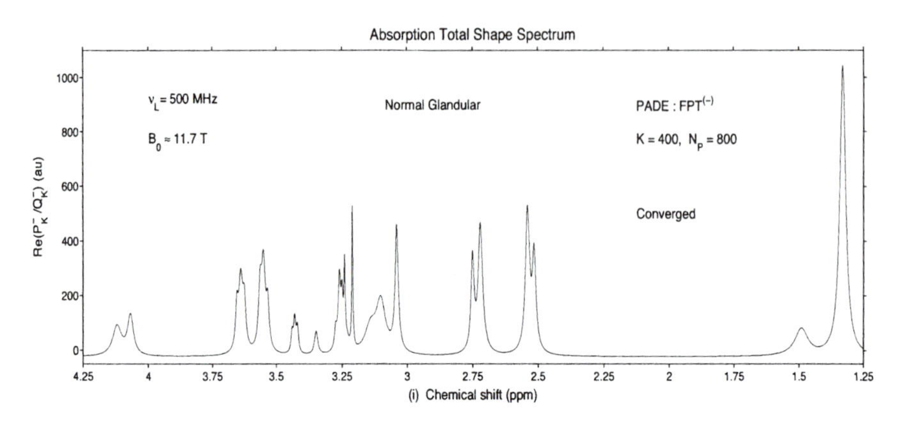

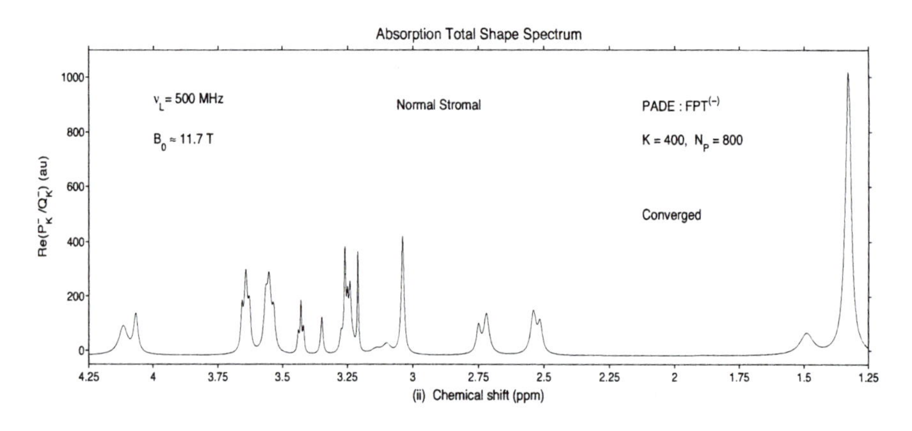

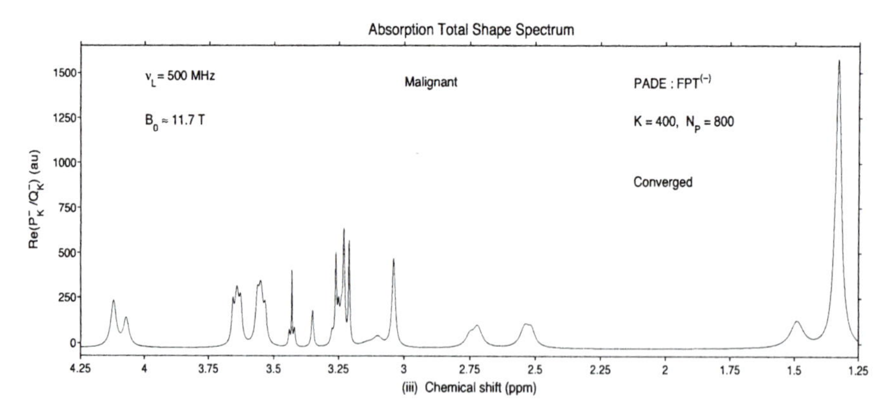

FIGURE 11.11
Converged Padé absorption total shape spectra at $N_P = 800$ for normal glandular prostate (panel (i)), normal stromal prostate (panel (ii)) and malignant prostate (panel (iii)) using *in vitro* MRS data derived from Ref. [443].

On the absorption component shape spectrum, the two polyamine peaks centered at 3.10 ppm and 3.14 ppm are broad, large and clearly distinguished. However, on the total shape spectrum for the glandular prostate, there is only a possible leftward shoulder at around 3.14 ppm which might suggest a second polyamine component.

The spectra from normal glandular and normal stromal prostate differ strikingly. The intensity of most of the spectral structures is lower for stromal tissue. The doublets of citrate doublets are clearly smaller, and the polyamines are difficult to detect for stromal prostate tissue. The creatine peak at 3.04 ppm is the most prominent after lactate at 1.33 ppm. The myoinositol triplet centered around 3.55 ppm is also smaller in the stromal prostate compared to the normal glandular tissue.

For the case of prostate cancer, the lactate peaks at 1.33 ppm and 4.12 ppm are larger than for the two normal prostate tissues. The choline components at 3.21 ppm to 3.24 ppm (i.e., total choline) are altogether more abundant than creatine at 3.04 ppm. The spectrum for the malignant prostate is most sharply distinguished from that of normal glandular prostate, particularly in that for prostate cancer the citrate doublet peaks and the two polyamine resonances are much smaller than the components of choline.

The converged Padé-reconstructed absorption component and total shape spectra for the normal glandular prostate (top panels (i)), normal stromal prostate (middle panel (ii)) and malignant prostate data (bottom panel (iii)) within the zoomed region from 2.40 ppm to 3.70 ppm using the partial length $N_P = 800$ are shown in Figs. 11.12 and 11.13, respectively.

In these figures, the component spectra clearly delineate the phosphocholine and glycerophosphocholine peaks that are not seen on the total shape spectra. The present analysis shows that the FPT unequivocally resolves multiplet resonances, as well as providing exact quantification. This is true for regions of otherwise very high spectral density.

Using the FFT and fitting via the Levenberg-Marquardt algorithm, the authors of Ref. [443] noted that spectral overlap compromised the accuracy of quantification. Procedures that require integrating the areas under the peaks in the Fourier absorption spectra are vulnerable to subjectivity due to the uncertainty about lower and upper integration limits. Such techniques for reconstructing metabolite concentrations are especially difficult with peak overlap [346].

Padé-based reconstruction yields not only the possibility to exactly extract all the spectral frequencies and amplitudes of all the resonances, but also provides certainty about their true number. This ensures unique and maximally reliable quantification of all the physical metabolite concentrations, including when there is an ample number of multiplet resonances, as seen in the present analyses.

As we have discussed in the preceding chapters, the "spectral crowding" problem does not obstruct the FPT, which via parametric analysis, without any fitting or numerical integration of peak areas, reconstructed all the mul-

PADE COMPONENT SHAPE SPECTRA with ZOOMING into MULTIPLETS : PARTIAL SIGNAL LENGTH N_P = 800
THREE TYPES of PROSTATE TISSUE : NORMAL GLANDULAR (TOP), NORMAL STROMAL (MIDDLE) and MALIGNANT (BOTTOM)

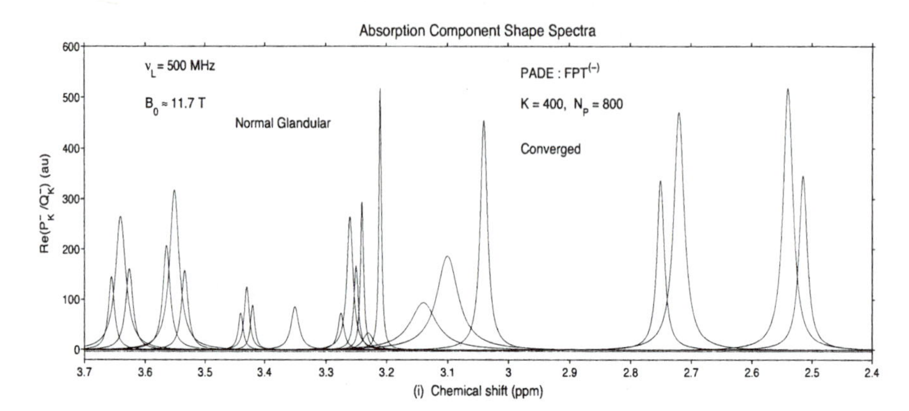

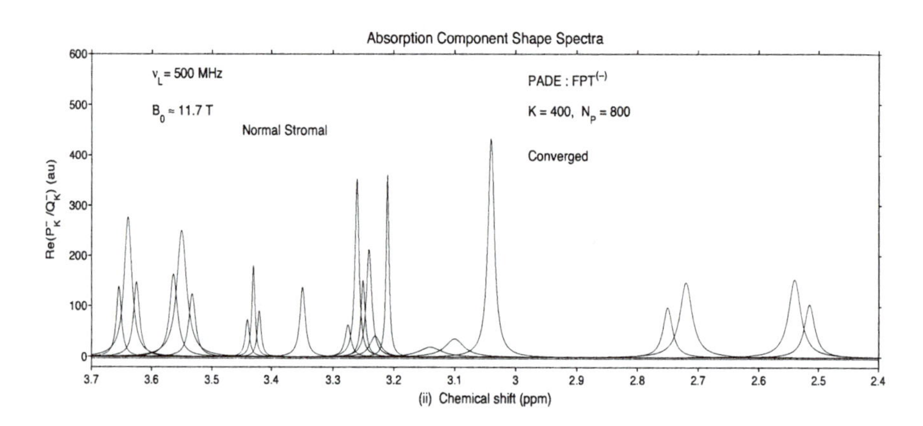

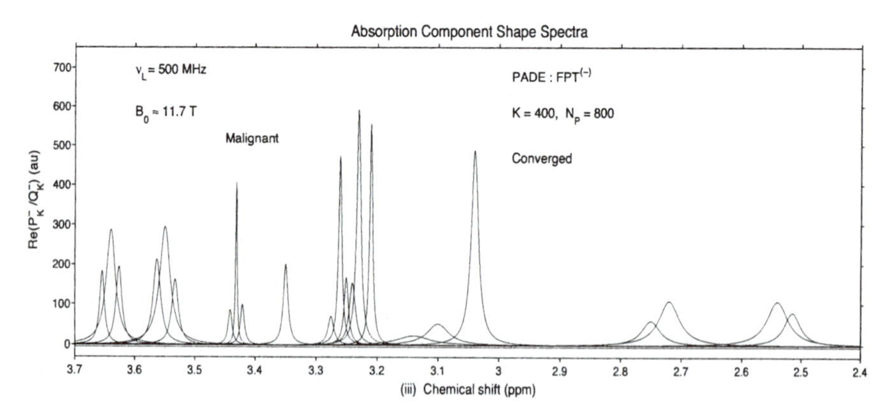

FIGURE 11.12

Converged Padé absorption component spectra at $N_P = 800$ for normal glandular prostate (panel (i)), normal stromal prostate (panel (ii)) and malignant prostate (panel (iii)) within the zoomed region from 2.4 ppm to 3.7 ppm using *in vitro* MRS data derived from Ref. [443].

PADE TOTAL SHAPE SPECTRA with ZOOMING into MULTIPLETS : PARTIAL SIGNAL LENGTH N_P = 800
THREE TYPES of PROSTATE TISSUE : NORMAL GLANDULAR (TOP), NORMAL STROMAL (MIDDLE) and MALIGNANT (BOTTOM)

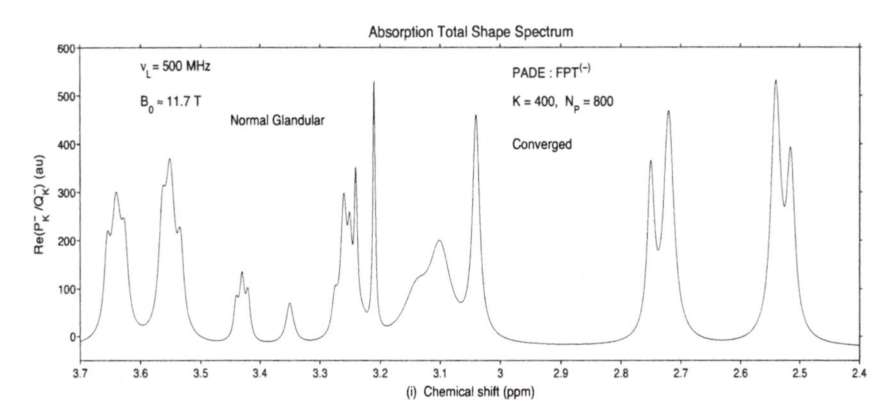

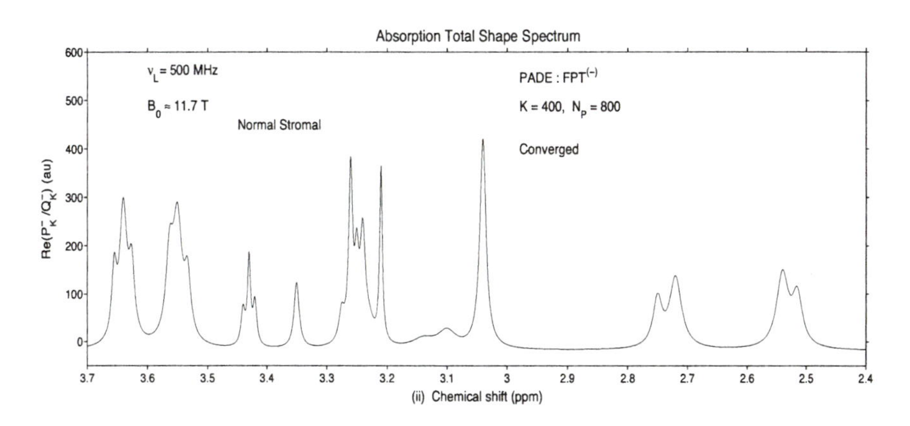

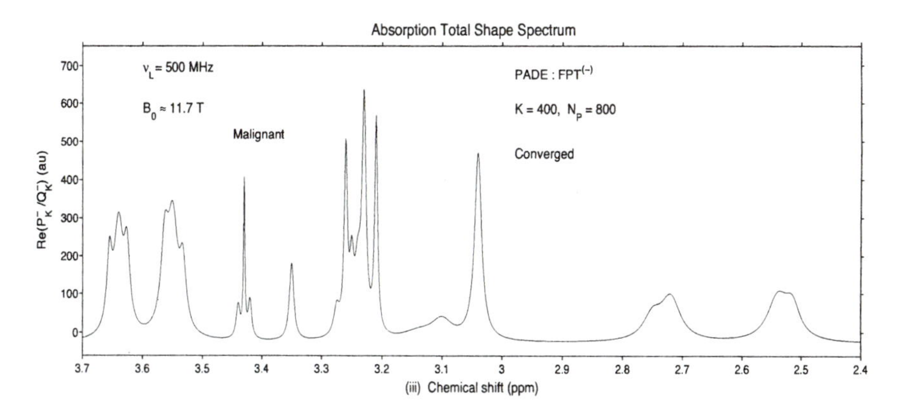

FIGURE 11.13

Converged Padé absorption total shape spectra at $N_P = 800$ for normal glandular prostate (panel (i)), normal stromal prostate (panel (ii)) and malignant prostate (panel (iii)) within the zoomed region from 2.4 ppm to 3.7 ppm using *in vitro* MRS data derived from Ref. [443].

tiplets and closely overlying resonances of different metabolites for all three problems under study here, namely for normal glandular and normal stromal prostate, as well as for the spectra from prostate cancer.

For the normal glandular prostate data convergence was attained at $N_P = 700$ and for the normal stromal and the malignant data at $N_P = 600$ for the given total signal length $N = 1024$. At those partial signal lengths, the FPT exactly reconstructed all the spectral parameters for the time signals corresponding to the normal glandular and stromal prostate, and to the prostate cancer.

Since $2K = N_P$, at these convergence lengths N_P there were 350, 300 and 300 resonances, respectively for the normal glandular prostate, normal stromal tissue and for the malignant case. Of these, only twenty-seven were genuine, the other 323 and 273, respectively were spurious, and were identified as such by zero amplitudes and the pole-zero coincidences. Thus, there were over eleven times and over ten times more spurious resonances than those that were genuine, respectively.

The number of spurious resonances (and their percentage in relation to those that are genuine) for the present prostate cancer MRS problem was much larger than for the ovarian cancer problem described to which we applied the FPT [27, 29], but fewer than for the breast cancer problem [32], as presented in the preceding chapters.

This is a reflection of some of the more subtle aspects of spectral processing, notably the smallest distance among the poles and zeros, the density of poles and zeros in the complex plane, as well as inter-separations among poles and zeros [34].

For the problems addressed in this chapter, it would be theoretically possible to retrieve all twenty-seven input resonances at $N_P = 54$ because there are 54 unknowns (27 complex frequencies and 27 complex amplitudes) with 54 linear equations and 54 signal points needed.

At $N_P = 54$ twenty-seven resonances were indeed retrieved, but fifteen of these resonances were spurious for the normal glandular case and sixteen were spurious for the normal stromal and prostate cancer cases. The pole-zero coincidences of the FPT (yielding zero-valued amplitudes) allowed us to identify these spurious resonances, and to thereby determine that only the remaining twelve or eleven were genuine.

Convergence seems to commence at the outermost spectral regions, with lactate and alanine at about 1.33 ppm and 1.49 ppm, respectively and with myoinositol and lactate at about 4.07 ppm and 4.12 ppm, respectively. The inner, denser spectral region converged later.

The last resonances to converge were those closest to the real axis with the smallest imaginary part of ν_k^-. It should be pointed out that at a very few signal points prior to convergence, there could still be an admixture of the absorption and dispersive modes, since the latter appear intermittently.

At convergence and beyond, the dispersive mode completely disappeared, so that the absorption spectrum was purely in the absorption mode, i.e., the real part of the complex-valued spectrum. The stability of convergence was confirmed at longer partial signal lengths and at the total signal length for the three problems under study. This has been true as well for all our other applications of the FPT to MRS time signals as described in chapters 3, 6, 7, 9 and 10.

Prior to but fairly near convergence, at $N_P = 600$ for the normal glandular data and at $N_P = 500$ for the normal stromal data, the phosphocholine resonance had not yet been detected and the computed concentration of glycerophosphocholine was over-estimated. For the malignant prostate data, even though phosphocholine was resolved prior to convergence, the concentration of GPC was still over-estimated by about a factor of two.

In this light, mention should also be made here concerning the ratio of GPC to PCho, in relation to the so-called "glycerophosphocholine to phosphocholine switch" which has been suggested to be a marker of malignant transformation for other tissues, notably breast [307], as discussed in the previous chapter. These inaccuracies that appeared at intermediate stages of calculation completely disappeared once convergence was attained. It therefore can be considered of potential clinical importance for prostate cancer diagnostics via MRSI to determine the exact number of true resonances and the exact point of convergence.

In the present chapter, we have focused upon prostate cancer, which is the most common cancer among men in the U.S. and in much of Europe, and a major cause of cancer deaths worldwide. We have discussed the dilemmas concerning screening, early detection and aspects of clinical decision making regarding prostate cancer.

We have seen that relying upon conventional Fourier-based data analysis, *in vivo* MRSI has made important strides in improving the accuracy with which prostatic tumor and extracapsular extension are identified, as well as helping to distinguish cancerous prostate from benign prostatic hypertrophy [413].

In vivo MRSI has also contributed to various aspects of clinical management of this malignancy. Nevertheless, major challenges and difficulties remain, and as of today, none of the existing non-invasive methods for prostate cancer diagnostics have sufficient sensitivity and specificity to be broadly recommended for screening and other aspects of surveillance.

It would obviously be premature to render definitive conclusions about the role the Padé-optimized MRSI in solving these dilemmas. From the input data of Ref. [443] for a fairly small number of prostate cancer samples, it cannot be stated with certainty which metabolites are best for identifying prostate cancer and distinguishing this from normal stromal and glandular prostate. Moreover, unfortunately, only the means and standard deviations of the computed concentrations were given in Ref. [443] for the three types of tissue.

Data from the individual patients, or the minimum and maximum of the computed concentrations would have been valuable. Notwithstanding these limitations, it is still clear that there is a far richer source of spectral information with which to identify prostate cancer than is currently used with Fourier-based *in vivo* MRS.

Padé-optimized MRSI with its capability to unequivocally resolve and quantify multiplet resonances and otherwise exceedingly challenging spectra with many overlapping resonances could undoubtedly provide valuable information for improving prostate cancer diagnostics.

11.4 Prospects for Padé-optimized MRSI within prostate cancer diagnostics

As in the other two problem-areas within cancer diagnostics, we have used noise-free FIDs, in order to set up the fully-controlled standard for the FPT. The methodological rationale for this strategy has been elaborated [10]. We are now applying this analysis to both noise-corrupted synthesized data for the prostate and to encoded FIDs similar to those from Ref. [443] as well as *in vivo* MRS and MRSI data from the prostate. These results will be reported soon.

It should be emphasized that since the time signals from MRSI are precisely of the same nature as those from MRS, the FPT can also be applied with equal success to MRSI. We have performed initial applications of the FPT to MRSI and demonstrated improvements in the resolution of MRSI and mitigation of Gibbs phenomena [290].

We conclude that the FPT is optimally suited to resolve and quantify the numerous overlapping resonances, including multiplets of metabolites in this very difficult area of signal processing in MRS within the realm of prostate cancer diagnostics.

Only short time signals were needed for this achievement, and, as discussed, this is a major advantage because free induction decay data become heavily corrupted with noise at the long total acquisition times as required by the FFT, which lacks interpolation and extrapolation features.

Herein, we have once again demonstrated that the FPT can unequivocally disentangle physical from spurious content of the studied time signals.

The many multiplets characteristic of the spectra of healthy and cancerous prostate represent a major challenge for signal processing. Padé optimization has been shown here to meet this challenge.

This line of investigation will continue with encoded data from normal, hypertrophic and cancerous tissue, *in vitro* and *in vivo.* We expect that Padé-optimized MRSI will improve the diagnostic accuracy of MR-based modalities. This could certainly contribute to a more timely and accurate diagnosis of prostate. Solutions to therapeutic dilemmas might also be forthcoming.

12

Recapitulation of Padé-optimized processing of biomedical time signals

In this chapter, we shall discuss the main aspects and place of Padé approximants in a general framework of rational functions in the mathematical literature through a large branch known as theory of approximations. As such, we shall recapitulate how one is unavoidably led to Padé approximants as the optimal rational function for many fields, and especially for solving the spectral analysis or quantification problem in various fields such as ICRMS, NMR, MRS, etc. The salient features of the Padé methodology will be illuminated, such as convergence acceleration of slowly converging series and induced convergence of divergent series as well as sequences. Also summarized will be the astounding precision of Padé approximants in reconstructing the machine accurate, true values of all the input spectral parameters via complex frequencies and amplitudes. The novel concept of exact signal-noise separation will be highlighted, as well, with reference to Froissart doublets. Finally, a special focus will be placed upon the relevance of the Padé-guided MRS for tumor diagnostics in clinical oncology. Not all these multifaceted issues will be covered in this recapitulation with equal weight. Some pertinent aspects are left for the next chapter with our concluding remarks and prospects for future developments with emphasis on applications to diagnostics in medicine, where Padé approximants find its new home.

12.1 The central role of rational functions in the theory of approximations

General rational functions $R(z^{-1})$ are defined as quotients of two other functions $f(z^{-1})$ and $g(z^{-1})$ via[1]

$$R(z^{-1}) = \frac{f(z^{-1})}{g(z^{-1})}. \tag{12.1}$$

[1]Here, for convenience, the independent variable z^{-1} is chosen as the inverse of a general complex variable z.

They represent the leading class of functions from mathematics that find rich areas of applications in many research fields, ranging from physics to life sciences. This is primarily due to the main features of rational approximations, as they apply to analysis and interpretation of data that can come from either experimental measurements or from theory by means of numerical computations. Such features are interpolation and extrapolation. By interpolation, one attempts to reliably generate the values of the observables in certain ranges or points where measured data are unavailable. By extrapolation, one tries to faithfully predict the values that could have been measured had the experiment continued beyond the last recorded data point of the studied physical quantity. Both cases are of great practical importance, since a reliable method would save extra measurements or possibly very time consuming numerical computations. Yet, the goal of an optimal theory is not to try in vain to achieve physically adequate interpolation and extrapolation by simple-minded fitting with its non-uniqueness, subjectivity and bias. Rather, the aim is to ingrain these two features into adequate mathematical models without adjustable parameters for physics theories. The most suitable framework for solving this challenging simultaneous interpolation-extrapolation problem is provided by rational functions of the general type (12.1).

12.2 The dominant role of Padé approximant among all rational functions

The simplest, and crucially, the most powerful rational functions, are Padé approximants $R^{(\mathrm{PA})}(z^{-1})$, introduced by ratios of two polynomials $P_L^-(z^{-1})$ and $Q_K^-(z^{-1})$ of degrees L and K

$$R^{(\mathrm{PA})}(z^{-1}) = \frac{P_L^-(z^{-1})}{Q_K^-(z^{-1})} \tag{12.2}$$

$$P_L^-(z^{-1}) = \sum_{r=0}^{L} p_r^- z^{-r} \qquad Q_K^-(z^{-1}) = \sum_{s=0}^{K} q_r^- z^{-s} \tag{12.3}$$

where $\{p_r^-, q_s^-\}$ are the expansion coefficients of $P_L^-(z^{-1})$ and $Q_K^-(z^{-1})$. The most stable are the diagonal and para-diagonal PA as obtained for $L = K$ and $L - 1 = K$, respectively. The polynomial ratio from (12.2) becomes unique if it is taken to approximate a given Maclaurin series

$$F(z^{-1}) = \sum_{n=0}^{\infty} c_n z^{-n} \tag{12.4}$$

where the elements c_n of infinite set $\{c_n\}$ are the known expansion coefficients. In applications to signal processing in many different fields, the expansion coefficients $\{c_n\}$ from the input Maclaurin series (12.4) are time signal points or FIDs, or equivalently, auto-correlation functions. These are given by linear combinations of decaying trigonometric functions that are complex-valued damped exponentials called fundamental harmonics (transients)

$$c_n = \sum_{k=0}^{K} d_k z_k^n \qquad z_k = \mathrm{e}^{i\omega_k \tau} \qquad \mathrm{Im}(\omega_k) > 0. \tag{12.5}$$

Here, τ is the sampling or dwell time, whereas $\{\omega_k, d_k\}$ are the nodal angular frequencies and the associated amplitudes, respectively. By inserting (12.5) into $F(z^{-1})$ from (12.4), the infinite sum over n can be carried out using the exact result for the geometric series $\sum_{n=0}^{\infty}(z_k/z)^{-n} = 1/(1 - z_k/z) = z/(z - z_k)$. The obtained fraction $z/(z - z_k)$ is the simplest 1st order diagonal ($L = K = 1$) rational polynomial in the variable $z^{+1} \equiv z$. To cohere with (12.5), general variable z can also be written in the harmonic form $z = \exp(i\omega\tau)$, where ω is a running complex angular frequency. Obviously, a sum of K elementary fractions $z/(z - z_k)$, as implied by $F(z^{-1}) = \sum_{n=0}^{\infty} c_n z^{-n} = \sum_{k=1}^{K} d_k \sum_{n=0}^{\infty}(z_k/z)^n = \sum_{k=1}^{K} z d_k/(z - z_k)$ is the Kth order diagonal ($L = K$) rational polynomial $P_K^+(z)/Q_K^+(z)$

$$F(z^{-1}) = \sum_{k=1}^{K} \frac{z d_k}{z - z_k} \equiv \frac{P_K^+(z)}{Q_K^+(z)} \tag{12.6}$$

$$P_L^+(z) = \sum_{r=1}^{L} p_r^+ z^r \qquad Q_K^+(z) = \sum_{s=0}^{K} q_s^+ z^s. \tag{12.7}$$

Therefore, for the expansion coefficients $\{c_n\}$ in the form of geometric progression (12.5), the exact result for the infinite sum in (12.4) is given precisely by the rhs of (12.6), which can alternatively be rewritten as

$$F(z^{-1}) = R^{(\mathrm{P}z\mathrm{T})}(z) \tag{12.8}$$

$$R^{(\mathrm{P}z\mathrm{T})}(z) \equiv \frac{P_K^+(z)}{Q_K^+(z)}. \tag{12.9}$$

Here, the acronym PzT stands for the so-called Padé z-transform. Distinguishing PA from PzT is essential due to the subtle, but critical differences (i) and (ii) between these two methods:

• (i) The standard Padé approximant is invariably introduced in the literature on this method as the rational polynomial $P_L^-(z^{-1})/Q_K^-(z^{-1})$ from (12.2) in the same variable z^{-1} as the original function $F(z^{-1})$ from (12.4). On the other hand, we can alternatively interpret (12.4) as the usual z-transform in variable z^{-1}. As such, subsequently using geometric progression (12.5) for the c_n's, the resulting rational function $R^{(\mathrm{P}z\mathrm{T})}(z) = P_K^+(z)/Q_K^+(z)$ from (12.9) becomes the exact Padé polynomial quotient, but in the new variable z relative to the initial z–transform $F(z^{-1})$. Thus, given $F(z^{-1})$, the first key difference between the PA and PzT is that the former and the latter are defined in variables z^{-1} and z, respectively. Of course, since the PzT is also a rational polynomial, the PzT and PA both belong to the same family of Padé approximants, albeit with two different tasks. To specify these tasks, given (12.4), we can consider two regions $|z| > 1$ and $|z| < 1$ in the complex z−plane. For $|z| > 1$ and $|z| < 1$, the series $F(z^{-1})$ from (12.4) will converge (say, slowly) and diverge, respectively. Therefore, the rational polynomial $P_K^-(z^{-1})/Q_K^-(z^{-1})$ from the usual PA in the same variable z^{-1} with respect to $F(z^{-1})$ accelerates the already existing convergence of (12.4) for $|z| > 1$. For the opposite case $|z| < 1$, the input series (12.4) diverges. However, for the same case $|z| < 1$, the rational polynomial $P_K^+(z)/Q_K^+(z)$ from the standard PzT converges, as it is defined in terms of the variable z as opposed to z^{-1} from $F(z^{-1})$. In this way, by means of the Cauchy analytical continuation, the PzT effectively induces convergence into the originally divergent series $F(z^{-1})$ for $|z| < 1$. This is how the same Padé methodology can achieve two opposite mappings via transforming divergent series into convergent ones, and converting slowly into faster converging series (hence acceleration).

• (ii) The numerator polynomial $P_K^-(z^{-1})$ in, e.g., the diagonal PA generally possesses the free, constant expansion coefficient ($p_0^- \neq 0$), such that the sum over r in (12.3) can start from $r = 0$ yielding

$$P_K^-(z^{-1}) = p_0^- + p_1^- z^{-1} + p_2^- z^{-2} + \cdots + p_K^- z^{-K}. \tag{12.10}$$

However, by definition, the corresponding expansion coefficient p_0^+ of the numerator polynomial $P_K^+(z)$ in the diagonal PzT is zero. Hence, this time, the sum over r in (12.7) for $P_K^+(z)$ begins with $r = 1$ with no free, z−independent term, thus producing

$$P_K^+(z) = p_1^+ z + p_2^+ z^2 + \cdots + p_K^+ z^K. \tag{12.11}$$

The mentioned uniqueness of the Padé approximant for the given input Maclaurin series (12.4) presents a critical feature of this method. In other words, the ambiguities encountered in other mathematical modelings are eliminated from the outset already at the level of the definition of the PA. Moreover, this definition contains its "figure of merit" by revealing how well the PA can really describe the function $F(z^{-1})$ to be approximated. More precisely, given the infinite sum $F(z^{-1})$ via (12.4), the key question to ask is whether could it be possible to determine the value of a positive integer M for which the PA would be able to exactly reproduce the Maclaurin polynomial $F_M(z^{-1})$ term by term? Here, $F_M(z^{-1})$ is a partial, finite sum from (12.4)

$$F(z^{-1}) = F_M(z^{-1}) + \sum_{n=M+1}^{\infty} c_n z^{-n} \qquad F_M(z^{-1}) = \sum_{n=0}^{M} c_n z^{-n}. \tag{12.12}$$

The answer to the posed question is in the affirmative and can be found by expanding $R^{(\mathrm{PA})}(z^{-1})$ from (12.2) as an infinite sum in powers of z^{-1} around the point z_∞ located at infinity, $z = z_\infty \equiv \infty$. Then, the ensuing first $L+K$ expansion coefficients of the generated series for $R^{(\mathrm{PA})}(z^{-1})$ would exactly coincide with $F_M(z^{-1})$ from (12.4)

$$\frac{P_L^-(z^{-1})}{Q_K^-(z^{-1})} = \sum_{n=0}^{\infty} b_n^- z^{-n} \tag{12.13}$$

$$\therefore \quad \sum_{n=0}^{L+K} b_n^- z^{-n} = \sum_{n=0}^{L+K} c_n z^{-n} \tag{12.14}$$

$$\therefore \quad b_n^- = c_n \qquad 0 \le n \le L+K. \tag{12.15}$$

Hence $M = L+K$ (QED). Relationship (12.15) follows from the uniqueness theorem of power series expansions. Thus, the PA from (12.2) and the original infinite sum $F(z^{-1})$ from (12.4) are in the best contact since they exhibit exact agreement to within the first $L+K$ terms of the Maclaurin expansion (12.13) of $P_L^-(z^{-1})/Q_K^-(z^{-1})$. Moreover, the result $M = L+K$ simultaneously provides the Padé estimate of the difference between (12.4) and (12.2), as expressed symbolically by

$$F(z^{-1}) - \frac{P_L^-(z^{-1})}{Q_K^-(z^{-1})} = \mathcal{O}^-(z^{-L-K-1}) \qquad z \longrightarrow \infty \tag{12.16}$$

where $\mathcal{O}^-(z^{-L-K-1})$ is the remainder of power series expansions around $z = z_\infty = \infty$. The function $\mathcal{O}^-(z^{-L-K-1})$, as an explicit error of the approximation $F(z^{-1}) \approx P_L^-(z^{-1})/Q_K^-(z^{-1})$, itself represents a power series with expansion coefficients $\{a_n^-\}$ multiplied by z^{-L-K-m} $(m = 1, 2, 3, ..., \infty)$

$$\mathcal{O}^-(z^{-L-K-1}) = \sum_{n=L+K+1}^{\infty} a_n^- z^{-n} \qquad a_n^- = c_n - b_n^- \tag{12.17}$$

where, in general, $a_n^- \neq c_n$ for $L+K+1 \le n \le \infty$. In other words, the mentioned "figure of merit" is explicitly given by the easily obtainable error term $\mathcal{O}^-(z^{-L-K-1})$, which is an infinite sum with higher-order expansion terms than those retained in the Maclaurin series (12.13) for the polynomial quotient $P_L^-(z^{-1})/Q_K^-(z^{-1})$ from the PA.

The definition (12.16) of the PA is reminiscent of a variance-type estimate for the difference between the input data (observed, measured), $F(z^{-1})$, and output (modeled) function, $P_L^-(z^{-1})/Q_K^-(z^{-1})$

$$\begin{aligned} \text{Padé Variance} &= \mathcal{O}^-(z^{-L-K-1}) \\ &= F(z^{-1})\,\{\text{input data (observed, measured)}\} \\ &- \frac{P_L^-(z^{-1})}{Q_K^-(z^{-1})}\,\{\text{output data (modeled, objective function)}\}. \end{aligned} \tag{12.18}$$

However, and this is what sets the PA apart from other methods, this "built in" Padé variance is not an adjustable function, which can be used as a fitting recipe for producing anything which one subjectively decides to be "sufficiently good". Rather, the variance in the PA is free from any adjustable parameter and, as such, it is fully and objectively controlled solely by the structure of the input data $F(z^{-1})$.

The possibility of explicitly computing the difference term $\mathcal{O}^-(z^{-L-K-1})$ in the Padé estimate $F(z^{-1}) \approx P_L^-(z^{-1})/Q_K^-(z^{-1})$ is the basis of the error analysis of proven validity in the PA. If desired, such an evaluated variance-type remainder $\mathcal{O}^-(z^{-L-K-1})$ may be fed back into the PA which, in turn, can undergo iterative and systematic improvements with the possibility of reproducing the original, input function $F(z^{-1})$ with any prescribed accuracy. However, this is not even necessary because with, e.g., the diagonal PA, the numerically exact relationship $\sum_{n=0}^{\infty} c_n z^{-n} = P_K^-(z^{-1})/Q_K^-(z^{-1})$ to literally hundreds of decimal places, if an application of this unprecedented accuracy would ever be needed, can be achieved by systematically increasing the common degree K in polynomials $P_K^-(z^{-1})$ and $Q_K^-(z^{-1})$.

Padé approximants can be computed through many different numerical algorithms, including the most stable numerical computations via continued fractions. Moreover, unlike any other related method, for the known $F(z^{-1})$, both Padé polynomials $P_L^-(z^{-1})$ and $Q_K^-(z^{-1})$ in the PA can be extracted by purely analytical means in their simple and concise closed forms [5, 17]. This represents the gold standard against which all the corresponding numerical algorithms should be benchmarked for their stability and robustness.

Outside mathematics, *per se*, theoretical physicists are most appreciative of the power and usefulness of the PA, which they began to use more than half a century ago in many problems ranging from the Brillouin-Wigner perturbation series to divergent expansions in quantum chromodynamics in the theory of strong interactions of elementary particles. The reason for such a widespread usage of this method in theoretical physics is that, in fact, the most interesting and also the most important series expansions emanating from realistic problems are divergent. Other frequently encountered series, although convergent in principle, often converge so slowly that they become virtually impractical in any exhaustive application. Here, the PA comes to rescue the situation in both cases by converting divergent into convergent series and accelerating slowly converging series.

The reason that the same method is able to tackle these diametrically opposing difficulties is in the non-linearity of the PA, as is obvious from the definition (12.16). The condition for linearity of a given function $F(z^{-1})$ is, for example, $F(az_1^{-1}+bz_2^{-1}) = aF(z_1^{-1})+bF(z_2^{-1})$ for any two constants $\{a,b\}$ and any two values $\{z_1^{-1},z_2^{-1}\}$ of independent variable z^{-1} from the domain of definition of $F(z^{-1})$. In general, rational functions including the PA from (12.2) do not satisfy this latter linearity condition. Therefore, general rational functions $R(z^{-1})$ from (12.1) as well as the Padé approximant, $R^{(\mathrm{PA})}(z^{-1})$, from (12.2) are non-linear. It is due to its non-linearity that the PA gains its versatile and powerful capability of performing analytical continuations and convergence accelerations. These two features of the PA are not limited to series via power expansions as in (12.4), but are equally applicable to any given sequence of numbers for which, e.g., the limiting values are sought when the number of included terms becomes infinitely large.

The discussed two main features of the Padé functions via its two wings, the PA (convergence rate enhancement of slowly convergent series or sequences), and the PzT (forced convergence of originally divergent series) are jointly embodied into the fast Padé transform, FPT. In the fast Padé transform, the PzT and PA are relabeled as $\text{FPT}^{(+)}$ and $\text{FPT}^{(-)}$, respectively, where the superscripts $\pm$ refer to the employed independent variables, $z^{+1} \equiv z$ and $z^{-1} = 1/z$, respectively. By definition, the $\text{FPT}^{(+)}$ accomplishes analytical continuation through the forced convergence of divergent series. Likewise, the $\text{FPT}^{(-)}$ achieves acceleration of slowly converging series or sequences. Given a Maclaurin series (12.4), the $\text{FPT}^{(+)}$ and $\text{FPT}^{(-)}$ are aimed at approximating the same function $F(z^{-1})$

$$F(z^{-1}) \approx R^{(\text{FPT})\pm}(z^{\pm 1}). \tag{12.19}$$

Functions $R^{(\text{FPT})\pm}(z^{\pm 1})$ are explicitly defined as rational polynomials

$$R^{(\text{FPT})\pm}(z^{\pm 1}) \equiv \frac{P_L^{\pm}(z^{\pm 1})}{Q_K^{\pm}(z^{\pm 1})} \tag{12.20}$$

$$P_L^{\pm}(z^{\pm 1}) = \sum_{r=1,0}^{L} p_r^{\pm} z^{\pm r} \qquad Q_K^{\pm}(z^{\pm 1}) = \sum_{s=0}^{K} q_r^{\pm} z^{\pm s} \tag{12.21}$$

where $r = 0$ and $r = 1$ correspond to $P_L^{-}(z^{-1})$ and $P_L^{+}(z)$, respectively. As in (12.16), the qualities of the $\text{FPT}^{(\pm)}$, i.e., the adequacy of the two approximations in (12.19), are governed by the explicit definitions

$$\sum_{n=0}^{\infty} c_n z^{-n} - \frac{P_L^{\pm}(z^{\pm 1})}{Q_K^{\pm}(z^{\pm 1})} = \mathcal{O}(z^{\pm(L+K+1)}). \tag{12.22}$$

The remainders $\mathcal{O}^{\pm}(z^{\pm(L+K+1)})$ follow by developing $R^{(\text{FPT})\pm}(z^{\pm 1})$ as

$$\frac{P_L^{\pm}(z^{\pm 1})}{Q_K^{\pm}(z^{\pm 1})} = \sum_{n=0}^{\infty} b_n^{\pm} z^{\pm n} \tag{12.23}$$

where the expansion coefficients $\{b_n^{\pm}\}$ can be computed from the previously extracted polynomial coefficients $\{p_r^{\pm}, q_s^{\pm}\}$. Like the earlier reasoning with (12.14) and (12.15), the explicit calculations and comparisons between (12.4) and (12.23) reveal the significance of the error terms $\mathcal{O}(z^{\pm(L+K+1)})$. This implies that both rational functions $P_L^{\pm}(z^{\pm 1})/Q_K(z^{\pm 1})$ would be able to exactly reproduce the first $L + K$ terms from the infinite set $\{c_n\}$ of the input Maclaurin series (12.19). According to (12.23), the rational polynomials

$P_L^+(z)/Q_K^+(z)$ and $P_L^-(z^{-1})/Q_K^-(z^{-1})$ from the $\text{FPT}^{(+)}$ and $\text{FPT}^{(-)}$ yield the series expansions in powers of z and $1/z$. Therefore, remainders $\mathcal{O}(z^{L+K+1})$ and $\mathcal{O}(z^{-L-K-1})$ from (12.23) are themselves power series expansions around the points $z = 0$ and $z = \infty$. When z is a harmonic variable as in MRS, where $z = \exp(i\omega\tau)$ and $\text{Re}(\omega) > 0$ then, by definition, the $\text{FPT}^{(+)}$ and $\text{FPT}^{(-)}$ converge inside and outside the unit circle $|z| < 1$ and $|z| > 1$, respectively. However, being in the family of Padé approximants, the $\text{FPT}^{(+)}$ and $\text{FPT}^{(-)}$ (just like the PzT and PA) converge, as well, in the complementary regions $|z| > 1$ and $|z| < 1$, respectively, by virtue of their analytical continuations. In other words, the $\text{FPT}^{(\pm)}$ are defined throughout the complex $z-$plane with the exception of K poles $z = z_k^\pm$ $(1 \leq k \leq K)$ of $P_K^\pm(z^{\pm 1})/Q_K^\pm(z^{\pm 1})$, where $z_k^\pm$ are zeros of the denominator polynomials, $Q_K^\pm(z_k^\pm) = 0$.

The recapitulations from sections 12.1 and 12.2 emphasize the universal importance of rational functions and their most powerful proponent – Padé approximants. This led to a straightforward identification of the origin of the fast Padé transforms, via the $\text{FPT}^{(-)}$ and $\text{FPT}^{(+)}$, as the standard PA (acceleration of slowly converging series) and the PzT (transformation of diverging into converging series), respectively. However, in addition to convergence acceleration and induced convergence, the FPT can be applied to signal processing, where the main task is to carry out spectral analysis, i.e., to solve the quantification problem. In this latter research area, a sampled time signal $\{c_n\}$ is available either from computations or measurements. Particularly in MRS, magnetic resonance physics dictates that each c_n is a sum of K complex damped exponentials as in (12.5). Here, we are given a set of N sampled time signal points $\{c_n\}$ $(0 \leq n \leq N-1)$, where the dwell time τ and the total signal length N are also known. The principal goal in parametric signal processing is to solve the quantification problem as an inverse problem. For the mentioned input data $\{c_n\}$ $(0 \leq n \leq N-1)$ as well as N and τ, this inverse problem amounts to finding the unique solutions for the three types of unknown quantities that are the complex fundamental frequencies $\{\omega_k\}$ and the corresponding complex amplitudes $\{d_k\}$, as well as their true number K.

In the preceding chapters, the main features of the $\text{FPT}^{(\pm)}$ are expounded and illustrated through detailed analysis and computations while solving typical quantification problems encountered in MRS. The overall outcome of the presented results from these two versions of the FPT is an impressive list of the main features in terms of:

• stability; stable convergence to the true results (reconstructed spectral parameters and component as well as total shape spectra) with increasing degrees of Padé polynomials,

• robustness; algorithmic robustness even against computational round-off errors, as well as robustness of extracting the physical content in spite of the presence of noise in the process of reconstructions of fundamental frequencies, amplitudes and the number of genuine components,

• accuracy; machine accuracy for the reconstructed spectral parameters in the case of all genuine resonances from synthesized time signals, as well as optimal accuracy for noise-corrupted input data that can be either theoretically generated or experimentally measured,

• super-resolution; the highest resolution beyond the Rayleigh-Fourier bound $2\pi/T$ with T being the duration of the investigated time signal, and

• signal-noise separation; unequivocal disentangling of genuine from spurious information by means of Froissart doublets (pole-zero coincidences).

Of paramount importance is to single out the milestone achievement of the FPT in solving the noise problem which evaded an exact treatment for more than half a century. The FPT provides the first exact separation of genuine (physical) from spurious (unphysical, noise and/or noise-like) information encountered either in theory or measurements involving time signals. This is accomplished by means of Froissart doublets that are coincident pairs of poles $z_k^\pm \equiv z_{k,Q}^\pm$ and zeros $z_{k,P}^\pm$ in the response functions $P_K^\pm(z^{\pm 1})/Q_K^\pm(z^{\pm 1})$ from the FPT$^{(\pm)}$. Here, $z_{k,P}^\pm$ and $z_{k,Q}^\pm$ are the solutions of the numerator and denominator characteristic or secular equations

$$P_K^\pm(z_{k,P}^\pm)=0 \quad z_{k,P}^\pm=\mathrm{e}^{\pm i\omega_{k,P}^\pm\tau} \qquad Q_K^\pm(z_{k,Q}^\pm)=0 \quad z_{k,Q}^\pm=\mathrm{e}^{\pm i\omega_{k,Q}^\pm\tau}. \tag{12.24}$$

Froissart pole-zero confluences are synchronized with the corresponding zero values obtained for Froissart amplitudes

$$\text{Spurious}: \quad z_{k,Q}^\pm = z_{k,P}^\pm \quad \therefore \quad \{d_k^\pm\}_{z_{k,Q}^\pm = z_{k,P}^\pm} = 0. \tag{12.25}$$

By changing the degree K of the polynomials in the diagonal FPT$^{(\pm)}$ from $P_K^\pm(z^{\pm 1})/Q_K^\pm(z^{\pm 1})$, Froissart doublets unpredictably and uncontrollably alter their positions in the complex $z^{\pm 1}-$planes. They never converge (stabilize) even when the whole signal length is exhausted. Therefore, these latter resonances that roam around in the complex planes are considered as spurious or unphysical. As such, unstable resonances are identified by their twofold signature: pole-zero coincidences and zero amplitudes for noise-free time signals. The same type of signature is also operative for noise-corrupted time signals (theoretically generated or experimentally measured), but with the approximations $z_{k,Q}^\pm \approx z_{k,P}^\pm$ and $d_k^\pm \approx 0$. Crucially, however, although Froissart doublets are unstable against even the smallest external perturbation (e.g., altering the degree of the Padé polynomial, adding noise, etc.), they nevertheless consistently preserve the relationships in (12.25).

By contrast, there are the retrieved resonances with spectral parameters that converge, and these are viewed as stable, genuine or physical resonances. The signatures of all such genuine resonances are

$$\text{Genuine}: \quad z_{k,Q}^\pm \neq z_{k,P}^\pm \quad \therefore \quad \{d_k^\pm\}_{z_{k,Q}^\pm \neq z_{k,P}^\pm} \neq 0. \tag{12.26}$$

Amplitudes $d_k^\pm$ are the Cauchy residues of quotients $P_K^\pm(z^{\pm 1})/Q_K^\pm(z^{\pm 1})$ taken at the poles $z_{k,Q}^\pm$. Here, the word residues has the transparent meaning of the residual differences which remain after the values of the poles are subtracted from the zeros, implying that $d_k^{\pm}$ are proportional to $z_{k,Q}^{\pm} - z_{k,P}^{\pm}$ via

$$d_k^{\pm} \propto (z_{k,Q}^{\pm} - z_{k,P}^{\pm}). \tag{12.27}$$

Thus, the distances between poles $z_{k,Q}^{\pm}$ and zeros $z_{k,P}^{\pm}$ are proportional to the amplitudes $d_k^{\pm}$. Distances[2] have the meaning of a metric; they are related to the so-called measure in a given vector space. Indeed, as shown in Ref. [5], amplitudes $d_k^{\pm}$ are present in the complex-valued Lebesgue measures $\mathrm{d}\sigma^{\pm}(z)$

$$\mathrm{d}\sigma^{\pm}(z) = \rho^{\pm}(z)\mathrm{d}z \qquad \sigma^{\pm}(z) = \sum_{k=1}^{K} d_k^{\pm}\vartheta(z^{\pm 1} - z_{k,Q}^{\pm}) \tag{12.28}$$

$$\rho^{\pm}(z) = \sum_{k=1}^{K} d_k^{\pm}\delta(z^{\pm 1} - z_{k,Q}^{\pm}) \tag{12.29}$$

$$\vartheta(z^{\pm 1} - z_{k,Q}^{\pm}) = \begin{cases} 1, & z^{\pm 1} \in \mathcal{K}_{\mathrm{G}}^{\pm} \\ 0, & z^{\pm 1} \notin \mathcal{K}_{\mathrm{G}}^{\pm} \end{cases} \tag{12.30}$$

with $\mathrm{d}\vartheta(z)/\mathrm{d}z = \delta(z)$, where $\delta(z)$ and $\vartheta(z)$ are the Dirac and Heaviside complex functions, respectively. Here, $\mathcal{K}_{\mathrm{G}}^{\pm}$ denotes the set of K_{G} genuine signal poles $\{z_{k,Q}^{\pm}\}$ $(1 \leq k \leq K_{\mathrm{G}})$ from the investigated FID. The differentials $\mathrm{d}\sigma^{\pm}(z)$, i.e., the measures, are complex-valued and, therefore, the corresponding norms are also complex numbers.

Despite the appearance of complex measures and norms, the orthogonality relationships of the Padé denominator polynomials in the $\mathrm{FPT}^{(\pm)}$ are still preserved. This is important, since distributions and locations of spurious poles depend on features of the orthogonality relations satisfied by the Padé denominator polynomials. Clearly, it is vital to have full control over the locations of all the zeros of $Q_K^{\pm}(z^{\pm 1})$ in the Padé quotients $P_K^{\pm}(z^{\pm 1})/Q_K^{\pm}(z^{\pm 1})$ from the $\mathrm{FPT}^{(\pm)}$. Such a control is possible in the $\mathrm{FPT}^{(+)}$ and $\mathrm{FPT}^{(-)}$, because all the genuine zeros of $Q_K^{+}(z)$ and $Q_K^{-}(z^{-1})$ are inside and outside the unit circle, respectively. However, despite our prior knowledge about such precise locations before reconstructing these zeros, as soon as the systematically increased degree K of $Q_K^{\pm}(z^{\pm 1})$ surpasses the unknown true order K_{G}, spurious roots $\{z_{k,Q}^{\pm}\}$ of the characteristic equations $Q_K^{\pm}(z^{\pm 1}) = 0$ would inevitably appear. For the same reason, spurious zeros $\{z_{k,P}^{\pm}\}$ will also emerge from the accompanying secular equations of the numerator polynomials $P_K^{\pm}(z^{\pm 1}) = 0$. This is where the Froissart concept comes into play to take advantage of

[2]In complex vector spaces, distances should be taken in a generalized sense which, of course, need not necessarily be reduced to a literal distance in units of physical length.

the spuriousness in the set $\{z^{\pm}_{k,P}\}$. Namely, the two types of spuriousness from the two sources $Q^{\pm}_K(z^{\pm 1}) = 0$ and $P^{\pm}_K(z^{\pm 1}) = 0$ are strongly coupled together. As a result, spurious Froissart poles $\{z^{\pm}_{k,Q}\}$ and zeros $\{z^{\pm}_{k,P}\}$ are always born out as pairs. It is in this way that Froissart doublets manifest themselves through pole-zero coincidences, $\{z^{\pm}_{k,Q}\} = \{z^{\pm}_{k,P}\}$, as in (12.25). Such an occurrence cancels the entire spuriousness from the polynomial quotients $P^{\pm}_K(z^{\pm 1})/Q^{\pm}_K(z^{\pm 1})$. This becomes particularly apparent when these ratios are written in their canonical forms

$$\frac{P^{\pm}_K(z^{\pm 1})}{Q^{\pm}_K(z^{\pm 1})} = \frac{p^{\pm}_K}{q^{\pm}_K} \prod_{k=1}^{K} \frac{(z^{\pm 1} - z^{\pm}_{k,P})}{(z^{\pm 1} - z^{\pm}_{k,Q})}. \tag{12.31}$$

If the running degree K is larger than the number of genuine resonances $K_{\rm G}$, then all the terms $(z^{\pm 1} - z^{\pm}_{k,P})/(z^{\pm 1} - z^{\pm}_{k,Q})$ from (12.31) for $K - K_{\rm G} > 0$ would contain spurious Froissart poles $z^{\pm}_{k,Q}$ and zeros $z^{\pm}_{k,P}$. Hence pole-zero cancellations leading to $(z^{\pm 1} - z^{\pm}_{k,P})/(z^{\pm 1} - z^{\pm}_{k,Q}) = 1$ for Froissart doublets $z^{\pm}_{k,Q} = z^{\pm}_{k,P}$, as per (12.25).

The ensuing consequence of these pole-zero cancellations onto the corresponding amplitudes of Froissart resonances can be seen at once from the explicit formulae for $d^{\pm}_k$ in terms of all the recovered poles and zeros

$$d^{\pm}_k = \frac{p^{\pm}_K}{q^{\pm}_K} \prod_{k'=1}^{K} \frac{(z^{\pm}_{k,Q} - z^{\pm}_{k',P})}{(z^{\pm}_{k,Q} - z^{\pm}_{k',Q})_{k' \neq k}}. \tag{12.32}$$

Here, in the numerator, it is permitted to have $k' = k$, in which case every Froissart doublet from (12.25) would produce zero-valued terms $(z^{\pm}_{k,Q} - z^{\pm}_{k,P})$ and, thus, the whole product in (12.32) will become zero. This yields $d^{\pm}_k = 0$ for $z^{\pm}_{k,Q} = z^{\pm}_{k,P}$, as in (12.25). Note that in our computations, we never use (12.32) to obtain the amplitudes $d^{\pm}_k$ in the FPT$^{(\pm)}$. This is because formula (12.32) employs the whole set of the reconstructed amplitudes to compute $d^{\pm}_k$ for the kth resonance. Therefore, even the slightest inaccuracy, such as near cancellations of poles and zeros, rather than the theoretically exact cancellations, could spoil the precision of the sought $d^{\pm}_k$ for the given k. Instead, we use the alternative expressions

$$d^{\pm}_k = \frac{P^{\pm}_K(z^{\pm}_{k,Q})}{Q^{\pm\,\prime}_K(z^{\pm}_{k,Q})} \qquad Q^{\pm\,\prime}_K(z^{\pm}_{k,Q}) \neq 0 \tag{12.33}$$

where the prime denotes the first derivative. Here, each kth amplitude on the lhs depends only on one, i.e., the kth value of the rhs of Eq.(12.32) and, hence, no other resonance can deteriorate the accuracy of the retrieved $d^{\pm}_k$.

Overall, it is clear from these remarks that the FPT$^{(\pm)}$ possesses a very elegant, simple and powerful solution for the exact identification of all spurious Froissart resonances. When these are discarded, only genuine resonances are left and this yields the exact solution of the quantification problem.

12.3 Relevance of Padé-optimized MRS for diagnostics in clinical oncology

Both MRI and *in vivo* MRS are becoming the non-invasive diagnostic methods of choice for a wide range of medical applications, particularly in oncology. Technological advances, most notably the use of very fast gradient variations (echo-planar scanning) [1, 4] have allowed high quality imaging to be performed rapidly. Here, MRS yields biochemical information which often improves the specificity of MRI. Further, MRSI combines the chemical specificity of MRS with the spatial localization techniques that have been developed for MRI to obtain multiple MRS signals over a volume of tissue.

There are many advantages of MR-based modalities. These include multiplanar capabilities, excellent contrast among tissues, as well as the potentially rich array of spectral information that can help distinguish malignant from benign lesions, as well as from healthy tissue. The lack of exposure to ionizing radiation renders MR-based modalities very attractive for early detection, especially for screening surveillance of persons at high risk for specific cancers.

The combination of anatomic localization and insight into metabolic characteristics from spectral information can be decisive for accurate and timely identification of cancers. This is especially true in the exceedingly difficult and critically important differential diagnostic dilemmas such as distinguishing recurrent tumor from radiation necrosis or from post-operative changes.

These advantages have become particularly clear in neuro-oncology, where MRS and MRSI are becoming a key modality for nearly all aspects of brain tumor diagnostics, as reviewed in chapter 8. Another area in which MRI and MRS have made an important impact is prostate cancer, as reviewed in chapter 11, where these methods in combination have provided diagnostic clarity unmatched by literally any other non-invasive approach, and are critical in improving the yield of biopsies. Further, MRS and MRSI have also helped improve the diagnostic accuracy of a number of other malignancies, most notably head and neck cancers, non-Hodgkin's lymphoma as well as breast cancer [7].

Notwithstanding these achievements, the full potential of MRI and MRS for cancer diagnostics has not been realized. As we have intimated, in actual clinical practice, all too often the interpretation of automatically generated MR spectra is shrouded by confusion and ambiguity. In order for MRS and MRSI to become a standard diagnostic tool in the area where it is needed the most, for clinical oncology, including especially cancer screening and surveillance, it is necessary to go beyond technical (hardware) improvements, as important as these are.

We will now briefly recapitulate why many of the major limitations of current applications of MRS and MRSI are due to the almost exclusive reliance upon the conventional Fourier-based mode of data analysis, i.e., the FFT, and what the FPT can offer to help solve these problems that are critical for cancer diagnostics. As presented in chapters 9 through 11, the advantages of Padé-optimization have clearly been demonstrated for MRS data from three problem areas within oncology: ovarian, breast and prostate cancer. We chose these problem areas because of their urgent clinical importance. For the first time, we have applied the FPT to time signals that were generated according to *in vitro* MRS data as encoded from (a) malignant and benign ovarian lesions, (b) breast cancer, fibroadenoma and normal breast tissue and (c) for cancerous prostate, normal stromal and glandular prostate.

The FFT is a low-resolution spectral/image estimator. One of the major problems for Fourier-based MR imaging and spectroscopy is the need for long imaging times. For MRI, image artifacts with edge distortions occur due to truncations in the FFT, which cannot supply images that are simultaneously bright and sharp. This is because sharpness of images stems from improved resolution which in the FFT is obtained by increasing the number of grid points per axis.

However, larger data sets unavoidably contain more noise, which reduces brightness. Due to its extrapolation feature, the FPT allows sampling at shorter acquisition time, and this enhances both sharpness and brightness, as well as diminishing the Gibbs phenomenon. Envelopes of time signals, such as those observed in MRS, decay exponentially, so that the larger signal intensities are found early in the recording. It is, therefore, advantageous to encode the time signal as rapidly as possible, i.e., to avoid long acquisition times at which mainly noise is recorded.

Within the FFT, attempts to improve resolution can only be made by increasing acquisition time T, thus usually leading to a worsening of the SNR, since clinical MRS time signals encoded at 1.5T become heavily corrupted with background noise at larger acquisition times. For breast cancer diagnostics using ^{1}H MRS, this is especially troublesome, due to the need for lipid suppression. One of the current strategies has been to increase the echo time, TE, which diminishes the overlap with the lipid signal, but this is achieved by a diminution in signal intensity. Poor SNR was cited as one of the major reasons for false negative findings using ^{1}H MRS to detect malignant breast lesions [116]. Moreover, some of the potentially informative metabolites for identifying breast cancer have short T_2 relaxation times, and will have decayed at longer TE [22]. Problems related to Gibbs phenomena are particularly important for MRSI.

Poor resolution and low SNR represent a major limitation for current applications of MRS in oncology. This has been a key obstacle for the use of *in vivo* MRS to identify cancers located in deep-seated abdominal/pelvic organs that are also in motion due to respiration and/or peristalsis.

Ovarian cancer is a case in point, with major public health implications. In many developed countries it is the leading cause of death from gynecological malignancies. Non-invasive early detection of ovarian cancer would confer a major survival advantage, but yet as of today, there still is no diagnostic method which has achieved this aim. The few published studies [334]–[339] to date applying *in vivo* MRS to benign and malignant adnexal lesions have been mired most notably by poor resolution and SNR, such that the results are inconclusive. Thus, for this very small, moving pelvic organ, attempts to glean diagnostically meaningful information from *in vivo* MRS using the conventional Fourier analysis have heretofore been modest. Yet, *in vitro* analysis [340]–[343] reveals that there is a rich store of MR-visible metabolic information which distinguishes benign from malignant ovarian lesions.

Taken together, these considerations spurred us to perform the comparative study [27, 29] of the resolution performance of the FPT and FFT for MRS data as encoded in Ref. [343] from cancerous and non-cancerous ovarian cyst fluid. As reviewed in chapter 9, for these ovarian data, remarkably, all the input spectral parameters were reconstructed exactly by the FPT using only sixty-four signal points, over two orders of magnitude fewer than with the conventional FFT. Specifically, using the FFT, some 32768 time signal data points were scanned (512 times more than needed in the FPT), and the FIDs were zero-filled to 65536 data points, in order to obtain this same Padé level of resolution in the Fourier absorption spectra. This provides direct evidence of the powerful extrapolation features of the FPT.

For the benign and malignant ovarian cases, direct visual comparisons of the reconstructed total shape spectra via the FPT versus the FFT at various partial signal lengths are also telling. For example, at a partial signal length of $N/32 = 32$, the main spectral features are already well apparent with Padé-reconstruction. In contrast, only the most rudimentary total shape spectrum is generated by the FFT at that partial signal length. When the total shape spectrum is fully converged at $N/16 = 64$ with the FPT, the Fourier-reconstructed spectrum is still completely uninformative.

Improved resolution and SNR as provided by the FPT could be of crucial help for improving the diagnostic yield of *in vivo* MRS in ovarian cancer diagnostics and for malignancies of other deep-seated, moving organs. Applications of *in vivo* MRS and MRSI for neuro-oncology, breast cancer diagnostics and elsewhere would also benefit greatly from this improvement in resolution and SNR. Moreover, the capabilities of the FPT to clearly resolve overlapping resonances would provide the opportunity to use short echo times, thereby further enhancing resolution and SNR.

Regarding 2D MRS, our initial applications [439] indicate that the Padé-optimized 2D MRS also yield marked improvements in resolution relative to the 2D FFT. We are currently exploring further this avenue via detailed computations on clinical 2D FIDs. This is an area which is attracting increasing attention in clinical oncology, including brain tumors, breast and prostate cancer [89, 387, 438].

The difficulties of attempts at quantification when using Fourier-based methods with post-processing fitting, have often lead to the use of either a dichotomous variable of the Bernoulli type (e.g., presence or absence of a composite choline peak) or to reliance upon metabolite ratios. Applications of *in vivo* ^{1}H

MRS for breast cancer detection have frequently been based upon the presence or absence of a composite choline peak. This can compromise diagnostic accuracy, since choline may also be observed in benign breast lesions, and in the normal breast during lactation. Furthermore, choline is often undetected in small tumors that are then misclassified as benign [22]. Metabolite ratios are also problematic for a number of reasons, including their variability according to different echo times [252]. Several other confounding factors including the cancer treatment itself, can affect metabolite ratios. Overall, the cut-points used to define malignancy vary widely from author to author [7].

One of the key obstacles to a greater use of MRS in clinical oncology has been the lack of a uniform approach to data analysis. Data compatibility is needed in order to make multi-center comparisons, indispensable for wider routine application of these methods [306]. *Differences in data processing methods, rather than real differences, have been considered as the major contributor to deviations among various clinical results* [259].

At odds with the need for data compatibility is the requirement within the FFT for fitting. As discussed, this can lead both to over-fitting (spurious peaks) as well as under-fitting (true metabolites being undetected). This is not only unacceptable to diagnosticians, but renders inter-study comparisons tenuous, at best, unless the same *in vitro* basis set is used in some fitting codes to predetermine the number of metabolites. We have cited several examples of contradictory findings with respect to brain tumor diagnostics directly related to whether or not a given metabolite was included in a basis set of a linear combination of model *in vitro* spectra from signals encoded separately [144, 145].

The problems of Fourier-based signal processing with post-processing fitting are most pronounced with respect to overlapping resonances. As seen repeatedly, closely-lying or overlapping metabolites are often the most important for clinical oncology. Sharply counter-posed to these limitations of Fourier-based data analytical techniques applied to MRS, is Padé-optimized MRS. As demonstrated, the FPT unequivocally indicates the number of metabolites, including those that are overlapping, and provides accurate and precise parametric information needed to determine metabolite concentrations. For the two other clinical areas in oncology, breast cancer and prostate cancer, it was the comparison between the MR total shape spectra and the component spectra, that most dramatically demonstrated the advantages of the FPT. For both these areas, in order to clearly distinguish between normal/benign and malignant tissue, absolute certainty is required about the number of resonances and their chemical shifts.

As presented in chapter 10, the FPT provided exact reconstruction of all the input spectral parameters for the time signals corresponding to the normal, benign as well as to the malignant breast lesions. The Padé absorption spectra yielded unequivocal resolution of all the extracted physical metabolites, even of those that were completely overlapping (phosphocholine and phosphoethanolamine at 3.22 ppm). The capacity of the FPT to resolve the physical resonances as encountered in normal breast versus fibroadenoma versus malignant breast was thereby demonstrated. In particular, the FPT unambiguously delineated diagnostically important metabolites such as lactate, as well as choline, phosphocholine and glycerophosphocholine that are very closely-lying and may represent MR-retrievable molecular markers of breast cancer. Within a very narrow spectral band, there were seven resonances, including phosphocholine which was completely buried underneath a much larger phosphoethanolamine peak.

Because of the so-called "glycerophosphocholine to phosphocholine switch" associated with malignant transformation of the breast [307]–[309], it is vital to identify and precisely quantify phosphocholine, as well as the other resonances lying within this narrow frequency band. This difficult task was achieved by the FPT. Moreover, due to this region of high spectral density, there were approximately 80 times more spurious resonances than those which were genuine. The FPT unequivocally separated out all of these spurious peaks, so that the true spectral information that distinguished malignant breast tissue from fibroadenoma and normal breast tissue could be evaluated with full confidence.

Particular note should be made that for all three cases of breast tissue, the total shape spectrum converged at a shorter signal length than did the component shape spectra. The peak at 3.22 ppm on the total shape spectrum appeared completely symmetrical without any hint that there were, in fact, two components: a larger phosphoethanolamine and a smaller, underlying phosphocholine peak at that position along the chemical shift axis. Because the FPT exactly determines the number of resonances and the spectral parameters of each of these, there is no speculation whatsoever.

Spectral overlap with multiplet resonances is recognized to be exceedingly troublesome for MRS of the prostate [443]. Exact determination of the number of genuine resonances becomes particularly vital in this setting. Our results as presented in chapter 11 illustrate that the FPT can unequivocally resolve and quantify a large number of overlapping resonances, including multiplets of metabolites that distinguish normal glandular prostate, normal stromal prostate and prostate cancer.

For all the presented applications of Padé-optimized MRS to problem areas within oncology, the metabolite concentrations were exactly and unequivocally reconstructed. With the standard Fourier approach, metabolite concentrations are estimated from the shape spectra by integrating the areas under the peaks or fitting the peaks to a subjectively chosen number of Lorentzians and/or Gaussians. Even for clearly delineated peaks, as noted, this procedure of numerical quadrature is subjective due to the uncertainty about the lower and upper integration limits.

With respect to breast cancer, notwithstanding the need to expand the number of metabolites upon which the diagnosis is made, accurate quantification of total choline via the FPT would represent a major breakthrough for early detection of this malignancy. This could also be important for therapeutic decision-making, in light of preliminary data [118] using changes in total choline concentration to gauge the response to chemotherapy. One of the main current problems with attempts at quantifying total choline has been subjectivity due to the uncertainty with lower and upper integration limits when using the procedure of numerical quadratures to determine the concentration via peak integration, as typically done [343, 443]. As mentioned, this problem does not occur with the FPT, since this processor yields the spectral parameters uniquely, thereby obviating any subjective assessment about where a given peak begins and ends.

Together with its stability and high resolution, these advantages indicate that the fast Padé transform appears as the optimal processor for applications of MRS and MRSI in clinical oncology. In summary, we enumerate the major advantages of the FPT relative to other methods for processing MR spectra. Specifically, the FPT:

- Uses only the originally encoded signal to extract the entire unique spectral information, in sharp contrast to the subjectivity of fitting recipes, some of which require additional measurements merely to initialize the least-square algorithms, only to end up with a biased pre-selection for the very resonances that are sought,

- Provides efficient numerical algorithms and closed analytical formulae for parametric and non-parametric signal processing (both can be used simultaneously for cross-validation),

- Truly embodies several powerful estimators with rigorous equivalence to the auto-regressive moving average, the Shanks transform, Padé-Lanczos approximant and continued fractions,

- Can efficiently compute the shape of a spectrum without prior extraction of any of the spectral parameters $\{\omega_k, d_k\}$,

- Greatly enhances resolution and signal-to-noise ratio compared to the FFT,

- Yields precise numerical data for all the peak parameters, i.e., the complex frequencies $\{\omega_k\}$ and complex amplitudes $\{d_k\}$ that define the position $\mathrm{Re}(\omega_k)$, height $|d_k|$, width $\mathrm{Im}(\omega_k)$ and phase $\arg(d_k)$ for every genuine (physical) resonance,

- Unequivocally extracts the exact number K of resonances directly from the encoded time signal $\{c_n\}$ $(0 \leq n \leq N-1)$ by means of the two simultaneous conditions for the Hankel determinants, in sharp contrast to guessing done by other processors for which K is either under-estimated (missing genuine peaks) or over-estimated (producing spurious peaks),

- Identifies with fidelity all the spurious (unphysical) resonances by numerical and analytical procedures via pole-zero cancellation (Froissart doublets) to unequivocally distinguish true from spurious resonances, thus establishing the exact signal-noise-separation,

- Unambiguously resolves overlapping peaks and retrieves weak resonances with small peak areas (low concentrations in MRS),

- Obtains the amplitude d_k of each resonance separately by using only *one* frequency (ω_k) at a time, rather than including *all* the found frequencies $\{\omega_m\}(m = 1, 2, ..., k-1, k, k+1, ...)$ as done by the HLSVD in which inadequate estimates for $\omega_m(m \neq k)$ can partially or totally undermine the accuracy of the computed d_k,

- Derives each amplitude d_k from an analytical expression, in contrast to the HLSVD where this is performed numerically by solving a system of linear equations that invokes additional round-off errors,

- Handles Lorentzian and non-Lorentzian spectra on the same footing by modeling the signal as a sum of K damped complex exponentials with either stationary/constant amplitudes (distinct ω_k's only) or time-dependent polynomial amplitudes (equal *and* distinct ω_k's), and thereby surpasses the HLSVD which is limited exclusively to pure Lorentzian spectra,

- Provides strikingly robust and completely stable convergence for varying fractions of the full signal length, yielding reasonable estimations of the main resonances even for severely truncated signals, as opposed to other parametric estimators that oscillate in an unwieldy manner before eventually converging,

- Can cross-validate all the found estimates by its two variants, $\text{FPT}^{(+)}$ and $\text{FPT}^{(-)}$, with the two complementary convergence regions, inside and outside the unit circle, respectively,

- Undergoes rigorous validation and error analysis by generating the variational estimates with unique upper and lower bounds for the spectral parameters as well as the lines-shape,

- Extends naturally to multi-dimensional signal processing since the data are treated as a coherent whole, in contrast to the corresponding Fourier sequential one-dimensional estimations,

- Has been validated in direct applications for oncology to MRS data as encoded *in vitro* from ovarian, breast and prostate cancer.

The next and urgently needed step is to more widely apply the fast Padé transform to a variety of time signals encoded via MRS and MRSI, especially those emanating from patients with cancers whose detection and characterization remain a major clinical challenge.

13

Conclusion and outlooks

This book is on theory and practice of spectral analysis as it applies to quantification of a wide class biomedical time signals. The presented methodology is general and can be applied to many other inter-disciplinary fields which need not have an overlap with biomedicine. Our principal method selected for this challenging task of solving inverse synthesis-type problems in data interpretation is the fast Padé transform, FPT. This method, which can autonomously pass from the time to the frequency domain with no recourse to Fourier integrals, represents a novel unification of the customary Padé approximant and the causal Padé$-z-$transform. The FPT automatically and simultaneously performs interpolation and extrapolation of the examined data. The idea of synthesis of time signals in a search for an adequate explanation of the observed variation in studied phenomena consisting of composite effects, is to find a sub-class of simpler constituent elements related to the fundamental structure of the examined system which produces a response to external perturbations. Such decompositions of complicated into simpler effects is in the heart of quantification of time signals through their parametrizations.

Finding a relatively small number of fundamental parameters, eliciting poles and zeros, that could capture the main features of the investigated system associated with a given time signal is of paramount theoretical and practical importance. In this way, theoretical explanations of phenomena involving time signals exhibit a great potential in simplifying the stated initial task and coordinating its different parts by the decomposition analysis of observed composite phenomena. As such, the theory of time signals becomes an essential complement of the corresponding measurements. This complementarity does not stop with theoretical explanations and interpretations, but also provides practical tools that enable interpolation where measured data have not been recorded, and extrapolation to the ranges where predictions could be made about the possible behavior of the system under study. Measurements in this field yield time signals, but it is theory which provides frequency spectra and decomposition of encoded data into their constituent fundamental elements. Such inverse problems are difficult due to mathematical ill-conditioning and the possible solutions are further hampered by inevitable noise.

The present book shows how this type of important problem, known as spectral analysis, quantification or harmonic inversion, can be solved by the FPT either with machine accuracy for theoretically generated/simulated time signals or with the best possible precision for the corresponding experimentally measured data. We also provide the optimally reliable solution of the ubiquitous noise problem. This is done by unequivocal disentangling of the genuine from spurious information using the concept of Froissart doublets. By this strategy, spurious information is precisely identified by strong coupling of unphysical poles and zeros through their strict coincidences. Such pole-zero confluences are totally absent for physical, genuine resonances. Hence exact noise separation or SNS as a novel paradigm in data analysis.

13.1 Leading role of Padé approximants in the theory of rational functions and in MRS

General rational functions $R(u)$ are defined as quotients $R(u) = f(u)/g(u)$ of two other functions $f(u)$ and $g(u)$ of a complex-valued independent variable u. They play by far the most prominent role in the mathematical theory of approximations. Importantly, it is this latter practical theory by which mathematics make their most significant and useful bridges towards other disciplines across different research fields. The key mathematical features that determine any function used in mathematical modeling across inter-disciplinary applications are the possible singularities (poles, cuts, branch points) and zeros. The former and the latter are tightly connected, respectively, with the potential existence of maximae (peaks) and minimae (valleys between adjacent peaks) of the given function. The principal reason for the central role of rational functions of the general type $R(u) = f(u)/g(u)$ in the theory of approximations is in their mathematical form by which the numerator $g(u)$ can provide adequate descriptions of singularities, whereas the denominator $f(u)$ is suitable for description of zeros. Poles and zeros can fully describe any system.

Notwithstanding its importance, generality does not necessarily always lead to practical mathematical models for real-life systems. Thus, choosing some very general form for $f(u)$ in $R(u) = f(u)/g(u)$ might not be so useful in many practical situations, since $f(u)$ besides having zeros could also possess singularities on its own. In such a case, there will be a twofold source of singularities of $R(u)$, one due to the denominator $g(u)$ and the other stemming from the numerator $f(u)$. A more restricted, but also a more useful choice, would be a sub-class of general rational functions $R(u)$ with a clearer separation of the roles of the constituents $f(u)$ and $g(u)$. One such sub-class would be the set of pairs of functions $\{f(u), g(u)\}$ for construction of $R(u) = f(u)/g(u)$ in which the only role of $f(u)$ would be to describe zeros of $R(u)$ whose singularities would then be left exclusively to $g(u)$. In this way, we can be sure that the sole singularities of $R(u)$ are due to the denominator function $g(u)$. This choice would also imply that the only zeros of $R(u)$ are those due to zeros of the numerator $f(u)$. For many applications, even this latter, restricted class of rational functions could still be too general, because cuts and branch point singularities might not be so often encountered outside of physics. Even when cuts and branch points are inherently present in a given problem (e.g., in particle scattering and in any other systems with interactive dynamics leading to continuous spectra), they could be closely approximated by a sequence of poles. This leads to a specialized sub-class of rational functions called meromorphic functions whose only singularities are exclusively due to poles of the denominator function $g(u)$. This presumes analyticity of the numerator function $f(u)$ which can have zeros, but is otherwise free of singularities.

The sub-class of functions that optimally meets these natural demands is comprised of two different polynomials for the numerator $f(u)$ and denominator $g(u)$. With this choice of pure polynomials for both $f(u)$ and $g(u)$, function $R(u) = f(u)/g(u)$ becomes a quotient of two polynomials, i.e., a rational polynomial called the Padé approximant, PA. Such a ratio is unique whenever the function to be modeled is represented by a known Taylor or Maclaurin series in powers of the given independent variable. This eliminates ambiguities from the outset in the definition of the PA. Padé approximants can also be viewed as a generalization of Taylor or Maclaurin polynomials to the field of rational functions. These rational Padé polynomials are of great practical usefulness. Among their most important features is acceleration of slowly converging series or sequences and transformation of divergence into convergence by the powerful Cauchy concept of analytical continuation.

This versatile Padé strategy represents an excellent approximation methodology and, moreover, rational functions are no more difficult to employ than ordinary polynomials like those of Taylor, Maclaurin or Laurent. However, Padé rational polynomials are infinitely more powerful than any single polynomial approximation via Taylor, Maclaurin or Laurent truncated series. Being polynomials, $f(u)$ and $g(u)$ are both analytic and so is their quotient $R(u) = f(u)/g(u)$, with the exception of poles which are zeros of the denominator. For instance, in magnetic resonance spectroscopy, MRS, which is the main subject of the present book, zeros $\{u_k\}$ of $g(u)$ are complex-valued numbers related to fundamental energies $\{E_k\}$, or frequencies $\{\omega_k\}$, such that the Padé approximant $R(u) = f(u)/g(u)$ can be used to provide the most adequate model of the system response function for the given external perturbation. In such a case, $R(u) = f(u)/g(u)$ constitutes a complex-valued spectrum whose real value is the physical, absorption spectrum. In practice, we sweep across the real energies or frequencies at which these spectra can safely be computed. This makes the Padé approximant $R(u) = f(u)/g(u)$ a well-defined function throughout, even at the zeros $\{u_k\}$ of $g(u)$, since for any real-valued frequency, we have $u \neq u_k$, i.e., the values $g(u_k) = 0$ never occur, thus making the potential singularity $R(u_k) = \infty$ unattainable.

As stated, the importance of rational functions is not limited to mathematics alone. Rather, they have unparalleled applications in all other branches of science and technology, including industry. This can be easily understood by reference to the main characteristics of versatile systems considered in isolation from their environment and/or when they are perturbed by external excitations or fields. These major characteristics of general physical, chemical or biological systems of small as well as large dimensions or degrees of freedom are universally embodied in their spectra that can be comprehended and adequately interpreted in terms of a relatively limited number of leading parameters. Such parameters are the two sets of physically measurable quantities (observables) called the characteristic numbers or eigen-energies $\{E_k\}$ or eigen-frequencies $\{\omega_k\}$ of the system's operator $\hat{\Omega}$, and the corresponding intensities $\{I_k\}$ or amplitudes $\{d_k\}$ of the spectral lines or line-shapes.

The so-called system's operator $\hat{\Omega}$ is the generator of the system interactive dynamics. In quantum physics $\hat{\Omega} = \hat{\text{H}}$ where $\hat{\text{H}}$ is the system's Hamiltonian operator consisting of kinetic energy operators for describing kinematics, and potential operators responsible for interactions. Physically, eigen-frequencies $\{\omega_k\}$ are the fundamental or nodal frequencies of the system intrinsic oscillations that are supposed to be determined while applying any external perturbations. These latter disturbances can be, e.g., a radio-frequency pulse combined with a static and a gradient magnetic field, as implemented in MRS. The studied system first absorbs the external radiation or impinging particle, and then dispenses the received excess of energy through some discrete internal transitions among the quantum-mechanically possible energy levels leading to, e.g., excitation, particle emission, etc. We infer about such intrinsic transitions only later when they are completed and this can be manifested in various ways, as can be exemplified as follows.

• (i) Emission spectra of one or more ejected particles can be measured in experiments for the case of a decay of an unstable transient system comprised of the target and an external perturbation via the projectile beam. There are many examples of this mechanism throughout physics, e.g., nuclear transmutations, involving nuclei as projectiles and targets. Here, a large number of peaks is routinely observed in the measured cross sections. This is interpreted as formation of a transient compound system (projectile plus target) with a number of positive energies $\{E_k\}$ that produce metastable states of this decaying aggregate. If the impact energy E_i is swept across these quasi-bound energies $\{E_k\}$ of the compound system, enhanced cross sections are detected whenever $E_i \approx E_k$. This leads to resonance peaks in measured cross section spectra. The mathematical form of these so-called Breit-Wigner resonances coincides with the PA, which then can be used to determine, e.g., positions, lifetimes and areas of such peaks. The found peak areas are proportional to peak heights (resonance amplitudes) and these reconstructed data are directly related to the number of projectiles that undergo resonant collisions with the target. Such resonance spectra are usually plotted as counts per channel, where the ordinate gives sticks proportional to the number of the incident particles (counts) that hit the detector at the given resonant energy (channel). We can see here how the Padé approximant is perfectly suited to describe the well-known Breit-Wigner mechanism of formation and subsequent decay of compound systems comprised of the projectile and target particles. It is through this major picture and mechanism of resonant collision physics in the Breit-Wigner-Padé setting that many elementary particles in nature were detected and accurately quantified for their masses and lifetimes. This is a continuing story of physics – it was like this decades ago and it is likely to persevere for a long time to come.

• (ii) Absorption spectra for general systems in magnetic fields can also be studied to uncover the hidden internal structure of such systems. For example, a quantitative proton MRS in clinical diagnostics can, in principle, determine the intrinsic structure of the scanned tissue yielding the precise biochemical content and concentrations or abundance of the involved molecules or metabolites. This is done by *in vivo* proton MRS through detection of time signals emanating from the selected tissue of examined patients. The sought absorption spectrum in the frequency domain is not actually measured, but rather it is computed by applying certain mathematical transformations (Fourier, Padé, etc.) to the encoded time signals. However, such spectra cannot, on their own, give the sought unequivocal internal biochemical structure of the scanned tissue. This can unambiguously be obtained only by an adequate spectral analyses (quantification) of the measured time signals to reconstruct the concentrations of the main diagnostically interpretable metabolites. Such concentrations are related to the abundance of these resonating protons from the tissue. On the other hand, it is the number of resonating protons which gives the intensity of the tissue response to the external excitations. Thus, the intensity I_k or amplitude $|d_k|$ of each of the constituents of the ensuing time signal becomes crucial for extraction of every concrete metabolite concentrations. This presumes that the number of resonances has also been retrieved. Finally, resonances are assigned to metabolites. Each metabolite can be associated with more than one resonance. Resonating protons bound in these molecules respond differently to the same applied frequency due to unequal shielding from the surrounding electronic clouds. These small differences (chemical shifts) in the resonant frequencies of various protons in different chemical environments are the very basis of MRS. Such tiny differences in chemical shifts permit differentiation among various molecules into which protons are bound, and this gives the possibility for metabolite assignments. Hence, reconstruction of chemical shifts, as the real part of the corresponding complex-valued fundamental frequencies $\{\omega_k\}$, followed by retrieval of the associated moduli of amplitudes $\{|d_k|\}$ is critical to the diagnostic task of MRS. This amounts to extraction of biochemical information from the examined tissue via metabolite assignments and concentrations. The imaginary parts of $\{\omega_k\}$ and phase of $\{d_k\}$ are also useful in providing the lifetimes of metastable transients in the time signal and its phase, respectively.

Whether exemplified by topic (i) or (ii) or by a myriad of other related phenomena from interactive dynamics, we see that the fundamental frequencies $\{\omega_k\}$ of the given system can be identified as the poles of the corresponding response function. Rational polynomials via the unique Padé approximant as the response function of the studied system, are optimally suitable for this task, since they are meromorphic functions whose poles are the zeros of the denominator polynomial. Moreover, the zeros of the numerator polynomial are the zeros of this response function in the form of the polynomial quotient. The locations of these poles in complex planes give metabolite chemical shifts and lifetimes in MRS from example (ii). Further, the Cauchy residues of the rational polynomial taken at the found nodal frequencies $\{\omega_k\}$ are the sought amplitudes $\{d_k\}$ whose absolute values and phases yield metabolite concentrations and phases of the associated time signal components. From these remarks, particularly in MRS as the central theme of the present book, it is clear how theory (via mathematics, physics, chemistry, biology) comes into play. It is the experimental measurement which encodes the time signal, as a response of the examined system to external perturbations. The theory, however, explains and interprets the measured experimental data. Therefore, theories driven by experiments are an indispensable and inseparable part of measurements. Especially, spectral analysis, as an inverse problem, starts from the encoded time signal and builds the whole theory of measurement.

In this book, we show both through a theoretical development and illustrations as they apply to clinical diagnostics in the realm of MRS, how Padé approximants can be used to find the unique solutions $\{\omega_k, d_k\}$ to the difficult quantification problem of critical importance in medicine. Moreover, within the Padé approximant, we present a novel and long awaited procedure in signal processing via the denoising Froissart filter for exact signal-noise separation. This is implemented through the powerful concept of Froissart doublets, manifested as pole-zero coincidences in the system response function. Such a denoising filter can clearly distinguish genuine (physical) from spurious (unphysical) resonances by reliance upon the stability criterion. We can vary different quantities, e.g., signal length, or we may change the degrees of the numerator and denominator Padé polynomials. These variations would leave a certain number of pairs of spectral parameters $\{\omega_k, d_k\}$ stable, once they converged, and they indeed do, as soon as the Padé approximant exactly reconstructs the true number of metabolites. Such stable resonances represent genuine components from the input time signal. However, the same variations of the polynomial degrees would never lead to convergence of the remaining sub-set from the whole retrieved collection $\{\omega_k, d_k\}$. These non-converged resonances fluctuate irregularly with the changes in the degrees of Padé polynomials and they never stabilize. Hence, they are categorized as spurious, as they are absent from the input time signal. These spurious resonances are easily spotted through their unmistakable twofold signature identified as pole-zero confluences and the associated zero amplitudes in the Padé response function. Because they appear as pairs, such spurious structures are called Froissart doublets after Froissart who 40 years ago discovered this phenomenon by computer experiments when adding random noise to a deterministic synthesized signal with a single genuine component.

We significantly expand and generalize the concept of Froissart doublets as the main part of the comprehensive strategy of signal-noise separation, which can be used in the whole field of signal processing as the most reliable method to date for disentangling genuine from spurious information. In system theory as well as in control theory from engineering, stability of the system's parameters is the prerequisite for the system's overall performance. Both zeros and poles of response functions play their crucial roles in the sought stability of systems. Not unexpectedly, the Padé approximant has been firmly established over the years as the optimal response function in both system and control theory. This fully coheres with the like experience from quantum physics as emphasized throughout this book and mentioned in the outlined example (i) and (ii). Moreover, Froissart doublets are the proof that zeros of the Padé response function are critical to finding stability of the examined system. Genuine zeros remain in the final output of the performed data analysis as stable structures of the studied system, while spurious ones as unstable Froissart zeros are washed out by being canceled by the corresponding spurious poles at the coincident positions in the complex-valued frequency spectra.

Equipped with the exact quantification and accompanied by signal-noise separation via identification as well as through subsequent discarding of Froissart doublets leading *de facto* to noise suppression, the Padé approximant emerges as the theory of choice for spectral analysis of quantification of general time signals. Our implementations of such pivotal developments within magnetic resonance spectroscopy give a fresh and new platform to the otherwise unprecedentedly rich theory and practice of Padé approximants as they apply to life sciences in general, and clinical diagnostics, in particular.

13.2 Outlooks for Padé-optimized MRS and MRSI from a clinical perspective

Diagnostics obviously play a central role in medical practice, particularly in oncology. Clinical decision-making vitally depends upon the knowledge of whether or not a malignancy is present, and if so, its nature and extent. This information is crucial for guiding therapy as well as for subsequent surveillance. Early cancer detection holds the promise of improved prognosis and the need for less radical treatment. This is the motivation for screening programs aimed at malignancies of major public health importance such as breast and prostate cancer. These programs have succeeded in increasing survival. Ovarian cancer remains a major cause of death among women precisely because of the lack of early detection. Primary brain tumors are less common than the other three mentioned malignancies. Nevertheless, they draw a great deal of attention, often have poor prognosis and it is here that the most delicate of clinical decisions are made, requiring maximal information of the highest possible reliability. These four types of cancers are those upon which we have focused in this book on spectral analyses of the pertinent data.

Well over a century ago it was physics through the discovery of diagnostic X-rays, which provided the possibility of non-invasively detecting a number of tumors. Further innovations through computerized tomography, as well as positron emission tomography, single photon emission tomography, ultrasound, together with magnetic resonance modalities were all made possible through basic research from physics. Each of these is now an indispensable part of the armamentarium of cancer diagnostics. The greatest promise is held by molecular imaging which examines pathways of disease through cellular and molecular biology in conjunction with diagnostic imaging.

While at its outset molecular imaging was almost synonymous with clinical X-ray-based imaging, it now has a far wider scope, including magnetic resonance imaging, MRI, magnetic resonance spectroscopy, MRS, and magnetic resonance spectroscopic imaging, MRSI. Each of the mentioned molecular imaging methods is complementary. Especially in the role of screening and whenever there is a need for heightened surveillance frequency of persons at increased cancer risk, MRI, MRS and MRSI are advantageous as they are free from ionizing radiation. Other advantages of MRI and MRSI include submillimeter resolution and high contrast among tissues as well as multi-planar capabilities. In particular, MRI is very sensitive, and being non-invasive and free from ionizing radiation, has become indispensable for timely cancer detection. Its major drawback is poor specificity. On the other hand, MRS enhances specificity by detecting metabolic features of malignancy. Since molecular changes often precede morphological alterations, sensitivity can also be further improved by MRS. As such, MRSI brings together the rich biochemical information of MRS with the spatial localization methods of MRI to provide multiple spectra over a volume of tissue. Therefore, MRSI holds particular promise for screening and surveillance.

The critical complementary information provided by MRS relative to anatomical imaging via MRI is metabolic information and, hence, the overall status of the scanned tissue. Thus far, MRS and MRSI have made great progress in cancer diagnostics by relying upon at most a handful of metabolites. However, these methods still are not in the widespread use. Indeed, as a research field and a clinical diagnostic modality, MRS is considered to be on the verge of a veritable renaissance. Starting from its status as the well-established nuclear magnetic resonance in physics and analytical chemistry, MRS has developed to such a point in medicine that it is currently being viewed by experts as the diagnostic modality which could potentially revolutionize not only cancer diagnostics, but also guided surgery and target delineation for radiotherapy. In order to achieve this potential, it is necessary to expand the often prevailing mindset within cancer diagnostics which has emphasized the qualitative, morphological characteristics of the tissue under examination.

The first glimpse with a rough metabolic information from MRS and MRSI is given in the form of spectra via composite, total line-shapes as a function of frequencies at which various metabolites respond or resonate to the applied external perturbations. Such information can be given by the fast Fourier transfer, FFT. However, the ultimate goal is to extract the complete information from individual components as the constituent parts of these composite spectral envelopes. This cannot be done by FFT. The present book suggests that the optimal alternative method is the fast Padé transform, FPT, which can provide both the qualitative (envelopes) and quantitative (metabolite concentrations) information.

Thus, notwithstanding the need to broaden this mindset, what has mainly prevented MRS and MRSI from realizing their potential in cancer diagnostics and beyond, is the reliance upon total shape spectra without trustworthy quantification. In other words, there is a vital need to reconstruct component spectra and the spectral parameters from which metabolite concentrations are computed. This requires a multi-disciplinary approach. The key role here is played by mathematics by which the measured time signals are transformed into spectra. Spectroscopic methods are in fact the key strategy for studying the structure and content of matter. The basic sciences, physics and chemistry, developed their full potential using these methods. It is now time for medicine to take advantage of such developments.

Clinical scanners used in MRI all rely upon the FFT. By inertia, this has continued for MRS, even though the FFT lacks the necessary power to reveal the true information content. The more appropriate spectroscopic methods from physics and chemistry should have been transferred to MRS. Attempts to compensate for the quantification inadequacy of the FFT in MRS have used fitting, which is merely a form of patching which involves guesswork. The fundamental issue here is not which fitting algorithm should be used, but rather, that fitting is an inadequate, naive strategy for such a serious problem as quantification in clinical diagnostics where underestimation via under-fitting (missing true metabolites) and overestimation through over-fitting ("prediction" of false, non-existent metabolites) are completely anathema. Fitting can never unambiguously decipher the genuine, physical components hidden under a given peak. This will invariably be uncertain and subjective in all fitting techniques which are overwhelmingly in use in MRS and MRSI.

What are the alternatives? The first step is to properly define the task of quantification in MRS. This amounts to the determination of metabolite concentrations which are the norms, and then to find the potential correspondence between various patterns of deviation from normal and specific disease processes. This is the role of clinical interpretation. The proper definition of this task will be immensely facilitated by seeing the larger context. How has spectral analysis, which is quantification, been achieved in neighboring sciences? It is eminently clear that the key is mathematics, with rational response functions being the leading tool. Within these, the Padé approximant is the method of choice because it uniquely solves the quantification problem when the time signals are fully controlled. In the present book, we have taken this road by developing the fast Padé transform. Through the FPT, we tackled fully controlled problems with synthesized time signals that were reminiscent of the related measured data. We demonstrated that the FPT can also handle encoded time signals. Further, we have shown that the FPT can adequately solve the quantification problem in the presence of noise.

The clinically-relevant difficulties within MRS are noise, as well as obtaining the needed spectral parameters. These include the complex frequencies and amplitudes, from which metabolite concentrations are reliably deduced. In addition, there is a crucial fifth unknown parameter which is the number of genuine resonances that are constituents of a given spectrum. Unless the latter is determined unequivocally, under-fitting or over-fitting will inevitably result, with peaks either being missing or falsely detected. Both of these severe failures are totally unacceptable for clinical diagnostics and, moreover, they are the ones that made physicians skeptical of all fitting-type data analyses in MRS and MRSI. In sharp contrast, exploiting the powerful principle of Froissart doublets, we have shown that exact signal-noise-separation can be achieved, such that all spurious resonances are unequivocally identified and separated from the genuine information content.

In the thorough applications of the FPT presented in this book, with direct relevance to cancer diagnostics, we used noise-free and noise-corrupted theoretically-generated synthesized/simulated time signals to set up the standard with a high level of control for obtaining the unambiguous solution of the quantification problem. Having passed this most stringent test with an unprecedented accuracy, robustness and clinical reliability, the FPT was also applied with great success to clinically encoded time signals and noise-corrupted spectra that are abundant with overlapping resonances. We foresee the subsequent step as application of the FPT in combined studies of malignant versus benign lesions, in which *in vitro* and *in vivo* time signals encoded by MRS and MRSI are directly and comparatively analyzed, together with histopathology for cross-validation. This type of comprehensive and complementary image-histopathology correlation is considered to be particularly promising for improving the diagnostic accuracy of MRS in oncology.

It will be vital to apply the FPT to encoded *in vivo* MR time signals associated with various cancers and non-malignant tissue that have presented differential diagnostic dilemmas, notably benign tumors, infectious or inflammatory lesions. The FPT would yield unambiguous quantitative physical and biochemical information. This could facilitate the development of normative data bases for metabolite concentrations versus the corresponding findings seen in malignancy. Such an opportunity would, in turn, provide needed standards to aid in cancer diagnostics, identifying malignant versus benign disease with specific patterns of departures from normal metabolite concentrations.

Overall, we anticipate that Padé-based optimization suggested in this book will be important for realizing the full potential of magnetic resonance spectroscopy and imaging in early cancer diagnostics and various aspects of cancer treatment. Such a potential of these two modalities from magnetic resonance physics is appreciated by biomedical researchers and clinical practitioners, as a way to greatly improve both diagnostics and therapy.

List of acronyms

Ala:	Alanine
AMARES:	Advanced method for accurate, robust and efficient spectral fitting
AR:	Auto-regressive
au:	arbitrary units
BPH:	Benign prostatic hypertrophy
BW:	Breit-Wigner
CCF:	Contracted continued fractions
CDP:	Cytosine diphosphate
CF:	Continued fractions
CHESS:	CHEmical Shift Selective
Cho:	Choline
CI:	Confidence interval
Cit:	Citrate
CNS:	Central nervous system
Cr:	Creatine
Crn:	Creatinine
CSI:	Chemical shift imaging
CT:	Computerized tomography
DCE:	Dynamic contrast enhancement
DFF:	Denoising Froissart filter
DOS:	Density of states
DPA:	Decimated Padé approximant
DSD:	Decimated signal diagonalization
DWI:	Diffusion weighted imaging
FD:	Filter diagonalization
FDG-PET:	Fluorodeoxyglucose positron emission tomography
FFT:	Fast Fourier transform
FID:	Free induction decay
FPT:	Fast Padé transform
$\mathrm{FPT}^{(+)}$:	Fast Padé transform inside the unit circle
$\mathrm{FPT}^{(-)}$:	Fast Padé transform outside the unit circle
FWHM:	Full width at the half maximum
Glc:	Glucose
Gln:	Glutamate
Glu:	Glutamine
GPC:	Glycerophosphocholine
HLSVD:	Hankel–Lanczos singular value decomposition
HRMAS:	High-resolution magic-angle spinning
Hz:	hertz
ICRMS:	Ion cyclotron resonance mass spectroscopy
IDOS:	Integrated density of states
IFPT:	Inverse fast Padé transform
IMRT:	Intensity modulated radiation therapy
Ins:	Inositol
Iso:	Isoleucine
Lac:	Lactate
LCF:	Lanczos continued fractions
LCModel:	Linear combination of model *in vitro* spectra
LFS:	Limited family structure
lhs:	left-hand side
Lip:	Lipid
LS:	Least Square
Lys:	Lysine

MATLAB:	Mathematical Advanced Toolkit Laboratory
Met:	Methionine
m-Ins:	Myoinositol
MPA:	Matrix Padé approximant
MR:	Magnetic resonance
MRI:	Magnetic resonance imaging
MRS:	Magnetic resonance spectroscopy
MRSI:	Magnetic resonance spectroscopic imaging
NAA:	N-Acetyl Aspartate
N/D:	Numerator/Denominator
NMR:	Nuclear magnetic resonance
ODE:	Ordinary difference equation
OΔE:	Ordinary differential equation
OPA:	Operator Padé approximant
PA:	Padé approximant
PAm:	Polyamine
PCho:	Phosphocholine
PD:	Product-difference
PCr:	Phosphocreatine
PE:	Phosphoethanolamine
PET:	Positron emisson tomography
$\mathrm{pFPT}^{(+)}$:	Poles of the $\mathrm{FPT}^{(+)}$
$\mathrm{pFPT}^{(-)}$:	Poles of the $\mathrm{FPT}^{(-)}$
PLA:	Padé–Lanczos approximant
PLCO:	Prostate, Lung, Colorectal and Ovarian
ppm:	parts per million
PSA:	Prostate specific antigen
PzT:	Padé z-transform
QD:	Quotient-difference
QED:	That which was set to prove
rhs:	right-hand side
RMS:	Root mean square
ROPEM:	Recursive orthogonal polynomial expansion method
RT:	Radiation therapy
s-Ins:	Scylloinositol
SCS:	Statistical classification strategy
SD:	Standard deviation
SNR:	Signal to noise ratio
SNS:	Signal-noise separation
SPECT:	Single photon emission computerized tomography
ST:	Shank Transform
SVD:	Singular value decomposition
Tau:	Taurine
TE:	Echo time
Thr:	Threonine
TOCSY:	Total correlation spectroscopy
TR:	Repetition time
TROSY:	Transverse relaxation-optimized spectroscopy
TRUS:	Transurethral ultrasound
TVUS:	Transvaginal ultrasound
US:	Ultrasound
Val:	Valine
VARPRO:	variable projection method
ww:	Wet weight
$\mathrm{zFPT}^{(+)}$:	Zeros of the $\mathrm{FPT}^{(+)}$
$\mathrm{zFPT}^{(-)}$:	Zeros of the $\mathrm{FPT}^{(-)}$
1D:	One-dimensional
2D:	Two-dimensional
3D:	Three-dimensional

References

[1] Liang, Z.-P., and Lauturber, P.C. *Principles of Magnetic Resonance Imaging: a Signal Processing Prospective*, IEEE Press Series in Biomedical Engineering, New York, 2000.

[2] Günther, H., *NMR Spectroscopy: Basic Principles, Concepts and Applications in Chemistry*, 2nd edn, Wiley, Chichester, 1992.

[3] Freeman, R., *A Handbook of Nuclear Magnetic Resonance* Addison Wesley Longman, 2nd edn, Edinburgh, 1997.

[4] McRobbie, D.W., Moore, E.A., Graves, M.J., and Prince, M.R., *MRI from picture to proton*, Cambridge University Press, Cambridge, U.K., 2003.

[5] Belkić, Dž., *Quantum-Mechanical Signal Processing and Spectral Analysis*, Institute of Physics Publishing, Bristol, 2004.

[6] Belkić, Dž., *Quantum Theory of Resonant Scattering, Spectroscopy and Signal Processing*, Taylor & Francis, London, 2010 (ISBN: 9781584887744).

[7] Belkić, K., *Molecular Imaging through Magnetic Resonance for Clinical Oncology*, Cambridge International Science Publishing, Cambridge, 2004.

[8] Belkić, Dž., and Belkić, K., The fast Padé transform in magnetic resonance spectroscopy for potential improvements in early cancer diagnostics, *Phys. Med. Biol.*, 50, 4385 – 4408, 2005.

[9] Belkić, Dž., and Belkić, K., *In vivo* magnetic resonance spectroscopy by the fast Padé transform, *Phys. Med. Biol.*, 51, 1049 – 1075, 2006.

[10] Belkić, Dž., Exact quantification of time signals in Padé-based magnetic resonance spectroscopy, *Phys. Med. Biol.*, 51, 2633 – 2670, 2006.

[11] Belkić, Dž., Exponential convergence rate (the spectral convergence) of the fast Padé transform for exact quantification in magnetic resonance spectroscopy, *Phys. Med. Biol.*, 51, 6483 – 6512, 2006.

[12] Belkić, Dž., Fast Padé transform for magnetic resonance imaging and computerized tomography, *Nucl. Instr. Meth. Phys. Res. A*,471, 165 – 169, 2001.

[13] Belkić, Dž., Dando, P.A., Main, J., and Taylor, H.S., Three novel high-resolution nonlinear methods for fast signal processing, *J. Chem. Phys.*, 133, 6542 – 6556, 2000.

[14] Belkić, Dž., Dando, P.A., Taylor, H.S., Main, J., and Shin, S.-K., Decimated signal diagonalization for Fourier transform spectroscopy, *J. Phys. Chem. A*, 104, 11677 – 11684, 2000.

[15] Deschamps, M., Burghardt, I., Derouet, C., Bodenhausen, G., and Belkić, Dž., Nuclear magnetic resonance study of xenon-131 interacting with surfaces: effective Liouvillian and spectral analysis, *J. Chem. Phys.*, 113, 1630 – 1640, 2000.

[16] Belkić, Dž., Non-Fourier based reconstruction technique, *Magn. Reson. Mater. Phys. Biol. Med.*, 15 (Suppl. 1), 36 – 37, 2002.

[17] Belkić, Dž., Exact analytical expressions for any Lorentzian spectrum in the Fast Padé transform (FPT), *J. Comp. Meth. Sci. Eng.*, 3, 109 – 186, 2003.

[18] Belkić, Dž., Strikingly stable convergence of the Fast Padé transform (FPT) for high-resolution parametric and non-parametric signal processing of Lorentzian and non-Lorentzian spectra, *Nucl. Instr. Meth. Phys. Res. A*, 525, 366 – 371, 2004.

[19] Belkić, Dž., Analytical continuation by numerical means in spectral analysis using the fast Padé transform (FPT), *Nucl. Instr. Meth. Phys. Res. A*, 525, 372 – 378, 2004.

[20] Belkić, Dž., Error analysis through residual frequency spectra in the fast Padé transform (FPT), *Nucl. Instr. Meth. Phys. Res. A*, 525, 379 – 386, 2004.

[21] Belkić, K., MR spectroscopic imaging in breast cancer detection: possibilities beyond the conventional theoretical framework for data analysis, *Nucl. Instr. Meth. Phys. Res. A.*, 525, 313 – 321, 2004.

[22] Belkić, K., Current dilemmas and future perspectives for breast cancer screening with a focus on optimization of magnetic resonance spectroscopic imaging by advances in signal processing, *Isr. Med. Assoc. J.*, 6, 610 – 618, 2004.

[23] Belkić, Dž., and Belkić, K., Fast Padé transform for optimal quantification of time signals from magnetic resonance spectroscopy, *Int. J. Quantum Chem.*, 105, 493 – 510, 2005.

[24] Belkić, Dž., Fast Padé transform for exact quantification of time signals in magnetic resonance spectroscopy, *Adv. Quantum Chem.*, 51, 157 – 233, 2006.

[25] Belkić, Dž., and Belkić, K., Mathematical optimization of *in vivo* NMR chemistry through the fast Padé transform: Potential relevance for early breast cancer detection by magnetic resonance spectroscopy, *J. Math. Chem.*, 40, 85 – 103, 2006.

[26] Belkić, Dž., and Belkić, K., Decisive role of mathematical methods in early cancer diagnostics: Optimized Padé-based magnetic resonance spectroscopy, *J. Math. Chem.*, 42, 1 – 35, 2007.

[27] Belkić, K., Resolution performance of the fast Padé transform: Potential advantages for magnetic resonance spectroscopy in ovarian cancer diagnostics, *Nucl. Instr. Meth. Phys. Res A.*, 580, 874 – 880, 2007.

[28] Belkić, Dž., Machine accurate quantification in magnetic resonance spectroscopy, *Nucl. Instr. Meth. Phys. Res. A*, 580, 1034 – 1040, 2007.

[29] Belkić, Dž., and Belkić, K., Mathematical modeling applied to an NMR problem in ovarian cancer detection, *J. Math. Chem.*, 43, 395 – 425, 2008.

[30] Belkić, Dž., and Belkić, K., Unequivocal disentangling genuine from spurious information in time signals: Clinical relevance in cancer diagnostics through magnetic resonance spectroscopy, *J. Math. Chem.*, 44, 887 – 912, 2008.

[31] Belkić, Dž., and Belkić, K., The general concept of signal-noise separation (SNS): mathematical aspects and implementation in magnetic resonance spectroscopy, *J. Math. Chem.*, 45, 563 – 597, 2009.

[32] Belkić, Dž., and Belkić, K., Exact quantification of time signals from magnetic resonance spectroscopy by the fast Padé transform with applications to breast cancer diagnostics, *J. Math. Chem.*, 45, 790 – 818, 2009.

[33] Belkić, Dž., and Belkić, K., Unequivocal resolution of multiplets in MR Spectra for prostate cancer diagnostics achieved by the fast Padé transform, *J. Math. Chem.*, 45, 819 – 858, 2009.

[34] Belkić, Dž., Exact signal-noise separation by Froissart doublets in the fast Padé transform for magnetic resonance spectroscopy, *Adv. Quantum Chem.*, 54, 95 – 179, 2009.

[35] Belkić, Dž., Noise identification in comparative analysis of Froissart doublets by analytical and numerical means, *J. Math. Chem.*, 46, 23 – 61, 2009.

[36] Callaghan, M.F., Larkman, D.J., and Hajnal, J.V., Padé methods for reconstruction and feature extraction in magnetic resonance imaging, *Magn. Reson. Med.*, 54, 1490 – 1502 , 2005.

[37] Williamson, D.C., Hawesa, H., Thacker, N.A., and Williams, S.R., Robust quantification of short echo time ^{1}H magnetic resonance spectra using the Padé approximant, *Magn. Reson. Med.*, 55, 762 – 771, 2006.

[38] Gamow, G., Zur Quantentheorie des Atomkernes, *Z. Phys.*, 51, 204 – 212, 1928.

[39] Gamow, G., and Houtermans, F.G., Zur Quantenmechanik des radiaktiven Kerns, *Z. Phys.*, 52, 496 – 509, 1928.

[40] Gamow, G., Zur Quantentheorie der Atomzertrümmerung, *Z. Phys.*, 52, 510 – 515, 1928.

[41] Gurney, R.W., and Condon, E.U., Quantum mechanics and radiative desintegration, *Nature*, 122, 439 – 440, 1928.

[42] Gurney, R.W., and Condon, E.U., Quantum mechanics and radiative disintegration, *Phys. Rev.*, 33, 127 – 140, 1929.

[43] Peierls, R.E., Complex eigenvalues in scattering theory, *Proc. Roy. Soc. London, Ser. A*, 253, 16 – 36, 1959.

[44] Froissart, M., Approximation de Padé: application à la physique des particules élémentaires *Centre Nacional de la Récherche Scientifigue (CNRS), Recherche Cooperative sur Programme (RCP) N° 25, Strasbourg*, Eds. Carmona, J., Froissart, M., Robinson, D.W., and Ruelle, D., 9, 1 – 13, 1969.

[45] Cadzow, J.A., *Discrete-Time Systems*, Prentice-Hall, Inc., Englewod Cliffs, New Jersey, 1973.

[46] Basdevant, J.L., Padé approximants, in: *Methods in Subnuclear Physics*, ed. M. Nikolić, Gordon & Breach, 1970 pp. 129 – 168.

[47] Basdevant, J.L., The Padé approximation and its physical applications, *Fortsch. d. Phys.*, 20, 283 – 331, 1972.

[48] Gammel, J.L., and Nuttall, J., Convergence of Padé approximant to quazi-analytic functions beyond natural boundaries, *J. Math. Anal. Appl.*, 43, 694 – 696, 1973.

[49] Gammel, J.L., Effect of random errors (noise) in the terms of a power series on the convergence of the Padé approximant, in: *Padé approximants*, ed. Graves-Morris, P.R., Institute of Physics Publishing, Bristol, 1973, pp. 132 – 133.

[50] Gammel, J.L., *Rocky Mount. J. Math.*, 4, 203 – 206, 1974.

[51] Padé, H., Sur la représentation approchée d'une fonction par des fractions rationnelles, *Annales Scientifiques de l'Ecole Normale Supérieure, Paris*, 9 (Suppl.), S1 – S93, 1892.

[52] Gilewicz, J., *Approximants de Padé*, Lecture Notes in Mathematics, Nr. 667, Springer-Verlag, Berlin, Section "Doublets de Froissart et foncions quazi-analytique", 1978, pp. 306 – 313.

[53] Gilewicz, J., and Truong-Van, B., Froissart doublets in the Padé approximation and noise, in: *Constructive Theory of Functions*, ed. Sendov, B., Petrushev, P., Ivanov, K., and Maleev, R., Publishing House of the Bulgarian Academy of Sciences, Sofia, Bulgaria, 1988, pp. 145 - 151.

[54] Gilewicz, J., and Pindor, M., Padé approximants and noise: a case of geometric series, *J. Comput. Appl. Math.*, 87, 199 - 214, 1997.

[55] Gilewicz, J., and Pindor, M., Padé approximants and noise: rational functions, *J. Comput. Appl. Math.*, 105, 285 - 297, 1999.

[56] Gilewicz, J., Zeros of Froissart polynomials, *J. Comput. Appl. Math.*, 133, 687 - 681, 2001.

[57] Gilewicz, J., and Kryakin, Y., Froissart doublets in Padé approximation in the case of polynomial noise, *J. Comput. Appl. Math.*, 153, 235 - 242, 2003.

[58] Carleman, T., *Les Fonctions Quasi Analytiques*, Gauthier-Villars, Paris, 1926.

[59] Steinhaus, H., Über die Wahrscheinlichkeit dafür dass der Konvergenzreis einer Potenzreihe ihre natürliche Grence ist, *Math. Z.*, 31, 408 - 416, 1929.

[60] Endrei, A., The Padé table of functions having a finite number of essential singularities, *Pac. J. Math.*, 56, 429 - 453, 1975.

[61] Pommerenke, C., Padé approximant and convergence in capacity, *J. Math. Anal. Appl.*, 41, 775 - 780, 1973.

[62] Barone, P., On the distribution of poles of Padé approximants to the Z-transform of complex Gaussian white noise, *J. Approx. Theory*, 132, 224 - 240, 2005.

[63] O'Sullivan, E.A., and Winfrey, W.R., and Cowan, C.F.N., Padé-Fourier methods for music transposition, *15th International Conference on Mathematics in Digital Signal Processing*, The Institute of Mathematics and its Applications, 1, 543 - 546, 2007.

[64] O'Sullivan, E.A., and Cowan, C.F.N., Modeling room transfer functions using the decimated Padé approximant, *Signal Processing IET*, 2, 49 - 58, 2008.

[65] Becuwe, S., and Cuyt, A., On the Froissart phenomenon in multivariate homogeneous Padé approximation, *Adv. Comput. Math.*, 11, 21 - 40, 1999.

[66] Basdevant, J.L., Bessis, D., and Zinn-Justin, J., Padé approximants in strong interactions: two-body pion and kaon systems, *N. Cimento A*, 60, 185 - 238, 1969.

[67] Zinn-Justin, J., Strong interactions dynamics with Padé approximants *Phys. Rep. C*, 1, 55 – 102, 1971.

[68] Zinn-Justin, J., *Quantum Field Theory and Critical Phenomena*, Cambridge University Press, Cambridge, 4th edn., 1996.

[69] Masjuan, P., and Peris, S., Large N_c QCD and Padé approximant theory, *arXiv:hep-ph:* /0704.12471v1, 2007.

[70] Masjuan, P., and Peris, S., Padé theory applied to the vacuum polarization of a heavy quark, *arXiv:hep-ph* /0903.0294v1, 2008.

[71] Sanz-Cillero, J.J., Vector meson dominance as a first step of a sequence of Padé approximants, *arXiv:hep-ph* /0809.1863v1, 2008.

[72] Masjuan, P., Peris, S., and Sanz-Cillero, J.J., Vector meson dominance as a first step in a systematic approximation: The pion vector form factor, *Phys. Rev. D*, 78, 074028, 2008.

[73] van der Veen, J.W.C., de Beer, R., P.R. Luyten, P.R., and van Ormondt, D., Accurate quantification of *in vivo* ^{31}P NMR signals using the variable projection method and prior knowledge, *Magn. Reson. Med.*, 6, 92 – 98, 1988.

[74] Pijnappel, W.W.F., van den Boogaart, A., de Beer, R., and van Ormondt, D., SVD-based quantification of magnetic resonance signals, *J. Magn. Reson.*, 97, 122 – 134, 1992.

[75] Vanhamme, L., van den Boogaart, A., and van Haffel, S., Improved method for accurate and efficient quantification of MRS data with use of prior knowledge, *J. Magn. Reson.*, 129, 35 – 43, 1997.

[76] Provencher, S.W., Estimation of metabolite concentrations from localized *in vivo* proton NMR spectra, *Magn. Reson. Med.*, 30, 672 – 679, 1993.

[77] Press, W.H., Teukolsky, S.A., Vetterling, W.T. and Flannery, B.P., *Numerical Recipes in Fortran 77: The Art of Scientific Computing*, 2nd edn, Cambridge University Press, Cambridge, 1992.

[78] Istratov, A.A., and Vyvenko, O.F., Exponential analysis in physical phenomena, *Rev. Sci. Instr.*, 70, 1233 – 1257, 1999.

[79] Govindaraju, V., Young, K., and Maudsley, A.A., Proton NMR chemical shifts and coupling constants for brain metabolites, *NMR Biomed.*, 13, 129 – 153, 2000.

[80] Zandt, H.J.A., van der Graaf, M., and Heerschap, A., Common processing of *in vivo* MR spectra, *NMR Biomed.*, 14, 224 – 232, 2001.

[81] Vanhamme, L., Sundin, T., van Hecke, P., and van Huffel, S., MR spectroscopy quantitation: a review of time-domain methods, *NMR Biomed.*, 14, 233 – 246 , 2001.

[82] Miwerisová, Š., and Ala-Korpela, M., MR spectroscopy quantitation: a review of frequency domain methods, *NMR Biomed.*, 14, 247 – 259, 2001.

[83] Pullet, J.-B., Sima, D.M., and van Huffel, S., MRS signal quantitation: A review of time- and frequency-domain methods, *J. Magn. Reson.*, 195, 134 – 144, 2008.

[84] Porat, B., *A Course in Digital Signal Processing*, Wiley, New York, 1997.

[85] Sidi, A., *Practical Extrapolation Methods: Theory and Applications*, Cambridge University Press, Cambridge, England, 2003.

[86] McEliece, R.J., and Shearer, J.B., A property of Euclid's algorithm and an application to Padé approximation, *SIAM J. Appl. Math.*, 34, 611 – 615, 1978.

[87] Palmer, D.R., and Cruz, J.R., An ARMA spectral analysis technique based on the fast Euclid algorithm, *IEEE Trans. Acoust. Speech Sign. Process.*, 37, 1532 – 1536, 1989.

[88] Frahm, J., Bruhn, H., Gyngell, M.L., Merboldt, K.D., Hänicke, W., and Sauter, R., Localized high-resolution proton NMR spectroscopy using stimulated echoes: initial applications to human brain *in vivo*, *Magn. Reson. Med.*, 9, 79 – 93, 1989.

[89] Thomas, M.A., Ryner, L.N., Mehta, M.P., Turski, P.A., and Sorenson, J.A., Localized 2D J-resolved ^{1}H MR spectroscopy of human brain tumors *in vivo*, *J. Magn. Reson. Imag.*, 6, 453 – 459, 1996.

[90] Breit, G., and Wigner, E., Capture of slow neutrons, *Phys. Rev.*, 49, 519 – 531, 1936.

[91] Siegert, A.J.F., On the derivation of the dispersion formula for nuclear reactions, *Phys. Rev.*, 56, 750 – 752, 1939.

[92] Belkić, Dž., Asymptotic convergence in quantum scattering theory, *J. Comput. Meth. Sci. Eng.*, 4, 353 – 496, 2001.

[93] Belkić, Dž., *Principles of Quantum Scattering Theory*, Institute of Physics Publishing, Bristol, 2003.

[94] Lorentz, H.A., The width of spectral lines, *Proc. Roy. Acad. Amsterdam*, 13, 134 – 150, 1914.

[95] Voigt, W., Über das Gasetz der Intensitätsverteilung innerhalb der Linien eines Gasspektrums, *Münch. Ber.*, 603 – 620, 1912.

[96] Unsöld, A., *Physik der Sternatmosphären*, Springer, Berlin, 1968.

[97] Kilekopf, J.F., New approximation to the Voigt function with application to spectral-line profile analysis, *J. Opt. Soc. Am.*, 63, 987 - 995, 1973.

[98] Freise, A., Spencer, E., Marshall, I., and Higinbotham, J., A comparison of frequency and time domain fitting of ^{1}H MRS metabolic data, *Bull. Magn. Reson.*, 17, 302 – 303, 1995.

[99] Bruce, S.D., Higinbotham, J., Marshall, I., and Beswick, P.H., An analytical derivation of a popular approximation of the Voigt function for quantification of NMR spectra, *J. Magn. Reson.*, 142, 57 – 63, 2000.

[100] Gautschi, W., Efficient computation of the complex error function, *SIAM J. Numer. Anal.*, 7, 187 – 198, 1970.

[101] Kwock, L., Smith, J.K., Castillo, M., Ewend, M.G., Collichio, F., Morris, D.E., Bouldin T.W., and Cush, S., Clinical role of proton magnetic resonance spectroscopy in oncology: brain, breast and prostate cancer, *Lancet Oncol.*, 7, 859 – 868, 2006.

[102] Hričak, H., MR imaging and MR spectroscopic imaging in the pre-treatment evaluation of prostate cancer, *Br. J. Radiol.*, 78 (Suppl.), 103 – 111, 2005.

[103] Huzjan, R., Sala, E., and Hričak, H., Magnetic resonance imaging and magnetic resonance spectroscopic imaging of prostate cancer, *Nature Clin. Practice Urology*, 2, 434 – 442, 2005.

[104] Katz, S., and Rosen, M., MR imaging and MR spectroscopy in prostate cancer management, *Radiol. Clin. N. Am.* 44, 723 – 734, 2006.

[105] Bezabeh, T., Odlum, O., Nason, R., Kerr, P., Sutherland, D., Pael, R., and Smith, I.C.P., Prediction of treatment response in head and neck cancer by magnetic resonance spectroscopy, *Am. J. Neuroradiol.*, 26, 2108–2113, 2005.

[106] Bolan, P.J., Nelson, M.T., Yee, D., and Garwood, M., Imaging in breast cancer: Magnetic resonance spectroscopy, *Breast Cancer Res.*, 7, 149 – 152, 2005.

[107] Evelhoch, J., Garwood, M., Vigneron, D., Knopp, M., Sullivan, D., Menkens, A., Clarke, L., and Liu, G., Expanding the use of magnetic resonance in the assessment of tumor response to therapy, *Cancer Res.*, 65, 7041 – 7044, 2005.

[108] Hollingworth, W., Medina, L.S., Lenkinski, R.E., Shibata, D.K., Bernal, B., Zurakowski, D., Comstock, B., and Jarvik, J.G., A systematic literature review of magnetic resonance spectroscopy for the characterization of brain tumors, *Am. J. Neuroradiol.*, 27, 1404 – 1411, 2006.

[109] King, A.D., Yeung, D.K.W., Ahuja, A.T., Tse, G.M.K., Yuen, H.Y., Wong, K.T., and van Hasselt, A.C., Salivary gland tumors at *in vivo* MR spectroscopy, *Radiology*, 237, 563 – 569, 2005.

[110] King, A.D., Yeung, D.K.W., Ahuja, A.T., Tse, G.M.K., Chan, A.B.W., Lam, S.S.L., and van Hasselt, A.C., *In vivo* ^{1}H MR spectroscopy of thyroid cancer, *Eur. J. Radiol.*, 54, 112 – 117, 2005.

[111] Mankoff, D., Imaging in breast cancer – revisited, *Breast Cancer Res.*, 7, 276 – 278, 2005.

[112] Payne, G.S., and Leach, M.O., Applications of magnetic resonance spectroscopy in radiotherapy treatment planning, *Br. J. Radiol.*, 79 (Suppl.), 16 – 26, 2006.

[113] Sibtain, N.A., Howe, F.A., and Saunders, D.E., The clinical value of proton magnetic resonance spectroscopy in adult brain tumors, *Clin. Radiol.*, 62, 109 – 119, 2007.

[114] Belkić, K., and Belkić, Dž., Spectroscopic imaging through magnetic resonance for brain tumor diagnostics, *J. Comp. Meth. Sci. Eng.*, 4, 157 – 207, 2004.

[115] Nelson, S., Multi-voxel magnetic resonance spectroscopy of brain tumors, *Mol. Cancer Ther.*, 2, 497 – 507, 2003.

[116] Katz-Brull, R., Lavin, P.T., and Lenkinski, R.E., Clinical utility of proton magnetic resonance spectroscopy in characterizing breast lesions, *J. Natl. Cancer Inst.*, 9, 1197 – 1203, 2002.

[117] Bartella, H.L., Morris, E.A., Dershaw, D.D., Liberman, L., Thakur, S.B., Moskowitz, C., Guido, J., and Huang, W., Proton MR spectroscopy with choline peak as malignancy marker improves positive predictive value for breast cancer diagnosis: preliminary study, *Radiology*, 239, 686 – 692, 2006.

[118] Meisamy, S., Bolan, P.J., Baker, E.H., Bliss, R.L., Gulbahce, E., Everson, L.I., Nelson, M.T., Emory, T.H., Tuttle, T.M., Yee D., and Garwood M., Neoadjuvant chemotherapy of locally advanced breast cancer: predicting response with *in vivo* ^{1}H MR spectroscopy – a pilot study at 4T, *Radiology*, 233, 424 – 431, 2004.

[119] Maudsley, A., Can MR spectroscopy ever be simple and effective? *Am. J. Neuroradiol.*, 69, 2167, 2005.

[120] Wirestam, R., and Ståhlberg, F., Wavelet-based noise reduction for improved deconvolution of time-series data in dynamic susceptibility-contrast MRI, *Magn. Reson. Mater. Phys. Biol. Med.*, 18, 113 – 118, 2005.

[121] Bessis, D., and Talman, J.D., Variational approach to the theory of operator Padé approximant, *Rocky Mount. J. Math.*, 4, 151 – 158, 1974.

[122] Bessis, D., Mary, P., and Turchetti, G., Variational bounds from matrix Padé approximants in potential scattering, *Phys. Rev. D*, 15, 2345 – 2353, 1977.

[123] Hu, N., On the application of Heisenberg's theory of S-matrix to the problems of resonance scattering and reactions in nuclear physics, *Phys. Rev.*, 74, 131 – 140, 1948.

[124] Hylleraas, E.A., and Undheim, B., Numerische Berechnung der 2S-Terme von Ortho- und Par-Helium, *Z. Phys.*, 65 759 – 772, 1930.

[125] Mandelshtam, V.A., and Taylor, H.S., Spectral projection approach to the quantum scattering calculations, *J. Chem. Phys.*, 102, 7390 – 7399, 1995.

[126] Mandelshtam, V.A., and Taylor, H.S., A simple recursion polynomial expansion of the Green's function with absorbing boundary conditions: Application to the reactive scattering, *J. Chem. Phys.*, 103, 2903 – 2907, 1995.

[127] Chen, R., and Guo, H., The Chebyshev propagator for quantum systems, *Comp. Phys. Commun.*, 119, 19 – 31, 1999.

[128] Tanaka, H., Kunishima, W., and Itoh, M., Efficient scheme to calculate Green functions by recursive polynomial expansion, *RIKEN Rev.*, 29, 20 – 24, 2000.

[129] Kunishima, W., Itoh, M., and Tanaka, H., A new method to calculate the Green function by polynomial expansion, *Progr. Theor. Phys.*, 138 (Suppl.), 149 – 150, 2000.

[130] Kunishima, W., Itoh, M., and Tanaka, H., Generalized polynomial expansion of Green's function with applications to electronic structure calculations, *AIP Conf. Proc.*, 519, 350 – 352, 2000.

[131] Kunishima, W., Tokihiro T., and Tanaka, H., Error estimation of recursive orthogonal polynomial expansion method for large Hamiltonian system, *Comput. Phys. Commun.*, 148, 171 – 181, 2002.

[132] Arfken, G.B., and Weber, H.J., *Mathematical Methods for Physicists*, 5th edn, Academic, New York, 2000.

[133] Gradshteyn, I.S. and Ryzhik, I.M., *Tables of Integrals, Series and Products*, Academic, New York, 1980.

[134] Mandelshtam, V.A, and Taylor, H.S., Harmonic inversion of time signals, *J. Chem. Phys.* 107, 6756 – 6769, 1997 [Erratum, *ibid.*, 109, 4128, 1998].

[135] Mandelstham, V.A., FDM: the filter diagonalization method for data processing in NMR experiments, *Progr. Nucl. Magn. Reson. Spectrosc.*, 38, 159 - 196, 1999.

[136] Schneider, B.I., Accurate basis sets for the calculation of bound and continuum wave functions of the Schrödinger equation, *Phys. Rev. A*, 55, 3417 - 3421, 1997.

[137] Schneider, B.I., and Feder, D.L., Numerical approach to the ground and excited states of a Bose-Einstein condensed gas confined in a completely anisotropic trap, *Phys. Rev. A*, 59, 2232 - 2242, 1999.

[138] Prony, R., Des suites récurrentes, in: *Suite de Leçons d'Analyse de Prony*, Firmin Didot, Libraire pour le Génie, l'Architecture et les Mathématiques, Paris, N° XXIII - XXXII, 1790, pp. 459 - 569.

[139] Prony, R., Essai expérimental et analytique sur les lois de la dilatabilité des fluids élastiques et sur celles de la force expansive de la vapeur d'eau et de la vapeur de l'alkool à différentes températures, *J. de l'École Polytechnique, Paris*, 1, 24 - 76, 1795.

[140] Hazi, A.U., and Taylor, H.S., Stabilization method of calculating resonance energies: model problem, *Phys. Rev. A*, 1, 1109 - 1120, 1970.

[141] Tkáč, I., Andersen, P., Adriany, G., Merkle, H., Uğurbil, K., and Gruetter, R., *In vivo* ^{1}H NMR spectroscopy of the human brain at 7T, *Magn. Reson. Med.*, 46, 451 - 456, 2001.

[142] Cabanes, E., Confort-Gouny, S., Le Fur, Y., Simond, F., and Cozzone, P.J., Optimization of residual water signal by HLSVD on simulated short echo time proton MR spectra of the human brain, *J. Magn. Reson.*, 150, 116 - 125, 2001.

[143] Drost, D.J., Riddle, W.R., and Clarge, G.D., Proton magnetic resonance spectroscopy in the brain: report of AAPM MR task group # 9, *Med. Phys.*, 29, 2177 - 2197, 2002.

[144] Cho, Y.-D., Choi, G.-H., Lee, S-P., and Kim, J.-K., ^{1}H MRS metabolic patterns for distinguishing between meningiomas and other brain tumors, *Magn. Reson. Imag.*, 21, 663 - 672, 2003.

[145] Opstad, K.S., Provencher, S.W., Bell, B.A., Griffiths, J.R., and Howe, F.A., Detection of elevated glutathione in meningiomas by quantitative *in vivo* ^{1}H MRS, *Magn. Reson. Med.*, 49, 632 - 637, 2003.

[146] Lanczos, C., An iteration method for the solution of the eigenvalue problem of linear differential and integral operators, *J. Res. Nat. Bur. Stand.*, 45, 255 - 282, 1950.

[147] Lanczos, C., Solution of systems of linear equations by minimized iterations, *J. Res. Nat. Bur. Stand.*, 49, 33 – 53, 1952.

[148] Lanczos, C., *Applied Analysis*, Prentice-Hall, Inglewood Cliffs, U.S.A., 1956.

[149] Löwdin, P.-O., On the non-orthogonality problem connected with the use of atomic wave functions in the theory of molecules and crystals, *J. Chem. Phys.*, 18, 365 – 375, 1950.

[150] Löwdin, P.-O., Studies of atomic self-consistent fields: I. Calculation of Slater functions, *Phys. Rev.*, 90, 120 – 125, 1953.

[151] Löwdin, P.-O., Studies of atomic self-consistent fields: II. Interpolation problem, *Phys. Rev.*, 94, 1600 – 1609, 1954.

[152] Appel, K., and Löwdin, P.-O., Studies of self-consistent fields: III. Analytic wave functions for the argon-like ions and for the first row of the transition metals, *Phys. Rev.*, 103, 1746 – 1755, 1956.

[153] Löwdin, P.-O., Exchange, correlation and spin effects in molecular and solid-state theory, *Rev. Mod. Phys.*, 34, 80 – 97, 1962.

[154] Löwdin, P.-O., The normal constants of motion in quantum mechanics treated by projection technique, *Rev. Mod. Phys.*, 34, 520 – 530, 1962.

[155] Löwdin, P.-O., On the state of the art of quantum chemistry, *Int. J. Quantum Chem.*, 29, 1651 – 1683, 1986.

[156] Rutishauser, H., *Der Quotienten-Differenzen Algorithmus*, Birkhäuser, Basel & Stuttgart, 1957.

[157] Henrici, P., Some applications of the quotient-difference algorithm, *Proc. Symp. Appl. Math.*, 15, 159 – 183, 1963.

[158] Gordon, R.G., Error bounds in equilibrium statistical mechanics, *J. Math. Phys.*, 9, 655 – 663, 1968.

[159] Wheeler, J.C., Modified moments and Gaussian quadratures, *Rocky Mountain J. Math.*, 4, 287 – 296, 1974.

[160] Wheeler, J.C., and Gordon, R.G., in *The Padé Approximant in Theoretical Physics*, Eds. Baker G.A., and Gammel, J.L. Academic, New York, 1970, pp. 99 – 128.

[161] Reid, C.E., Transformation of perturbation series into continued fractions, with application to an anharmonic oscillator, *Int. J. Quantum Chem.*, 1, 521 – 534, 1967.

[162] Goscinski, O., and Brändas, E., Dispersion forces, second- and third-order energies, *Chem. Phys. Lett.*, 2, 299 – 302, 1968.

[163] Brändas, E., and Goscinski, O., Variation-perturbation expansions and Padé approximants to energy, *Phys. Rev. A*, 1, 552 – 560, 1970.

[164] Goscinski, O., and Brändas, E., Padé approximants to physical properties via inner projections, *Int. J. Quantum Chem.*, 5, 131 – 156, 1971.

[165] Brändas, E., and Bartlett, R.J., Reduced partitioning technique for configuration interaction calculations using Padé approximants and inner-projections, *Chem. Phys. Lett.*, 8, 153 – 156, 1971.

[166] Micha, D.A., and Brändas, E., Variational methods in wave operator formalism – unified treatment for bound and quasi-bound electronic and molecular states, *J. Chem. Phys.*, 55, 4792 – 4797, 1971.

[167] Brändas, E., and Micha, D.A., Variational methods in wave operator formalism – applications in variation-perturbation theory and theory of energy bounds, *J. Math. Phys.*, 13, 155 – 160, 1972.

[168] Bartlett, R.J., and Brändas, E., Reduced partitioning procedure in configuration interaction studies: 1. Ground-states, *J. Chem. Phys.*, 56, 5467 – 5477, 1972.

[169] Brändas, E., and Goscinski, O., Darboux functions and power series expansions with examples from isoelectronic sequences, *Int. J. Quantum Chem.*, 6, 59 –72, 1972.

[170] Bartlett, R.J., and Brändas, E., Reduced partitioning procedure in configuration interaction studies: 2. Excited states, *J. Chem. Phys.*, 59, 2032 – 2042 , 1973.

[171] Baker, G.A., and Gammel, J.L., *The Padé Approximant in Theoretical Physics*, Academic, New York, 1970.

[172] Baker, G.A., *Essentials of the Padé Approximants*, Academic, New York, 1975.

[173] Baker, G.A., and Graves-Morris, P., *Padé Approximants* Cambridge University Press, 2nd edn, Cambridge, 1996.

[174] Longman, I.M., Computation of Padé table, *Int. J. Comput. Math. B*, 3, 53 – 64, 1971.

[175] Levin, D., Development of nonlinear transformations for improving convergence of sequences, *Int. J. Comput. Math.*, 3, 371 – 388, 1973.

[176] Belkić, Dž., New hybrid non-linear transformations of divergent perturbation series for quadratic Zeeman effects, *J. Phys. A*, 22, 3003 – 3010, 1989.

[177] Weniger, E.J., Nonlinear sequence transformations for the acceleration of convergence and the summation of divergent series, *Comput. Phys. Rep.*, 10, 191 – 371, 1989.

[178] Chisholm, J.S.R., Genz, A.C., and Pusterla, M., A method for computing Feynman amplitudes with branch cuts, *J. Comput. Appl. Math.*, 2, 73 – 76, 1976.

[179] Claverie, P., Denis, A., and Yeramian, E., The representation of functions through the combined use of integral transforms and Padé approximants – Padé-Laplace analysis of functions as sums of exponentials, *Comput. Phys. Rep.*, 9, 249 – 299, 1989.

[180] Feldman P., and Freund, R., Efficient linear circuit analysis by Padé approximation via the Lanczos process, *IEEE Trans. Computer-Aided Design Integr. Circuits Syst.*, 14, 639 – 649, 1995.

[181] Freund, R, and Feldman, P., Structure-preserving model order reduction of RCL circuit equations, *IEEE Trans. Circuits Syst. Anal. Digit. Sign. Process.*, 43, 577 – 585, 1996.

[182] Levin, D., and Sidi, A., Extrapolation methods for infinite multiple series and integrals, *J. Comp. Meth. Sci. Eng.*, 1, 167 – 184, 2001.

[183] Barone, P., and March, R., A novel class of Padé based methods in spectral analysis, *J. Comp. Meth. Sci. Eng.*, 1, 185 – 211, 2001.

[184] Alejos, Ó., de Francisco, C., Muñoz, J.M., Hernández-Gómez, P., and Torres, C., Overcoming noise sources in multi-exponential fitting: A comparison of different algorithms, *J. Comp. Meth. Sci. Eng.*, 1, 213 – 228, 2001.

[185] Grotendorst, J., Maple programs for converting series expansions to rational functions using the Levin transformation – automatic generation of Fortran functions for numerical applications, *Comput. Phys. Commun.*, 55, 325 – 335, 1989.

[186] Grotendorst, J., Approximating functions by means of symbolic computation and a general extrapolation method, *Comput. Phys. Commun.*, 59, 289 – 301, 1990.

[187] Grotendorst, J., A Maple package for transforming series, sequences and functions, *Comput. Phys. Commun.*, 67, 325 – 342, 1991.

[188] Driscoll, T.A., and Fornberg, B., A Padé-based algorithm for overcoming the Gibbs phenomenon, *Num. Algor.*, 26, 77 – 92, 2001.

[189] Neuhauser, D., The application of optical potentials for reactive scattering – a case study, *J. Chem. Phys.*, 93, 2611 – 2616, 1990.

[190] Wall, M.R., and Neuhauser, D., Extraction, through filter diagonalization, of general quantum eigenvalues or classical normal mode frequencies from a small number of residues or a short time segment of a signal: I. Theory and application to a quantum dynamics model, *J. Chem. Phys.*, 102, 8011 – 8022, 1995.

[191] Roy, P.-N., and Carrington, T. Jr., An evaluation of methods designed to calculate energy levels in a selected range and application to a (one-dimensional) Morse oscillator and (3-dimensional) HCN/HNC, *Chem. Phys.*, 103, 5600 - 5612, 1995.

[192] Huang, S.-W., and Carrington, T. Jr., A comparison of filter diagonalization methods with the Lanczos method for calculating vibrational energy levels, *Chem. Phys. Lett.*, 312, 311 - 318, 1999.

[193] Chen, R., and Guo, H., Symmetry-enhanced spectral analysis via the spectral method and filter diagonalization, *Phys. Rev. E*, 57, 7288 - 7293, 1998.

[194] Beck, M.H., and Meyer, H.-D., Extracting accurate bound-state spectra from approximate wave packet propagation using the filter diagonalization method, *J. Chem. Phys.*, 109, 3730 - 3741, 1998.

[195] Vijay, A., and Wyatt, R.E., Spectral filters in quantum mechanics: A measurement theory perspective, *Phys. Rev. E*, 62, 4351 - 4364, 2000.

[196] Mandelshtam, V.A., and Carrington, T. Jr., Comment on Spectral filters in quantum mechanics: A measurement theory perspective, *Phys. Rev. E*, 65, 028701, 2002.

[197] Vijay, A., Reply to Comment on Spectral filters in quantum mechanics: A measurement theory perspective, *Phys. Rev. E*, 65, 028702, 2002.

[198] Nyman, G., and Yu, H.-Y., Iterative diagonalization of a large sparse matrix using spectral transformation and filter diagonalization, *J. Comp. Meth. Sci. Eng.*, 1, 229 - 250, 2001.

[199] Guo, H., Lanczos approach to molecular spectroscopy without explicit calculation of eigenfunctions, *J. Comp. Meth. Sci. Eng.*, 1, 251 - 265, 2001.

[200] Levitina, T., and Brändas, E.J., Perturbed ellipsoidal wave functions for quantum scattering, *Int. J. Quantum Chem.*, 70, 1017 - 1022, 1998.

[201] Larsson, B., Levitina, T., and Brändas, E.J., On prolate spheroidal wave functions for signal processing, *Int. J. Quantum Chem.*, 85, 392 - 397, 2001.

[202] Levitina, T.V., and Brändas, E.J., Computational techniques for prolate spheroidal wave functions in signal processing, *J. Comp. Meth. Sci. Eng.*, 1, 287 - 313, 2001.

[203] Levitina, T., and Brändas, E.J., Filter diagonalization with finite Fourier transform eigenfunctions, *J. Math. Chem.*, 40, 43 - 47, 2006.

[204] Taswell, C., Experiments in wavelet shrinkage denoising, *J. Comp. Meth. Sci. Eng.*, 1, 315 - 326, 2001.

[205] Antoine, J.P., and Coron, A., Time-frequency and time-scale approach to magnetic resonance spectroscopy, *J. Comp. Meth. Sci. Eng.*, 1, 327 – 352, 2001.

[206] Palumbo, B., Brain tumor recurrence: brain single-photon emission computerized tomography, PET and proton magnetic resonance spectroscopy, *Nucl. Med. Comm.*, 29, 730 – 735, 2008.

[207] Howe, F.A., and Opstad, K.S., ^{1}H spectroscopy of brain tumors and masses, *NMR Biomed.*, 16, 123 – 131, 2003.

[208] Li, A.E., and Bluemke, D.A., Magnetic resonance imaging in: de Vita, V.T., Hellman, S., and Rosenberg, S.A., *Cancer Principles & Practice of Oncology*, 6th edn, Lippincott Williams & Wilkins, Philadelphia, U.S.A, 2001, pp. 669 – 679.

[209] Levin, A., Leibel, S.A., and Gutin, P.H., Neoplasms of the central nervous system, in: de Vita, V.T., Hellman, S., and Rosenberg, S.A., *Cancer Principles & Practice of Oncology*, 6th edn, Lippincott Williams & Wilkins, Philadelphia, U.S.A, 2001, pp. 2100 – 2160.

[210] Tamraz, J.C., Outin, C., Forjaz Secca, M., and Soussi, B., *MRI Principles of the Head, Skull Base and Spine: A Clinical Approach*, Springer, Paris, France, 2003.

[211] Brandão, L.A., and Domingues, R.C., *MR Spectroscopy of the Brain*, Lippincott Williams & Wilkins, Philadelphia, U.S.A, 2004.

[212] Gudowius, S., Engelbrecht, V., Messing-Jünger, M., Reifenberger, G., and Gärtner J., Diagnostic difficulties in childhood bilateral thalamic astrocytomas, *Neuropediatrics*, 33, 331 – 335, 2002.

[213] Walecki, J., Tarasow, E., Kubas, B., Czemicki, Z., Lewko, J., Podgórski, J., Sokól, M., and Grieb, P., ^{1}H MR spectroscopy of the peritumoral zone in patients with cerebral glioma: assessment of the value of the method, *Acta Radiol.*, 10, 145 – 153 2003.

[214] Kadota, O., Kohno, K., Ohue, S., Kumon, Y., Sakaki, S., Kikuchi, K., and Miki, H., Discrimination of brain abscess and cystic tumor by *in vivo* proton magnetic resonance spectroscopy, *Neurol. Med. Chir. (Tokyo)*, 41, 121 – 126, 2001.

[215] McKnight, T.R., von dem Bussche, M.H., Vigneron, D.B., Lu, Y., Berger, M.S., McDermott, M.W., Dillon, W.P., Graves, E.E., Pirzkall, A., and Nelson, S.J., Histopathological validation of a three-dimensional magnetic resonance spectroscopy index as a predictor of tumor presence, *J. Neurosurg.*, 97, 794 – 802, 2002.

[216] Vigneron, D., Bollen, A., McDermott, M., Wald, L., Day, M., Moyher-Noworolski, S., Henry, R., Chang, S., Berger, M., Dillon, W.,

and Nelson, S., Three-dimensional magnetic resonance spectroscopic imaging of histologically confirmed brain tumors, *Magn. Reson. Imag.*, 19, 89 – 101, 2001.

[217] Nakaiso, M., Uno, M., Harada, M., Kageji, T., Takimoto, O., and Nagahiro, S., Brain abscess and glioblastoma identified by combined proton magnetic resonance spectroscopy and diffusion-weighted magnetic resonance imaging – two case reports, *Neurol. Med. Chir. (Tokyo)*, 42, 346 – 348, 2002.

[218] Bulakbasi, N., Kocaoglu, M., Ors, F., Tayfun, C, and Uçöz, T., Combination of single-voxel proton MR spectroscopy and apparent diffusion coefficient calculation in the evaluation of common brain tumors, *Am. J. Neuroradiol.*, 24, 255 – 233, 2003.

[219] Sijens, P.E., and Oudkerk, M., ^{1}H chemical shift imaging characterization of human brain tumor and edema, *Eur. Radiol.*, 12, 2056 – 2061, 2002.

[220] Utriainen, M., Komu, M., Vuorinen, V., Lehikoinen, P., Sonninen, P., Kurki, T., Utriainen, T., Roivainen, A., Kalimo, H., and Minn, H., Evaluation of brain tumor metabolism with ^{11}C choline PET and ^{1}H MRS, *J. Neuro-Oncol.*, 62, 329 – 338, 2003.

[221] Croteau, D., Scarpace, L., Hearshen, D., Gutierrez, J., Fisher, J.L., Rock, J.P., and Mikkelsen, T., Correlation between magnetic resonance spectroscopy imaging and image-guided biopsies: Semi-quantitative and qualitative histopathological analyses of patients with untreated glioma, *Neurosurgery*, 49, 823 – 829, 2001.

[222] Dowling, C., Bollen, A.W., Noworolski, S.M., McDermott, M.W., Barbaro, N.M., Day, M.R., Henry, R.G., Chang, S.M., Dillon, W.P., Nelson, S.J., and Vigneron, D.B., Preoperative proton MR spectroscopic imaging of brain tumors: correlation with histopathologic analysis of resection specimens, *Am. J. Neuroradiol.*, 22, 604 – 612, 2001.

[223] Nagar, V.A. Ye, J., Xu, M., Ng, W.H., Yeo, T.T., Ong, P.L., and Lim, C.C., Multi-voxel MR spectroscopic imaging – distinguishing intracranial tumors from non-neoplastic disease, *Ann. Acad. Med. Singapore*, 36, 309 – 313, 2007.

[224] Porto, L., Hattingen, E., Pilatus, U., Kieslich, M., Yan, B., Schwabe, D., Zanella, F.E., and Lanfermann, H., Proton magnetic resonance spectroscopy in childhood brainstem lesions, *Childs Nerv. Syst.*, 23, 305 – 314, 2007.

[225] Tarnawski, R., Sokol, M., Pieniazek, P., Maciejewski, B., Walecki, J., Miszczyk, L., and Krupska, T., ^{1}H MRS *in vivo* predicts the early

treatment outcome of postoperative radiotherapy for malignant gliomas, *Int. J. Radiat. Oncol. Biol. Phys.*, 52, 1271 – 1276, 2002.

[226] Lin, A.P., and Ross, B.D., Short-echo time proton MR spectroscopy in the presence of gadolinium, *J. Comput. Assist. Tomogr.*, 25, 705 – 712, 2001.

[227] Hourani, R., Brant, L.J., Rizk, T., Weingart, J.D., Barker, P.B., and Horska, A., Can proton MR spectroscopic and perfusion imaging differentiate between neoplastic and non-neoplastic brain lesions in adults? *Am. J. Neuroradiol.*, 29, 366 – 372, 2008.

[228] Majós, C., Aguilera, C., Alonso, J., Julia-Sape, M., Castaner, S., Sanchez, J.J., Samitier, A., Leon, A., Rovira, A., and Arús, C., Proton MR spectroscopy improves discrimination between tumor and pseudo-tumoral lesion in solid brain masses, *Am. J. Neuroradiol.*, 30, 544 – 551, 2009.

[229] Hall, W.A., Martin, A., Liu, H., and Truwit, C.L., Improving diagnostic yield in brain biopsy: coupling spectroscopic targeting with real-time needle placement, *J. Magn. Reson. Imag.*, 13, 12 – 15, 2001.

[230] Hall, W.A., Liu, H., Maxwell, R.E, and Truwit, C.L., Influence of 1.5 tesla intraoperative MR imaging on surgical decision-making, *Acta Neurochir.*, 85 (Suppl.), 29 – 37, 2003.

[231] Hermann, E.J., Hattingen, E., Krauss, J.K., Marquardt, G., Pilatus, U., Franz, K., Setzer, M., Gasser, T., Tews, D.S., Zanella, F.E., Seifert, V., and Lanfermann, H., Stereotactic biopsy in gliomas guided by 3 tesla ^{1}H chemical-shift imaging of choline, *Stereotact. Funct. Neurosurg.*, 86, 300 – 307, 2008.

[232] Ng, W.H., and Lim, T., Targeting regions with highest lipid content on MR spectroscopy may improve diagnostic yield in stereotactic biopsy, *J. Clin. Neurosci.*, 15, 502 – 506, 2008.

[233] Saindane, A.M., Cha, S., Law, M., Xue, H., Knopp, E.A., and Zagzag, D., Proton MR spectroscopy of tumefactive demyelinating lesions, *Am. J. Neuroradiol.*, 23, 1378 – 1386, 2002.

[234] Smith, J.K., Londono, A., Castillo, M., and Kwock, L., Proton magnetic resonance spectroscopy of brain-stem lesions, *Neuroradiology*, 44, 825 – 829, 2002.

[235] Fan, G., Sun, B., Wu, Z., Guo, Q., and Guo, Y., *In vivo* single-voxel proton MR spectroscopy in the differentiation of high-grade gliomas and solitary metastases, *Clin. Radiol.*, 59, 77 – 85, 2004.

[236] Barba, I., Moreno, A., Martinez-Pérez, I., Tate, A.R., Cabañas, M.E., Baquero, M., Capdevila, A., and Arús, C., Magnetic resonance

spectroscopy of brain hemangiopericytomas: high myoinositol concentrations and discrimination from meningiomas, *J. Neurosurg.*, 94, 55 – 60, 2001.

[237] Sener, R.N., Astroblastoma: diffusion MRI, and proton MR spectroscopy, *Comput. Med. Imag. Graph.*, 26, 187 – 191, 2002.

[238] Hattingen, E., Raab, P., Franz, K., Zanella, F.E., Lanfermann, H., and Pilatus, U., Myo-inositol: a marker of reactive astrogliosis in glial tumors? *NMR Biomed.*, 21, 233 – 241, 2008.

[239] Möller-Hartmann, W., Herminghaus, S., Krings, T., Marquardt, G., Lanfermann, H., Pilatus, U., and Zanella, F.E., Clinical application of proton magnetic resonance spectroscopy in the diagnosis of intracranial mass lesions, *Neuroradiology*, 44, 371 – 381, 2002.

[240] Kubas, B., Tarasów, E., Dzienis, W., Łebkowski, W., Zimnoch, L., Dzieciol, J., Siergiejczyk, L., Walecki, J., and Lewko, J., Magnetic resonance proton spectroscopy in neuro-oncology – preliminary report, *Neurol. Neurochir. Pol.*, 35 (Suppl. 5), 90 – 100, 2001.

[241] Lai, P.H., Weng, H.H., Chen, C.Y., Hsu, S.S., Ding, S., Ko, C.W., Fu, J.H., Liang, H.L., and Chen, K.H., *In vivo* differentiation of aerobic brain abscesses and necrotic glioblastomas multiforme using proton MR spectroscopic imaging, *Am. J. Neuroradiol.*, 29, 1511–1518, 2008.

[242] Li, X., Lu, Y., Pirzkall, A., McKnight, T., and Nelson, S.J., Analysis of the spatial characteristics of metabolic abnormalities in newly diagnosed glioma patients, *J. Magn. Reson. Imag.*, 16, 229 – 237, 2002.

[243] Toyooka, M., Kimura, H., Uematsu, H., Kawamura, Y., Takeuchi, H., and Itoh, H., Tissue characterization of glioma by proton magnetic resonance spectroscopy and perfusion-weighted magnetic resonance imaging: glioma grading and histological correlation, *Clin. Imag.*, 32, 251 – 258, 2008.

[244] Spampinato, M.V., Smith, J.K. Kwock, L., Ewend, M., Grimme, J.D., Camacho, D.L., and Castillo, M., Cerebral blood volume measurements and proton MR spectroscopy in grading of oligodendroglial tumors, *Am. J. Roentgenol.*, 188, 204 – 212, 2007.

[245] Fayed, N., Davila, J., Medrano, J., and Olmos, S., Malignancy assessment of brain tumors with magnetic resonance spectroscopy and dynamic susceptibility contrast MRI, *Eur. J. Radiol.*, 67, 427 – 433, 2008.

[246] Nelson, S.J., McKnight, T.R., and Henry, R.G., Characterization of untreated gliomas by magnetic resonance spectroscopic imaging, *Neuroimaging Clin. N. Am.*, 12, 599 – 613, 2002.

[247] Rijpkema, M., Schuuring, J., van der Meulen, Y., van der Graaf, M., Bernsen, H., Boerman, R., van der Kogel, A., and Heerschap, A., Characterization of oligodendrogliomas using short echo time ^{1}H MR spectroscopic imaging, *NMR Biomed.*, 16, 12 – 18, 2003.

[248] Zakrzewski, K., Kreisel, J., Polis, L., Nowosławska, E., Liberski, P.P., and Biegański, T., Clinical application of proton magnetic resonance spectroscopy for differential diagnosis of pediatric posterior fossa tumors, *Neurol. Neurochir. Pol.*, 35 (Suppl. 5), 19 – 25, 2001.

[249] Herminghaus, S., Dierks, T., Pilatus, U., Möller-Hartmann, W., Wittsack, J., Marquardt, G., Labisch, C., Lanfermann, H., Schlote, W., and Zanella, F.E., Determination of histopathological tumor grade in neuroepithelial brain tumors by using spectral pattern analysis of *in vivo* spectroscopic data, *J. Neurosurg.*, 98, 74 – 81, 2003.

[250] Nafe, R., Herminghaus, S., Raab, P., Wagner, S., Pilatus, U., Schneider, B., Schlote, W., Zanella, F., and Lanfermann, H., Preoperative proton-MR spectroscopy of gliomas – correlation with quantitative nuclear morphology in surgical specimen, *J. Neuro-Oncol.*, 63, 233 – 245, 2003.

[251] Murphy, M., Loosemore, A., Clifton, A.G., Howe, F.A., Tate, A.R., Cudlip, S.A., Wilkins, P.R., Griffiths, J.R., and Bell, B.A., The contribution of proton magnetic resonance spectroscopy (^{1}H MRS) to clinical brain tumor diagnosis, *Br. J. Neurosurg.*, 16, 329 – 334, 2002.

[252] Kaminogo, M., Ishimaru, H., Morikawa, M., Ochi, M., Ushijima, R., Tani, M., Matsuo, Y., Kawakubo, J., and Shibata, S., Diagnostic potential of short echo time MR spectroscopy of gliomas with single-voxel and point-resolved spatially localized proton spectroscopy of brain, *Neuroradiol.*, 43, 353 – 363, 2001.

[253] Murphy, P.S., Rowland, I.J., Viviers, L., Brada, M., Leach, M.O., and Dzik-Jurasz, A.S., Could assessment of glioma methylene lipid resonance by *in vivo* ^{1}H MRS be of clinical value? *Br. J. Radiol.* 76, 459 – 463, 2003.

[254] Panigrahy, A., Nelson, M.D. Jr., Finlay, J.L., Sposto, R., Krieger, M.D., Gilles, F.H., and Bluml, S., Metabolism of diffuse intrinsic brainstem gliomas in children, *Neuro-Oncology*, 10, 32 – 44, 2008.

[255] Tzika, A.A., Astrakas, L.G., Zarifi, M.K., Petridou, N., Young-Poussaint, T., Goumnerova, L., Zurakowski, D., Anthony, D.C., and Black, P.M., Multi-parametric MR assessment of pediatric brain tumors, *Neuroradiol.*, 45, 1 – 10, 2003.

[256] Tzika, A.A., Zurakowski, D., Poussaint, T.Y., Goumnerova, L., Astrakas, L.G., Barnes, P.D., Anthony, D.C., Billett, A.L., Tarbell,

N.J., Scott, R.M., and Black, P.M., Proton magnetic spectroscopic imaging of the child's brain: the response of tumors to treatment, *Neuroradiol.*, 43, 169 – 177, 2001.

[257] Di Costanzo, A., Scarabino, T., Trojsi, F., Popolizio, T., Catapano, D., Giannatempo, G.M., Bonavita, S., Portaluri, M., Tosetti, M., d'Angelo, V.A., Salvolini, U., and Tedeschi, G., Proton MR spectroscopy of cerebral gliomas at 3T: spatial heterogeneity, and tumor grade and extent, *Eur. Radiol.*, 18, 1727 – 1735, 2008.

[258] Del Sole, A., Falini, A., Ravasi, L., Ottobrini, L., De Marchis, D., Bombardieri, E., and Lucignani, G., Anatomical and biochemical investigation of primary brain tumors, *Eur. J. Nucl. Med.*, 28, 1851 – 1872, 2001.

[259] Danielsen, E.R., and Ross, B., *Magnetic Resonance Spectroscopy Diagnosis of Neurological Diseases*, Marcel Dekker, Inc., New York, U.S.A, 1999.

[260] Majós, C., Alonso, J., Aguilera, C., Serrallonga, M., Acebes, J.J., Arús, C., and Gili, J., Adult primitive neuroectodermal tumor: proton MR spectroscopic findings with possible application for differential diagnosis, *Radiology*, 225, 556 – 566, 2002.

[261] Majós, C., Alonso, J., Aguilera, C., Serrallonga, M., Coll, S., Acebes, J.J., Arús, C., and Gili, J., Utility of proton MR spectroscopy in the diagnosis of radiologically atypical intracranial meningiomas, *Neuroradiology*, 45, 129 – 136, 2003.

[262] Majós, C., Alonso, J., Aguilera, C., Serrallonga, M., Pérez-Martín J, Acebes, J.J., Arús, C., and Gili, J., Proton magnetic resonance spectroscopy (^{1}H MRS) of human brain tumors: assessment of differences between tumor types and its applicability in brain tumor categorization, *Eur. Radiol.*, 13, 582 – 591, 2003.

[263] Howe, F.A., Barton, S.J., Cudlip, S.A., Stubbs, M., Saunders, D.E., Murphy, M., Wilkins, P., Opstad, K.S., Doyle, V.L., McLean, M.A., Bell, B.A., and Griffiths, J.R., Metabolic profiles of human brain tumors using quantitative *in vivo* ^{1}H magnetic resonance spectroscopy, *Magn. Reson. Med.*, 49, 223 – 232, 2003.

[264] Krishnamoorthy, T., Radhakrishnan, V.V., Thomas, B., Jeyadevan, E.R., Menon, G., and Nair, S., Alanine peak in central neurocytomas on proton MR spectroscopy, *Neuroradiology*, 49, 551 – 554, 2007.

[265] Tate, A.R., Majós, C., Moreno, A., Howe, F.A., Griffiths, J.R., and Arús, C., Automated classification of short echo time in *in vivo* ^{1}H brain tumor spectra: a multi-center study, *Magn. Reson. Med.*, 49, 29 – 36, 2003.

[266] Law, M., Cha, S., Knopp, E.A., Johnson, G., Arnett, J., and Litt, A.W., High-grade gliomas and solitary metastases: differentiation by using perfusion and proton spectroscopic MR imaging, *Radiology*, 222, 715 – 721, 2002.

[267] Ricci, R., Bacci, A., Tugnoli, V., Battaglia, S., Maffei, M., Agati, R., and Leonardi, M., Metabolic findings on 3T ^{1}H MR spectroscopy in peritumoral brain edema, *Am. J. Neuroradiol.*, 28, 1287 – 1291, 2007.

[268] Davies, N.P., Wilson, M., Harris, L.M., Natarajan, K., Lateef, S., Macpherson, L., Sgouros, S., Grundy, R.G., Arvanitis, T.N., and Peet, A.C., Identification and characterization of childhood cerebellar tumors by *in vivo* proton MRS, *NMR Biomed.*, 21, 908 – 918, 2008.

[269] Wilson, M., Davies, N.P., Brundler, M.A., McConville, C., Grundy, R.G., and Peet, A.C., High resolution magic angle spinning ^{1}H NMR of childhood brain and nervous system tumors, *Molec. Cancer* 8, 2009 (doi:10.1186/1476-4598-8-6).

[270] Harris, L.M., Davies, N., Macpherson, L., Foster, K., Lateef, S., Natarajan, K., Sgouros, S., Brundler, M.A., Arvanitis, T.N., Grundy, R.G., and Peet, A.C., The use of short-echo-time ^{1}H MRS for childhood cerebellar tumors prior to histopathological diagnosis, *Ped. Radiol.*, 37, 1101 – 1109, 2007.

[271] Pirzkall, A., Nelson, S.J., McKnight, T.R., Takahashi, M.M., Li, X., Graves, E.E., Verhey, L.J., Wara, W.W., Larson, D.A., and Sneed, P.K., Metabolic imaging of low-grade gliomas with three-dimensional magnetic resonance spectroscopy, *Int. J. Radiat. Oncol. Biol. Phys.*, 53, 1254 – 1264, 2002.

[272] Nelson, S.J., Graves, E., Pirzkall, A., Li, X., Chan, A.A., Vigneron, D.B., and McKnight, T.R., *In vivo* molecular imaging for planning radiation therapy of gliomas: an application of ^{1}H MRSI, *J. Magn. Reson. Imag.*, 16, 464 – 476, 2002.

[273] Park, I., Tamai, G., Lee, M.C., Chuang, C.F., Chang, S.M., Berger, M.S., Nelson, S.J., and Pirzkall, A., Patterns of recurrence analysis in newly diagnosed glioblastoma multiforme after three-dimensional conformal radiation therapy with respect to pre-radiation therapy magnetic resonance spectroscopic findings, *Int. J. Radiat. Oncol. Biol. Phys.*, 69, 381 – 389, 2007.

[274] Chang, J., Thakur, S., Perera, G., Kowalski, A., Huang, W., Karimi, S., Hunt, M., Koutcher, J., Fuks, Z., Amols, H., and Narayana, A., Image-fusion of MR spectroscopic images for treatment planning of gliomas, *Med. Phys.*, 33, 32–40, 2006.

[275] Narayana, A., Chang, J., Thakur, S., Huang, W., Karimi, S., Hou, B., Kowalski, A., Perera, G., Holodny, A., and Gutin, P,H., Use of MR

spectroscopy and functional imaging in the treatment planning of gliomas, *Br. J. Radiol.*, 80, 347 – 354, 2007.

[276] Zeng, Q.S., Li, C.F., Liu, H., Zhen, J.H., and Feng, D.C., Distinction between recurrent glioma and radiation injury using magnetic resonance spectroscopy in combination with diffusion-weighted imaging, *Int. J. Radiat. Oncol. Biol. Phys.*, 68, 151 – 158, 2007.

[277] Weybright, P., Sundgren, P.C., Maly, P., Gomez, D., Hassan, B., Nan, S., Rohrer, S., and Junck, L., Differentiation between brain tumor recurrence and radiation injury using MR spectroscopy, *Am. J. Radiol.*, 185, 1471 – 1476, 2005.

[278] Smith, E.A., Carlos, R.C., Junck, L.R., Tsien, C.I., Elias, A., and Sundgren, P.C., Developing a clinical decision model: MR spectroscopy to differentiate between recurrent tumor and radiation change in patients with new contrast-enhancing lesions, *Am. J. Roentgenol.*, 192, W45 – W52, 2009.

[279] Czernicki, T., Szeszkowski, W., Marchel, A., and Golebiowski, M., Spectral changes in postoperative MRS in high-grade gliomas and their effect on patient prognosis, *Folia Neuropathologica*, 47, 43 – 49, 2009.

[280] Sankar, T., Caramanos, Z., Assina, R., Villemure, J.G., Leblanc, R., Langleben, A., Arnold, D.L., and Preul, M.C., Prospective serial proton MR spectroscopic assessment of response to tamoxifen for recurrent malignant glioma, *J. Neuro-Oncology*, 90, 63–76, 2008.

[281] Sijens, P.E., Heesters, M.A., Enting, R.H., van der Graaf, W.T., Potze, J.H., Irwan, R., Meiners, L.C., and Oudkerk, M., Diffusion tensor imaging and chemical shift imaging assessment of heterogeneity in low grade glioma under temozolomide chemotherapy, *Cancer Invest.*, 25, 706 – 710, 2007.

[282] Alimenti, A., Delavelle, J., Lazeyras, F., Yilmaz, H., Dietrich, P.Y., de Tribolet, N., and Lovblad, K.O., Mono-voxel ^{1}H magnetic resonance spectroscopy in the progression of gliomas, *Eur. Neurol.*, 58, 198 – 209, 2007.

[283] Guillevin, R., Menuel, C., Duffau, H., Kujas, M., Capelle, L., Aubert, A., Taillibert, S., Idbaih, A., Pallud, J., Demarco, G., Costalat, R., Hoang-Xuan, K., Chiras, J., and Vallee, J.N., Proton magnetic resonance spectroscopy predicts proliferative activity in diffuse low-grade gliomas, *J. Neuro-Oncology*, 87, 181 – 187, 2008.

[284] Laprie, A., Catalaa, I., Cassol, E., McKnight, T.R., Berchery, D., Marre, D., Bachaud, J.M., Berry, I., and Moyal, E.C., Proton magnetic resonance spectroscopic imaging in newly diagnosed glioblastoma: predictive value for the site of post-radiotherapy relapse

in a prospective longitudinal study, *Int. J. Radiat. Oncol. Biol. Phys.*, 70, 773 – 781, 2008.

[285] Chernov, M.F., Kawamata, T., Amano, K., Ono, Y., Suzuki, T., Nakamura, R., Muragaki, Y., Iseki, H., Kubo, O., Hori, T., and Takakura, K., Possible role of single-voxel ^{1}H MRS in differential diagnosis of suprasellar tumors, *J. Neuro-Oncology*, 91, 191 – 198, 2009.

[286] Tzika, A.A., Astrakas, L.G., Zarifi, M.K., Zurakowski, D., Poussaint, T.Y., Goumnerova, L., Tarbell, N.J., and Black, P.M., Spectroscopic and perfusion magnetic resonance imaging predictors of progression in pediatric brain tumors, *Cancer*, 15, 1246 – 1256, 2004.

[287] Marcus, K.J., Astrakas, L.G., Zurakowski, D., Zarifi, M.K., Mintzopoulos, D., Poussaint, T.Y., Anthony, D.C., De Girolami, U., Black, P.M., Tarbell, N.J., and Tzika, A.A., Predicting survival of children with CNS tumors using proton magnetic resonance spectroscopic imaging biomarkers, *Int. J. Oncol.*, 30, 651 – 657, 2007.

[288] Harris, L.M., Davies, N.P., Macpherson, L., Lateef, S., Natarajan, K., Brundler, M.A., Sgouros, S., English, M.W., Arvanitis, T.N., Grundy, R.G., and Peet, A.C., Magnetic resonance spectroscopy in the assessment of pilocytic astrocytomas, *Eur. J. Cancer*, 44, 2640 – 2647, 2008.

[289] Huang,Y., Lisboa, P.J.G., and El-Deredy, W., Tumor grading from magnetic resonance spectroscopy: a comparison of feature extraction with variable selection, *Statist. Med.*, 22, 147 – 164, 2003.

[290] Belkić, K., and Belkić, Dž., The fast Padé transform (FPT) for magnetic resonance spectroscopic imaging (MRSI) in oncology, *Medical Imaging Conference IEEE (MIC)* , Abstract Number, 1918 (CD), Portland (Oregon, USA), October 22 – 25, 2003.

[291] Rabinov, J.D., Lee, P.L., Barker, F.G., Louis, D.N., Harsh, G.R., Cosgrove, G.R., Chiocca, E.A., Thornton, A.F., Loeffler, J.S., Henson, J.W., and Gonzalez, R.G., *In vivo* 3T MR spectroscopy in the distinction of recurrent glioma versus radiation effects: Initial experience, *Radiology*, 225, 871 – 879, 2002.

[292] Osorio, J.A., Ozturk-Isik, E., Xu, D., Cha, S., Chang, S., Berger, M.S., Vigneron, D.B., and Nelson, S.J., 3D ^{1}H MRSI of brain tumors at 3.0 tesla using an eight-channel phased-array head coil, *J. Magn. Reson. Imag.*, 26, 23 – 30, 2007.

[293] Hattingen, E., Pilatus, U., Franz, K., Zanella, F.E., and Lanfermann, H., Evaluation of optimal echo time for ^{1}H spectroscopic imaging of brain tumors at 3 tesla, *J. Magn. Reson. Imag.*, 26, 427 – 431, 2007.

[294] Peet, A.C., Lateef, S., MacPherson, L., Natarajan, K., Sgouros, S., and Grundy, R.G., Short echo time ^{1}H magnetic resonance spectroscopy of childhood brain tumors, *Childs Nerv. Syst.*, 23, 163 – 169, 2007.

[295] Bottomley, P.A., The trouble with spectroscopy papers, *J. Magn. Reson. Imag.*, 2, 1 – 8, 1992.

[296] Ciskowska-Lyson, B., Krolicki, L., Teska, A., Janowicz-Webrowska, A., Zajda, K., Krzakowski, M., and Tacikowska, E., Proton magnetic resonance spectroscopy investigations in brain metabolic changes after first doses of chemotherapy, *Magn. Reson. Mater. Phys. Biol. Med.*, 15 (Suppl. 1), 149, 2002.

[297] Isobe, T., Matsumura, A., Anno, I., Yoshizawa, T., Nagatomo, Y., Itai, Y., and Nose, T., Quantification of cerebral metabolites in glioma patients with proton MR spectroscopy using T_2 relaxation time correction, *Magn. Reson. Imag.*, 20, 343 – 349, 2002.

[298] Auer, D.P., Gössl, C., Schirmer, T., and Czisch, M., Improved analysis of ^{1}H MR spectra in the presence of mobile lipids, *Magn. Reson. Med.*, 46, 615 – 618, 2001.

[299] Tosi, M.R., Bottura, G., Lucchi, P., Reggiani, A., Trinchero, A., and Tugnoli, V., Cholesteryl esters in human malignant neoplasms, *Int. J. Mol. Med.*, 11, 95 – 98, 2003.

[300] Tugnoli, V., Tosi, M.R., Tinti, A., Trinchero, A., Bottura, G., and Fini, G., Characterization of lipids from human brain tissues by multi-nuclear magnetic resonance spectroscopy, *Biopolymers*, 62, 297 – 306, 2001.

[301] Cheng, L.L., Anthony, D.C., Comite, A.R., Black, P.M., Tzika, A.A., and Gonzalez, R.G., Quantification of micro-heterogeneity in glioblastoma multiforme with *ex vivo* high-resolution magic-angle spinning (HRMAS) proton magnetic resonance spectroscopy, *Neuro-Oncol.*, 2, 87 – 95, 2000.

[302] Tzika, A.A., Cheng, L.L., Goumnerova, L., Madsen, J.R., Zurakowski, D., Astrakas, L.G., Zarifi, M.K., Scott, R.M., Anthony, D.C., Gonzalez, R.G., and Black, P.M., Biochemical characterization of pediatric brain tumors by using *in vivo* and *ex vivo* magnetic resonance spectroscopy, *J. Neurosurg.*, 96, 1023 – 1031, 2002.

[303] Opstad, K.S., Bell, B.A., Griffiths, J.R., and Howe, F.A., Taurine: a potential marker of apoptosis in gliomas, *Br. J. Cancer*, 100, 789 – 794, 2009.

[304] Novotny, E.J., Fulbright, R.K., Pearl, P.L., Gibson, K.M., and Rothman, D.L., Magnetic resonance spectroscopy of neurotransmitters in human brain, *Ann. Neurol.*, 54 (Suppl.), 25 – 31, 2003.

[305] Kuznetsov, Y.E., Caramanos, Z., Antel, S.B., Preul, M.C., Leblanc, R., Villemure, J.G., Pokrupa., R., Olivier, A., Sadikot, A., and Arnold, D.L., Proton magnetic resonance spectroscopic imaging can predict length of survival in patients with supratentorial gliomas, *Neurosurgery*, 53, 565 – 576, 2003.

[306] Arús, C., Tumors and spectroscopy, *Magn. Reson. Mater. Phys. Biol. Med.*, 15 (Suppl. 1), 38, 2002.

[307] Aboagye, E.O., and Bhujwalla, Z.M., Malignant transformation alters membrane choline phospholipid metabolism of human mammary epithelial cells, *Cancer Res.*, 59, 80 – 84, 1999.

[308] Glunde, K., Jie, C., and Bhujwalla, Z.M., Molecular causes of the aberrant choline phospholipid metabolism in breast cancer, *Cancer Res.*, 64, 4270 – 4276, 2004.

[309] Katz-Brull, R., Margalit, R., and Degani, H., Differential routing of choline in implanted breast cancer and normal organs, *Magn. Reson. Med.*, 46, 31 – 38, 2001.

[310] Ackerstaff, E., Pflug, B.R., Nelson, J.B., and Bhujwalla, Z.M., Detection of increased choline compounds with proton nuclear magnetic resonance spectroscopy subsequent to malignant transformation of human prostatic epithelial cells, *Cancer Res.*, 61, 3599 – 3603, 2001.

[311] Pecorelli, S., Favalli, G., Zigliani, L., and Odicino, F., Cancer in women, *Int. J. Gynaecol. Obstet.*, 82, 369 – 379, 2003.

[312] Brewer, M.A., Johnson, K., Follen, M., Gershenson, D., and Bast, R., Prevention of ovarian cancer: intraepithelial neoplasia, *Clin. Cancer Res.*, 9, 20 – 30, 2003.

[313] Ashworth, A., Balkwill, F., Bast, R.C., Berek, J.S., Kaye, A., Boyd, J.A., Mills, G., Weinstein, J.N., Woolley, K., and Workman, P., Opportunities and challenges in ovarian cancer research, a perspective from the 11th Ovarian cancer action/HHMT Forum, Lake Como, March 2007, *Gynecol. Oncol.*, 108, 652 – 657, 2008.

[314] Woodward, E.R., Sleightholme, H.V., Considine, A.M., Williamson, S., McHugo, J.M., and Cruger, D.G., Annual surveillance by CA125 and transvaginal ultrasound for ovarian cancer in both high-risk and population risk women is ineffective, *Br. J. Obstet. Gynaecol.*, 114, 1500 – 1509, 2007.

[315] Bhoola, S., and Hoskins, W.J., Diagnosis and management of epithelial ovarian cancer, *Obstet. Gynecol.*, 107, 1399 – 1410, 2006.

[316] Einhorn, N., Bast, R., Knapp, R., Nilsson, B., Zurawski, V., and Sjövall K., Long-term follow-up of the Stockholm screening study on ovarian cancer, *Gynecol. Oncol.*, 79, 466 – 470, 2000.

[317] Kurman, R.J., Visvanathan, K., Roden, R., Wu, T.C., and Shih, I.M., Early detection and treatment of ovarian cancer: shifting from early stage to minimal volume of disease based on a new model of carcinogenesis, *Am. J. Obstet. Gynecol.*, 198, 351 – 356, 2008.

[318] Dearking, A.C., Aletti, G.D., McGree, M.E., Weaver, A.L., Sommerfield, M.K., and Cliby, W.A., How relevant are ACOG and SGO guidelines for referral of adnexal mass? *Obstet. Gynecol.*, 110, 841 – 848, 2007.

[319] Duffy, M.J., Bonfrer, J.M., Kulpa, J., Rustin, G.J., Soletormos, G., Torre, G.C., Tuxen, M.K., and Zwirner, M., CA125 in ovarian cancer: European Group on Tumor Markers Guidelines for Clinical Use, *Int. J. Gynecol. Cancer*, 15, 679 – 691, 2005.

[320] Garner, E.I.O., Advances in the early detection of ovarian carcinoma, *J. Reprod. Med.*, 50, 447 – 453, 2005.

[321] Kong, F., Nicole White, C., Xiao, X., Feng, Y., Xu, C., He, D., Zhang, Z., and Yu, Y., Using proteomic approaches to identify new biomarkers for detection and monitoring of ovarian cancer, *Gynecol. Oncol.*, 100, 247 – 253, 2006.

[322] Liu, Y., Serum proteomic pattern analysis for early cancer detection, *Technol. Cancer Res. Treat.*, 5, 61 – 66, 2006.

[323] Nossov, V., Amneus, M. Su, F., Lang, J., Janco, J.M., Reddy, S.T., and Farias-Eisner, R., The early detection of ovarian cancer: from traditional methods to proteomics: Can we really do better than serum CA-125? *Am. J. Obstet. Gynecol.*, 199, 215 – 223, 2008.

[324] Buys, S.S., Partridge, E., Greene, M.H., Prorok, P.C., Reding, D., Riley, T.L., Hartge, P., Fagerström, R.M., Ragard, L.R., Chia, D., Izmirlian, G., Fouad, M., Johnson, C.C., Gohagan, J.K., and the PLCO Project Team, Ovarian cancer screening in the Prostate, Lung, Colorectal and Ovarian (PLCO) cancer screening trial: Findings from the initial screen of a randomized trial, *Am. J. Obstet. Gynecol.*, 193, 1630 – 1639, 2005.

[325] Kobayashi, H., Yamada, Y., Sado, T., Sakata, M., Yoshida, S., Kawaguchi, S., Kanayama, S., Shigetomi, H., Haruta, S., Tsuji, Y., Ueda, S., and Kitanaka, T., A randomized study of screening for ovarian cancer: a multi-center study in Japan, *Int. J. Gynecol. Cancer*, 18, 414 – 420, 2008.

[326] Taylor, K.L., Shelby, R., Gelmann, E., and McGuire, C., Quality of life and trial adherence among participants in the prostate, lung, colorectal, and ovarian cancer screening trial, *J. Natl. Cancer Inst.*, 96, 1083 – 1094, 2004.

[327] McGovern, P.M., Gross, C.R., Krueger, R.A., Engelhard, D.A., Cordes, J.E., and Church, T.R., False-positive cancer screens and health-related quality of life, *Cancer Nurs.*, 27, 347 – 352, 2004.

[328] U.S. Preventive Services Task Force, Screening for ovarian cancer: recommendation statement, *Ann. Fam. Med.*, 2, 260 – 262, 2004.

[329] Imaoka, I., Wada, A., Kaji, Y., Hayashi, T., Hayashi, M., Matsuo, M., and Sugimura, K., Developing an MR imaging strategy for diagnosis of ovarian masses, *Radiographics*, 26, 1431 – 1448, 2006.

[330] Spencer, J.A., A multi-disciplinary approach to ovarian cancer at diagnosis, *Br. J. Radiol.*, 78 (Suppl.), 94 – 102, 2005.

[331] Kinkel, K., Lu, Y., Mehdizade, A., Pelte, M-F., and Hričak, H., Indeterminate ovarian mass at US: Incremental value of second imaging test for characterization – Meta-analysis and Bayesian analysis, *Radiology*, 236, 85 – 94, 2005.

[332] Harlap, S., Olson, S.H., Barakat, R.R., Caputo, T.A., Forment, S., Jacobs, A.J., Nakraseive, C., and Xue, X., Diagnostic X-rays and risk of epithelial ovarian carcinoma in Jews, *Ann. Epidemiol.*, 12, 426 – 434, 2002.

[333] Hill, D.A., Preston-Martin, S., Ross, R.K., and Bernstein, L., Medical radiation, family history of cancer and benign breast disease in relation to breast cancer risk in young women, *Cancer Causes Control*, 13, 711 – 718, 2002.

[334] Booth, S.J., Pickles, M.D., and Turnbull, L.W., *In vivo* magnetic resonance spectroscopy of gynaecological tumors at 3.0 tesla, *Br. J. Obstet. Gynaecol.*, 116, 300 – 303, 2009.

[335] Cho, S.W., Cho, S.G., Lee, J.H., Kim, H.-J., Lim, M.H., Kim, J.H., and Suh, C.H., *In vivo* proton magnetic resonance spectroscopy in adnexal lesions, *Korean J. Radiol.*, 3, 105 – 112, 2002.

[336] Hascalik, S., Celik, O., and Erdem, G., Magnetic resonance spectral analysis of ovarian teratomas, *Int. J. Gynecol. Obstet.*, 90, 152 – 152, 2005.

[337] Hascalik, S., Celik, O., Sarak, K., Meydanli, M.M., Alkan, A., and Mizrak, B., Metabolic changes in pelvic lesions: findings at proton MR spectroscopic imaging, *Gynecol. Obstet. Invest.*, 60, 121 – 127, 2005.

[338] Okada, T., Harada, M., Matsuzaki, K., Nishitani, H., and Aono, T.J., Evaluation of female intrapelvic tumors by clinical proton MR spectroscopy, *Magn. Reson. Imag.*, 13, 912 – 917, 2001.

[339] Stanwell, P., Russell, P., Carter, J., Pather, S., Heintze, S., and Mountford, C., Evaluation of ovarian tumors by proton magnetic resonance spectroscopy at three tesla, *Invest. Radiol.*, 43, 745 – 751, 2008.

[340] Wallace, J. C., Raaphorst, G.P., Somorjai, R.L., Ng, C.E., Fung Kee Fung, M., Senterman, M., and Smith, I.C., Classification of ^{1}H MR spectra of biopsies from untreated and recurrent ovarian cancer using linear discriminant analysis, *Magn. Reson. Med.*, 38, 569 – 576, 1997.

[341] Smith, I.C., and Blandford, D.E., Diagnosis of cancer in humans by ^{1}H NMR of tissue biopsies, *Biochem. Cell. Biol.*, 76, 472 – 476, 1998.

[342] Massuger, L.F.A.G., van Vierzen, P.B.J., Engelke, U., Heerschap, A., and Wevers, R., ^{1}H magnetic resonance spectroscopy. A new technique to discriminate benign from malignant ovarian tumors, *Cancer*, 82, 1726 – 1730, 1998.

[343] Boss, E.A., Moolenaar, S.H., Massuger, L.F., Boonstra, H., Engelke, U.F., de Jong, J.G., and Wevers, R.A., High-resolution proton nuclear magnetic resonance spectroscopy of ovarian cyst fluid, *NMR Biomed.*, 13, 297 – 305, 2000.

[344] Mountford, C.E., Doran, S., Lean, C.L. and Russell, P.L., Proton MRS can determine the pathology of human cancers with a high level of accuracy, *Chem. Rev.*, 104, 3677 – 3704, 2004.

[345] Gluch, L., Magnetic resonance in surgical oncology: II Literature review, *ANZ. J. Surg.*, 75, 464 – 470, 2005.

[346] Nicholson, J.K., and Wilson, I.D., High resolution proton magnetic resonance spectroscopy of biological fluids, *Prog. NMR Spectrosc.*, 21, 449 – 501, 1989.

[347] Althuis, M.D., Dozier, J.M., Anderson, W.F., Devesa, S.S., and Brinton, L.A., Global trends in breast cancer incidence and mortality 1973 – 1997, *Int. J. Epidemiol.*, 34, 405 – 412, 2005.

[348] Love, R.R., Love, S.M., and Laudico, A.V., Breast cancer from a public health perspective, *Breast J.*, 10, 136 – 140, 2004.

[349] Masood, S., Coming together to conquer the fight against breast cancer in countries of limited resources: the challenges and the opportunities, *Breast J.*, 13, 223 – 225, 2007.

[350] Parkin, D.M., Bray, F., and Pisani, P., Global cancer statistics, *CA Cancer J. Clin.*, 55, 74 – 108, 2005.

[351] Perry, H.N., Broeders, M., de Wolf, C., Törnberg, S., Holland, R., and von Karsa, L.H., European guidelines for quality assurance in breast cancer screening and diagnosis fourth edition – summary document, *Ann. Oncol.*, 19, 614 – 622, 2008.

[352] Armstrong, K., Moye, E., Williams, S., Berlin, J.A., and Reynolds, E.E., Screening mammography in women 40 to 49 years of age: a systematic review for the American College of Physicians, *Ann. Intern. Med.*, 146, 516 – 526, 2007.

[353] Berg, W.A., Blume, J.D., Cormack, J.B., Mendelson, E.B., Lehrer, D., Böhm-Vélez, M., Pisano, E.D., Jong, R.A., Evans, W.P., Morton, M.J., Mahoney, M.C., Larsen, L.H., Barr, R.G., Farria, D.M., Marques, H.S., Boparai, K., and ACRIN 6666 Investigators, Combined screening with ultrasound and mammography vs mammography alone in women at elevated risk of breast cancer, *JAMA*, 299, 2151 – 2163, 2008.

[354] Saslow, D., Boetes, C., Burke, W., Harms, S., Leach, M.O., Lehman, C.D., Morris, E., Pisano, E., Schnall, M., Sener, S., Smith, R.A., Warner, E., Yaffe, M., Andrews, K.S., Russell, C.A., and American Cancer Society Breast Cancer Advisory Group, American Cancer Society guidelines for breast screening with MRI as an adjunct to mammography, *CA Cancer J. Clin.*, 57, 75 – 89, 2007.

[355] Nemec, C.F., Listinsky, J., and Rim, A., How should we screen for breast cancer: Mammography, ultrasonography, MRI? *Cleveland Clin. J. Med.*, 74, 897 – 904, 2007.

[356] Berrington de González, A., and Reeves, G., Mammographic screening before age 50 years in the UK: comparison of the radiation risks with the mortality benefits, *Br. J. Cancer*, 93, 590 – 596, 2005.

[357] Cardenas, K., and Frisch, K., Comprehensive breast cancer screening. Programs now include individual risk assessment, *Postgrad. Med.*, 113, 34 – 46, 2003.

[358] Surbone, A., Ethical implications of genetic testing for breast cancer susceptibility, *Crit. Rev. Oncol. Hematol.*, 40, 149 – 157, 2001.

[359] Márquez, M., Belkić, K., Nilsson, S., and Holmberg, A.R., Genetic testing in patients with hereditary cancer risk: social, ethical and legal considerations, *Instit. Investig. Jurid.*, 383, 295 – 340, 2007.

[360] Weitzel, J.N., Lagos, V.I., Callinane, C.A., Gambol, P.J., Culver, J.O., Blazer, K.R., Palomares, M.R., Lowstuer, K.J., and MacDonald, D.J., Limited family structure and BRCA gene mutation status in single cases of breast cancer, *JAMA*, 297, 2587 – 2595, 2007.

[361] Belkić, K., Cohen, M., Márquez, M., Berman, A.H., Mints, M., Wilczek, B., Castellanos, E., and Castellanos, M., Screening of high risk groups for breast and ovarian cancer in Europe (Manuscript under editorial review).

[362] Leu, M., Czene, K., and Reilly, M., The impact of truncation and missing family links in population-based registers on familial risk estimates, *Am. J. Epidemiol.*, 166, 1461 – 1467, 2007.

[363] Smith, J. A., and Andreopoulou, E., An overview of the status of imaging screening technology for breast cancer, *Ann. Oncol.*, 15, (Suppl. 1), i18 – i26, 2004.

[364] Kuni, H., Schmitz-Feuerhake, I., and Dieckmann, H., Mammography screening – neglected aspects of radiation risks, *Gesundheitswesen*, 65, 443 – 446, 2003.

[365] Laderoute, M.P., Improved safety and effectiveness of imaging predicted for MR mammography, *Br. J. Cancer*, 90, 278 – 279, 2004.

[366] Kriege, M., Brekelmans, C.T., Peterse, H., Obdeijn, I.M., Boetes, C., Zonderland, H.M., Muller, S.H., Kok, T., Manoliu, R.A., Besnard, A.P., Tilanus-Linthorst, M.M., Seynaeve, C., Bartels, C.C., Meijer, S., Oosterwijk, J.C., Hoogerbrugge, N., Tollenaar, R.A., de Koning, H.J., Rutgers, E.J., and Klijn, J.G., Tumor characteristics and detection method in the MRISC screening program for the early detection of hereditary breast cancer, *Breast Cancer Res. Treat.*, 102, 357 – 363, 2007.

[367] Greendale, G.A., Reboussin, B.A., Slone, S., Wasilauskas, C., Pike, M.C., and Ursin, G., Postmenopausal hormone therapy and change in mammographic density, *J. Natl. Cancer Inst.*, 95, 30 – 37, 2003.

[368] Kopans, D.B., *Breast Imaging*, 2nd edn, Lippincott-Raven Publishers, Philadelphia, Pennsylvania, 1998.

[369] Baek, H.M., Yu, H.J., Chen, J.H., Nalcioglu, O., and Su, M.Y., Quantitative correlation between ^{1}H MRS and dynamic contrast-enhanced MRI of human breast cancer, *Magn. Reson. Imag.*, 26, 523 – 531, 2008.

[370] Nass, S.J., Henderson, C., and Lashof, J.C., (Eds.), *Mammography and Beyond: Developing Technologies for the Early Detection of Breast Cancers*, National Academy Press, Washington, D.C., 2001.

[371] Morris, E.A., Breast cancer imaging with MRI, *Radiol. Clin. N. Am.*, 40, 443 – 466, 2002.

[372] Yu, J., Park, A., Morris, E., Liberman, L., Borgen, P.I., and King, T.A., MRI screening in a clinic population with a family history of breast cancer, *Ann. Surg. Oncol.*, 15, 452 – 461, 2008.

[373] Schrading, H.S., and Kuhl, C.K., Mammographic, US, and MR imaging phenotypes of familial breast cancer, *Radiology*, 246, 58 – 70, 2008.

[374] Essink-Bot, M.L., Rijnsburger, A.J., van Dooren, S., de Koning, H.J., and Seynaeve, C., Women's acceptance of MRI in breast cancer surveillance because of a familial or genetic predisposition, *Breast*, 15, 673 – 676, 2006.

[375] Robson, M., Breast cancer surveillance in women with hereditary risk due to BRCA1 or BRCA2 mutations, *Clin. Breast Cancer*, 5, 260 – 268, 2004.

[376] Houssami, N., and Wilson, R., Should women at high risk of breast cancer have screening magnetic resonance imaging (MRI)? *J. Breast*, 16, 2 – 4, 2007.

[377] Bartella, H.L., Thakur, S.B., Morris, E.A., Dershaw, D.D., Huang, W., Chough, E., Cruz, M.C., and Liberman, L., Enhancing non-mass lesions in the breast: evaluation with proton ^{1}H MR spectroscopy, *Radiology*, 245, 80 – 87, 2007.

[378] Bartella, L., and Huang, W., Proton ^{1}H MR spectroscopy of the breast, *Radiographics*, 27 (Suppl. 1), 241 – 252, 2007.

[379] Meisamy, S., Bolan, P.J., Baker, E.H., Le, C.T., Kelcz, F., Lechner, M.C., Luikens, B.A., Carlson, R.A., Brandt, K.R., Amrami, K.K., Nelson, M.T., Everson, L.I., Emory, T.H., Tuttle, T.M., Yee, D., and Garwood, M., Adding *in vivo* quantitative ^{1}H MR spectroscopy to improve diagnostic accuracy of breast MR imaging: preliminary results of observer performance study at 4.0T, *Radiology*, 236, 465 – 475, 2005.

[380] Sardanelli, H. F., Fausto, A., and Podo, F., MR spectroscopy of the breast, *Radiol. Med.* (*Torino*), 113, 56 – 64, 2008.

[381] Tse, G.M., Yeung, D.K., King, A.D., Cheung, H.S., and Yang W.T., *In vivo* proton magnetic resonance spectroscopy of breast lesions: an update, *Breast Cancer Res. Treat.*, 104, 249 – 255, 2007.

[382] Huang, W.H., Fisher, P.R., Dulaimy, K.H., Tudorica, L.A.H., O'Hea, B.H., and Button, T.H., Detection of breast malignancy: diagnostic MR protocol for improved specificity, *Radiology*, 232, 585 – 591, 2004.

[383] Kim, J.-K., Park, S.-H., Lee, H.M., Lee, Y.-H., Sung, N.-K., Chung, D.-S., and Kim, O.-D., *In vivo* ^{1}H MRS evaluation of malignant and benign breast diseases, *Breast*, 12, 179 – 182, 2003.

[384] Okamoto, K., Kurihara, Y., Imamura, K., Kanemaki, Y., Nakajima, Y., Fukuda, M., and Maeda, I., Desmoid tumor of the breast: The role

of proton magnetic resonance spectroscopy for a benign breast lesion mimicking a malignancy, *Breast J.*, 14, 376 – 378, 2008.

[385] Das, C.J., and Medhi, K., Proton magnetic resonance spectroscopy of tubercular breast abscess: Report of a case, *J. Comput. Assist. Tomogr.*, 32, 599 – 601, 2008.

[386] Jagannathan, N. R., Singh, M., Govindaraju, V., Raghunathan, P., Coshic, O., Julka, P.K., and Rath, G.K., Volume localized *in vivo* proton MR spectroscopy in breast carcinoma: variations of water-fat ratios in patients receiving chemotherapy, *NMR Biomed.*, 11, 414 – 422, 1998.

[387] Thomas, M.A., Wyckoff, N., Yue, K., Binesh, N., Banakar, S., Chung, H.-K., Sayre, J., and DeBruhl, N., Two-dimensional MR spectroscopic characterization of breast cancer *in vivo*, *Technol. Cancer. Res. Treat.*, 4, 99 – 106, 2005.

[388] Jacobs, M.A., Barker, P.B., Bottomley, P.A., Bhujwalla, Z., and Bluemke, D.A., Proton magnetic resonance spectroscopic imaging of human breast cancer: a preliminary study, *J. Magn. Reson. Imag.*, 19, 68 – 75, 2004.

[389] Bolan, P.J., Meisamy, S., Baker, E.H., Lin, J., Emory, T., Nelson, M., Everson, L.I., Yee, D., and Garwood, M., *In vivo* quantification of choline compounds in the breast with ^{1}H MR spectroscopy, *Magn. Reson. Med.*, 50, 1134 – 1143, 2003.

[390] Bolan, P.J., DelaBarre, L., Baker, E.H., Merkle, H., Everson, L.I., Yee, D., and Garwood, M., Eliminating spurious lipid sidebands in ^{1}H MRS of breast lesions, *Magn. Reson. Med.*, 48, 215 – 222, 2002.

[391] Hu, H.J., Yu, Y., Kou, Z., Huang, W., Jiang, Q., Xuan, Y., Li, T., Sehgal, V., Blake, C., Haacke, E.M., and Soulen, R.L., A high spatial resolution ^{1}H magnetic resonance spectroscopic imaging technique for breast cancer with a short echo time, *Magn. Reson. Imag.*, 26, 360 – 366, 2008.

[392] Stanwell, H.P., and Mountford, C., *In vivo* proton MR spectroscopy of the breast, *Radiographics*, 27 (Suppl. 1), 253 – 266, 2007.

[393] Tozaki, M., Proton MR spectroscopy of the breast, *Breast Cancer*, 15, 218 – 223, 2008.

[394] Tozaki, M., Sakamoto, M., Oyama, Y., Ouchi T., Kawano, N., Suzuki, T., Yamashiro, N., Ozaki, S., Sakamoto, N., Higa, K., Abe, S., Ogawa, T., and Fukuma, E., Monitoring of early response to neoadjuvant chemotherapy in breast cancer with ^{1}H MR spectroscopy: Comparison to sequential 2 ^{18}F fluorodeoxyglucose positron emission tomography, *J. Magn. Reson. Imag.*, 28, 420 – 427, 2008.

[395] Gribbestad, I.S., Sitter, B., Lundgren, S., Krane, J., and Axelson, D., Metabolite composition in breast tumors examined by proton nuclear magnetic resonance spectroscopy, *Anticancer Res.*, 19, 1737 – 1746, 1999.

[396] Lippman, M.E., Breast cancer, in: Braunwald, E., Fauci, A., Kasper, D.L., Hauser, S.L., Longo, D.L., and Jameson, J.L., *Harrison's Principles of Internal Medicine*, 15th edn, McGraw-Hill, New York, 2001, pp. 571 – 578.

[397] Katz-Brull, R., Seger, D., Rivenson-Segal, D., Rushkin, E., and Degani, H., *Cancer Res.*, 62, 1966 – 1970, 2002.

[398] Sharma, U., Mehta, A., Seenu, V., and Jagannathan, N.R., Biochemical characterization of metastatic lymph nodes of breast cancer patients by *in vitro* ^{1}H magnetic resonance spectroscopy: a pilot study, *Magn. Reson. Imag.*, 22, 697 – 706, 2004.

[399] Rivenzon-Segal, D., Margalit, R., and Degani, H., Glycolysis as a metabolic marker in orthotopic breast cancer, monitored by *in vivo* ^{13}C MRS, *Am. J. Physiol. Endocrinol. Metab.*, 283, E623 – E630, 2002.

[400] Loening, N.M., Chamberlin, A.M., Zepeda, A.G., Gilberto Gonzalez, R., and Cheng, L.L., Quantification of phosphocholine and glycerophosphocholine with ^{31}P edited ^{1}H NMR spectroscopy, *NMR Biomed.*, 18, 413 – 420, 2005.

[401] Horwich, A., Parker, C., and Kataja, V., ESMO Guidelines Working Group, Prostate cancer: ESMO clinical recommendations for diagnosis, treatment and follow-up, *Ann. Oncol.*, 19 (Suppl. 2), ii45 – ii46, 2008.

[402] Kundra, V., Silverman, P.M., Matin, S.F., and Choi, H., Imaging in oncology from the University of Texas M. D. Anderson Cancer Center: diagnosis, staging, and surveillance of prostate cancer, *Am. J. Roentgenology*, 189, 830 – 844, 2007.

[403] Lim, L.S., Sherin, K., and ACPM Prevention Practice Committee, Screening for prostate cancer in U.S. men: ACPM position statement on preventive practice, *Am. J. Prev. Med.*, 34, 164 – 170, 2008.

[404] Scher, H.L., Hyperplastic and malignant diseases of the prostate, in: Braunwald E., Fauci A., Kasper D.L., Hauser, S.L., Longo, D.L., and Jameson, J.L., *Harrisons Principles of Internal Medicine*, 15th edn, McGraw-Hill, New York, 2001, pp. 608 – 616.

[405] Signorello, L.B., and Adami, H.-O., Prostate cancer, in: Adami, H.-O., Hunter D., and Trichopoulos, D., *Textbook of Cancer Epidemiology*, Oxford University Press, Oxford, 2002, pp. 400 – 428.

[406] Bouchardy, C., Fioretta, G., Rapiti, E., Verkooijen, H., Rapin, C., Schmidlin, F., Miralbell, R., and Zanetti, R., Recent trends in prostate cancer mortality show a continuous decrease in several countries, *Int. J. Cancer*, 123, 421 – 429, 2008.

[407] Bosetti, C., Bertuccio, P., Levi, F., Lucchini, F., Negri, E., and La Vecchia, C., Cancer mortality in the European Union, 1970-2003, with a joinpoint analysis, *Ann. Oncology*, 19, 631 – 640, 2008.

[408] Thompson, I.M., Pauler, D.K., Goodman, P.J. Tangen, C.M., Lucia, M.S., Parnes, H.L., Minasian, L.M., Ford, L.G., Lippman, S.M., Crawford, E.D., Crowley, J.J., and Coltman, C.A., Prevalence of prostate cancer among men with a prostate-specific antigen level ≤ 4.0 ng per milliliter, *N. Engl. J. Med.*, 350, 2239 – 2246, 2004.

[409] U.S. Preventive Services Task Force Screening for prostate cancer, U.S. Preventive Services Task Force Recommendation Statement, *Ann. Intern. Med.*, 149, 185 – 191, 2008.

[410] Lin, K., Lipsitz, R., Miller, T., Janakiraman, S., and U.S. Preventive Services Task Force, Benefits and harms of prostate-specific antigen screening for prostate cancer: an evidence update for the U.S. Preventive Services Task Force, *Ann. Intern. Med.*, 149, 192 – 199, 2008.

[411] Ornstein, D.K., and Kang, J., How to improve prostate biopsy detection of prostate cancer, *Curr. Urology Rep.*, 2, 218 – 223, 2001.

[412] Mazaheri, Y., Shukla-Dave, A., Hričak, H., Fine, S.W., Zhang, J., Inurrigarro, G., Moskowitz, C.S., Ishill, N.M., Reuter, V.E., Touijer, K., Zakian, K.L., and Koutcher, J.A., Prostate cancer: identification with combined diffusion-weighted MR imaging and 3D ^{1}1H MR spectroscopic imaging – correlation with pathologic findings, *Radiology*, 246, 480 – 488, 2008.

[413] Kurhanewicz, J., Swanson, M.G., Nelson, S.J., and Vigneron, D.B., Combined magnetic resonance imaging and spectroscopic imaging approach to molecular imaging of prostate cancer, *J. Magn. Reson. Imag.*, 16, 451 – 463, 2002.

[414] Casciani, E., and Gualdi, G.F., Prostate cancer: value of magnetic resonance spectroscopy 3D chemical shift imaging, *Abdom. Imag.* 31, 490 – 499, 2006.

[415] Spielman, D., *In vivo* proton MR spectroscopy: basic principles and clinical applications, *Magn. Section for Magnetic Resonance Technologists Educational Seminars*, 3, 19 – 37, 2000.

[416] Costello, L.C., Franklin, R.B., and Narayan, P., Citrate in the diagnosis of prostate cancer, *Prostate*, 38, 237 – 245, 1999.

[417] Garcia-Segura, J.M., Sanchez-Chapado, M., Ibarburen, C., Viano, J., Angulo, J.C., Gonzalez J., and Rodriguez-Vallejo, J.M., *In vivo* proton magnetic resonance spectroscopy of disease prostate: spectroscopic features of malignant versus benign pathology, *Magn. Reson. Imag.*, 17, 755 – 765, 1999.

[418] Yu, K.K., Scheidler, J., Hričak, H., Vigneron, D.B., Zaloudek, C.J., Males, R.G., Nelson, S.J., Carroll, P.R., and Kurhanewicz, J., Prostate cancer: prediction of extracapsular extension with endorectal MR imaging and three-dimensional proton MR spectroscopic imaging, *Radiology*, 213, 481 – 488, 1999.

[419] Wu, X., Dibiase, S.J., Gullapalli, R., and Yu, C.X., Deformable image registration for the use of magnetic resonance spectroscopy in prostate treatment planning, *Int. J. Radiat. Oncol. Biol. Phys.*, 58, 1577 – 1583, 2004.

[420] Chen, M., Dang, H.D., Wang, J.Y., Zhou, C., Li, S.Y., Wang, W.C., Zhao, W.F., Yang, Z.H., Zhong, C.Y., and Li, G.Z., Prostate cancer detection: comparison of T2-weighted imaging, diffusion-weighted imaging, proton magnetic resonance spectroscopic imaging, and the three techniques combined, *Acta Radiol.*, 49, 602 – 610, 2008.

[421] Cirillo, S., Petracchini, M., Della Monica, P., Gallo, T., Tartaglia, V., Vestita, E., Ferrando, U., and Regge, D., Value of endorectal MRI and MRS in patients with elevated prostate-specific antigen levels and previous negative biopsies to localize peripheral zone tumors, *Clin. Radiol.*, 63, 871 – 979, 2008.

[422] Lawrentschuk, N., and Fleshner, N., The role of magnetic resonance imaging in targeting prostate cancer in patients with previous negative biopsies and elevated prostate-specific antigen levels, *Br.J. Urol. Int.*, 103, 730 – 733, 2009.

[423] Kumar, R., Nayyar, R., Kumar, V., Gupta, N., Hemal, A., Jagannathan, R., Dattagupta, S., and Thulkar, S., Potential of magnetic resonance spectroscopic imaging in predicting absence of prostate cancer in men with serum prostate-specific antigen between 4 and 10 ng/ml: a follow-up study, *Urology*, 72, 859 – 863, 2008.

[424] Cabrera, A.R., Coakley, F.V., Westphalen, A.C., Lu, Y., Zhao, S., Shinohara, K., Carroll, P.R., and Kurhanewicz, J., Prostate cancer: is in-apparent tumor at endorectal MR and MR spectroscopic imaging a favorable prognostic finding in patients who select active surveillance? *Radiology*, 247, 444 – 450, 2008.

[425] Shukla-Dave, A., Hričak, H., and Scardino, P.T., Imaging low-risk prostate cancer, *Curr. Opin. Urol.*, 18, 78 – 86, 2008.

[426] Coakley, F.V., Qayyum, A., and Kurhanewicz, J., Magnetic resonance imaging and spectroscopic imaging of prostate cancer, *J. Urology*, 170 (Suppl.), 69 – 76, 2003.

[427] Kurhanewicz, J., Swanson, M.G., Wood, P.J., and Vigneron, D.B., Magnetic resonance imaging and spectroscopic imaging: improved patient selection and potential for metabolic intermediate endpoints in prostate cancer chemoprevention trials, *Urology*, 57 (Suppl. 4A), 124 – 128, 2001.

[428] Mueller-Lisse, U.G., Swanson, M.G., Vigneron, D.B., Hričak, H., Bessette, A., Males, R.G., Wood, P.J., Noworolski, S., Nelson, S.J., Barken, I., Carroll, P.R., and Kurhanewicz, J., Time-dependent effects of hormone-deprivation therapy on prostate metabolism as detected by combined magnetic resonance imaging and 3D magnetic resonance spectroscopic imaging, *Magn. Reson. Med.*, 46, 49 – 57, 2001.

[429] Westphalen, A.C., McKenna, D.A., Kurhanewicz, J., and Coakley, F.V., Role of magnetic resonance imaging and magnetic resonance spectroscopic imaging before and after radiotherapy for prostate cancer, *J. Endourol.*, 22, 789 – 794, 2008.

[430] Cirillo, S., Petracchini, M., D'Urso, L., Dellamonica, P., Illing, R., Regge, D., and Muto, G., Endorectal magnetic resonance imaging and magnetic resonance spectroscopy to monitor the prostate for residual disease or local cancer recurrence after transrectal high-intensity focused ultrasound, *Br. J. Urol. Int.*, 102, 452 – 458, 2008.

[431] Pucar, D., Sella, T., and Schoder, H., The role of imaging in the detection of prostate cancer local recurrence after radiation therapy and surgery, *Curr. Opinion Urol.*, 18, 87 – 97, 2008.

[432] Pickett, B., Kurhanewicz, J., Pouliot, J., Weinberg, V., Shinohara, K., Coakley, F., and Roach, M., Three-dimensional conformal external beam radiotherapy compared with permanent prostate implantation in low-risk prostate cancer based on endorectal magnetic resonance spectroscopy imaging and prostate-specific antigen level, *Int. J. Radiat. Oncol. Biol. Phys.*, 65, 65 – 72, 2006.

[433] Pouliot, J., Kim, Y., Lessard, E., Hsu, I.C., Vigneron, D.B., and Kurhanewicz, J., Inverse planning for HDR prostate brachytherapy used to boost dominant intraprostatic lesions defined by magnetic resonance spectroscopy imaging, *Int. J. Radiat. Oncol. Biol. Phys.*, 59, 1196 – 1207, 2004.

[434] Joseph, T., McKenna, D.A., Westphalen, A.C., Coakley, F.V., Zhao, S., Lu, Y., Hsu, I.C., Roach, M., and Kurhanewicz, J., Pre-treatment endorectal magnetic resonance imaging and magnetic resonance spectroscopic imaging features of prostate cancer as predictors of

response to external beam radiotherapy, *Int. J. Radiat. Oncol. Biol. Phys.*, 73, 665 – 671, 2009.

[435] Weinreb, J.C., Blume, J.D., Coakley, F.V., Wheeler, T.M., Cormack, J.B., Sotto, C.K., Cho, H., Kawashima, A., Tempany-Afdhal, C.M., Macura, K.J., Rosen, M., Gerst, S.R., and Kurhanewicz, J., Prostate cancer: sextant localization at MR imaging and MR spectroscopic imaging before prostatectomy – results of ACRIN prospective multi-institutional clinicopathologic study, *Radiology*, 251, 122 – 133, 2009.

[436] Weis, J., Ahlstrom, H., Hlavčak, P., Haggman, M., Ortiz-Nieto, F., and Bergman, A., Two-dimensional spectroscopic imaging for pretreatment evaluation of prostate cancer: comparison with the step-section histology after radical prostatectomy, *Magn. Reson. Imag.*, 27, 87 – 93, 2009.

[437] Kim, C.K., and Park, B.K., Update of prostate magnetic resonance imaging at 3T, *J. Comput. Assist. Tomogr.*, 32, 163 – 172, 2008.

[438] Thomas, M.A., Binesh, N., Yue, K., Banakar, S., Wyckoff, N., Huda, A., Marumoto, A., and Raman, S., Adding a new spectral dimension to localized ^{1}H MR spectroscopy of human prostates using an endorectal coil, *Spectroscopy*, 16, 521 – 527, 2003.

[439] Belkić, Dž., High-resolution parametric estimation of two-dimensional magnetic resonance spectroscopy, 20th Annual Meeting of European Soc. Magn. Res. Med. Biol. (ESMRMB), Abstract Number 365 (CD), Rotterdam (Netherlands), September 18 – 21, 2003.

[440] Swindle, P., McCredie, S., and Russell, P., Pathologic characterization of human prostate tissue with proton MR spectroscopy, *Radiology*, 228, 144 – 151, 2003.

[441] Koutcher, J.A., Zakian, K., and Hričak, H., Magnetic resonance spectroscopic studies of the prostate, *Mol. Urol.*, 4, 143 – 153, 2000.

[442] Swanson, M.G., Vigneron, D.B., Tabatabai, Z.L., Simko, J., Jarso, S., Keshari, K.R., Schmitt, L., Carroll, P.R., Shinohara, K., Vigneron, D.B., and Kurhanewicz, J., Proton HRMAS spectroscopy and quantitative pathologic analysis of MRI/3D-MRSI-targeted postsurgical prostate tissues, *Magn. Reson. Med.*, 50, 944 – 954, 2003.

[443] Swanson, M.G., Zektzer, A.S., Tabatabai, Z.L., Simko, J., Jarso, S., Keshari, K.R., Schmitt, L., Carroll, P.R., Shinohara, K., Vigneron, D.B., and Kurhanewicz, J., Quantitative analysis of prostate metabolites using ^{1}H HRMAS spectroscopy, *Magn. Reson. Med.*, 55, 1257 – 1264, 2006.

[444] Swanson, M.G., Keshari, K.R., Tabatabai, Z.L., Simko, J.P., Shinohara, K., Carroll, P.R., Zektzer, A.S., and Kurhanewicz, J., Quantification of choline- and ethanolamine-containing metabolites in human prostate tissues using ^{1}H HRMAS total correlation spectroscopy, *Magn. Reson. Med.*, 60, 33 – 40, 2008.

[445] Tessem, M.B., Swanson, M.G., Keshari, K.R., Albers, M.J., Joun, D., Tabatabai, Z.L., Simko, J.P., Shinohara, K., Nelson, S.J., Vigneron, D.B., Gribbestad, I.S., and Kurhanewicz, J., Evaluation of lactate and alanine as metabolic biomarkers of prostate cancer using ^{1}H HRMAS spectroscopy of biopsy tissue, *Magn. Reson. Med.*, 60, 510 – 516, 2008.

[446] Belkić, Dž., Critical validity assessment of theoretical models: charge-exchange at intermediate and high energies, *Nucl. Instr. Meth. Phys. Res. B*, 154, 220 – 246, 1999.

[447] Fullerton, G.D., The development of technologies for molecular imaging should be driven by biological questions to be addressed rather than simply modifying existing imaging technologies, *Med. Phys.*, 32, 1231 – 1232, 2005.

[448] Hazle, J.D., Comment on: The development of technologies for molecular imaging should be driven by biological questions to be addressed rather than simply modifying existing imaging technologies, *Med. Phys.*, 32, 1232 – 1233, 2005.

[449] Barrio, J.R., Editorial, *Mol. Imag. Biol.*, 4, 267 – 273, 2002.

[450] Rollo, F.D., Molecular imaging: an overview and clinical applications, *Radiol. Manag.*, 25, 28 – 32, 2003.

[451] Jones, T., The spectrum of medical imaging, *Eur. J. Cancer*, 38, 2067 – 2069, 2002.

[452] Wüthrich, K., The second decade – into the third millenium, *Nat. Struct. Biol.*, 5 (Suppl.) 492 – 495, 1998.

[453] Bessis, D., and Talman, J.D., Variational approach to the theory of operator Padé approximations, *Rocky Mountain J. Math.*, 4, 151 – 158, 1974.

[454] Tani, S., Padé approximant in potential scattering, *Phys. Rev.*, B1011 – B1020, 1965.

[455] Coleman, J.P., Iteration-variation methods and the $\varepsilon-$algorithm, *J. Phys. B*, 9, 1079 – 1093, 1976.

Index